Fundamentals of Low Dimensional Magnets

A low-dimensional magnet is a key to the next generation of electronic devices. In some respects, low-dimensional magnets refer to nanomagnets (nanostructured magnets) or single-molecule magnets (molecular nanomagnets). They also include the group of magnetic nanoparticles, which have been widely used in biomedicine, technology, industries, and environmental remediation.

Low-dimensional magnetic materials can be used effectively in the future in powerful computers (hard drives, magnetic random-access memory, ultra-low power consumption switches, etc.). The properties of these materials largely depend on the doping level, phase, defects, and morphology. This book covers various nanomagnets and magnetic materials. The basic concepts, various synthetic approaches, characterizations, and mathematical understanding of nanomaterials are provided. Some fundamental applications of 1D, 2D, and 3D materials are covered.

This book provides the fundamentals of low-dimensional magnets along with synthesis, theories, structure-property relations, and applications of ferromagnetic nanomaterials. This book broadens our fundamental understanding of ferromagnetism and mechanisms for realization and advancement in devices with improved energy efficiency and high storage capacity.

Series in Materials Science and Engineering

The series publishes cutting-edge monographs and foundational textbooks for interdisciplinary materials science and engineering. It is aimed at undergraduate and graduate-level students, as well as practicing scientists and engineers. Its goal is to investigate the relationships between material qualities, structure, synthesis, processing, characterization, and performance.

Automotive Engineering: Lightweight, Functional, and Novel Materials
Brian Cantor, P. Grant, C. Johnston

Multiferroic Materials: Properties, Techniques, and Applications
Junling Wang, Ed.

2D Materials for nanoelectronics
Michel Houssa, Athanasios Dimoulas, Alessandro Molle

Skyrmions: Topological Structures, Properties, and Applications
J. Ping Liu, Zhidong Zhang, Guoping Zhao, Eds.

Computational Modeling of Inorganic Nanomaterials
Stefan T. Bromley, Martijn A. Zwijnenburg, Eds.

Physical Methods for Materials Characterisation, Third Edition
Peter E. J. Flewitt, Robert K. Wild

Conductive Polymers: Electrical Interactions in Cell Biology and Medicine
Ze Zhang, Mahmoud Rouabhia, Simon E. Moulton, Eds.

Silicon Nanomaterials Sourcebook, Two-Volume Set
Klaus D. Sattler, Ed.

Advanced Thermoelectrics: Materials, Contacts, Devices, and Systems
Zhifeng Ren, Yucheng Lan, Qinyong Zhang

Fundamentals of Ceramics, Second Edition
Michel Barsoum

Flame Retardant Polymeric Materials, A Handbook
Xin Wang and Yuan Hu

2D Materials for Infrared and Terahertz Detectors
Antoni Rogalski

Fundamentals of Fibre Reinforced Composite Materials
A. R Bunsell. S. Joannes, A. Thionnet

Series Preface

The series publishes cutting-edge monographs and foundational textbooks for interdisciplinary materials science and engineering.

Its purpose is to address the connections between properties, structure, synthesis, processing, characterization, and performance of materials. The subject matter of individual volumes spans fundamental theory, computational modeling, and experimental methods used for design, modeling, and practical applications. The series encompasses thin films, surfaces, and interfaces, and the full spectrum of material types, including biomaterials, energy materials, metals, semiconductors, optoelectronic materials, ceramics, magnetic materials, superconductors, nanomaterials, composites, and polymers.

This book in the series is aimed at undergraduate and graduate-level students, as well as practicing scientists and engineers in the field of nanomaterials.

Proposals for new volumes in the series may be directed to Carolina Antunes, Commissioning Editor at CRC Press, Taylor & Francis Group (Carolina.Antunes@tandf.co.uk).

Fundamentals of Low Dimensional Magnets

Edited by
Ram K. Gupta
Sanjay R. Mishra
Tuan Anh Nguyen

CRC Press is an imprint of the
Taylor & Francis Group, an **informa** business

First edition published 2023
by CRC Press
6000 Broken Sound Parkway NW, Suite 300, Boca Raton, FL 33487–2742

and by CRC Press
4 Park Square, Milton Park, Abingdon, Oxon, OX14 4RN

CRC Press is an imprint of Taylor & Francis Group, LLC

© 2023 selection and editorial matter Ram K. Gupta, Sanjay R. Mishra and Tuan Anh Nguyen, individual chapters, the contributors

Reasonable efforts have been made to publish reliable data and information, but the author and publisher cannot assume responsibility for the validity of all materials or the consequences of their use. The authors and publishers have attempted to trace the copyright holders of all material reproduced in this publication and apologize to copyright holders if permission to publish in this form has not been obtained. If any copyright material has not been acknowledged, please write and let us know so we may rectify it in any future reprint.

Except as permitted under U.S. Copyright Law, no part of this book may be reprinted, reproduced, transmitted, or utilized in any form by any electronic, mechanical, or other means, now known or hereafter invented, including photocopying, microfilming, and recording, or in any information storage or retrieval system, without written permission from the publishers.

For permission to photocopy or use material electronically from this work, access www.copyright.com or contact the Copyright Clearance Center, Inc. (CCC), 222 Rosewood Drive, Danvers, MA 01923, 978–750–8400. For works that are not available on CCC please contact mpkbookspermissions@tandf.co.uk

Trademark notice: Product or corporate names may be trademarks or registered trademarks and are used only for identification and explanation without intent to infringe.

ISBN: 978-1-032-04872-7 (hbk)
ISBN: 978-1-032-05421-6 (pbk)
ISBN: 978-1-003-19749-2 (ebk)

DOI: 10.1201/9781003197492

Typeset in Times
by Apex CoVantage, LLC

Contents

Editors ix

1 **Nanomagnets: Basics, Applications, and New Prospectives** 1
  Biswanath Bhoi and Mangesh Diware

2 **Nanostructured Magnetic Semiconductors** 23
  Alessandra S. Silva, Éder V. Guimarães, Tasso O. Sales, Wesley S. Silva, Elisson A. Batista, Carlos Jacinto, Anielle C.A. Silva, Noelio O. Dantas, and Ricardo S. Silva

3 **Nanowire Magnets: Synthesis, Properties, and Applications** 41
  Daljit Kaur

4 **Synthesis Techniques for Low Dimensional Magnets** 59
  Kalyani Chordiya, Gergely Norbert Nagy, and Mousumi Upadhyay Kahaly

5 **2D Magnetic Systems: Magnetic Properties, Measurement Techniques, and Device Applications** 73
  Daljit Kaur and Shikha Bansal

6 **3D Magnonic Structures as Interconnection Element in Magnonic Networks** 93
  A. A. Martyshkin, S. A. Nikitov, and A. V. Sadovnikov

7 **Nanostructured Hybrid Magnetic Materials** 111
  Sha Yang and Wei Liu

8 **Methods for the Syntheses of Perovskite Magnetic Nanomagnets** 125
  Xinhua Zhu

9 **Design of Room Temperature d^0 Ferromagnetism for Spintronics Application: Theoretical Perspectives** 161
  Ravi Trivedi and Brahmananda Chakroborty

10 **Crystal Structures and Properties of Nanomagnetic Materials** 183
  Mirza H. K. Rubel and M. Khalid Hossain

11 **Nanomagnetic Materials: Structural and Magnetic Properties** 207
  P. Maneesha, Suresh Chandra Baral, E. G. Rini, and Somaditya Sen

12 **Magnetism in Monoatomic and Bimetallic Clusters: A Global Geometry Optimization Approach** 225
  J. L. Morán-López, A. P. Ponce-Tadeo, and J. L. Ricardo-Chávez

13 Nanoscale Characterization 245
Arvind Kumar, Swati, Manish Kumar, Neelabh Srivastava, and Anadi Krishna Atul

14 Mathematical Modeling and Simulation of Exchange Coupling Constant (*J*) and Zero-Field Splitting Parameters (*D*) 269
Satadal Paul

15 Novel Magnetism in Ultrathin Films With Polarized Neutron Reflectometry 289
Saibal Basu and Surendra Singh

16 Magnetosomes: Biological Synthesis of Magnetic Nanostructures 309
Marta Masó-Martínez, Paul D Topham, and Alfred Fernández-Castané

17 Theory and Modeling of Spintronics of Nanomagnets 325
Mehmet C. Onbaşlı, Ahmet Avşar, Saeedeh Mokarian Zanjani, Arash Mousavi Cheghabouri, and Ferhat Katmis

18 Research Trends and Statistical-Thermodynamic Modeling the α''-Fe$_{16}$N$_2$-Based Phase for Permanent Magnets 343
Taras M. Radchenko, Olexander S. Gatsenko, Vyacheslav V. Lizunov, and Valentyn A. Tatarenko

Index 367

Editors

Dr. Ram K. Gupta is Associate Professor at Pittsburg State University. Dr. Gupta's research focuses on conducting polymers and composites, green energy production and storage using biowastes and nanomaterials, optoelectronics and photovoltaics devices, organic-inorganic hetero-junctions for sensors, bio-based polymers, flame-retardant polymers, bio-compatible nanofibers for tissue regeneration, scaffold and antibacterial applications, corrosion inhibiting coatings, and bio-degradable metallic implants. Dr. Gupta has published over 250 peer-reviewed articles, made over 300 national, international, and regional presentations, chaired many sessions at national/international meetings, edited many books, and written several book chapters. He has received several million dollars for research and educational activities from many funding agencies. He is serving as Editor-in-Chief, Associate Editor, and editorial board member of numerous journals.

Dr. Sanjay Mishra joined the Department of Physics at the University of Memphis in 1999. He has been consistently productive in research, instruction, and service to the University of Memphis (UoM) since 1999. Dr. Mishra initiated an active multidisciplinary Materials Research program at the UoM. Before receiving postdoctoral experience from the Lawrence Berkeley National Laboratory, the University of California-Berkeley at the Advanced Light Source Synchrotron Facility, he received his PhD in Physics from the Missouri University of Science and Technology—Rolla; his MS from Pittsburg State University, Pittsburg, KS; his MSc from the South Gujarat University, Surat, India; and his postgraduate diploma in Space Sciences from Gujarat University, Ahmedabad, India. Dr. Mishra's research work focuses on magnetic nanomaterials and nanocomposites for energy applications, including alloys, ferrites, and magnetocaloric materials. Dr. Mishra has published more than 300 peer-reviewed journal articles and has given numerous presentations at national and international conferences. He has been successful in securing federal grants of more than two and a half million dollars over the years and has received numerous awards from the UoM for his outstanding research accomplishments.

Dr. Tuan Anh Nguyen completed his BSc in Physics from Hanoi University in 1992 and his PhD in Chemistry from Paris Diderot University (France) in 2003. He was Visiting Scientist at Seoul National University (South Korea, 2004) and the University of Wollongong (Australia, 2005). He then worked as Postdoctoral Research Associate and Research Scientist at Montana State University (U.S.), 2006–2009. In 2012, he was appointed Head of the Microanalysis Department at the Institute for Tropical Technology (Vietnam Academy of Science and Technology). He has managed four PhD theses as thesis director, and three are in progress. He is Editor-In-Chief of Kenkyu Journal of Nanotechnology & Nanoscience and Founding Co-Editor-In-Chief of Current Nanotoxicity & Prevention. He is the author of 4 Vietnamese books and the editor of 32 Elsevier books in the Micro & Nano Technologies Series.

Nanomagnets
Basics, Applications, and New Prospectives

1

Biswanath Bhoi[1,2] and Mangesh Diware[3]

1 National Creative Research Initiative Center for Spin Dynamics and Spin-Wave Devices, Nanospinics Laboratory, Department of Materials. Science and Engineering, Seoul National University, Seoul, South Korea

2 Department of Physics, Indian Institute of Technology (Banaras Hindu University) Varanasi, Varanasi, India

3 CeNSCMR and Institute of Applied Physics, Department of Physics and Astronomy, Seoul National University, Seoul, South Korea

Contents

1.1	Introduction	2
1.2	Theoretical Background	3
	1.2.1 Magnetism at Nanoscale	3
	1.2.2 Magnetization Dynamics	5
1.3	Magnetic Nanostructures	7
1.4	Synthesis of MNMs	8
	1.4.1 Chemical Methods	8
	1.4.2 Physical Methods	10
	1.4.3 Biological Methods	11
	1.4.4 Advanced Synthesis Method	11
1.5	Characterization of MNMs	13
1.6	Applications of MNMs	13
	1.6.1 High-Density Data Storage	13
	1.6.2 Nano-Electronics and Spintronic Applications	14
	1.6.3 Biomedical Applications	15
	1.6.4 Biosensing Applications	16
	1.6.5 Environmental and Agricultural Applications	16
	1.6.6 Other Applications	17
1.7	Emerging Research Areas in MNMs	17
1.8	Summary	19
1.9	Acknowledgments	19
References		19

1.1 INTRODUCTION

Nano-scale magnetism has been a subject of intense research within the last few decades, not only for technology development but also for understanding material science and fundamental physics. With the advancements in nanoscience and nanotechnology, studies have been centered on designing and manipulating the properties of magnetic nanomaterials (MNMs) by controlling their size, morphology, and composition. As MNM research has progressed, the prime focus has shifted from the chemistry of the materials (synthesis routes, morphology control, and characterization) to the investigation of functionality and integration with the physical, electrical, and biomedical fields. The notable potential applications of MNMs are in a range of multidisciplinary fields like magnetic memory, magnetic resonance imaging, biomedicine and health science, spintronics, and other areas [1–3], as shown in Figure 1.1. Recently, MNMs have increasingly gained attention in emerging research fields such as spin-torque nano-oscillators (STNOs), spin logic, two-dimensional (2D) ferromagnetism, and quantum magnonic-based on magnetization dynamics in nanomagnets [4–6].

The MNMs are an interdisciplinary subject, with researchers from both basic sciences and engineering equally interested in developing novel materials with controlled morphology and properties able to perform multiple functions. Further, magnetic properties like saturation magnetization (M_S), coercivity

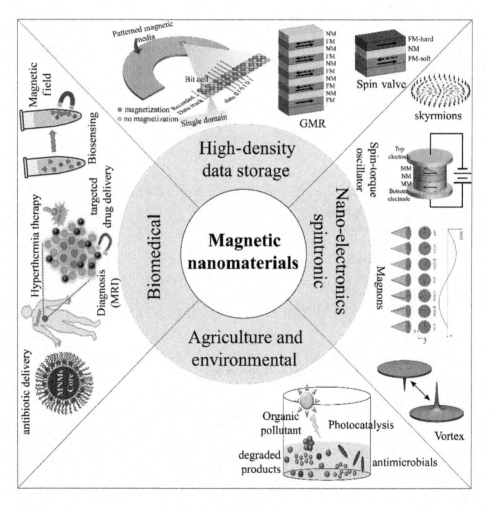

FIGURE 1.1 Schematic illustration of magnetic nanomaterial for different applications.

(H_C), and anisotropy vary significantly with the morphology and composition of the MNMs within the nano-scale regime. There are several kinds of MNMs available based on iron (Fe), cobalt (Co), nickel (Ni), and their alloys and oxides (e.g., ferro- and antiferromagnets), rare-earth metals (e.g., gadolinium [Gd], terbium [Tb], dysprosium [Dy]), and multicomponent compounds for various applications [7–11]. More efforts are needed to uncover the intrinsic relationship in MNMs between size/morphology/structure and magnetic properties. Apart from superparamagnetism and spin-glass phenomena, MNMs offer an exciting platform for investigating the interplay among various competing magnetic and electronic order and magnetic phenomena involving the quantum confinement effect. Recent studies on magnetization dynamics of MNMs have been extremely vigorous owing to the accelerating miniaturization of magnetic units in spintronic-based devices and other spin logic devices.

Moreover, the new technologies require structuring magnetic materials in all three dimensions (3D) at various length scales and the utilization of novel phenomena ranging from magnetic vortex to spin-wave propagation. This introduces multiple magnetic structures at different length scales, such as nanodots, -wires, -stripes, -discs, and 2D materials. We attempt to summarize the underlying mechanism and issues that have yet to be realized in the various emerging nanomagnetism research fields.

1.2 THEORETICAL BACKGROUND

Magnetic nanomaterials are a class of engineered particles (typically ≤ 150 nm) that can be controlled by applying an external magnetic field. Change in the surface-to-volume ratio helps discover novel properties in nanomaterials compared to the corresponding bulk materials. Factors that influence these properties of MNMs include crystal structure, morphology, chemical composition, and, finally, how they interact with neighboring particles or the surrounding matrix.

1.2.1 Magnetism at Nanoscale

It is well known that magnetic (ferro or ferri) materials minimize their magnetostatic energy below their ordering temperature by forming magnetic domains separated by domain walls, i.e., interfaces where a gradual reorientation of individual magnetic moments occurs. However, on decreasing the particle volume, a critical domain size will be reached below which the proximity of domain walls is no longer energetically favored, and the single domain (SD) configuration is adopted. Frenkel and Dorfman formulated the criteria of critical radius for a magnetic particle which acts as an SD is given by [12]

$$D_s = 18 \frac{(AK)^{1/2}}{\mu_0 M_s^2} \quad (1)$$

where A is the exchange stiffness constant in J/m, K, is the magnetocrystalline anisotropy constant in J/m^3, μ_0 is the permeability of free space (4 π × 10^{-7} H/m), and M_S is the saturation magnetization. Depending on the size and material composition, the magnetic moments of single-domain particles can be 10^3-10^5 μ_B.

Coercivity (H_C) is extremely sensitive to particle size, unlike saturation magnetization, which is size-independent in principle, as shown in Figure 1.2a. In the multi-domain (MD) region, the H_C gradually increases with decreasing particle diameter (subdivided into domains). In contrast, it rapidly decreases to zero in the SD region with decreasing particle size. At the limiting condition $d = D_S$, the SD particle magnetized uniformly along the anisotropic easy axes, leading to a substantial H_C enhancement. Below D_S, H_C value decreases with the particle size due to the decrease of the magnetic anisotropy energy ($E_a = KV$, V is the particle volume). As the size is reduced further, the anisotropy energy value decreases and becomes comparable or even lower than the thermal energy (k_BT, k_B is Boltzmann constant). As a result, thermal

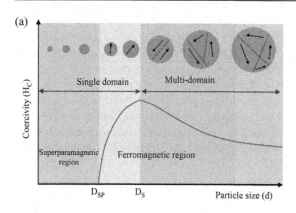 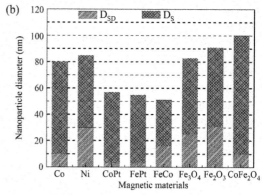

FIGURE 1.2 (a) Schematic illustration of the coercivity-size ($H_C - d$) relations of small ferromagnetic particles. As the size of ferromagnetic particles decreases, the H_C initially increases and reaches a maximum value at the critical single domain size (D_S). Furthermore, in the single domain regime ($d < D_S$), the H_C decreases as the particle size decreases until it reaches zero at ($d = D_{SP}$), known as a superparamagnetic regime. (b) Variation of critical size for single domain D_S and superparamagnetic limit D_{SP} for common ferro and ferri materials.

energy dominates the energy barrier for magnetization reversal, which spontaneously randomizes the magnetization of a particle from one easy axis to other directions even in the absence of a magnetic field. In the most extreme case, when the particle size is reduced further, the H_C becomes zero, resulting in a superparamagnetic state. Figure 1.2 depicts how the critical size for superparamagnetic transition (D_{SP}) and from a single domain to a multi-domain state (D_s) for several common magnetic materials depends on the type of material. Unlike the ferro- to paramagnetic transition in bulk magnets, the ferro- to superparamagnetic in nanoparticles is entirely from the size effect.

In the limit, $k_B T \gg KV$ the superparamagnetic particle can be considered freely fluctuating with a frequency f or a relaxation time, $\tau = (2\pi f)^{-1}$, which can be modeled by the Néel-Brown theory, expressed as [13]

$$\tau = \tau_0 \exp\left(\frac{KV}{k_B T}\right) \quad (2)$$

where τ_0 is a material-specific relaxation time in the range of 10^{-9} to 10^{-13} s. The τ increases as the samples are cooled to lower temperatures, which means fluctuations slow down. When $\tau \gg \tau_m$ (experimental measuring time), the system becomes static. Therefore, magnetization is measurable only if $\tau < \tau_m$. However, $\tau \approx \tau_m$ the particle is said to be blocked and the magnetic properties are characterized as "blocking" temperature, T_B, below which the particle moments appear to be frozen for τ_m. The T_B can be obtained from Eq. 2 as [13]

$$T_B \approx \frac{KV}{k_B \ln(\tau_m/\tau_0)} \quad (3)$$

It is clear from Eq. (3) that for the same size non-interacting particles, T_B depends on the magnetocrystalline anisotropy and the volume of magnetic nanoparticles. However, interparticle interaction modifies the energy barrier and produces collective properties in the case of multicore nanoparticles. The Vogel-Fulcher model describes the magnetic properties of interacting MNMs with reasonable accuracy, given by [13]

$$\tau = \tau_0 \exp\left[\frac{E_a}{k_B(T-T_0)}\right] \quad (4)$$

where T_0 is the Vogel-Fulcher temperature (a measure of the interaction strength) and activation energy (E_a/k_B) is required to overcome the energy barrier for magnetization reversal. With increased interparticle interaction, a spin-glass-like collective state, called super-spin glass, is formed. Also, in the super-ferromagnet state, long-range ferromagnetic order is occurred due to strong enough interparticle interaction. Recently, super-spin glass and the super-ferromagnetic state have been realized in Fe_3O_4 and Co nanoparticles. The inter-/intraparticle interactions can be controlled to further manipulate the magnetic properties, apart from size, shape, composition, and morphology.

1.2.2 Magnetization Dynamics

Nanomaterials' quasistatic and dynamic magnetic properties are different from their bulk counterparts. The time scale is very wide for magnetization dynamics. For example, exchange interaction occurs within tens of femtoseconds while domain wall movement takes a few nanoseconds to microseconds. Similarly, the spin waves (or SW, i.e., collective excitations of the precessions of individual spins) in ferromagnetic (FM) material can propagate in time scales ranging from a few hundreds of picoseconds to tens of nanoseconds before it dies out. The precessional motion of magnetization in the presence of a time-dependent magnetic field can be described mathematically by the Landau-Lifshitz-Gilbert equation of motion, given as [14]

$$\frac{d\mathbf{m}}{dt} = -\gamma \mathbf{m} \times \mathbf{H}_{eff} + \alpha \mathbf{m} \times \frac{d\mathbf{m}}{dt} \quad (5)$$

where $\mathbf{m} = \mathbf{M}/M_s$, M_s, γ, α, $\mathbf{H}_{eff}$ are the saturation magnetization, gyromagnetic ratio, Gilbert damping, and effective magnetic field, respectively. The first term on the right of Eq. (5) corresponds to the precessional motion of the magnetization vector $\mathbf{m}$ about the effective magnetic field direction. The second term is the damping term responsible for the magnetization vector's alignment in the direction of $\mathbf{H}_{eff}$. Schematic representations of magnetization precession with and without damping are shown in Figure 1.3.

Under the macro-spin model and for a uniform ellipsoidal particle with demagnetizing tensor axes N_x, N_y, and N_z, where $N_x + N_y + N_z = 4\pi$, Kittel [14] derived the frequency for the uniform precessional mode, also known as ferromagnetic resonance (FMR) mode as

$$\omega_r = \gamma \sqrt{\left[H + \mu_0 M_s \left(N_x - N_z\right)\right]\left[H + \mu_0 M_s \left(N_y - N_z\right)\right]} \quad (6)$$

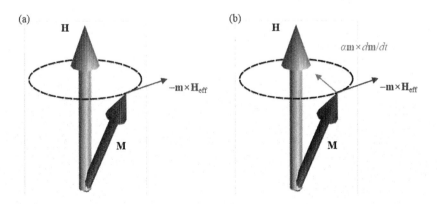

FIGURE 1.3 Sketch of magnetization precession around an applied magnetic field (H) (a) without damping and (b) with Gilbert damping.

For a thin film magnetized in the film plane ($N_x = 1$, $N_y = N_z = 0$ in SI unit), this formula reduces to [15] $\omega_r = \gamma\sqrt{H(H + \mu_0 M_s)}$, while for magnetizations oriented in the axis normal to the film plane ($N_x = N_y = 0$, $N_z = 1$), it turns out to be $\omega_r = \gamma(H - \mu_0 M_s)$. For a spherical particle ($N_x = N_y = N_z$), the FMR can be written as $\omega_r = \gamma H$. These equations are valid for the case of a vanishing wave vector, $k_{sw} = 0$.

For the cases of $\vec{k}_{sw} \neq 0$, the phase of the precessions of the neighboring spins differs, and thus the spins are no longer parallel with each other. Thus, the exchange interaction cannot be neglected. Furthermore, the dynamic part of the magnetization is a function of the position. Herrings and Kittel derived the dispersion relation of dipole-exchange spin waves in a ferromagnetic material of infinite size as given by the formula:

$$\omega(k_{sw}) = \gamma\sqrt{\left(H_{eff} + Dk_{sw}^2\right)\left(H_{eff} + Dk_{sw}^2 + \mu_0 M_s \sin^2\varphi\right)} \tag{7}$$

where φ is the angle between the direction of the spin-wave vector and the static magnetization; H_{eff} denotes effective magnetic field, which is as defined by the vector sum of externally applied magnetic field and the demagnetization field in the magnetic film, and $D = 2g_L\mu_B A/M_s$ is the exchange stiffness constant, where g_L is the Landé factor and μ_B is the Bohr magneton.

However, the dispersion relation cannot be explained by Eq. (7) for ultrathin films. R.W. Damon and J.R. Eshbach solved the Landau-Lifshitz-Gilbert (LLG) equation by considering Maxwell's equations in the magnetostatic limit for an in-plane magnetized thin film [14]. They discovered two types of solutions: the surface (Damon-Eshbach or DE) and the volume mode. In general, for an in-plane magnetized film, the surface mode called magnetostatic surface wave (MSSW) propagates perpendicularly to the magnetization, while the volume modes, called backward volume magnetostatic wave (BVMSW), propagate along the magnetization direction. Considering negligible anisotropy, the dispersion relations of MSSW and BVMSW modes are given by [15]

$$f_{MSSW} = \frac{\gamma}{2\pi}\sqrt{\left(H_{eff} + \frac{\mu_0 M_s}{2}\right)^2 - \left(\frac{\mu_0 M_s}{2}\right)^2 e^{-2k_{sw}d}} \tag{8a}$$

$$f_{BVMSW} = \frac{\gamma}{2\pi}\sqrt{H_{eff}\left(H_{eff} + \mu_0 M_s \frac{1 - e^{-k_{sw}d}}{k_{sw}d}\right)} \tag{8b}$$

The negative slope of the dispersion of BVMSW implies that the group and phase velocity are in opposite directions. On the other hand, a forward volume magnetostatic wave (FVMSW) mode can be excited in normally magnetized films, whose dispersion is described as

$$f_{FVMSW} = \frac{\gamma}{2\pi}\sqrt{H_{eff}\left[H_{eff} + \mu_0 M_s\left(1 - \frac{1 - e^{-k_{sw}d}}{k_{sw}d}\right)\right]} \tag{8c}$$

The amplitude of the magnetization precession has a cosinusoidal distribution across the film thickness for both volume modes (FVMSWs and BVMSWs). On the other hand, MSSWs are localized to one film surface on which they propagate. The distribution of precessional amplitude across the film thickness is exponential, with a maximum at one surface of the film, and can be switched to the other surface by reversal (i.e., 180° rotation) of either the field or the propagation direction. The SWs can be quantized in the plane of the film of confined magnetic structures. For a nanostructure of width w, the quantized SW vectors can be written as $k_n = 2\pi \lambda_n = n\pi/w$. However, solving the LLG equation (5) within the framework of micromagnetism is a convenient alternative for calculating the quantized SWs [16].

1.3 MAGNETIC NANOSTRUCTURES

Small magnetic particles of various types exist in nature or are created artificially. As illustrated in Figure 1.4, magnetic nanostructures have a fascinating variety of geometries, including a wide range of low-dimensional systems. The rough criteria for the existence of individuality of dots, rods, and particles of different shapes and sizes are the strength of the exchange and magnetostatic interparticle interactions. Particles properties are also of interest in the study of dot arrays, nanowires, thin film, and composites. Exciting use of small magnetic particles (< 10 nm size) is known as ferrofluids, where a variety of materials such as Fe_3O_4, $BaFe_{12}O_{19}$, Fe, Co, Ni, and their alloys form stable colloidal suspensions in organic liquids or hydrocarbons [7]. Water-based ferrofluids, on the other hand, are more challenging to produce.

The core-shell type magnetic nanoparticles have received special attention due to their physical and chemical properties, which strongly depend on the core, shell, and interface structure. Also, the core/shell interface microstructure is a point of interest in controlling the surface strain anisotropy on the magnetic core. The core-shell nanoparticles contain three parts: (i) FM core with a nonmagnetic (NM) shell, (ii) FM core with an FM shell, and (iii) FM core with an antiferromagnetic (AFM) shell or vice versa. So far, the nonmagnetic coating has been demonstrated to be useful in stabilizing magnetic cores and functionalizing surfaces for biological application [17–18]. The AFM shell coated on the FM core induces the so-called exchange bias effect. In this effect, under-compensated spins at the interface induce the unidirectional anisotropy that appears as a shift in the hysteresis loop along the magnetic field axis. Exchange bias in core-shell nanoparticles (NPs) has been observed in various systems, including Co/CoO, Co/MnO, CoPt/CoO, NiCo/NiCoO, FeNi/oxide, and so on [19]. Similarly, the magnetic core was encapsulated with an organic coating in biological applications to stabilize the entire particle and provide biocompatible support for biomolecules. Several approaches have been developed to functionalize magnetic NPs, including in situ coatings and post-synthesis coating with different organic materials, such as dextran, polyethylene glycol, starch, poly D, L-lactide, and polyethyleneimine, particularly hydrophilic organic materials [8].

Many magnetic thin films and multilayers can be considered nanostructures and exhibit exciting properties required for many applications. For example, nanostructured thin films with intermediate or high coercivities [9, 20] have been studied in the context of permanent magnetism and magnetic recording. These thin films or multilayers can be polycrystalline or epitaxial using FM, AFM, or combining both

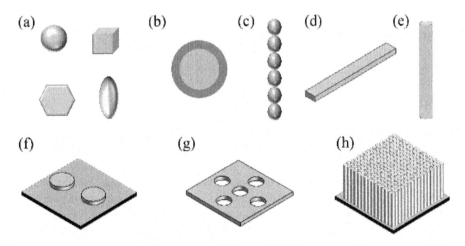

FIGURE 1.4 Nanostructure geometries: (a) nanoparticles of different shapes, (b) core-shell spherical nanoparticle, (c) chain of fine particles, (d) striped nanowire, (e) cylindrical nanowire, (f) nanodots, (g) antidots, and (h) self-assembly of nanorods.

or with nonmagnetic metal and insulating materials. Recently, there has been an increasing demand for fabricating epitaxial thin films in the nanostructured form to miniaturize integrated devices and extend the potential of establishing full nano-scale architectures crucial for modern and future spin-electronic-based devices.

The research on MNM progressed smoothly from elongated dots or thin-film patches to nanowires form, which is scientifically exciting due to their future applications in advanced nanotechnology [20]. Initially, the research on magnetic nanowires focused on exploratory issues, including the synthesis process to control or establish magnetic easy axis and magnetostatic interactions between the wire and the significance of shape anisotropy over magnetocrystalline. More recently, attention has shifted towards understanding the magnetization dynamic in nanowires or patterned magnetic media for magneto-optical, microwave nano-electronics applications [6].

Finally, the MNM in granular and composites form has significant importance in nanotechnology and science. The structural correlation lengths of typical nanocomposite materials range from 1 nm in amorphous X-ray structures to several 100 nm in submicron structures. Some well-known nanocomposite materials have been used in devices including Nd–Fe–B alloy as permanent magnets, Fe–Cu–Nb–Si–B amorphous alloy as soft magnets, and Co–Ag granular composites as magnetoresistive materials.

1.4 SYNTHESIS OF MNMS

There are several important issues for the synthesis of MNMs, including (i) obtaining materials with satisfactory high crystallinity and desirable crystal structure; (ii) obtaining monodisperse nanoparticles of controlled size and shape in a reproducible manner; and (iii) long-term stability of the nanoparticles. Several techniques for preparing MNMs, including physical, chemical, and biological methods, can be broadly classified into top-down and bottom-up approaches [1, 8] as outlined schematically in Figure 1.5. A comprehensive review of different synthesis methods is beyond the scope of this chapter. However, some of the key methods that have provided excellent control over the shape and size of MNMs are discussed briefly in this section. Apart from these conventional methods, we also briefly describe modern-day synthesis methods such as template-assisted fabrication, self-assembly, and lithography for fabrications of MNMs [3, 8].

1.4.1 Chemical Methods

The chemical synthesis of MNMs is a widely used method due to its simplicity and ability to mass-produce the desired material. Furthermore, in the chemical method, a short burst of nucleation followed by slow growth resulted in well-controlled monodisperse particles with sizes ranging from a few nm to μm. Although a wide range of chemical methods has been developed to date, we present some of the most commonly used methods in this section.

The thermal decomposition method involves the disintegration of organometallic compounds (e.g., carbonyls, acetylacetonates, or cuproferronates) in organic solvents in the presence of surfactants (e.g., hexadecyl amine or oleic acid) to produce monodisperse MNMs [1]. The shape and size of MNMs can be controlled by optimizing the proportions of different precursors during the reaction process. This technique has been successfully employed in the synthesis of a variety of MNMs, including magnetic oxides and metals (Co, Ni, Fe) with tunable shapes and sizes over a wide size range (3–50 nm) [21]. There are three essential parameters to be noted for achieving shape-controlled MNMs; (i) a suitable organometallic precursor with lower decomposing temperature than that of surfactants; (ii) the adsorption ability of the two surfactants must be different; and (iii) one of the surfactants should stimulate monomer exchange between particles to allow narrow size distribution.

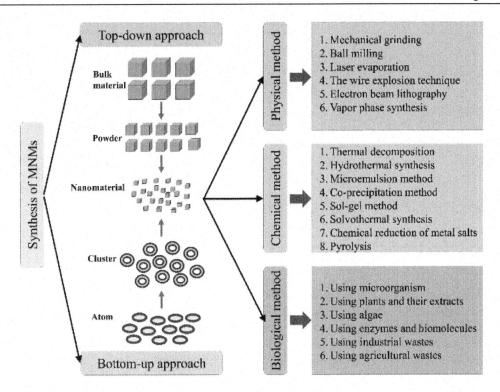

FIGURE 1.5 Different approaches and methods for synthesizing magnetic nanomaterials.

Hydrothermal synthesis, which employs the liquid-solid-solution (LSS) reaction, is another important chemical synthesis method that provides excellent control over the shape and size of the MNMs. The general strategy is based on the phase separation occurring at the interface of LSS phases present in the reaction at different temperature conditions. For example, metal nanomaterials were produced by reducing metal ions with ethanol at the interfaces of solid metal linoleate, the liquid phase of ethanol-linoleic acid, and water-ethanol solutions at different temperatures under hydrothermal conditions.

Another method to synthesize MNMs is the microemulsion approach, where two immiscible liquids are mixed to form a thermodynamically stable isotropic dispersion, and the interfacial film of surfactant molecules stabilizes the microdomain of one or both liquids. The microemulsion technique has been used to prepare a variety of MNMs, including metals and oxide, as detailed in several recent articles [3, 8].

Co-precipitation is a facile and convenient method for synthesizing oxides MNMs from aqueous metal salt solutions by adding a base in an inert atmosphere at room or elevated temperature. On the other hand, the shape, size, and composition of the MNMs are determined by the type of salts used (e.g., chlorides, nitrates, or sulfates), the reaction temperature, the pH value, and the ionic strength of the media. Significant progress has been made in preparing monodisperse MNMs using organic additives as a stabilizer or reducing agent. For example, Fe_3O_4 nanoparticles (4–10 nm) can be stabilized in 1 wt% polyvinyl alcohol (PVA) aqueous solution, but it forms chainlike clusters when PVA with 0.1 mol% carboxyl groups is used as the stabilizer. This finding suggests that selecting a suitable surfactant is critical for the stabilization of nanoparticles [3, 19].

The sol-gel is a proper wet chemical method that involves three steps. The first is to initiate a sol of nm-sized particles by hydroxylation and condensation of molecular precursors. Second, further condensation and inorganic polymerization result in forming a 3D metal oxide network known as a wet gel. Finally, heat treatment of gels results in fine crystalline MNMs. The shape, size, and composition of the MNMs are determined by the type of solvent, precursors, additives and catalysts, pH value, reaction temperature,

TABLE 1.1 Summary of the Comparison of the Synthetic Methods

METHOD	REACTION TEMPERATURE (°C)	REACTION PERIOD	SIZE DISTRIBUTION	SHAPE CONTROL	YIELD	DIFFICULTY
Thermal decomposition	100–320	Hours–days	Very narrow	Very good	Very high	Complicated, inert atmosphere
Microemulsions	20–50	Hours	Relatively narrow	Good	Low	Complicated, ambient condition
Hydrothermal	> 220	Hours	Very narrow	Very good	Medium	Simple, high pressure and temperature
Co-precipitation	20–90	Minutes–hours	Relatively narrow	Mode rate	Very High	Moderate
Sol-gel	> 40	Minutes–hours	Narrow	Good	High	Moderate

and mechanical agitation that affects the reaction rate, hydrolysis, and condensation process during the reaction. Compared with other methods, the sol-gel process offers several advantages for metal oxides compared to other methods. This includes excellent homogeneity, low cost, and high purity [2, 22].

Table 1.1 summarizes the benefits and drawbacks of the aforementioned synthesis methods. Asides from the methods listed prior, electrochemical reactions, solvothermal synthesis, atomic or molecular condensation, plasma or flame spraying synthesis, sputtering and thermal evaporation, chemical vapor deposition, and bio-assisted synthesis are all used to prepare MNMs [1, 3, 20, 23–25].

1.4.2 Physical Methods

The physical methods also include both top-down and bottom-up approaches. However, the bottom-up physical method produces more finely dispersed MNMs than the top-down approach. For example, a top-down approach is breaking bulk materials into irregularly shaped and sized nanoparticles by high-energy ball milling [23]. In contrast, bottom-up approaches like laser evaporation can produce controlled shape and size [17]. Other important physical methods used to prepare MNMs are inert-gas condensation and wire explosion.

Ball milling is a convenient and straightforward top-down approach that involves the mechanical grinding of coarse-textured particles into fine nanostructured particles [26]. In this method, the raw materials are enclosed in a small hollow cylindrical jar containing many steel balls as a grinding medium. The steel balls continuously collide with the solid materials, imparting kinetic energy to the solid material and resulting in nano/micro-sized powder. This method is widely used to prepare nanostructured magnetic alloys in powder form. The factors that need to be considered in ball milling are the ball-to-powder ratio, milling speed, and time, which affect the alloy formation process. In addition, a nonreactive organic solvent can be used to avoid agglomerations and reduce the milling temperature. Although the particles have wide size distribution compared to chemical methods, this technique is suitable for large-scale production when the shape of the MNMs is not essential.

Laser evaporation is another simple bottom-up technique in which a high-energy laser beam strikes the raw materials in a vacuum chamber. The electromagnetic energy is converted into electronic excitation, which is then converted to thermal, chemical, and mechanical energy causing evaporation of the target material, and MNMs are formed through condensation from the gaseous phase [3, 17]. This method is cost-effective for producing metallic magnetic nanoparticles without hazardous waste, as wet chemical methods do. On the other hand, wire explosion is a new attractive physiochemical technique for synthesizing MNMs. This method is a highly efficient one-step process that does not require additional steps such as separating NMs from solution or retreatment of byproducts. Although this method is environmentally friendly and requires minimal energy to produce less contaminated nanopowders, the produced MNMs are not monodispersed [3].

The wire explosion technique is a new physiochemical technique that is a safe and clean process for synthesizing MNMs. This method is a one-step, highly productive process that requires no additional steps like separating NMs from solution and retreatment of byproducts. This method was previously used to prepare iron oxide MNMs for the removal of arsenic from water. It is environmentally safe and requires minimum energy for making less contaminated nanopowders. However, the NMs produced through this method are not monodispersed [3].

1.4.3 Biological Methods

The biological method is a unique route for synthesizing MNMs that involves using living organisms like plants and microorganisms such as fungi, bacteria, viruses, and actinomycetes [21, 23–24]. The MNMs produced by this synthesis method are biocompatible and have valuable applications in the biological field. The advantages of this method are its efficiency and being environmentally friendly; however, the disadvantage is poor NPs distribution [21, 23]. Recently this method has become popular for preparing MNMs using plant tissue, exudates, extracts, and other plant parts. For example, MNMs of 20–45 nm size consisting of a magnetic core with a stable lipid membrane can be produced by microbial methods. Although the biological method is a promising synthesis technique, the mechanism of MNMs formation by microorganisms and plants is not fully understood and is still being investigated [23]. Despite some drawbacks, such as the yield and distributions of MNMs associated with biological methods, Fe_3O_4 prepared has been successfully used as a catalyst in photo-catalysis [25].

1.4.4 Advanced Synthesis Method

The current research in nanomagnetism focuses on materials-by-design to create engineered nanostructures that are not found in nature. The nanostructures can be of several types, including nanodots, nanoholes or antidots, nanostripes, nanowires, nanorings, NPs, granular media, nanojunctions, and multilayered nanostructures. In the early days, static and dynamic magnetization research was primarily focused on unpatterned magnetic films or multilayers. Recently with the advancement in the fabrication of well-defined MNM structures, the emphasis has gradually shifted to confined magnetic systems, including single nanostructures and their one- or two-dimensional (1D or 2D) arrays. Particularly, template-assisted synthesis, self-assembly, and lithography are the crucial methods for the fabrication of MNMs used for modern-day nanotechnology research [1, 3].

Self-assembly is a thermodynamically driven process that organizes structural units (such as atoms, or molecules or particles) into larger arrays of complex shapes that can withstand thermal fluctuations due to the system's inherent nano-scale forces. The process can be stimulated by either using a nanostructured surface known as template-assisted assembly or by applying external fields known as field-assisted assembly [1]. The external magnetic field guides the organization process for field-assisted assembles and controls their dimension and anisotropy. Self-assembly is a popular bottom-up approach for nanofabrication that has recently gained popularity.

On the other hand, lithography is a top-down microfabrication technique that allows for precise control of patterned nanostructure's shape, size, and alignment. However, the high cost and slow manufacturing process are a demerit to mass production. At the same time, the dimensional range of the nanostructure is determined by the types of lithography used. For example, electron beam lithography can reduce the sizes to tens of nanometers, but patterning takes a long time. On the other hand, photolithography saves time for patterning due to the large area exposure, whereas the wavelength of the light primarily determines the minimum size. The microfabrication involves two processes: (1) etching and

TABLE 1.2 Nanoparticle Characterization, Separation, and Quantification Techniques

TYPE	TECHNIQUE	SPECIFIC PURPOSE
Formation of nanoparticle	Ultraviolet-visible spectrophotometry	Size, stabilization, and aggregation of nanoparticles
Morphology and particle size	Transmission electron microscopy	Shape, size (10–10 m), morphology, and allographic structure
	High-resolution transmission electron microscopy	Arrangement of the atoms and their local microstructures, such as lattice fringe, glide plane, lattice vacancies and defects, screw axes, and surface atomic arrangement of crystalline nanoparticles
	Scanning electron microscopy	Morphology by direct visualization
	Atomic force microscopy	Length, width, height, and surface texture
	Dynamic light scattering	Particle size distribution
Surface charge	Zeta potential	Stability and surface charge of the colloidal nanoparticles, as well as the nature of core and shell
	Fourier transform infrared spectroscopy	Functional groups and determine the emission, absorption, and photoconductivity
	X-ray photoelectron spectroscopy	Bonding of different elements involved, as well as confirms the chemical composition
	Thermal gravimetric analysis	Confirm the formation of coatings such as surfactants or polymers to estimate the binding efficiency on the surface of magnetic nanoparticles
Crystallinity	X-ray diffraction	Crystal structure and related information such as space group, point group, bond length, etc.
Magnetic properties	Vibrating sample magnetometry	Magnetization and coercivity and confirm the type of magnetism present in the nanoparticles
	Superconducting quantum interference device	Determine the magnetic properties with better sensitivity
	Ferromagnetic resonance	Properties related to dynamic magnetization such as damping and spin wave
	The physical property measurement system	Magnetic susceptibility, magnetoresistance, Hall effect
Other techniques	Chromatography and related techniques	Separate nanoparticles based on their affinity towards the mobile phase
	Energy-dispersive X-ray spectra	Elemental composition of the nanoparticles
	Field flow flotation	Separate different nanoparticles based on their magnetic susceptibility
	Filtration and centrifugation techniques	Fractionate the preparative size of the nanoparticles
	Laser-induced breakdown detection	Analyze the concentration and size of colloids
	Small-angle X-ray scattering	Investigate the structural characterization of solid and fluid materials in the nanometer range
	X-ray fluorescence spectroscopy	Identify and determine the concentrations of elements present in solid, powdered, or liquid samples

(ii) lift-off. A film is first deposited on a substrate in the case of the etching process. The film is then spin-coated with a photoresist to form the desired resist pattern for lithographical patterning. Finally, the film is etched through the resist mask to create the desired pattern. However, in the case of the lift-off process, a resist is first spin-coated onto a substrate. After patterning the resist, a film is deposited on the substrate containing the patterned resist. After removing the resist from the substrate, the thin film deposited on the resist-free regions remains. Many reports are on the static and dynamic magnetic properties of nanostructured MNMs, including NiFe, Fe, Co, and Co/Pt [5, 16].

Another important approach for nanostructures fabrication is template-assisted techniques, with two significant advantages over other methods [26]. The first is the template determines the shape and the size of the nanostructures formed, and the second is the ease of fabrication of complex nanostructures with precise control of the composition along different branch lengths. On the other hand, this method has the inherent disadvantage of being a two-step process involving the production of high-quality templates followed by the deposition of magnetic material in the template. Furthermore, although several templates are commercially available, the options for pore size, thickness, and uniformity necessitate in-house template fabrication. Therefore, the method has usually employed the fabrication of highly oriented metal (Ni, Co) and metal oxide nanowires [1].

1.5 CHARACTERIZATION OF MNMS

The properties of MNMs depend primarily on the synthesis route, microstructure, and chemical structure of the involved materials. Therefore, proper characterization of MNMs is an essential part of research that helps understand the quality and properties essential for different technological applications. Therefore, different methods were applied to examine the structural, morphological, chemical, physical (magnetic, optical, mechanical, etc.), and biological properties of MNMs. Table 1.2 describes some of the important characterization techniques used to measure specific parameters.

1.6 APPLICATIONS OF MNMS

Advances in synthesis and characterization techniques have created novel opportunities for obtaining MNMs with controlled magnetic performances. Due to this versatility, these materials have been used successfully in a wide range of applications from biomedical to modern information technologies. The application of MNMs in various fields, including data storage, electronic, biological, environmental, agriculture, and catalysis, are briefly discussed in this section.

1.6.1 High-Density Data Storage

The unique ability of MNMs to control coercivity has led to many important technological applications, especially in data storage. As global data storage needs grow by 40% annually, it is important to increase the areal density in the hard disk drive (HDD) industry. Information storage density has reached a terabit per square inch (TB/in^2), which can be further increased by using small magnetic particles [17]. However, the fundamental issue for magnetic data storage is the stability of the bit information as the size of the area that holds the physical information is reduced. As the particle volume of the recording medium shrinks and bits are packed closer together, the medium must have high coercivity to maintain thermal stability beyond currently available magnetic write fields. Therefore, because of the superparamagnetic limit, the perpendicular magnetic recording (PMR) technology is unlikely to advance the areal density beyond 1 TB/in^2 [16].

Another promising method for increasing the areal density of data storage beyond 1 TB/in^2 is heat-assisted magnetic recording (HAMR) based on local heating through near-field optics [28]. Substantial progress has been made in HAMR, and companies in the HDD industry are looking to bring HAMR technology to products in the coming years. Seagate expects to have 36 TB HDDs by 2022, 48 TB drives before 2024, and 100 TB units in 2025. According to Seagate, HAMR can double the areal density every 2.5 years. The current PMR technology may be phased out in the future, and it appears that HDDs will remain a valuable asset for those seeking the lowest price per GB.

Researchers are interested in patterned magnetic media such as 2D dot arrays because of their potential applications in information storage or nonvolatile magnetic random access memory (MRAM) [5–6]. Microwave-assisted magnetic recording (MAMR) was also proposed as another energy-assisted recording approach [29]. MAMR reduces the switching field by an order of magnitude by using microwaves generated by a spin-transfer oscillator patterned in the write gap of the write head. Significant progress has been made lately on this technology. As a huge amount of data is generated and needs to be stored every day, advancement and progress for HDD, the leading candidate for next-generation storage, are required, while flash-based solid-state drives (SSDs) are proving to be a fierce competitor to conventional HDDs.

1.6.2 Nano-Electronics and Spintronic Applications

Magnetic nanoparticles open up new avenues for the development of nano-electronic devices. For example, single-electron devices can control the motion of a single electron or a group of electrons through the Coulomb blockade effect. However, when the magnetic properties of the nanoparticles are used, even more, intriguing effects can be observed. This starts the special subject of electronics based on spin named spintronics [30]. The giant magnetoresistance (GMR) phenomenon was the first example of using electron spin as a degree of freedom in nano-electronic devices [31]. GMR is the large change in electrical resistance in a magnetic multilayer system where the resistance depends on whether the orientation of magnetic moments is parallel or antiparallel in FM layers separated by a NM metallic layer, and the relative difference in resistance can reach 200% for some structures. A similar effect was also observed in granular magnetic composites consisting of FM nanoparticles embedded in an NM metallic matrix. Unfortunately, the magnetic fields required to align the moments are in the order of 10 kG, limiting granular material's applicability.

The next effect used in spintronics is tunneling magnetoresistance (TMR) [31], consisting of two FM layers (or nanoparticles) separated by a thin insulating barrier. The FM layer (or nanoparticle) polarizes the spins of electrons crossing the barrier by quantum mechanical tunneling before reaching the second FM layer (or nanoparticle). When the magnetic moments in FM layers (or particles) are parallel, electrons can tunnel the junction easier than the antiparallel orientation of magnetizations. Therefore, the magnitude of the current can be used to define two states, 1 and 0, in the TMR device and can be used to store digital information. Recently TMR has been used in MRAMs to combine the fast access time of the semiconductor RAM with the nonvolatility of the magnetic memories. Apart from standard FM materials and their alloys, spintronics can employ Heusler alloys (general composition X_2YZ) which exhibit a high (nearly 100%) level of spin polarization. For example, X = Co or Fe, Y = Mn, and Z = Al or Si or CrO_2 are popular for such applications [31].

While the initial application of spintronics was primarily related to data storage, increased efforts have recently been made to incorporate spintronic concepts into the logic operations of information technologies. In contrast to purely charge-based systems, information encoded in magnetization states is generally nonvolatile, reducing the power required for retaining data. However, as the size decreases, thermal stability becomes an issue that may necessitate the use of high anisotropic magnetic materials to develop ultra-miniaturized magnetic devices [30].

Recently the key developments in spintronics are poised to impact applications such as spin-orbitronics, spin-caloritronics, and magnonics. Spin-orbitronics is a new subfield of spintronics that exploits the spin-orbit coupling to generate and detect spin currents even in NM materials. The spin-orbit coupling also drives another phenomenon known as spin-caloritronics, which focuses on the interaction of charge and spin currents with thermal currents. This heat-driven phenomenon can be understood by considering the spin-wave excitations, which form the basis of a new spintronics concept called magnonics. The details are discussed in the section on emerging research in nanomagnetism. Furthermore, the discovery of spin-transfer torque (STT) and spin-orbit torque (SOT) of giant TMR in MgO-based magnetic tunnel junctions (MTJs) and large interfacial magnetic anisotropy at magnetic metal/oxide interfaces has led to the development of scalable nonvolatile MRAMs [5, 32]. The commercial STT-MRAMs are now used as a replacement for embedded flash (eFlash) memory or static RAM (SRAM) in embedded cache memories due to their easy integration with CMOS technology, low power consumption, and ultrafast switching with superior endurance [6].

1.6.3 Biomedical Applications

MNMs have been widely used in various biological applications due to their diverse physicochemical properties and biocompatibility. When an external magnetic field is applied to magnetic nanoparticles, it causes a torque on magnetic dipoles resulting in translation, rotation, and energy dissipation. This type of phenomenon has numerous applications, including cell separation/biomarker, targeted drug delivery, magneto-mechanical actuation of cell surface receptor, magnetic resonance imaging, drug release triggering, and hyperthermia [18, 23, 33–34]. However, the biocompatibility and toxicity of MNMs are the most important factors that need to consider for biomedical research. Over the last few decades, the fields of nanobiotechnology and molecular biology have emerged as novel approaches for the early detection and visualization of cancer cells by using MNMs for nano-scale imaging probes. In addition, MNMs are also being designed as vehicles to effectively deliver anticancer agents, genes, or proteins to the targeted tumor sites via enhanced infiltration and retention effect [21, 23, 33–34].

Magnetic hyperthermia is an important cancer therapy treatment by producing heat in suspensions of MNMs within the body under an external microwave magnetic field (MMF) [35]. The cancer cells have a lower pH than normal cells, making them more sensitive to magnetic hyperthermia. The general mechanism of heat release from MNMs depends on magnetization reversal processes that occur through hysteresis loss and susceptibility loss (Neel and Brownian relaxation) [35]. In the case of FM nanomaterials, the hysteresis loss contributes to magnetic hyperthermia, while in the case of superparamagnetic nanostructures, relaxation losses play a major role. The general strategy for magnetic hyperthermia is to raise the local temperature above 40°C by surface functionalizing MNMs with targeting agents that selectively accumulate at the tumor site and result in tumor suppression in the presence of external MMF. The most significant benefits of MNMs-based hyperthermia are the specific killing of cancer cells by penetration into deep tissue without damaging surrounding healthy tissues. This localized and selective heating dramatically improves the efficacy of cancer treatment, which has made it move from the lab to clinical trials [3].

The dissipation of thermal energy is measured in terms of a specific absorption rate or specific loss power (SAR/SLP), which can be defined as the ratio of thermal power dissipation to the mass of MNMs. SAR can be determined experimentally by measuring the rise in temperature over some time in a magnetic sample dispersed in a solvent when exposed to an MMF of particular amplitude and frequency and is given by [36]:

$$SAR = C_s \frac{\Delta T}{\Delta t} \frac{M_{sol}}{M_{MNM}} \qquad (9)$$

where C_s is the specific heat capacity of the solvent, M_{sol} is the mass of the solvent, M_{MNM} is the mass of the MNM, and $\Delta T/\Delta t$ is the temperature-time-dependent slope.

An important application of magnetic nanoparticles in biomedicine is MRI contrast agents, where superparamagnetic nanoparticles shorten the spin-spin relaxation time (T_2) of surrounding water protons in the presence of an external magnetic field, enabling these regions to darken in the T_2 weighted images

[8, 18, 37–39]. To ensure effective deployment as a contrast agent, MNM needs to target the desired tissue specifically. The specific targeting requires surface functionalization of MNM using various targeting agents. Iron oxide nanoparticles are used as the contrast agent with and without specific targeting [38]. Many other actively targeted magnetic nanostructures, including anti-carcinoembryonic antigen conjugated iron oxide and chlorotoxin conjugated PEG-coated iron oxide nanoparticles, have improved contrast in magnetic imaging [40]. In addition, spinel ferrites MFe_2O_4 (where M is +2 cation of Mn, Fe, Co, or Ni) were also investigated for in vivo MR targeted imaging [18, 40]. Herceptin conjugated Mn-doped iron oxide nanoparticles showed highly sensitive targeted in vivo mice imaging among all the targeting agents. Nevertheless, there are various reports of multifunctional nanoparticles consisting of magnetic and plasmonic materials, which can provide multifunctionality and will remain an active research area for biomedical applications.

1.6.4 Biosensing Applications

MNMs-based sensors have found widespread use in various fields, including food technology, lab testing, clinical diagnosis, and environmental monitoring, because of their biocompatibility, durability, and safety [3]. MNMs-based biosensors are superior to traditional biosensors due to their unique magnetic signaling and magnetic separation properties. MNMs can also be used as magnetic probes to detect analytes in biological samples due to their high signal-to-background ratio. The MNMs of Fe, Co, and Ni and their oxides were usually used for biosensors application. For example, a study reported a Fe_3O_4@Ag SER-based strip (SER: surface-enhanced Raman scattering) for detecting two respiratory viruses. Furthermore, a Pt-decorated Fe_3O_4 enzymes-based bioassay was prepared, where the magnetic properties enhance the magnetization in liquid samples, and their catalytic properties allow signal amplification via enzyme mimic reactions.

Magnetic-based microspheres or composites are made up of a variety of magnetic entities, porous polymeric structures, and specific metal cores, which offer higher affinities for targeted biomolecules [40–41]. Ferromagnetic NMs can significantly enhance the metal core's localized surface plasmon resonance (LSPR). Due to their high index of refraction and low molecular weight, these NMs are good candidates for enhancing the plasmonic response to biological binding and can detect even small molecules. In addition, their physicochemical properties and stability make them suitable for detection in vitro and in vivo without interfering with biological interactions. Therefore, much effort has been made to design MNMs-based biosensors that are easy to use and have high sensitivity for accurate detection.

1.6.5 Environmental and Agricultural Applications

The deterioration and contamination of water, soil, and atmosphere are becoming a central environmental issue due to the increased release of toxic and lethal chemicals from anthropogenic activities, which may pose serious health risks to humans if they enter the food chain. Nanotechnology has recently emerged as one of the reliable alternatives compared to conventional treatments. For example, Fe nanoparticles can degrade the chlorinated compounds producing eco-friendly trichloroethylene. This method has also been used to treat the immobilization of heavy metals and radionuclides [42].

Similarly, MNMs have also been used for water purification by targeting bacteria, dye degradation, and removing organic species. For example, functionalized Fe_3O_4@amino acid showed more extraordinary capturing ability (97%) for gram-positive and harmful bacteria [43]. In addition, magneto-catalysis is a well-known and highly effective method for degrading persistent organic pollutants into low-risk compounds. For example, cobalt ferrite–bismuth ferrite (CFO–BFO) core-shell nanoparticles were used to purify water under ac magnetic fields by catalytically degrading organic pollutant Rhodamine-B and other pharmaceutical compounds. Furthermore, the MNMs-based wastewater treatment also can reduce power and energy consumption with chemicals and residual wastes.

Another successful application of MNMs in the environment includes plant protection, seed germination, and soil quality improvement. For example, iron is an essential element that is required for many physiological activities in plants, including respiration, chlorophyll, biosynthesis, and redox reactions. Therefore, functionalized iron oxide MNMs were used as soil nutrients to increase production with minimal negative impacts [44]. The use of MNMs possesses great potential with unprecedented opportunities in improving water, soil, and environmental quality [45]

1.6.6 Other Applications

In the last few years, wave-absorbing materials have been proposed for use in communication technology and electromagnetic controlling systems. Wave-absorbing materials could weaken electromagnetic energy via absorption and then convert them into thermal energy within the materials while the attenuated reflective energy is reflected into the air. Different materials based on MNMs and their composites with natural and synthetic textiles have been proposed by different groups for EMI shielding [46]. It was reported that nanocomposites of carbon fiber with Ni and Fe_3O_4 nanoparticles could be a good candidate for EMI shielding [46]. Similarly, other nanocomposite fibers such as polyacrylonitrile coated Ni, Co, and their alloys and magnetic polyimide/Fe_3O_4 can be applicable for microwave absorption for broadband frequency applications.

The rapid development of intelligent electronic devices has led to the development of small-sized lithium batteries and has raised expectations for their performance. One of the key components that can be used to improve the performance of lithium batteries is electrode materials. Recent studies showed that a low-cost, high-quality electrode could be developed from magnetic materials (Fe_3O_4) and a carbon nanotube and graphene oxide hybrid that can maintain high stability and reversible specific capacity after hundreds of cycles [47]. Meanwhile, Co_3O_4/graphene and reduced graphene oxide/Fe_2O_3 electrodes also showed excellent performance in lithium-ion batteries.

1.7 EMERGING RESEARCH AREAS IN MNMS

Spintronics is a remarkable success story in transitioning fundamental science into real-world applications. Within a decade of its discovery, GMR has become an essential component of magnetic data storage redheads, allowing ever-increasing amounts of digital data to be stored. Thus, spintronics played a vital role in the digital revolution; digital data essentially replaced analog information to the point where information transmission, manipulation, and storage have become an integral part of our daily lives. In this section, we summarize recent key developments within spintronics.

It has recently been discovered that spin-orbit interactions can provide an efficient alternative route for generating spin currents from the flow of charge currents through nonmagnetic conductors. The phenomenon is called spin Hall effects (SHE) [5, 32, 48]. The spin current generated by the SHE is comparable to or even higher than that generated by direct electrical injection when the charge current from the FM layer passes through the interface to an adjacent nonmagnetic conductor with spin-polarized charges. The transverse geometry of SHEs allows for the generation of large spin currents from charge currents and allows for careful metrology of these effects using macroscopically generated spin currents. Such macroscopic spin currents can be achieved via spin pumping, which occurs when FMR is excited in the FM in direct contact with a nonmagnetic conductor [32].

Another recently discovered effect occurs at the interfaces between FM layers and heavy-element materials: chiral Dzyaloshinskii-Moriya interaction (DMI) stabilized by breaking inversion symmetry [6, 40]. DMI has the attractive property of stabilizing domain walls with well-defined chirality when combined with spin-transfer torques from SHEs. This has potential applications for racetrack memories, where information is encoded in the presence or absence of magnetic domain walls, which can be reversibly shifted using electric charge currents. Furthermore, the chirality of the domain walls can also produce more complex spin textures known as skyrmions, which can be used in information carriers [49]. Indeed, it has already been demonstrated that 2D patterns can be coherently moved in magnetic-field-driven skyrmion bubbles paving the way for massively parallel racetrack memories [49].

The SHE discussed prior has been instrumental in ushering in a new field of research, spin-caloritronics [32], wherein, in addition to purely charge current–driven effects, the interaction of spin and charge degrees of freedom with heat currents are also investigated. The key phenomenon that started this field is the spin Seebeck effect (SSE), where the generation of a spin current in a bilayer system made up of a ferromagnetic material (FM) and a normal metal (NM) via the magnetization dynamics induced by the application of a thermal gradient. The nontrivial connections between heat and spin transport have opened several fundamental questions and new application possibilities.

As discussed in Section 1.2.2, "Magnetization Dynamics", spin waves are the fundamental quasiparticle excitations of magnetically ordered systems and are also referred to as magnons. Spin waves have been proposed as information carriers for low-power data storage and processing, which has given rise to the field of magnonics [15, 50]. The utilization of magnonic approaches in spintronics gave birth to the field of magnon spintronics. Furthermore, magnonic crystals are critical components for magnon spintronic applications as they enable access to novel multifunctional magnonic devices. These devices can be used as spin-wave conduits and fitters, sensors, delay lines, phase shifters, auto-oscillators' components, frequency and time inverters, data-buffering elements, power limiters, nonlinear enhancers in a magnon transistor, and components of logic gates [4, 51].

The ongoing development of devices and experimental techniques that utilize the quantum nature of the system for the storage, transfer, and processing of quantum information is important to achieve the ambitious goals of workable quantum computation. In the last few years, an exciting strategy has been explored to access, through hybrid quantum systems, novel capabilities in existing quantum technologies by combining different physical systems (e.g., photons and magnons) possessing distinct characteristics [52]. Combining the superconducting resonator and nano-scale FM material opens a promising platform for investigating on-chip quantum magnonics and spintronics. It brings new potential for coherent manipulation and long-distance propagation of spin information.

Cleavable two-dimensional van der Waals (vdW) nanomaterials, which can thin down to monolayer, showed the multiple functionalities due to quantum confinement of the electrons in one plane, revolutionizing the research world. As noted earlier, that size can significantly modify the spin configuration. Therefore, for the completeness of the topic, we briefly review the advances in vdW MNMs. The 2D magnetism research field is rapidly growing since the long-range magnetic order recently added functionality in the vdW material's category, which can survive down to the monolayer limit. A wide variety of materials with different spin ordering were realized with a short period of the last five years, and many more are theoretically predicted. Namely, insulating antiferromagnets ($FePS_3$, $MnPS_3$, $NiPS_3$, $MnPSe_3$), ferromagnetic semiconductors (CrI_3, $CrGeTe_3$), and ferromagnetic metals ($MnSe_2$, VSe_2). The magnetic anisotropy, spin direction, interlayer magnetic ordering, exchange gap, and magneto-optical behavior can be easily tuned via just varying the halogen composition within the same material family. Different types of magnetic ordering behavior are observed in the same materials (CrI_3) depending on the odd number of layers used. The magnetic ordering temperature of the vdW MNMs ranges from room temperature down to liquid helium. The current synthesis status of the vdW MNMs is laboratory-based single-crystal growth using chemical vapor transport and self-flux methods, and the monolayer is exfoliated using mechanical means. There is a long way to go to achieve batch processing for practical devices. The biggest hurdle is the stability of vdW MNMs in the air, which severely affects performance. Investigating vdW MNMs aims to discover a stable semiconducting material with long-range FM order above room temperature with gate tenability, which means the magnetic order can be controlled by an applied electric field. This allows realizing low-power data storage and transport application that can synchronize with current CMOS technology.

1.8 SUMMARY

This introductory chapter discusses broad aspects of the research on magnetic nanomaterial MNMs, from their fundamental nanomagnetism to synthesis to essential applications. The synthesis of MNMs has made significant progress focusing on the geometry of nanoparticles and uniformity yield, particularly with chemical routes and advanced techniques such as nanolithography and laser ablation. Details of some of the important synthesis routes and different shapes of nanostructures are presented based on the available literature. Next, diverse applications of MNMs are summarized, ranging from biomedical to information technology. Biomedical is the most benefited field where MNMs are applied to diagnose the infected region through nanomedicine imaging or treatment. Various magnetic nanoparticle modalities are being studied in clinical trials for cancer cell imaging and therapy. The information technology sector is constantly searching for low-power, efficient, and high-density data storage and transfer devices to tackle energy crises. Some of the recent developments in nanomagnetism with new computational paradigms for information processing technologies have been proposed. The vdW MNMs are promising candidates for low-power devices; a short description is given here. This chapter provides a helpful introduction and different perspectives on nanomagnetism and its emerging research field for the researcher working on physics or chemistry or the beginner material scientist or biologist.

1.9 ACKNOWLEDGMENTS

One of the authors, B. Bhoi, acknowledges BK21 PLUS SNU Materials Education/Research Division for Creative Global Leaders for providing financial support while preparing this book chapter.

REFERENCES

1. Singamaneni, S., Bliznyuk, V.N., Binek, C., and Tsymbal, E.Y.: 'Magnetic nanoparticles: Recent advances in synthesis, self-assembly and applications', *Journal of Materials Chemistry*, 21 (2011) 16819–16845
2. Lu, A.-H., Salabas, E.L., and Schüth, F.: 'Magnetic nanoparticles: Synthesis, protection, functionalization, and application', *Angewandte Chemie International Edition*, 46 (2007) 1222–1244
3. Ali, A., Shah, T., Ullah, R., Zhou, P., Guo, M., Ovais, M., Tan, Z., and Rui, Y.: 'Review on recent progress in magnetic nanoparticles: Synthesis, characterization, and diverse applications', *Frontiers in Chemistry*, 9 (2021) 629054
4. Pirro, P., Vasyuchka, V.I., Serga, A.A., and Hillebrands, B.: 'Advances in coherent magnonics', *Nature Reviews Materials*, 6 (2021) 1114–1135
5. Hirohata, A., Yamada, K., Nakatani, Y., Prejbeanu, I.-L., Diény, B., Pirro, P., and Hillebrands, B.: 'Review on spintronics: Principles and device applications', *Journal of Magnetism and Magnetic Materials*, 509 (2020) 166711
6. Dieny, B., Prejbeanu, I.L., Garello, K., Gambardella, P., Freitas, P., Lehndorff, R., Raberg, W., Ebels, U., Demokritov, S.O., Akerman, J., Deac, A., Pirro, P., Adelmann, C., Anane, A., Chumak, A.V., Hirohata, A., Mangin, S., Valenzuela, S.O., Onbaşlı, M.C., d'Aquino, M., Prenat, G., Finocchio, G., Lopez-Diaz, L., Chantrell, R., Chubykalo-Fesenko, O., and Bortolotti, P.: 'Opportunities and challenges for spintronics in the microelectronics industry', *Nature Electronics*, 3 (2020) 446–459
7. Skomski, R.: 'Nanomagnetics', *Journal of Physics: Condensed Matter*, 15 (2003) R841–R896
8. Duan, M., Shapter, J.G., Qi, W., Yang, S., and Gao, G.: 'Recent progress in magnetic nanoparticles: Synthesis, properties, and applications', *Nanotechnology*, 29 (2018) 452001

9. Mohapatra, J., Xing, M., Elkins, J., and Liu, J.P.: 'Hard and semi-hard magnetic materials based on cobalt and cobalt alloys', *Journal of Alloys and Compounds*, 824 (2020) 153874
10. Krishnan, K.M., Pakhomov, A.B., Bao, Y., Blomqvist, P., Chun, Y., Gonzales, M., Griffin, K., Ji, X., and Roberts, B.K.: 'Nanomagnetism and spin electronics: Materials, microstructure and novel properties', *Journal of Materials Science*, 41 (2006) 793–815
11. Pereira, C., Pereira, A.M., Fernandes, C., Rocha, M., Mendes, R., Fernández-García, M.P., Guedes, A., Tavares, P.B., Grenèche, J.-M., Araújo, J.P., and Freire, C.: 'Superparamagnetic MFe2O4 (M = Fe, Co, Mn) nanoparticles: Tuning the particle size and magnetic properties through a novel one-step coprecipitation route', *Chemistry of Materials*, 24 (2012) 1496–1504
12. O'Handley, R.C.: *Modern Magnetic Materials: Principles and Applications* (1999). Wiley & Sons, Inc., New York
13. Rosensweig, R.E.: 'Heating magnetic fluid with alternating magnetic field', *Journal of Magnetism and Magnetic Materials*, 252 (2002) 370–374
14. Maksymov, I.S., and Kostylev, M.: 'Broadband stripline ferromagnetic resonance spectroscopy of ferromagnetic films, multilayers and nanostructures', *Physica E: Low-dimensional Systems and Nanostructures*, 69 (2015) 253–293
15. Serga, A.A., Chumak, A.V., and Hillebrands, B.: 'YIG magnonics', *Journal of Physics D: Applied Physics*, 43 (2010) 264002
16. Bedanta, S., Barman, A., Kleemann, W., Petracic, O., and Seki, T.: 'Magnetic nanoparticles: A subject for both fundamental research and applications', *Journal of Nanomaterials*, 952540 (2013) 1–22
17. Biehl, P., Von der Luhe, M., Dutz, S., and Schacher, F.H.: 'Synthesis, characterization, and applications of magnetic nanoparticles featuring polyzwitterionic coatings', *Polymers (Basel, Switz.)*, 10 (2018) 91
18. Williams, H.M.: 'The application of magnetic nanoparticles in the treatment and monitoring of cancer and infectious diseases', *Bioscience Horizons: The International Journal of Student Research*, 10 (2017) 1–10
19. Sun, S., Zeng, H., Robinson, D.B., Raoux, S., Rice, P.M., Wang, S.X., and Li, G.: 'Monodisperse MFe2O4 (M = Fe, Co, Mn) nanoparticles', *Journal of the American Chemical Society*, 126 (2004) 273–279
20. Bowden, G.J., Beaujour, J.M.L., Gordeev, S., Groot, P.A.J.d., Rainford, B.D., and Sawicki, M.: 'Discrete exchange-springs in magnetic multilayer samples', *Journal of Physics: Condensed Matter*, 12 (2000) 9335–9346
21. Zhang, D., Ma, XL, Gu, Y., Huang, H., and Zhang, G.W.: 'Green synthesis of metallic nanoparticles and their potential applications to treat cancer', *Frontiers in Chemistry*, 8 (2020) 799
22. Mosayebi, J., Kiyasatfar, M., and Laurent, S.: 'Synthesis, functionalization, and design of magnetic nanoparticles for theranostic applications', *Advanced Healthcare Materials*, 6 (2017) 1700306
23. Patra, JK, and Baek, K.-H.: 'Green nanobiotechnology: Factors affecting synthesis and characterization techniques', *Journal of Nanomaterials*, 417305 (2014) 1–12
24. Verma, R., Pathak, S., Srivastava, A.K., Prawer, S., and Tomljenovic-Hanic, S.: 'ZnO nanomaterials: Green synthesis, toxicity evaluation and new insights in biomedical applications', *Journal of Alloys and Compounds*, 876 (2021) 160175
25. Zhang, Q., Yang, X., and Guan, J.: 'Applications of magnetic nanomaterials in heterogeneous catalysis', *ACS Applied Nano Materials*, 2 (2019) 4681–4697
26. Ozin, G.A.: 'Nanochemistry: Synthesis in diminishing dimensions', *Advanced Materials*, 4 (1992) 612–649
27. Castro, C.M., and Mitchell, B.: 'Nanoparticles from mechanical attrition', in *Synthesis, functionalization and surface treatment of nanoparticles*, Baraton, M.-L. ed. (2003) American Scientific Publishers, Valencia, CA
28. Sharrock, M.P.: 'Time-dependent magnetic phenomena and particle-size effects in recording media', *IEEE Transactions on Magnetics*, 26 (1990) 193–197
29. Zhu, J., Zhu, X., and Tang, Y.: 'Microwave assisted magnetic recording', *IEEE Transactions on Magnetics*, 44 (2008) 125–131
30. Wolf, S.A., Awschalom, D.D., Buhrman, R.A., Daughton, J.M., Molnár, S.V., Roukes, M.L., Chtchelkanova, A.Y., and Treger, D.M.: 'Spintronics: A spin-based electronics vision for the future', *Science*, 294 (2001) 1488–1495
31. Zabel, H.: 'Progress in spintronics', *Superlattices and Microstructures*, 46 (2009) 541–553
32. Hoffmann, A., and Bader, S.D.: 'Opportunities at the frontiers of spintronics', *Physical Review Applied*, 4 (2015) 047001
33. Wu, W., Wu, Z., Yu, T., Jiang, C., and Kim, W.-S.: 'Recent progress on magnetic iron oxide nanoparticles: Synthesis, surface functional strategies and biomedical applications', *Science and Technology of Advanced Materials*, 16 (2015) 023501
34. Shabatina, T.I., Vernaya, O.I., Shabatin, V.P., and Melnikov, M.Y.: 'Magnetic nanoparticles for biomedical purposes: Modern trends and prospects', *Magnetochemistry*, 6 (2020) 30

35. Mohapatra, J., Xing, M., and Liu, J.P.: 'Inductive thermal effect of ferrite magnetic nanoparticles', *Materials*, 12 (2019) 3208
36. Rajan, A., and Sahu, NK: 'Review on magnetic nanoparticle-mediated hyperthermia for cancer therapy', *Journal of Nanoparticle Research*, 22 (2020) 319
37. Ansari, SAMK, Ficiarà, E., Ruffinatti, F.A., Stura, I., Argenziano, M., Abollino, O., Cavalli, R., Guiot, C., and D'Agata, F.: 'Magnetic iron oxide nanoparticles: Synthesis, characterization and functionalization for biomedical applications in the central nervous system', *Materials*, 12 (2019) 465
38. Ahmed, N., Jaafar-Maalej, C., Eissa, M.M., Fessi, H., and Elaissari, A.: 'New oil-in-water magnetic emulsion as contrast agent for in vivo magnetic resonance imaging (MRI)', *Journal of Biomedical Nanotechnology*, 9 (2013) 1579
39. Kole, M., and Khandekar, S.: 'Engineering applications of ferrofluids: A review', *Journal of Magnetism and Magnetic Materials*, 537 (2021) 168222
40. Sun, C., Veiseh, O., Gunn, J., Fang, C., Hansen, S., Lee, D., Sze, R., Ellenbogen, R.G., Olson, J., and Zhang, M.: 'In Vivo MRI detection of gliomas by chlorotoxin-conjugated superparamagnetic nanoprobes', *Small*, 4 (2008) 372–379
41. Legge, C.J., Colley, H.E., Lawson, M.A., and Rawlings, A.E.: 'Targeted magnetic nanoparticle hyperthermia for the treatment of oral cancer', *Journal of Oral Pathology & Medicine*, 48 (2019) 803
42. Belyanina, I., Kolovskaya, O., Zamay, S., Gargaun, A., Zamay, T., and Kichkailo, A.: 'Targeted magnetic nanotheranostics of cancer', *Molecules*, 22 (2017) 975
43. Mondal, P., Anweshan, A., and Purkait, M.K.: 'Green synthesis and environmental application of iron-based nanomaterials and nanocomposite: A review', *Chemosphere*, 259 (2020) 127509
44. Jin, Y., Liu, F., Shan, C., Tong, M., and Hou, Y.: 'Efficient bacterial capture with amino acid modified magnetic nanoparticles', *Water Research*, 50 (2014) 124–134
45. Mishra, S., Keswani, C., Abhilash, P.C., Fraceto, L.F., and Singh, H.B.: 'Integrated approach of agri-nanotechnology: Challenges and future trends', *Frontiers in Plant Science*, 8 (2017) 471
46. Shahidi, S.: 'Magnetic nanoparticles application in the textile industry – A review', *Journal of Industrial Textiles*, 50 (2019) 970–989
47. Piraux, L.: 'Magnetic nanowires', *Applied Sciences*, 10 (2020) 1832
48. Fergus, J.W.: 'Recent developments in cathode materials for lithium ion batteries', *Journal of Power Sources*, 195 (2010) 939–954
49. Sinova, J., Valenzuela, S.O., Wunderlich, J., Back, C.H., and Jungwirth, T.: 'Spin hall effects', *Reviews of Modern Physics*, 87 (2015) 1213–1260
50. Everschor-Sitte, K., Masell, J., Reeve, R.M., and Kläui, M.: 'Perspective: Magnetic skyrmions – Overview of recent progress in an active research field', *Journal of Applied Physics*, 124 (2018) 240901
51. Chumak, A.V., Serga, A.A., and Hillebrands, B.: 'Magnonic crystals for data processing', *Journal of Physics D: Applied Physics*, 50 (2017) 244001
52. Clerk, A.A., Lehnert, K.W., Bertet, P., Petta, J.R., and Nakamura, Y.: 'Hybrid quantum systems with circuit quantum electrodynamics', *Nature Physics*, 16 (2020) 257–267

Nanostructured Magnetic Semiconductors

2

Alessandra S. Silva[1], Éder V. Guimarães[1], Tasso O. Sales[2], Wesley S. Silva[2], Elisson A. Batista[3], Carlos Jacinto[2], Anielle C. A. Silva[4,5], Noelio O. Dantas[4], and Ricardo S. Silva[1]

1 Instituto de Ciências Exatas, Naturais e Educação (ICENE), Departamento de Física, Universidade Federal do Triângulo Mineiro, Uberaba, Minas Gerais, Brazil

2 Group of Nano-Photonics and Imaging, Instituto de Física, Universidade Federal de Alagoas, Maceió, Brazil

3 Instituto de Física, Universidade Federal de Uberlândia, Uberlândia MG, Brazil

4 Laboratório de Novos Materiais Nanoestruturados e Funcionais, Instituto de Física, Universidade Federal de Alagoas, Maceió, AL, Brazil

5 Programa de Pós-Graduação da Rede Nordeste de Biotecnologia (RENORBIO), Universidade Federal de Alagoas, Maceió, AL, Brazil

Contents

2.1	Introduction	24
2.2	Nanostructured Materials and Quantum Dots	24
2.3	Diluted Magnetic Semiconductor (DMS) Nanocrystals	25
	2.3.1 Exchange Interactions in DMS Nanocrystals	26
	2.3.2 Crystal Field Theory (CFT)	26
2.4	Experimental Procedures	26
	2.4.1 Growth of NCs in Glassy Matrices	26
	2.4.2 Growth of NCs by the Precipitation Method	27
2.5	Nanostructured Magnetic Semiconductors	27

DOI: 10.1201/9781003197492-2

2.5.1 $Zn_{1-x}A_x Te$ Nanocrystals (A = Mn; Cu) Grown in the P_2O_5–ZnO–Al_2O_3–BaO–PbO
         Glassy System   27
2.5.2   Nanostructured Glasses With $Zn_{1-x}Mn_xTe$ and Doped With Europium   30
2.5.3 $Pb_{1-x}Co_xS$ Nanocrystals Synthesized in SiO_2–Na_2CO_3–Al_2O_3–B_2O_3–PbO Host Glass   31
2.5.4   Optical and Magnetic Properties of $Zn_{1-x}Co_xO$ Nanopowders   35
2.6 Conclusion   36
2.7 Acknowledgments   37
References   37

2.1 INTRODUCTION

Nanoscience and nanotechnology are present in the world market in the form of technologically sophisticated products, such as state-of-the-art microprocessors, digital TVs, cell phones, and broadband Internet with optical fibers, among others. Advances in nanomaterials have allowed the development of new types of lasers, as well as an increase in densities and digital data storage capacities [1]. There is also the diagnosis of diseases using nanosensors and in the release and control of drug concentration in the body [2], more diversified and efficient nanometric catalysts [3], advanced materials for prostheses [4], and destroying viruses or cancer cells, where they are located in the body [5].

Due to the worldwide relevance of nanoscience and nanotechnology, several research groups are engaged in studying a variety of nanostructured compounds. The main motivation for the study of nanostructures/nanocrystals (NCs) is that, at nanometer size scale, the materials have unique and often surprising properties caused by effects predicted by quantum mechanics. That is, due to size reduction or shape modification, these NCs can present quantum confinement effects of their charge carriers, thus being called quantum dots (QDs). On the other hand, when these QDs are doped with magnetic impurities, their properties can be modified by interaction dependent on the size and concentration of the magnetic ions in the host environment.

Possible applications for materials with these properties include: dramatically increasing the data storage and processing capacity of computers; creating new mechanisms for drug delivery that are safer and less harmful to the patient; creating materials for buildings, cars, and airplanes that are lighter and more resistant than metals and plastics; and many more innovations aimed at saving energy, protecting the environment, and using less scarce raw materials. In the field of nanotechnology, these are very current and concrete possibilities.

2.2 NANOSTRUCTURED MATERIALS AND QUANTUM DOTS

A quantum dot semiconductor is a nanocrystal that exhibits quantum confinement of its charge carriers (electrons and holes) in three spatial dimensions. Due to this confinement, electrons and holes in a QD have their energy quantized into discrete values, as in an atom. For this reason, QDs are sometimes called "artificial atoms." In this way, energy levels can be controlled by changing the shape and size of these NCs [6]. For this reason, these semiconductor NCs exhibit unique electronic, optical, and photochemical properties that differ significantly from those observed in materials with bulk properties (without quantum confinement). One of the most exciting qualities is the ability to alter the properties of the material at specific wavelengths of the electromagnetic spectrum, resulting in a variety of optoelectronic applications. Among several semiconductor NCs, there are zinc oxide (ZnO), lead sulfide (PbS), and zinc telluride (ZnTe). The zinc oxide (ZnO) presents a bandgap of 3.44 eV and interesting physical and biological properties, such as ultraviolet absorption and bactericidal and antitumor properties [7], being suitable for technological applications of photonic devices operating in the blue and ultraviolet region and fabrication of nanodevices electronics, among others [8]. Lead sulfide (PbS) is a crystalline material that belongs to the narrow gap semiconductor class (bandgap of 0.41 eV). PbS has been used in optoelectronics, sensors, solar cells, diodes, and laser technologies. However, the structural, electronic, optical, and thermal properties of this semiconductor can be enhanced with changes in the gap energy. The relatively large exciton Bohr radius (20 nm) [9] provides the PbS quantum confinement regimes that keep the crystal structure (rock-salt) undisturbed. Different quantum confinement regimes change the bandgap to 5.2 eV in PbS quantum dots [10]. The breaking of translational symmetry, large surface area, structural defects, and anisotropic surface [11] are phenomena provided by quantum confinement that differ from such properties in PbS. PbS NCs are efficient sensitizers: that is, materials that can be applied as photodetectors and photovoltaic cells in the near-infrared region [12]. At room temperature, ZnTe NCs have very interesting physical properties, such as cubic zincblende-type structure, exciton Bohr radius a_B = 5.2 nm, and an energy gap around E_g = 2.26 eV, [13]. Depending on their size and shape, these quantum dots can absorb and emit light in the visible and near-ultraviolet (UV) electromagnetic spectrum. In this context, ZnTe QDs have been synthesized using several methods, such as: molecular beam epitaxy (MBE), in which ZnTe NCs are usually self-assembled in layers of another semiconductor, such as ZnSe, and grown on a GaAs substrate [14]; colloidal, in which the ZnTe QDs are obtained from organic solutions and it is possible to control the size and shape through the synthesis temperature, growth time, concentration of precursor reagents, and chemical nature of the ligands [15]; and mechanical alloying, in which the precursors in powder form are subjected to mechanical grinding, resulting in the formation of NCs with desired properties, after suitable thermal treatment of the nanopowders [13].

Although QDs have been synthesized by the aforementioned methods, some possible applications require these NCs to be incorporated into robust and transparent host materials. Thus, a way to produce high-quality QDs is to grow them in a host glassy system synthesized by the fusion method. Thus, from post-fusion heat treatments, precursor ions diffuse and form NCs. Two parameters are fundamental at this stage: the time and temperature of the heat treatment, which allow for the control of the size of the NCs. Another effective and viable way to produce these nanostructures is by the chemical precipitation method via aqueous solution. Among the characteristics of this method, it can be mentioned that it is a relatively simple, efficient, reproducible process, applicable on a large scale, with relatively low cost, environmentally friendly, and allows direct control over synthesis parameters, enabling greater control over the composition, shape, and size of NCs [16].

2.3 DILUTED MAGNETIC SEMICONDUCTOR (DMS) NANOCRYSTALS

In diluted magnetic semiconductor (DMS) nanocrystals, a part of the native cations of the host semiconductor is replaced by transition metals. Thereby, an exchange interaction between the host electronic subsystem and the partially filled d or f level electrons from the intentionally introduced magnetic atom may occur. This allows for the control of the physical properties of the DMS as function of magnetic field. For this reason, QDs doped with a variety of transition metals (TM) have been studied due to their significant physical and chemical properties [17]. These can be exploited for several technological applications, such as spintronic devices or tunable wavelength lasers. Due to these interesting DMS properties, $Zn_{1-x}A_x$Te (A = Mn; Cu), $Pb_{1-x}Co_x$S, and $Zn_{1-x}Co_x$O NCs have been synthesized via several methods, such as the hydrothermal [18], the solvothermal [19], the thermal evaporation deposition [20], combustion reaction [21], and microwave-assisted hydrothermal [22]. These materials can also be doped with rare-earth (RE) ions to improve the quantum efficiency of the luminescence of TM or RE ions from the energy transfer process between RE and TM ions.

2.3.1 Exchange Interactions in DMS Nanocrystals

The *sp-d* exchange interaction originates from the spatially extended electronic wave functions of carriers that overlap with a large number of local magnetic spin moments, aligned in the presence of an external magnetic field. Thus, the Hamiltonian for an exciton in a DMS under the action of an external magnetic field is given by [23]:

$$H = H_0 + H_{int} + H_{sp-d} + H_{d-d} \qquad (1)$$

in which H_0 describes the kinetic and potential energies of the exciton in a perfect crystal, H_{int} describes the intrinsic interaction of the exciton with the external magnetic field, H_{sp-d} describes the magnetic exchange interactions (*sp-d*) [electrons (e) and holes (h)] and the magnetic dopants, and H_{d-d} refers to the interaction between the neighboring TM ions, which interact through the so-called *d-d* double exchange interaction. Depending on the magnetic dopant concentration (low concentrations), this *d-d* interaction is weaker than the *sp-d* interaction, since, in this case, it can be disregarded. The term H_{int} is independent of the magnetic dopant concentration, and it can be considered as an intrinsic contribution to the total Zeeman splitting observed in doped semiconductors. Already the term H_{sp-d} is directly dependent on the concentration of the magnetic dopants.

2.3.2 Crystal Field Theory (CFT)

The crystal field theory (CFT) is a study that explains the electronic absorption and emission transitions obtained, respectively, by OA and PL, of TM ions incorporated at semiconductor NC sites. As proposed by Hans Bethe in 1929, it is a simple approach to the formation of coordination compounds, analyzing only the effects caused by the ligands on the *d* orbital of the central atom. These effects are very interesting in interpreting spectroscopic, magnetic, and thermochemical properties of compounds with electron configuration d^n (n = 1, 2, ... 10). This theory has as a basic presupposition that the metal-ligand interaction is, exclusively, of an electrostatic nature. Ligand species are understood as negative point charges that interact with *d* orbital electrons of the central atom, promoting their split into new groups of orbitals with different energies.

According to quantum mechanics, there are five *d* orbitals (d_{xy}, d_{xz}, d_{yz}, $d_{x^2-y^2}$, and d_{z^2}), which are the stationary states of the wave function of an electron, given by the Schrödinger equation $H\Psi = E\Psi$, where Ψ is the wave function [24]. In isolated atoms or ions, these five *d* orbitals are degenerate – that is, they have the same energy. However, when these atoms or ions coordinate with the ligands, the energies of those orbitals become different. Thus, in compounds formed in these combinations, crystal fields originate, determined from the difference in energy of the *d* orbitals.

2.4 EXPERIMENTAL PROCEDURES

2.4.1 Growth of NCs in Glassy Matrices

The glassy matrices studied in this work, called PZABP and SNABP, have the following nominal compositions: $65P_2O_5$, $14ZnO$, $1Al_2O_3$, $10BaO$, $10PbO$ (mol%), and $40SiO_2$, $30Na_2CO_3$, $1Al_2O_3$, $24B_2O_3$, and $5PbO$ (mol%). High purity compounds, obtained from the Sigma-Aldrich Company, were used in accordance with the specifications of the American Chemical Society: P_2O_5 (≥ 99.99%), ZnO (≥ 99%), Al_2O_3 (≥ 98%), BaO (≥ 99.99%), e PbO (≥ 99%), SiO_2 (99.9%), Na_2CO_3 (99.5%), Al_2O_3 (99.9%), B_2O_3 (99.98%), PbO (99%), S (99.5%), and e Co (99.9%).

The compounds in powder form were properly weighed, following the appropriate stoichiometry, and mixed and homogenized using sterilized alumina crucibles. Afterward, the composition of the glassy matrix PZABP underwent a fusion process at 1300°C for 30 minutes. Already the SNABP glassy matrix was subjected to 1200°C for 30 minutes. Soon after the fusion, the resulting melt was poured onto a metal plate, at a temperature of approximately 0°C, becoming glass.

The doping of the glassy matrices, synthesized by the fusion method, was carried out, remelting them already pulverized, with the addition of dopants and also at the same melting temperature and time. The compounds were homogenized and then subjected to fusion. Thus, adopting the same synthesis procedure of the glassy matrices, the *melt* of the doped glassy matrix was poured onto a metallic plate at 0°C, obtaining, in this way, glass sheets doped with precursor ions. Soon after fusion, followed by rapid cooling, the glassy samples were subjected to appropriate heat treatments to favor the nucleation and growth of NCs from different dopings.

2.4.2 Growth of NCs by the Precipitation Method

ZnO NCs were synthesized by the precipitation method. Under vigorous stirring at room temperature, 2 MNaOH was added into an aqueous 1M zinc nitrate ($Zn(NO_3)_2 \cdot 6H_2O$, 98%) solution, resulting in the formation of a white suspension. This suspension was centrifuged at 6000 rpm for 5 min and washed several times with distilled water until the pH of the solution was around 7. The obtained samples were dried at 100°C for 24 hours. To produce ZnO:xCo NCs (x = 0.0, 0.1, and 5.0), we use the same procedure as ZnO NCs, but with the cobalt (II) chloride ($CoCl_2$, 98%) solution during the synthesis process. All reagents are nearly pure and purchased from Sigma-Aldrich Company.

2.5 NANOSTRUCTURED MAGNETIC SEMICONDUCTORS

2.5.1 $Zn_{1-x}A_x$Te Nanocrystals (A = Mn; Cu) Grown in the P_2O_5–ZnO–Al_2O_3–BaO–PbO Glassy System

This study investigates some physical properties of $Zn_{1-x}Mn_x$Te and $Zn_{1-x}Cu_x$Te nanosized DMS embedded in the PZABP glassy matrix synthesized by the fusion method. To the chemical composition of this glassy matrix has been added 2Te (wt.%) and Mn or Cu at doping x content varying with Zn content from 0 to 10 (wt.%). The physical properties of the glass samples were studied by optical absorption (OA), Raman spectroscopy, X-ray diffraction (XRD), transmission electron microscopy (TEM), atomic/magnetic force microscopy (AFM/MFM), and electron paramagnetic resonance (EPR). OA spectra were recorded using a UV-VIS-NIR spectrometer (model UV-3600 Shimadzu) in the range from 190 to 3300 nm, with a resolution of 1 nm [25]. Raman spectra were collected using Raman spectrometer (JY-T64000 micro-Raman spectrometer) with a 514.5 nm line of an Argon laser [26]. XRD patterns were recorded using X-ray diffractometer (XRD-6000 Shimadzu diffractometer) with monochromatic Cu-$K_{\alpha 1}$ radiation (λ = 1.54056 Å) and a resolution of 0.02° [27]. TEM images were taken using transmission electron microscope (JEM-2100,JEOL) operated at 200 kV to investigate the formation of the NCs. TEM images were analyzed by Image J software [25]. A Shimadzu Scanning Probe Microscope (SPM-9600) was used to obtain AFM/MFM images in which the resolution in the vertical z-direction for topographic images was 0.01 nm [25]. The EPR spectra were recorded at X-band (~9 GHz) and Q-band (~9 GHz) microwave frequencies using a Bruker Elexis spectrometer.

Figure 2.1 (a) presents TEM, AFM and MFM images, XRD diffractograms, Raman spectra, and EPR spectra of $Zn_{1-x}Mn_x$Te NCs, with x = 0.00, 0.05, and 0.10. TEM images confirm the formation of

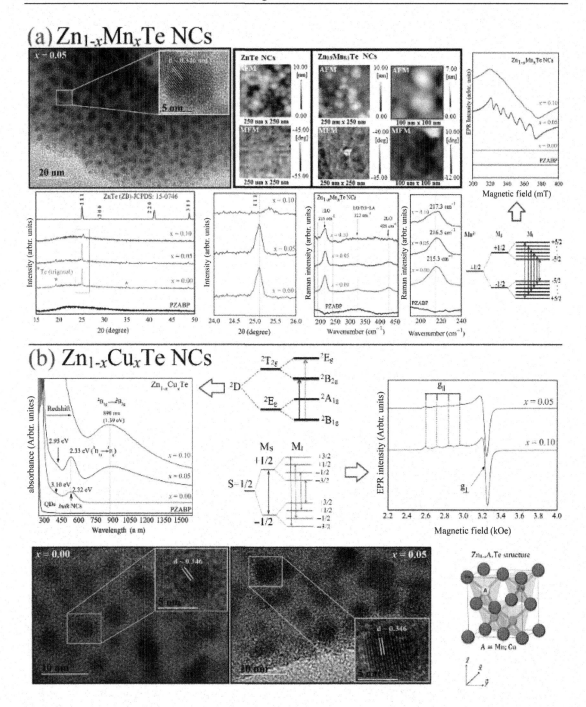

FIGURE 2.1 (a) TEM and AFM/MFM images, XRD diffractograms, Raman spectra, and EPR spectra of the samples containing $Zn_{1-x}Mn_xTe$ NCs, with Mn concentrations of $x = 0.00$, $x = 0.05$, and $x = 0.10$. Reproduced with permission from [29]. Copyright (2012) Elsevier. (b) OA and EPR spectra and TEM images of the samples containing $Zn_{1-x}Cu_xTe$ NCs, with Mn concentrations of $x = 0.00$, $x = 0.05$, and $x = 0.10$; Incorporation of Mn^{2+} or Cu^{2+} ions into the ZB crystal structure of ZnTe. Reproduced with permission from [30]. Copyright (2018) Elsevier.

the $Zn_{1-x}Mn_xTe$ spherical NCs (with $x = 0.05$) and an interplanar spacing d ~ 0.346 nm corresponds to the densest (111) plane of ZnTe (zincblende) [25]. AFM and MFM images refer to ZnTe NCs ($x = 0.00$) and $Zn_{0.95}Mn_{0.05}Te$ NCs ($x = 0.05$) embedded in the glass. Each panel shows a topographic image and a corresponding magnetic phase image, as well as the size distribution of NCs. The images corresponding to $Zn_{0.95}Mn_{0.05}Te$ NC are enlarged to better visualize the formation of NCs and magnetic contrasts. The mean radius range is from 2 to 10 nm, assigned to two groups of NCs with different sizes: a group that presents quantum confinement (QDs), with an average radius around ~ 2 nm, and another group without quantum confinement, with an average radius around ~ 10 nm, also called bulk NCs. This result has already been well discussed in our previous work [26].

XRD patterns were obtained to evaluate the crystallographic characteristics of the as-synthesized $Zn_{1-x}Mn_xTe$ NCs in terms of Mn x-content (0.00, 0.05, and 0.10), as well as of the pure PZABP sample (Figure 2.1a). The XRD pattern of the PZABP glassy matrix shows an amorphous band at around $20° < 2\theta° < 30°$, confirming the glassy characteristics [27]. $Zn_{1-x}Mn_xTe$ NC samples present the typical (111), (220), and (311) diffraction peaks of ZnTe with a cubic zincblende (ZB) structure (JCPDS: 15–0746) [27]. It is noted that the typical bulk ZnTe zincblende crystal structure is preserved for the $Zn_{1-x}Mn_xTe$ NC samples having an Mn-concentration of $x = 0.05$. Nevertheless, the characteristic XRD peaks are shifted towards higher diffraction angle values as the Mn^{2+} incorporation in the host ZnTe increases. This is a clear indication of decrease in the lattice constant with doping [28]. This decrease in the lattice constant is related to the replacement of Zn^{2+} ions in the zincblende ZnTe crystal structure by Mn^{2+} ions. At lower concentrations, the zincblende crystalline structure remains unchanged. On the other hand, for high concentrations of Mn^{2+}, a decrease in the crystallographic quality of the samples is expected.

Raman spectra revealing transitions at approximately 215 cm^{-1} (1LO), 323 cm^{-1} (TO/LO + LA), and 428 cm^{-1} (2LO) correspond to the normal phonon modes characteristic of the ZnTe phase (zincblende) [29]. It is observed that Raman scattering, refer to the 1LO phonon, presents a blue shift with increasing Mn concentration. This blue shift is related to the different atomic mass of ZnMnTe NCs where a heavier Zn^{2+} ion is replaced by a lighter Mn^{2+} ion: m_{Mn} (55) < m_{Zn} (65) [29]. This causes an increase in phonon mode frequencies with increasing Mn concentration.

In Figure 2.1a, the six absorption lines in the EPR spectrum result from the hyperfine interaction between the electron spin (S = 5/2) and the nuclear spin (I = 5/2) of Mn^{2+} ions and are due to transitions between electronic states MS = ±1/2 obeying the selection rules $\Delta M_S = \pm 1$ and $\Delta M_I = 0$ [31], as shown by the energy diagram. This result indicates that the paramagnetic Mn^{2+} ions are well incorporated into the $Zn_{1-x}Mn_xTe$ NCs.

Figure 2.1b presents OA and EPR spectra and TEM images of the samples containing $Zn_{1-x}Cu_xTe$ NCs, as a function of Mn concentration. When $x = 0.00$, are observed absorption bands centered at 3.10 eV (400 nm) and 2.33 eV (535 nm) that are attributed to ZnTe QDs and bulk NCs, respectively [26]. This result confirms the evidence observed from the AFM images in Figure 2.1a. An increase in Cu concentration (from 0.00 to 0.05) causes a redshift in the OA bands assigned to the QDs, from 3.10 eV (400 nm) to 2.95 eV (420 nm). This decrease is the strong evidence of Cu^{2+} being incorporated into ZnTe QDs, since the energy gap of ZnTe bulk (2.26 eV) tends to CuTe gap bulk (1.5 eV) [30]. This interesting behavior shows that the sp-d exchange interaction in the QDs is stronger than in the bulk-like NCs, when $x = 0.05$. However, when $x = 0.10$, a large redshift of the absorption bands of both QDs and bulk ZnTe NCs is observed. This result is related to the fact that at high concentrations of Cu^{2+}, it is not possible to distinguish between Cu^{2+} ions that are incorporated into the NCs and those that are dispersed in the PZABP glass matrix. The increase in the band centered at 535 nm (2.32 eV) with increasing copper concentration is due to two overlapping absorption bands: bulk NCs (around 535 nm) and a band around 531 nm (2.33 eV), which has been attributed to $^2B_{1g} \rightarrow {}^2E_g$ transition copper ions with monovalent nature (Cu^{1+}) [30], dispersed in the glass matrix. In addition is observed one band centered at 890 nm (1.39 eV), which becomes more intense with increasing Cu concentration. This band is attributed to the $^2B_{1g} \rightarrow {}^2B_{2g}$ transition of Cu^{2+} ions in the distorted octahedral sites [30]. In the octahedral crystal field, the free term for Cu^{2+} (d^9) ion is ^2D, which is divided into ^2E_g and $^2T_{2g}$, with ^2E_g being the state of lowest energy. Due to the Jahn-Teller effect, which causes distortions in octahedral symmetry, the ^2E_g ground state can be divided into

$^2B_{1g}$ and $^2A_{1g}$ and the $^2T_{2g}$ state into $^2B_{2g}$ and ^2E_g, with $^2B_{1g}$ being the ground state [30]. The energy diagram, containing these transitions, is represented in Figure 2.1b. It is well known that in a crystal structure of the zincblende type, e.g., the ZnTe lattice, a cation Zn^{2+} is bound to four Te^{2-} anions (i.e., the coordination number CN = 4), with tetrahedral coordination (Td). Thus, according to the results obtained from the OA spectra, one can infer that part of the Cu^{2+} ions are dispersed in the PZABP glass matrix, since, probably, these ions may have octahedral (Oh) symmetry with tetragonal distortion, and the other part of these are incorporated into ZnTe NCs, which further increases these tetragonal distortions in Oh symmetry.

TEM image confirms the formation of the $Zn_{1-x}Cu_xTe$ spherical NCs (with $x = 0.00$ and $x = 0.05$) in the PZABP glass matrix. The amplified region of the TEM image shows an interplanar spacing d ~ 0.346 nm corresponds to the densest (111) plane of the zincblende ZnTe crystal system [30]. The invariance of this interplanar spacing with the incorporation of Cu^{2+} is expected since there is a little difference between the Zn^{2+} (0.68 Å) and Cu^{2+} (0.73 Å) ionic radii [30]. Thus, it is expected that the lattice parameter of the NCs is not modified with the incorporation of Cu^{2+} ions in ZnTe NCs.

The EPR spectra of the PZABP glass samples and this containing $Zn_{1-x}Cu_xTe$ NCs with concentrations of 0.00, 0.05, and 0.10 are shown in Figure 2.1b. In these spectra, one can observe four parallel weak components in the lower magnetic field region and a corresponding intense resonance signal to the four components perpendicular to the higher magnetic field region. The parallel components are due to the hyperfine interaction between the electron spin (S = 1/2) and nuclear spin (I = 3/2) of ^{63}Cu and/or ^{65}Cu, resulting from transitions between electronic states $M_S = \pm 1/2$ and $M_I = \pm 3/2, \pm 1/2$ [32], according to the selection rules $\Delta M_S = \pm 1$ and $\Delta M_I = 0$, as shown in the inset of Figure 2.1b. This hyperfine interaction, due to Cu^{2+} ions, results from the presence of a distorted axially octahedral crystal field [32]. Since the symmetry is smaller than octahedral, anisotropy occurs mainly in the values of g and A tensors. Thus, this anisotropy observed in EPR spectra may be related to Cu^{2+} ions incorporated into sites with tetrahedral symmetry (Td) of ZnTe NCs, replacing Zn^{2+} ions (following the redshift observed in the OA spectra), as well as dispersed in PZABP glass matrix, especially in high Cu concentrations. The hyperfine interaction, due to Cu^{2+} ions incorporated in the zincblende lattice of ZnTe NCs, results from the presence of a crystalline field in these NCs, which favors the strong interactions of exchange between the d sublevel of electrons of Cu^{2+} ions and electrons in the sp sublevel of host semiconductor (ZnTe).

Figure 2.1b shows the ZB crystal structure of ZnTe with the incorporation of A^{2+} ions (A = Mn or Cu). As discussed in the introduction of this chapter, this incorporation enables an exchange interaction between the host electronic subsystem and the electrons of the partially filled d or f levels of the magnetic ions. This allows for the control of several physical and chemical properties, such as those mentioned in this study. In turn, these can be explored for various technological applications, such as spintronic devices and new lasers.

2.5.2 Nanostructured Glasses With $Zn_{1-x}Mn_xTe$ and Doped With Europium

In recent decades, increased studies on RE ion doped glass systems are reported [33]. In particular, phosphate-based oxide glasses have proven to be a very suitable system for the advancement of new optoelectronic devices due to their high transparency, low melting temperature, high thermal stability, low viscosity, and low phonon energy, besides easy preparation and low-cost production [34]. These RE-doped glasses have interesting optical properties such as sharp absorption and emission lines and high emission efficiencies besides emission of intense radiation from visible (Vis) to infrared (IR) spectral regions under suitable excitation conditions [34]. An important activator is Eu^{3+}, which can be used in photonic applications as orange/red light emitting devices because to narrowband almost monochromatic of the $^5D_0 \rightarrow {}^7F_2$ transition around 612 nm (4f-4f transitions) [34]. It has been shown that to increase the efficiency of Eu^{3+} ion emission in the red spectral region, to reduce the laser threshold, and to extend the pumping range, it is necessary to have sensitizers that favor the energy transfer process.

Mn^{2+} ions have often been used as sensitizers via energy transfer to RE ions such as Nd^{3+}, Er^{3+}, Pr^{3+}, and Eu^{3+}. As is well known, the transition metal Mn^{2+} ion–doped luminescent materials, especially the DMS NCs, are very interesting as they feature a wide range of emissions from green to infrared, depending on the crystalline environment of said hosts [35]. In this context, the motive of the present study is to investigate the energy transfer (ET) process between Mn^{2+} and Eu^{3+} ions (Mn^{2+} → Eu^{3+}) on a PZABP phosphate glass system containing Zn$_{1-x}$Mn$_x$Te NCs, with x ranging from 0.0 to 0.05, and doped with Eu$_2$O$_3$ (1 wt.%). We use AO and PL at room temperature to study the effect of Mn^{2+} concentration on ET processes. OA spectra were recorded with a UV-VIS-NIR spectrometer model UV-3600 Shimadzu operating between 190–3300 nm, with a resolution of 1 nm. The emissions were obtained using the Fluorimeter NanoLog™ (HORIBA) armed with a Xenon lamp (CW 450 W) as the excitation source and a photomultiplier detector (model R928P). The emissions of the sample containing Zn 0.99Mn 0.01Te NCs of Zn$_{0.99}$Mn$_{0.01}$Te were taken with a 410 nm (~3.02 eV) continuous-wave laser focused to a ~ 200 μm ray with an excitation power of 13 mW [36].

Figure 2.2 presents OA spectra of PZABP glass matrix doped with 1.0 wt.% of Eu$_2$O$_3$, named PZABP:1Eu, and PL of another PZABP glass matrix doped with 1.0 wt.% of Mn, using a 410 nm (3.02 eV) excitation line; simplified energy levels diagrams with all the transitions relative to absorption and emission observed for Mn^{2+} ions and Eu^{3+} ions, as well as the energy levels likely involved in the ET processes between Mn^{2+} and Eu^{3+} ions; PL spectra, obtained under excitation at 410 nm (3.02 eV), for glass samples containing Zn$_{0.99}$Mn$_{0.01}$Te NCs ($x = 0.01$) and doped with 1.0 wt.% of Eu$_2$O$_3$ (named Zn$_{0.99}$Mn$_{0.01}$Te:1Eu); and illustrative scheme of the energy transfer process from Mn^{2+} to Nd^{3+}/Eu^{3+} ions.

The OA spectrum shows characteristic transitions of Eu^{3+} ions centered at 365, 380, 392, 415, 465, 530, and 583 nm, assigned respectively to the following transitions: $^5D_0 \rightarrow {}^7F_0$ (580 nm), $^5D_0 \rightarrow {}^7F_1$ (592 nm), $^5D_0 \rightarrow {}^7F_2$ (612 nm), $^5D_0 \rightarrow {}^7F_3$ (653 nm), and $^5D_0 \rightarrow {}^7F_4$ (700 nm) [37]. The PL spectrum shows the $^4T_1(^4G) \rightarrow {}^6A_1(^6S)$ main emission of Mn^{2+} ions. The overlap between the broad emission band centered at around 605 nm, characteristic of Mn^{2+} ions, the $^7F_0 \rightarrow {}^5D_{0,1}$ absorption transition, and characteristic of Eu^{3+} ions suggest that an energy transfer can occur from Mn^{2+} to Eu^{3+} ions in Zn$_{1-x}$Mn$_x$Te NCs, as shown in the energy diagram beside. The PL spectra of Zn$_{0.99}$Mn$_{0.01}$Te and Zn$_{0.99}$Mn$_{0.01}$Te:1Eu samples show, in addition to the Mn^{2+} ions characteristic emission, the Eu^{3+} characteristic emission transitions ($^5D_0 \rightarrow {}^7F_0$ (580 nm), $^5D_0 \rightarrow {}^7F_1$ (592 nm), $^5D_0 \rightarrow {}^7F_2$ (612 nm), $^5D_0 \rightarrow {}^7F_3$ (653 nm), and $^5D_0 \rightarrow {}^7F_4$ (700 nm)) [37]. Comparing the PL spectra of both samples, there is a observed decrease in $^4T_1(^4G) \rightarrow {}^6A_1(^6S)$ emission intensity with Eu$_2$O$_3$ doping. This decrease is due to the efficient energy transfer from Mn^{2+} to Eu^{3+} ions since this ET efficiently competes with the spontaneous emission of Mn^{2+} ions. The illustration in Figure 2.2 shows that Mn^{2+} ions are excited at 410 nm and emit at 605 nm transferring energy to Eu^{3+} ions, which emit at 612 nm. These results indicate that the said glass system is a potential candidate for use in various technological applications, as in the development of fiber-optic amplifiers used in communication devices and solid-state laser systems.

2.5.3 Pb$_{1-x}$Co$_x$S Nanocrystals Synthesized in SiO$_2$–Na$_2$CO$_3$–Al$_2$O$_3$–B$_2$O$_3$–PbO Host Glass

Transition metal–doped PbS NCs have fascinating properties that define the magnetically diluted as consequences of the magnitude of the *sp-d* exchange interactions. Specifically, cobalt has properties that allow the manipulation of spins in PbS NCs [38]. In this work, the structural, morphological, electronic, optical, and magnetic properties were investigated by XRD, TEM with energy-dispersive X-ray spectroscopy (EDS), AFM with MFM, and OA spectroscopy. The characteristics of the equipment have already been well described in Sections 2.5.1 and 2.5.2.

Pb$_{1-x}$Co$_x$S NCs are highlighted in the TEM image of Figure 2.3a. On the right of Figure 2.3a, the rock-salt structure face-centered cubic unit cell (FCC) and the Pb$_{1-x}$Co$_x$S quantum dot are shown. Pb$_{1-x}$Co$_x$S NCs with a diameter smaller than the Bohr radius of the PbS exciton ($a_B = 20$ nm) are evidence of a

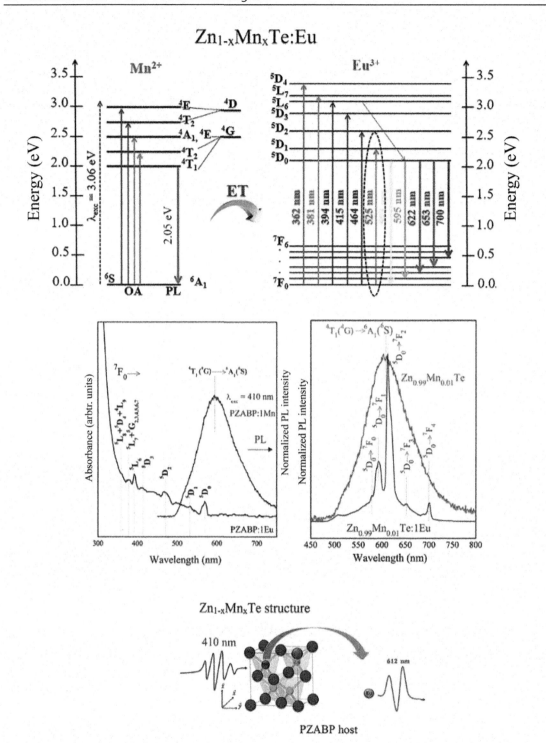

FIGURE 2.2 OA spectrum of the PZABP:1Eu glassy sample, and PL spectrum of the PZABP:1Mn glassy sample, using a 410 nm excitation line; simplified energy levels diagrams with all the absorption and emission transitions observed for Mn^{2+} and Eu^{3+} ions; PL spectra of the Zn$_{0.99}$Mn$_{0.01}$Te and Zn$_{0.99}$Mn$_{0.01}$Te:1Eu glass samples; and illustrative scheme of the energy transfer process from Mn^{2+} to Nd^{3+}/Eu^{3+} ions.

2 • Nanostructured Magnetic Semiconductors 33

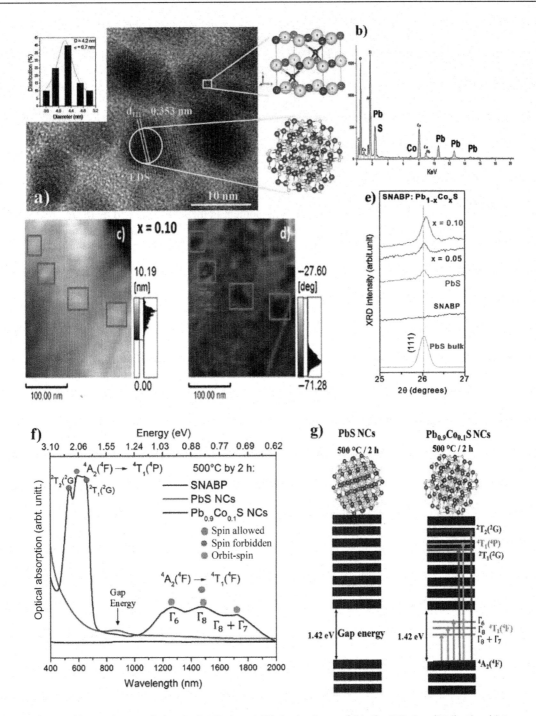

FIGURE 2.3 (a) TEM image of Pb$_{1-x}$Co$_x$S NCs (x = 0.10). In the inset of (a), the NC size distribution histogram. On the right of (a), FCC unit cell and Pb$_{1-x}$Co$_x$S quantum dot with Co^{2+} dopant at tetrahedral sites. (b) EDS spectrum of the yellow circle area in the MET images, suggesting the presence of the elements Co, Pb, and S. (c) MFA topographic image and (d) MFM magnetic image for Pb$_{1-x}$Co$_x$S NCs (x = 0.10). (e) DRX diffractograms show the peak (111), characteristic of PbS NCs. (f) Optical absorption spectra of PbS and Pb$_{0.9}$Co$_{0.1}$S NCs incorporated in the SNABP glassy matrix annealed at 500°C for 2 h. The absorption spectrum of the SNABP glassy matrix is represented in the black background line. (g) Power level diagram for PbS and Pb$_{1-x}$Co$_x$S NCs. Reproduced with permission from [29]. Copyright (2021) Elsevier.

strong quantum confinement regime and, consequently, the formation of quantum dots. Due to the similar growth kinetics, NCs with approximate spherical symmetry have a uniform mean diameter (D = 4.2 nm). The Gaussian fit performed on the size distribution histogram is shown in the inset of Figure 2.3a. In the EDS analysis of the circle area (Figure 2.3a), a characteristic Co peak at 6.91 KeV suggests the doping of PbS NCs with Co^{2+} ions. In Figure 2.3a the value of the interplanar space d_{111} = 0.353 nm of the crystal plane (111) of the PbS rock-salt structure was measured.

The FCC structure of the PbS semiconductor is confirmed by the XRD (111) peak compatible with the JCPDS card powder diffraction pattern No. 05–0592 (Figure 2.3e) [39]. The sharpness of the XRD peak shows the presence of crystalline PbS NCs. The small shift of the peak (111) is evidenced by the Co^{2+} incorporation into PbS NCs. According to Bragg's law, increasing the θ angle decreases the d_{111} distance of the crystal plane (111) and the lattice parameter a = 5.936 Å of the FCC structure.

The AFM (Figure 2.3c) and MFM (Figures 2.3d) images represent the *sp-d* electronic exchange interactions that contribute to the total magnetic spin moment in the NCs indicated by the red rectangles. The light/dark contrast in MFM images is due to the repulsion/attraction of the magnetized tip towards the NCs. MFM images are evidence of the presence of spin domain upon the incorporation of Co^{2+} in the PbS NCs.

Figure 2.3f shows the optical absorption spectrum of PbS and $Pb_{0.9}Co_{0.1}S$ NCs (nominal composition) grown in the SNABP glassy matrix and thermally annealed at 500°C for 2 h. The position of the PbS band around 860 nm (1.43 eV) is evidence of carrier mobility in a strong confinement regime. The SNABP matrix is transparent in the adsorption and photon emission spectral region of the PbS and $Pb_{0.9}Co_{0.1}S$ NCs.

In the crystal environment of $Pb_{0.9}Co_{0.1}S$ SMD NCs, the free Co^{2+} ion senses the presence of a ligand field of S^{2-} ions. The free Co^{2+} ion energy states represented by the fundamental terms ^4F and excited ^4P unfold into the spectral terms $^4A_2 + {}^4T_1 + {}^4T_2$ e $^4T_1$. Such a phenomenon occurs under the disturbing influence of the expansion of the electron cloud of the tetrahedral crystalline field (Td). These energy states of the Co^{2+} ion at crystal sites of the PbS host were identified with the OA spectroscopy technique UVVISNIR together with the crystal field theory (Figure 2.3f) [38].

The absorption bands in the VIS and NIR regions were analyzed and identified based on the Tanabe-Sugano energy diagram (C/B = 4.5) $3d^7$ (Td). The transition energies, the parameters of Racah B (B = 792 cm^{-1}), and crystal field division (Δ = 3897 cm^{-1}) confirm the high-spin state (weak field) of the Co^{2+} ion at tetrahedral coordination sites $[CoS_4]^{6-}$ in the crystal structure of the $Pb_{1-x}Co_xS$ NCs [38]. The covalent character of the Co–S bond in the $[CoS_4]^{6-}$ complex can be described with the term "electron cloud expansion" or the nephelauxetic effect. For the free Co^{2+} ion B = 1028 cm^{-1}. The ratio β = B$[CoS_4]^{6-}$/ BCo^{2+} for the $Pb_{1-x}Co_xS$ NCs (β = 0.73) evidences the reduction in the interelectronic repulsion in relation to the free Co^{2+} ion and, consequently, the covalence of the Co–S bond [40]. The excitation energy of a photoelectron in the NIR promotes the electron from the state $e^4t_2^3$ to $e^3t_2^4$. The transitions allowed by spin are identified by a green dot. The transition allowed by spin $^4A_2(^4F) \rightarrow {}^4T_1(^4F)$ around 1477 nm (NIR) can be explained by a strong spin-orbit coupling interaction. The Co^{2+} ions at Td sites divide the excited state $^4T_1(^4F)$ into three sub-energy states designated by points of symmetry $Γ_6$, $Γ_8$ e $Γ_8 + Γ_7$ [40]. The blue dot on the spectrum identifies the energy states of the spin-orbit coupling.

In the VIS region of the spectrum, the band around 593 nm is attributed to the transition allowed by spin $^4A_2(^4F) \rightarrow {}^4T_1(^4P)$. Such excitation energy consists of the electron in the first excited state of electron configuration $e^2t_2^5$. The absorption bands at 530 and 661 nm (red dot) are spin-prohibited transitions $^4A_2(^4F) \rightarrow {}^2T_2(^2G)$ and $^4A_2(^4F) \rightarrow {}^2T_1(^2G)$, respectively [41]. The energy level diagram (Figure 2.2g) depicts the PbS and $Pb_{1-x}Co_xS$ NCs heat-treated at 500°C for 2 h. The confinement energy (gap energy) of exciton charge carriers in NCs with quantum dot properties is much higher than the narrow bandgap (0.41 eV) of the corresponding bulk material (PbS). Cobalt doping provides characteristic d-d transition energies of the Co^{2+} ion under the influence of the crystal field Td in terms of Δ of S^{2-} ($[CoS4]^{6-}$) ligand ions. Therefore, *sp-d* potential and kinetic exchange interactions arise from the hybridization between the 3d orbitals of Co^{2+} and the *sp* orbitals of S^{2-}. That is, from the spectroscopic states of the Co^{2+} $3d^7$ ion to the PbS semiconductor conduction band [41]. Consequently, the broadband coming from the strong quantum confinement energy intensifies the *sp-d* exchange interactions.

The correlation of results obtained by OA, XRD, TEM, and MFM strongly confirm the incorporation of Co^{2+} ions in Pb$_{1-x}$Co$_x$S NCs. We can see that the fusion method is a low-cost synthesis for producing and tuning the properties of NCs. The investigated properties of Pb$_{1-x}$Co$_x$S NCs enhance technological applications in light-emitting diodes, lasers, solar cells, photodetectors, and possibly spintronic devices, among others.

2.5.4 Optical and Magnetic Properties of Zn$_{1-x}$Co$_x$O Nanopowders

In this study, we investigated the optical and magnetic properties of Zn$_{1-x}$Co$_x$O NCs with content, x, by Raman spectroscopy, OA, CFT, fluorescence (FL), and EPR. Raman spectra were obtained using a HORIBA Scientific Raman spectrometer, with an excitation line at $\lambda = 532$ nm and 1800 gr/mm grid. OA spectra were recorded using a UV-VIS-NIR spectrometer (Shimadzu) [16]. Fluorescence spectra were assessed using the LabRAM HR Evolution (HORIBA), with an excitation line of 325 nm [16]. EPR spectra were recorded on a Bruker-EMX spectrometer, operating around 9.5 GHz (X-band) [16].

Raman spectra are shown in Figure 2.4a, and the characteristics of wurtzite ZnO bands at 383 cm^{-1}, 408 cm^{-1}, 438 cm^{-1}, 575 cm^{-1}, and 581 cm^{-1} were observed [42]. Additional bands at 332 cm^{-1} and 540 cm^{-1} are related to multiple processes, while the bands at 492 cm^{-1}, 620 cm^{-1}, and 685 cm^{-1} reach peaks with secondary phase, probably ZnCo$_2$O$_4$ [43]. Figure 2.4b shows the absorption spectra of samples in which

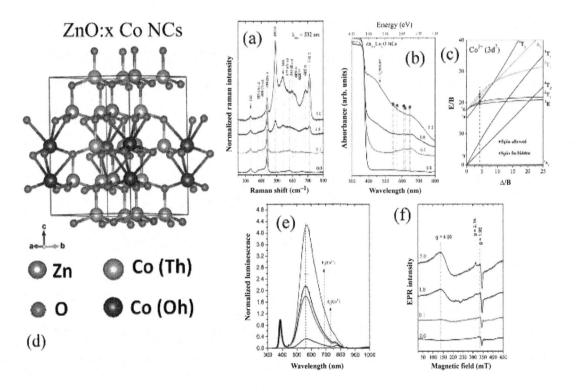

FIGURE 2.4 (a) Raman spectra, (b) optical absorption spectra, (c) Tanabe-Sugano diagram for the electronic configuration of the Co^{2+} ion (3d^7), (d) wurtzite structure of ZnO NCs with coordination geometry in which the Co^{2+} ions meet, represented by tetrahedral sites (Td) and octahedral sites (Od). (e) Fluorescence spectra and (f) EPR spectra of Zn$_{1-x}$Co$_x$O NCs with x-content. Reproduced with permission from [16]. Copyright (2021) Springer Nature.

a characteristic ultraviolet band and a redshift, as a function of the Co concentration due to the *sp-d* exchange interaction, are observed. Additionally, the absorption band in the blue spectral region is related to the secondary $ZnCo_2O_4$ phase (2.63 eV), which is in the excellent agreement with Raman spectra (Figure 2.4a).

The OA spectra were analyzed based on the CFC, in which it was possible to calculate the crystalline field strength parameters Δ and Racah-B (Figure 2.4c). Based on this resource, the energies of the characteristic electronic transitions of Co^{2+} are: $^4A_2(^4F) \rightarrow {}^2E\ (^2G)$ (646 nm), $^4A_2(^4F) \rightarrow {}^2T_1(^2G)$ (618 nm), $^4A_2(^4F) \rightarrow {}^4T_1(^4P)$ (610 nm), $^4A_2(^4F) \rightarrow {}^2A_1(^2G)$ (571 nm), and $^4A_2(^4F) \rightarrow {}^2T_2(^2G)$ (546 nm), with crystalline field splitting (Δ = 3480 cm^{-1}) and Racah-B parameter (802 cm^{-1}). Both parameters are calculated under the dashed vertical line at Δ/B = 4.34 on the Tanabe-Sugano energy level diagram (C/B = 4.50) for Co^{2+} (3d^7) (Figure 2.4c) [44].

Figure 2.4e shows FL spectra of the ZnO NCs, with two bands around 387 nm and 555 nm, assigned to the excitonic transition, and an emission overlap. The absence of the band peaks at 443 nm is due to Zn-related interstitial defects. This confirms the low density of interstitial Zn ions in the as-synthesized ZnO NCs [45]. In these spectrums, it is also possible to observe emission bands in orange (from 600 to 700 nm) and in red (from 730 to 900 nm) assigned to $^4T_1(^4P) \rightarrow {}^4A_2(^4F)$ and $^4T_1(^4P) \rightarrow {}^4T_2(^4F)$ tetrahedral transitions of Co^{2+} ions, respectively. The redshift and broadening of these bands with increasing concentration are related to increasing Co^{2+} ions density in the crystal structure of ZnO, replacing Zn^{2+} ions. Furthermore, the FL data show that the Co^{2+}-related fluorescence originates from strongly localized Co^{2+} ion levels, incorporated into the core and the surface of the ZnO nanocrystals. Figure 2.4d shows a representation of the wurtzite structure of ZnO NCs with coordination geometry in which the Co^{2+} ions meet, represented by tetrahedral sites (Td) and octahedral sites (Od).

The EPR spectra in Figure 2.4f show a resonant signal around g = 1.96, characteristic of the crystal structure of the ZnO NCs [46]. Two extra EPR signals observed in all recorded spectra are attributed to Co^{2+} ions under the influence of the hosting ZnO crystal field. The feature peaking at g = 4.00 is assigned to Co^{2+} ions in octahedral symmetry [47] whereas the feature appearing at g = 2.14 has been assigned to Co^{2+} ions in tetrahedral symmetry [48]. Indeed, it is worth mentioning that the EPR signal intensity increases as a function of Co content, *x*. EPR signals of Co^{2+} ions (3d^7) are due to the interaction between electron spin (S = 3/2) and nuclear spin (I = 7/2), since the substitution of Co^{2+} for Zn^{2+} occurs in both octahedral and tetrahedral sites of ZnO structure.

2.6 CONCLUSION

This chapter showed the development and applications of several nanostructured magnetic semiconductors, such as $Zn_{1-x}Co_xO$ nanopowders, $Zn_{1-x}Mn_xTe$, $Zn_{1-x}Cu_xTe$, and $Pb_{1-x}Co_xS$ embedded in pure glass systems or doped with europium. TEM, AFM, and MFM images confirm the formation of DMS NCs. The optical properties of these materials investigated by photoluminescence and UV-Vis spectroscopy techniques investigated the optical properties based on crystal field theory. These show the influence of TM^{2+} (Mn^{2+}, Cu^{2+}, and Co^{2+}) ions on the visible spectrum, altering the optical absorption and photoluminescence bands of undoped nanocrystals. On the other hand, when the glasses are europium doped, these techniques confirm energy transfer between Mn^{2+} and Eu^{3+}. XRD and Raman spectra show, respectively, the influence of the *x*Mn concentration on NC lattice parameters and vibrational modes. The EPR spectra confirm the incorporation of TM^{2+} ions in the doped nanocrystals' crystalline structure. Therefore, we believe that this work can contribute significantly to the study of new nanostructured magnetic semiconductors and their possible technological applications.

2.7 ACKNOWLEDGMENTS

This work was supported by PDJ/CNPQ (151349/2020–7) Project. We thank the Laboratório Multiusuário de Microscopia de Alta Resolução (LabMic), at the Universidade Federal de Goiás, for the use of facilities for the TEM and EDs measurements, and the Physics Institute of the Universidade Federal de Uberlândia for the Raman, AFM, and MFM measurements.

REFERENCES

1. Gu M, Zhang Q, Lamon S (2016) Nanomaterials for optical date storage. *Nature Reviews Materials* 12: 1–14
2. Arndt N, Tran H D N, Zhang R, Xu Z P, Ta H T (2020) Different approaches to develop nanosensors for diagnosis of diseases. *Advanced Science*, 2001476: 1–31
3. Mitchell S, Qin R, Zheng N, Pérez-Ramírez J (2021) Nanoscale engineering of catalytic materials for sustainable technologies. *Nature Nanotechnology* 16: 129–131
4. Sybil D, Guttal S, Midha S (2021) Nanomaterials in prosthetic rehabilitation of maxillofacial defects. In: Chaughule R S, Dachaoutra R (eds) *Advances in Dental Implantology using Nanomaterials and Allied Technology Applications*. Springer
5. Aghebati-Maleki A, Dolati S, Ahmadi M, Baghbanzhadeh A, Asadi M, Fotouhi A, Aghebati-Maleki L (2019) Nanoparticles and cancer therapy: Perspectives for application of nanoparticles in the treatment of cancers. *Journal of Cellular Physiology* 2019: 1–11
6. Dantas N O, Fanyao Q, Daud S P, Alcalde A M, Almeida C G, Diniz Neto O O, Morais P C (2003) The effects of external magnetic field on the surface charge distribution of spherical nanoparticles. *Microelectronics Journal* 34: 471–473
7. Borysiewicz, M A (2019) ZnO as a functional material, a review. *Crystals* 9: 505
8. Mishra Y K, Adelung R (2018) ZnO tetrapod materials for functional applications. *Mater Today* 21: 631–651
9. Dantas N O, de Paula P M N, Silva R S, Lopez-Richard V, Marques G E (2011) Radiative versus nonradiative optical processes in PbS nanocrystals. *Journal of Applied Physics* 109: 024308
10. Sukhovatkin V, Musikhin S, Gorelikov I, Cauchi S, Bakueva L, Kumacheva E, Sargent E H (2005) Room-temperature amplified spontaneous emission at 1300 nm in solution-processed PbS quantum-dot films. *Optics Letters* 30: 171–173
11. Mandal S K, Mandal A R, Banerjee S (2012) High ferromagnetic transition temperature in PbS and PbS:Mn nanowires. *ACS Applied Maters Interfaces* 4: 20–29
12. De Iacovo A, Venettacci C, Colace L, Scopa L, Foglia S (2016) PbS colloidal quantum dot photodetectors operating in the near infrared. *Scientific Reports* 6: 37913
13. Ersching K, Campos C E M, De Lima J C, Grandi T A, Souza S M, Silva D L, Pizani P S (2009) X-ray diffraction, Raman, and photoacoustic studies of ZnTe nanocrystals. *Journal of Applied Physics* 105: 123532
14. Kuo M C, Yang C S, Tseng P Y, Lee J, Shen J L, Chou W C, Shih Y T, Ku C T, Lee M C, Chen W K (2002) Formation of self-assembled ZnTe quantum dots on ZnSe buffer layer grown on GaAs substrate by molecular beam epitaxy. *Journal of Crystal Growth* 242: 533–537
15. Jiang F, Li Y, Ye M, Fan L, Ding Y, Li Y (2010) Ligand-tuned shape control, oriented assembly, and electrochemical characterization of colloidal ZnTe nanocrystals. *Chem Mater* 22: 4632–4641
16. Batista E A, Silva A C A, Rezende T K L, Guimarães E V, Pereira P A G, Souza P E N, Silva R S, Morais P C, Dantas N O (2021) Modulating the magnetic-optical properties of $Zn_{1-x}Co_xO$ nanocrystals with x-content. *Journal of Materials Research* 36: 1657–1665
17. Dantas N O, Freitas Neto E S, Silva R S (2010) Diluted magnetic semiconductor nanocrystals in glass matrix. In: Masuda Y (ed) *Nanocrystals*. Intechopen
18. Tang F L, Su H L, Chuang P Y, Wu J C, Huang A, Huang X L, Jin Y (2014) Dependence of magnetism on the doping level of $Zn_{1-x}Mn_xTe$ nanoparticles synthesized by a hydrothermal method. *RSC Advances* 4: 49308–49314

19. Zhao L, Zhang B, Pang Q, Yang S, Zhang X, Ge W, Wang J (2006) Chemical synthesis and magnetic properties of dilute magnetic ZnTe:Cr crystals. *Applied Physics Letters* 89: 092111
20. Gul Q, Zakria M, Khan T M, Mahmood A, Iqbal A (2014) Effects of Cu incorporation on physical properties of ZnTe thin films deposited by thermal evaporation. *Materials Science in Semiconductor Processing* 19: 17–23
21. Shankar D B, Bhagwat V R, Shinde A B, Jadhav K M (2016) Effect of Co^{2+} ions on structural, morphological and optical properties of ZnO nanoparticles synthesized by sol – gel auto combustion method. *Materials Science in Semiconductor Processing* 41: 441–449
22. Kwong T-L, Yung K-F (2015) Surfactant-free microwave-assisted synthesis of fe doped ZnO nanostars as photocatalyst for degradation of tropaeolin o in water under visible light. *Journal of Nanomaterials* 2015: 1–9
23. Beaulac R, Ochsenbein S T, Gamelin D R (2010) Colloidal transition-metal-doped quantum dots. In: Klimov V I (ed) *Nanocrystal Quantum Dots*. CRC Press
24. Orchin M, Macomber R S, Pinhas A, Wilson R M (2005) Atomic orbital theory. In: *The Vocabulary and Concepts of Organic Chemistry*. Wiley-Interceince.
25. Silva A S, Lourenço S A, Dantas N O (2016) Mn concentration-dependent tuning of Mn^{2+} d emission of $Zn_{1-x}Mn_xTe$ nanocrystals grown in a glass system. *Physical Chemistry Chemical Physics* 18: 6069–6076
26. Dantas N O, Silva A S, Silva S W, Morais P C, Pereira-da-Silva MA, Marques G E (2010) ZnTe nanocrystal formation and growth control on UV-transparent substrate. *Chemical Physics Letters* 500: 46–48
27. Silva A S, Silva S W, Morais P C, Dantas N O (2016) Solubility limit of Mn^{2+} ions in $Zn_{1-x}Mn_xTe$ nanocrystals grown within an ultraviolet-transparent glass template. *Journal of Nanoparticle Research* 18: 125
28. Dantas N O, Pelegrini F, Novak M A, Morais P C, Marques G E, Silva R S (2012) Control of magnetic behavior by $Pb_{1-x}Mn_xS$ nanocrystals in a glass matrix. *Journal of Applied Physics* 111: 064311
29. Dantas N O, Silva A S, Ayta W E F, Silva S W, Morais P C, Pereira-da-Silva M A, Marques G E (2012) Dilute magnetism in $Zn_{1-x}Mn_xTe$ nanocrystals grown in a glass template. *Chemical Physics Letters* 541: 44–48
30. Silva A S, Pelegrini F, Figueiredo L C, Souza P E N, Morais P C, Dantas N O (2018) Effects of Cu^{2+} ion incorporation into ZnTe nanocrystals dispersed within a glass matrix. *Journal of Alloys and Compounds* 749: 681–686
31. Silva A S, Franco Jr A, Pelegrini F, Dantas N O (2015) Paramagnetic behavior at room temperature of $Zn_{1-x}Mn_xTe$ nanocrystals grown in a phosphate glass matrix by the fusion method. *Journal of Alloys and Compounds* 647: 637–643
32. Hameed A, Ramadevudu G, Shareefuddin M, Chary M N (2014) EPR and optical absorption by Cu^{2+} ions in $ZnO–Li_2O–Na_2O–K_2O–B_2O_3$ glasses. *Solid State Physics AIP Conference Proceedings* 1591: 842–844
33. Yang D L, Gong H, Pun E Y B, Zhao X, Lin H (2010) Pr^{3+}-doped heavy metal germanium tellurite glasses for irradiative light source in minimally invasive photodynamic therapy surgery. *Optic Express* 18: 18997–19008
34. Chanthima N, Tariwong Y, Kim H J, Kaewkhao J, Sangwaranatee N (2018) Effect of Eu^{3+} Ions on the physical, optical and luminescence properties of aluminium phosphate glasses. *Key Engineering Materials* 766: 122–126
35. Wan M H, Wong P S, Hussin R, Lintang H O, Endud S (2014) Structural and luminescence properties of Mn^{2+} ions doped calcium zinc borophosphate glasses. *Journal of Alloys and Compounds* 595: 39–45
36. Silva A S, Sales T O, Silva W S, Jacinto C, Dantas N O (2020) Energy transfer from Mn^{2+} to Nd^{3+} ions embedded in a nanostructured glass system with $Zn_{1-x}Mn_xTe$ nanocrystals. *Journal of Luminescence* 226:117511
37. Morais R F, Serqueira E O, Dantas N O (2013) Effects of OH radicals and the silicon network on the lifetime of Eu^{3+}-doped sodium silicate glasses. *Optical Materials Express* 3: 853–867
38. Silva R S, Guimaraes E V, Melo R E S, Silva A S, Silva A C A, Dantas N O, Lourenço S A (2021) Investigation of structural and optical properties of $Pb_{1-x}Co_xS$ nanocrystals embedded in chalcogenide glass. *Materials Chemistry and Physics* 269: 124766
39. Bai R, Chaudhary S, Pandya D K (2018) Temperature dependent charge transport mechanisms in highly crystalline p-PbS cubic nanocrystals grown by chemical bath deposition. *Materials Science in Semiconductor Processing* 75: 301–310
40. Pasternak A, Goldschmidt Z B (1972) Spin-dependent interactions in the $3d^N$ configurations of the third specta of the iron group. *Physical Review A* 6: 55
41. Guimarães E V, Gonçalves E R, Lourenço S A, Oliveira L C, Baffa O, Silva A C A, Dantas N O, Silva R S (2018) Concentration effect on the optical and magnetic properties of Co^{2+}-doped Bi_2S_3 semimagnetic nanocrystals growth in glass matrix. *Journal of Alloys and Compounds* 740: 974–979
42. Gao Q, Dai Y, Li C, Yang L, Li X, Cui C (2016) Correlation between oxygen vacancies and dopant concentration in Mn-doped ZnO nanoparticles synthesized by co-precipitation technique. *Journal of Alloys and Compounds* 684: 669–676
43. Amini M N, Dixit H, Saniz R, Lamoen D, Partoens B (2014) The origin of p-type conductivity in ZnM_2O_4 (M = Co, Rh, Ir) spinels. *Physical Chemistry Chemical Physics* 16: 2588–2596

44. S Husain, L A Alkhtaby, I Bhat, E Giorgetti, A Zoppi, M Muniz Miranda (2014) Study of cobalt doping on structural and luminescence properties of nanocrystalline ZnO. *Journal of Luminescence* 154: 430–436
45. Tirupataiah C, Kumar A S, Narendrudu T, Ram G C, Sambasiva Rao M V, Veeraiah N, Rao D K (2019) Characterization, optical and luminescence features of cobalt ions in multi-component $PbO - Al_2O_3 - TeO_2 - GeO_2 - SiO_2$ glass ceramics. *Optical Materials (Amst)* 88: 289–298
46. Toby B H (2001) EXPGUI, a graphical user interface for GSAS. *Journal of Applied Crystallography* 34: 210–213
47. Rietveld H M (1969) A profile refinement method for nuclear and magnetic structures. *Journal of Applied Crystallography* 2: 65–71
48. Sakata M, Cooper M J (1979) An analysis of the Rietveld refinement method. *Journal of Applied Crystallography* 12: 554–563

Nanowire Magnets
Synthesis, Properties, and Applications

3

Daljit Kaur
Department of Physics, DAV University, Jalandhar, India

Contents

3.1	Introduction	42
3.2	Synthesis Techniques	43
	3.2.1 Electrodeposition	43
	3.2.1.1 Process of Electrodeposition	43
	3.2.2 Electroless Deposition	45
	3.2.3 Electron Beam Lithography	45
	3.2.4 Other Techniques	45
3.3	Magnetic Properties	45
	3.3.1 Magnetic Anisotropy	45
	3.3.1.1 Shape Anisotropy Energy	46
	3.3.1.2 Magnetocrystalline Anisotropy Energy	46
	3.3.2 Exchange Energy	46
	3.3.3 Zeeman Energy	47
	3.3.4 Dipolar Energy	47
	3.3.5 Micromagnetism	47
3.4	Magnetization Reversal/Switching in Nanowires and Their Arrays	48
	3.4.1 Coherent Rotation	49
	3.4.2 Curling-Vortex Domain Nucleation and Propagation	50
	3.4.3 Transverse Domain Nucleation and Propagation	51
	3.4.4 Other Reversal Modes – Buckling and Fanning	52
3.5	Applications	52
	3.5.1 Data Storage	52
	3.5.1.1 Perpendicular Recording and Patterned Media	52
	3.5.1.2 Racetrack Memory	53
	3.5.2 Magnonic Crystal Devices	53
	3.5.3 Spintronics Applications	54

DOI: 10.1201/9781003197492-3

3.5.4 Biomedical Applications 54
3.5.4.1 Biosensors 54
3.5.4.2 Cancer Cell Destruction 54
3.5.5 Catalysis 54
3.5.6 Neuromorphic Computation 55
3.6 Conclusion and Future Scope 55
References 55

3.1 INTRODUCTION

Nanowires (NWs) are wires with diameters in nanometers and are generally referred to as one-dimensional (1D) materials. NWs have many novel properties that are not seen in three-dimensional (3D) or bulk materials. This is because NWs have two quantum restricted directions, while the third direction is unrestricted, and thus, the occupancy of electrons in the energy levels is different than the energy levels found in three-dimensional bulk materials. Because of the unique density of electronic states in NWs, as shown in Figure 3.1, they exhibit intriguing electrical, magnetic, and optical properties different from their bulk 3D crystalline counterparts [1–2].

The magnetic NWs are cylindrical wires with diameter in nanometers and length in micrometers, made up of magnetic material like Co, Ni, Fe, Gd, or their alloys, while nanotubes are hollow cylindrical. Template-assisted electrodeposition and electroless chemical deposition are the most common methods employed to deposit the bunch of wires in an array. Lithography uses a focused electron or ion beam to fabricate 1D stripes with long lengths, while the thickness and breadth of a single stripe referred to as nanowire, is of few nanometers. The potential applications of magnetic NWs are high-density magnetic

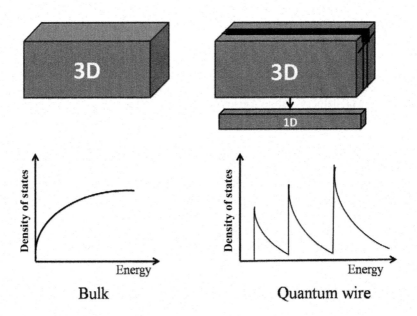

FIGURE 3.1 Density of states d(E) vs. E diagram for 3D bulk material and 1D NW [2]. Adapted with permission from [2]. Copyright the author (2013), Thesis IIT Delhi.

data storage, biomedical applications, racetrack memory devices, and magnonic crystal devices [3–4]. The magnetic NWs can be fabricated with different materials and different structures as follows:

1. Metallic and semiconducting nanowire: The electrical property of NW of magnetic material can be metallic (like Co, Fe, Ni, etc.), semiconducting (pristine or Co-doped ZnO, etc.), or insulating.
2. Multi-segmented NWs: When the NW has different material segments, e.g., Co/Ni, Cu-Co multilayered nanowires are called multi-segmented NWs. The alternating hard and soft magnetic elements or magnetic/nonmagnetic elements can be used for the application of spin-torque nano-oscillators and giant magnetoresistive devices. These multi-segmented NWs can be easily achieved by electrodeposition using multi-potential step potentiometry, which switches to the different reduction potentials of the metal ions to be reduced.
3. Core-shell NWs: The core-shell NWs are made up of two materials, one inside as the core and another material coated around the core in a tubular form known as a shell. A combination of soft and hard magnetic materials can be deposited. The ferromagnetic (FM) core with an antiferromagnetic (AFM) shell or vice versa can exhibit an exchange bias phenomenon useful for various spintronic applications. These structures require a combination of deposition methods to achieve well-defined cores and outer shells.

3.2 SYNTHESIS TECHNIQUES

3.2.1 Electrodeposition

Electrodeposition (ED) is a simple, precise, and cost-effective technique for the deposition of thin films and nanostructures of metals, metal oxide, semiconductors, and bio-minerals. ED is used in depositing Cu interconnects for the fabrication of integrated circuits [5], semiconductor thin films and nanostructures for solar cells [6–7], plasmonic sensor devices [8], and optoelectronic devices [9]. Also, composite material nanostructures can be deposited, and their bandgap tunability has been achieved by the control of deposition parameters. It has many advantages over other vacuum techniques like:

1. ED is a very simply controlled, fast, and mass-production process of material fabrication that is carried out without the requirement of evacuating the chamber like in sputtering, thermal evaporation, and other vacuum techniques.
2. The deposition conditions like supporting electrolyte concentration, temperature, pH value, and applied potential can control the growth rate and tune the crystallographic growth direction. Hence, shape, size, and crystallographic orientation can be effectively tailored.
3. ED can deposit on the selective area because a deposition is possible only on conducting site on the substrate and hence nanoparticles and NWs are easily deposited by this technique.

3.2.1.1 Process of Electrodeposition

The electrochemical process deposits pure metals or alloys on conducting surfaces by the application of an electric current. ED is carried out in a three-electrode cell composed of three terminals:

1. Anode or a counter electrode: The terminal at which oxidation of anions takes place and consists of a 2 cm × 1.5 cm platinum sheet in our case.

2. The cathode or a working electrode: The terminal at which reduction of the cations takes place. For NWs, it consists of commercially available polycarbonate (PCT) membrane or anodic alumina (AAO) membranes which are available in different pore diameters. These membranes/templates are coated on one side with a metallic layer of Au/Ag/Cu with thickness up to 200 nm. This layer acts as a cathode for deposition in the cylindrical pores. For nanotubes deposition, the thickness of the coating is less, like 25–50 nm, so the whole of the pore does not get covered up with the conducting layer.
3. Reference electrode: All the potentials are measured with respect to this electrode because it has a constant composition, e.g., saturated calomel Hg/HgCl$_2$ electrode, Ag/AgCl electrode.

The electrolyte solution called the ED bath contains the salts of the metals to be deposited. The working electrode is connected to the negative terminal of the battery (external current source) and the counter electrode to the positive terminal of the battery. When the circuit is completed, the current through the electrolytic solution is passed due to the movement of charged particles (ions) through it. The negative anions travel towards the anode and the positive cations towards the cathode. The cations accept electrons from the cathode surface and deposit as adatoms on the cathode [2]. Figure 3.2a shows the schematic of the three-electrode cell. The SEM images of unfilled PCT (before deposition) template and filled PCT (after deposition) of pore diameter of 50 nm are shown in Figure 3.2b and Figure 3.2c, respectively. There are numerous reports available on the electrodeposited magnetic NWs and nanotubes of Co, Ni, Fe, NiFe, CoFe, CoNiFe, multilayered Cu-Co, Co/Ni, etc. Nielsch *et al.*, Ounadjela *et al.*, Blondel *et al.*, and Piraux *et al.* were the first to publish on uniformly electrodeposited NWs or multilayered NWs and their magnetic and MR properties [10–13]. However, the demonstration of the spintronic application of these electrodeposited NWs is quite a few in comparison.

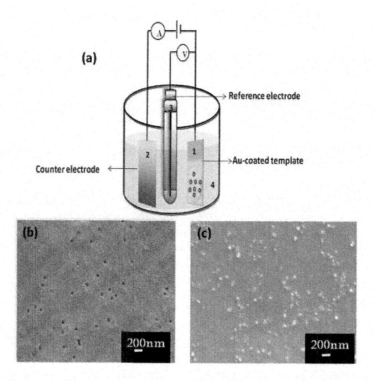

FIGURE 3.2 (a) Schematic diagram of a three-electrode cell, (b) SEM image of unfilled PCT before deposition, and (c) filled PCT after deposition [2]. Adapted with permission from [2]. Copyright the author (2013), Thesis IIT Delhi.

3.2.2 Electroless Deposition

Electroless deposition is similar to ED except that the use of voltage/current for deposition is replaced by reducing agents like boron, phosphorous, hydrazine, formaldehyde, etc. in the chemical bath solution. Hence there is no requirement of conducting substrate or layer. The reducing agent used depends on the desired material of deposition and the chemical resilience of the template. The pH and concentration of the reducing agent affect the structural, magnetic, and electrical properties of the deposited material. There are many reports on the electroless deposition of Co, Ni, CoFeB, NiFeB, NiB, etc. in templates and bio-templates [14].

3.2.3 Electron Beam Lithography

Focused electron beam lithography is a technique that allows the atoms to deposit at the preferred site/position with the help of scanning focused electron beam. It is equipped with a gas injection system that emits gaseous precursors to the substrate. The focused electron beam dissociates the adsorbed precursor, and metal atoms are deposited at the desired beam position with a resolution of a few tens of nanometers. The metallic content is generally 50% without purification and purification/annealing in a reducing environment; the metal content goes up to 95%. The planar nanostripes, vertical NWs, and core-shell NWs of magnetic materials have been successfully deposited, and Au/Pt/Ag contacts or protective coating helps in studying spin hall effect and magnetoresistance in these structures [15–16].

3.2.4 Other Techniques

The 1D magnetic NWs, nanotubes, nanorods, and nanostripes can also be deposited using the sol-gel method, electrospinning, chemical vapor deposition (CVD), atomic layer deposition (ALD), co-precipitation, etc. The crystal quality and growth of deposited NWs can be controlled by the deposition parameters in each technique. The NWs obtained from ALD have better crystallinity compared to other techniques [17–20].

3.3 MAGNETIC PROPERTIES

Magnetism is an inherently quantum mechanical phenomenon that arises due to the spin of the electrons. Most of the theoretical framework in magnetism is semi-classical, and as the magnetic materials are pushed to the nanoscale regime, it opens an interesting area of nanomagnetism with different novel phenomena like magnetic anisotropy, magnetization dynamics, superparamagnetism, exchange bias, giant magnetoresistance, tunneling magnetoresistance, and molecular magnetism, which can be the base of many new magnetic applications. The different energies involved in the resultant magnetization of NWs are discussed as follows:

3.3.1 Magnetic Anisotropy

There are different anisotropy energies associated with these magnetic NWs owing to the shape, crystallinity, and interwire spacing of the NW array. These different energies are as follows:

3.3.1.1 Shape Anisotropy Energy

The direction of the applied field does not produce any anisotropy in the magnetization of a spherical object; while for a cylindrical rod-shaped object, there is shape-induced anisotropy in the magnetization of the body. A magnetized body will produce magnetic poles at the ends of the surface, which gave rise to the demagnetizing field, H_{demag} [21]. It is called so because it demagnetizes or acts in opposition to the magnetization that produces it and hence is proportional to it:

$$H_{demag} = -K_d \cdot M$$

where K_d is the demagnetization factor. The magnetostatic energy E_D (erg/cm³) associated with a particular magnetization direction can be expressed as:

$$E_D = \frac{1}{2} K_d M_s^2$$

where M_s is the saturation magnetization of the material (emu/cm³) and K_d is the demagnetization factor along the magnetization direction. The demagnetization factor values along the easy axis (parallel to the nanowire axis) is 0 and along the hard axis (perpendicular to the nanowire axis) is 2π. Thus, the shape anisotropy energy difference is $E_D = -\pi M_s^2$. For the case of bulk cobalt, the E_D value comes out to be equal to 6×10^6 erg/cm³.

3.3.1.2 Magnetocrystalline Anisotropy Energy

The magnetocrystalline anisotropy arises in a magnetic material when the magnetization is favored along with any defined crystalline orientation/direction. This is caused by spin-orbit interaction when the electron spin gets coupled to the orbital moment of the electron, which is further connected to the crystal structure of the material. The magnetocrystalline anisotropy term can be mathematically expanded in a power series taking into account the crystal symmetry and using coefficients from experimentally observed values. The magnetocrystalline anisotropy energy for a hexagonal close-packed (hcp) crystal is a function of the angle ϕ between the magnetization, M, and the c-axis of the hcp crystal. The energy is symmetric with respect to the basal plane, which leads to the exclusion of odd powers of $sin\phi$ in the series expansion for the anisotropy energy density, E_{mc}. The first two terms of the energy density are:

$$E_{mc} = K_1 sin^2\phi + K_2 sin^4\phi$$

where K_1 and K_2 are first order and second order anisotropy constants. If K_1 is positive, then the c-axis is an easy axis of magnetization; if K_1 is negative, it is a hard axis with an easy plane perpendicular to the c-axis [22].

3.3.2 Exchange Energy

The exchange interaction tries to align the neighboring atomic dipole moments along the same direction in a magnetic material by direct exchange or indirect exchange. The direct exchange energy between two neighboring magnetic moments μ_i and μ_j is usually described by:

$$\varepsilon_{ex}^{i,j} = -JS_i \cdot S_j$$

The exchange integral J has its origin in the extent of overlap of the wave function of two electrons Ψ (r₁, r₂). S_i and S_j are the unit vectors of the direction of the neighboring magnetic moments. The positive value of J favors ferromagnetic order in a magnetic material with minimum exchange energy for parallel alignment of neighboring spins, while the negative J value favors antiferromagnetic ordering with minimum exchange energy when the spins are aligned antiparallel to each other.

3.3.3 Zeeman Energy

The Zeeman energy is the interaction energy of magnetization M of a body with an external applied magnetic field H_{ext}. The expression of this Zeeman energy is given by:

$$E_{Zeeman} = -M.H_{ext}$$

3.3.4 Dipolar Energy

The dipolar anisotropy generally arises in an array or a group of similar shape magnetic entities e.g. nanowire array, nanoparticles, etc. This anisotropy depends on the shape of the magnetic object and inter-object distance. In a highly dense magnetic array of NWs, there exists a magnetostatic long-range dipolar interaction among the neighboring NWs. The dipolar anisotropy energy density is difficult to calculate mathematically as it is an open boundary problem with one of its boundary conditions at infinity. The dipolar interaction between the NWs leads to rounding of the in-plane hysteresis loops and broadening of in-plane resonance peaks.

Magnetostatic interactions between the NWs depend on the inter-wire distance, length, and diameter of the NWs. These interactions increase with a decrease in inter-wire spacing and increase in the diameter of the wires [23]. The wires with a diameter d and inter-axis distance D in ordered alumina membranes yield the packing factor:

$$p = \left(\frac{\pi}{2\sqrt{3}}\right)\left(\frac{D}{d}\right)^2$$

The dipolar energy E_d in a composite medium for magnetic tape recording described is given as:

$$E_d = -\mu_0 \vec{M}.\vec{H_d} = K_d \left(\frac{3p-1}{2}\right)\cos^2\theta$$

3.3.5 Micromagnetism

The micromagnetics theory of ferromagnetic materials was put forth by W. F. Brown in the 1940s. It has been extensively used to model ferromagnetic behavior and hysteresis curves in any given magnetic material [24]. It is a continuum theory that has bridged the gap between Maxwell's electromagnetic theory and the microscopic quantum theory of spin structures.

Any system in thermodynamic equilibrium tries to attain its lowest energy state, which implies that its total free energy must attain a minimum value. The magnetic properties of a material depend on several energy terms in the micromagnetics theory. The total free energy is a result of the following energy terms: the dipolar energy E_d, the exchange energy E_{ex}, the Zeeman energy E_Z, and the anisotropy energy E_a is the resultant of magnetocrystalline and shape anisotropy energy when the system is at 0 K (i.e. in the of thermal energy spin fluctuations):

$$E_{tot} = E_d + E_a + E_{ex} + E_z$$

The normalized magnetization M(r), per unit volume, is determined at any value of the applied field, H, by minimizing the total free energy. The energy minimization calculation is done by the finite element method (FEM) or finite difference method in the magnetic structure. FEM divides or discretizes the magnetic structure into many small cells of nanometer size. The size of cells should be small enough to achieve linearity in the solution of the equation for each cell. The given equation is solved numerically by taking into account the continuity of solution at each cell boundary (solving boundary conditions). There are distinct advantages that lead to the successful development and applications of the micromagnetic theory:

1. The magnetic domains, domain walls, and spin singularities structures in nanoparticles and thin films can be identified easily.
2. The nucleation and magnetization reversal process can be analyzed completely.
3. The characteristic parameters of the hysteresis loops can be interpreted.

The mathematical computation of complex micromagnetic problems is now possible by applying various numerical methods with the help of supercomputers, which further helped in the successful application of the micromagnetic theory in nanomagnetic elements [25].

3.4 MAGNETIZATION REVERSAL/SWITCHING IN NANOWIRES AND THEIR ARRAYS

Magnetization reversal studies are critical to study and analyze the magnetic properties of nanostructures as they give precise information about the spin structures and domain nucleation phenomena which could be useful for spintronics applications. The microstructure and defects play a significant role in deciding the various physical processes that take place during magnetization reversal, like domain nucleation, its propagation, pinning and de-pinning of domain walls, etc. The various magnetic characterizations used to determine the magnetization reversal mechanism are temperature dependence of the coercivity (H_c), the angular dependence of H_c, magnetic viscosity, and the determination of dipolar fields by numerical modeling. The determination of coercivity and remanence can help to understand the magnetic ordering of the sample under study [26]. In the case of magnetic NWs, the angle of the applied magnetic field to the NW axes critically affects the process of magnetization reversal and H_c. It has been established that different magnetization reversal modes have different angular dependence variations with H_c, which has been explained further in this section.

Magnetization reversal studies in cobalt and nickel NWs have been studied in detail in the last few years with different models. Frei et al. [27] in 1957 were the first to give theoretical studies on the magnetization reversal of an ideal ferromagnetic cylinder. They have explained three modes of magnetization reversal, namely, coherent rotation, curling, and buckling. Aharoni et al. [28] in 1958 explained the nucleation modes of buckling and curling for all radii of an infinite cylinder and the transition from buckling to curling from lower to a higher value of radius. Wernsdorfer et al. [29] have proposed that magnetization reversal in a ferromagnetic NW of finite length results from the thermally activated nucleation and propagation reversal by the switching time measurements in 100 nm Ni NWs while the observed nucleation field values are close to the curling mode reversal. O'Barr et al. [30] have studied the magnetization reversal in Ni columns with radii varying from 20 nm to 500 nm by magnetic force microscopy using in situ applied magnetic field and found that curling mode applies to a narrow range of radii only. Ferre et al. [31] have studied the magnetization configuration and processes by 3D micromagnetics simulations and compared it with resistivity and magnetization measurements in Co and Ni NWs for diameters 35–100 nm and found that for Co NWs, the *c*-axis of hcp Co lies parallel to the NW axis for diameters smaller than 50 nm and its orientation becomes perpendicular to the NW axis at larger diameters. The Co NWs possess a single domain spin texture with an easy axis along the wires for diameters less than 25 nm generally. An irreversible jump in the resistivity measurements of Co NWs has been found when the reversal takes place in each one of the wires independently, which is initiated by domain nucleation as studied by micromagnetic calculations. Wegrowe *et al.* [32] have studied magnetization reversal in electrode-posited Ni NWs by anisotropic magnetoresistance (AMR) measurements. The magnetization reversal can be of the following types in NWs:

3.4.1 Coherent Rotation

Stoner and Wohlfarth, in 1948 [33] gave the simplest model of coherent rotation, in which all the atomic spins are collinear. All of them switch from one direction to the other coherently. The assumption of the same direction of all magnetic moments in the system makes the exchange energy value zero. Hence the only interaction is between the applied field and the anisotropic energy of the system. The total free energy is now the addition of only two energies, Zeeman energy, and demagnetization energy, viz.,

$$E = E_Z + E_{Demag} \tag{1.1}$$

The demagnetization energy for a prolate spheroid, which closely resembles a cylindrical nanowire, is given by the expression

$$E_{Demag} = K_u \sin^2(\theta - \theta_0) \tag{1.2}$$

where θ_0 is the angle between the applied field and easy axis of magnetization; θ is the angle between the magnetization, *M*, and the external field, *H*; and K_u is the uniaxial shape anisotropy constant given as $K_u = \frac{1}{2}(N_x - N_z)M_s^2$ (N_x and N_z are demagnetization factors along the short axis and the long axis of the wire respectively). The NW has a high aspect ratio, such that the values of $N_x = 2\pi$ and $N_z = 0$ and hence $K_u = \pi M_s^2$. As $E_Z = -H.M$, the total magnetic energy becomes

$$E = K_u \sin^2(\theta - \theta_0) - H.M \tag{1.3}$$

The component of the magnetization along the field axis is $M_s \cos\theta$. The energy minimization is done with respect to the angle θ and hence the hysteresis loop can be simulated numerically by solving for ***m*** as a function of *h*, where ***m*** = M/M_s and $h = H/2(K_u/M_s)$. The nucleation field or switching field at which the magnetization flips is given as

$$h_n = \left(\cos^{2/3}\theta_0 + \sin^{2/3}\theta_0\right)^{-3/2} \tag{1.4}$$

where $h_n = H_n/(2K_u/M_s)$.

In this model, there are two regions:

1. When the applied field is close to the direction of the easy magnetization axis, $0° < \theta < 45°$, the hysteresis loop is approximately square, as shown in Figure 3.3.

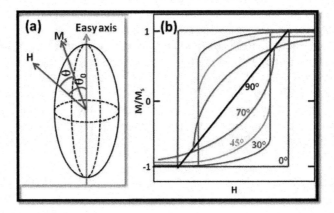

FIGURE 3.3 (a) Prolate spheroid and (b) Stoner-Wohlfarth simulated hysteresis loops for various angles θ (0°, 30°, 45°, 70°, and 90°) of the field H for prolate spheroid. Adapted with permission from [33]. Copyright (1948) Royal Society (U.K.).

2. When the applied field is oriented closer to the direction of the hard axis, $45° < \theta < 90°$, the hysteresis loop is sheared such that switching occurs after the magnetization changes sign [34]. The coercive field can be determined as follows:

$$H_c^{coherent}(\theta) = \begin{cases} H_c(\theta) = -H_n(\theta) = -2K_u \dfrac{\left(\sqrt{1-t^2+t^4}\right)}{\mu_0 M_s (1+t^2)}, & 0 \leq \theta \leq \pi/4 \\ H_c(\theta) = H_n(\pi/4) - H_n(\theta), & \pi/4 \leq \theta \leq \pi/2 \end{cases} \quad (1.5)$$

where $t = \tan^{1/3}\theta$.

The Stoner-Wohlfarth model applies to smaller systems having larger anisotropy contribution, thus leading to a single domain state, whereas, for larger systems, this approximation fails due to sizeable dipolar anisotropy contribution leading to different spin structures and magnetic textures than that of a single-domain state. Thus cobalt NWs with a diameter less than 20 nm are expected to have a single domain ground state exhibit magnetization reversal by coherent rotation.

3.4.2 Curling-Vortex Domain Nucleation and Propagation

With an increase in the size of nanostructures, the magnetization reversal mode changes from coherent rotation to an inhomogeneous magnetization reversal such as curling, buckling, or the creation-annihilation of a vortex-antivortex pair. In the curling model, the reversal of magnetization takes place when the spins rotate progressively via propagation of a vortex domain wall, and the H_c is very close to the nucleation field, H_n. The angular dependence of the normalized nucleation field for a prolate spheroid based on the curling model is given [35] by

$$H_c^{curling}(\theta) = -H_n(\theta) = \dfrac{\left(2N_z - \dfrac{k}{S^2}\right)\left(2N_x - \dfrac{k}{S^2}\right)}{\sqrt{\left(2N_z - \dfrac{k}{S^2}\right)^2 \sin^2\theta + \left(2N_x - \dfrac{k}{S^2}\right)^2 \cos^2\theta}} \quad (1.6)$$

where N_z and N_x are the demagnetizing factors along and perpendicular to the NW-axis, 0 and 2π respectively, k is the geometrical parameter whose value for an infinite cylinder is taken as 1.08, S is the reduced radius given by R/R_o, R_o being the threshold radius above which the curling takes place. Here $R_o = \dfrac{2\sqrt{A}}{M_s}$, where A is the exchange energy constant and M_S is saturation magnetization. In the curling reversal mode, the H_c of NWs is increased by decreasing the wire diameter. Thus, curling is found to be the dominant mode of reversal in cobalt NWs with a diameter > 50 nm [32, 36–37].

3.4.3 Transverse Domain Nucleation and Propagation

In coherent rotation mode, all the spins rotate concomitantly, while in curling mode, the spins rotate progressively via nucleation and propagation of a vortex domain wall; there is one other transverse wall reversal mode in which the spins rotate along the magnetized body via nucleation and propagation of a transverse domain wall. For the transverse mode, the angular dependence of the H_c can be studied by an adapted Stoner-Wohlfarth model in which the width of the transverse domain wall w_T is used as the length of the region undergoing coherent rotation as shown in Figure 3.4. The width of the domain wall depends on the wire geometry. The domain wall width increases with the radius of the wire and is about 50 nm larger than the wire's radii. Following this approach, the angular dependence of H_c in the transverse mode is given the same as that of coherent mode Eq. (1.5). The only difference is that this mode will always exhibit a lower H_c, independent of θ, because there would be a change in the measure of uniaxial anisotropy constant, K_u, which depends on the length of a transverse domain formed in the NW. Recently Vivas et al. [34, 38] have shown that transverse reversal mode is the magnetization reversal mechanism in cobalt and nickel NWs, which is similar to the coherent reversal except that the coherent rotation takes place in a small region along the NW which is the length of the transverse domain.

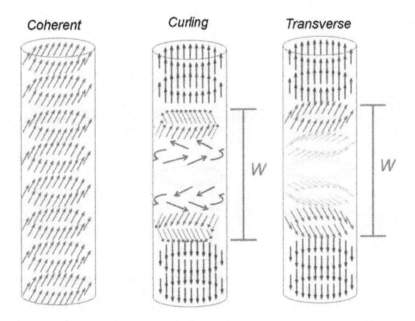

FIGURE 3.4 Pictorial presentation of movement of spins in coherent, curling, and transverse reversal modes of reversal with w showing the width of two types of domain walls, vortex, and transverse DW, in a cylindrical nanowire [2]. Adapted with permission from [2]. Copyright the author (2013), Thesis IIT Delhi.

3.4.4 Other Reversal Modes – Buckling and Fanning

There are other lesser-known modes, one of which is magnetization buckling, whose formulation was given by Frei et al. [27]. In this mode, the spin rotates only along the NW diameter and the spin deviation is a periodic function along the length of the NW. Formulism is complicated. The fanning reversal mode was suggested by Jacobs and Bean [39] for an assembly of elongated fine particles by approximating them as the "chain of spheres." In fanning mode, the magnetization of successive spheres in the chain rotates in a plane in alternate directions, while its magnetization rotates in unison when all the spheres are parallel to each other.

3.5 APPLICATIONS

3.5.1 Data Storage

3.5.1.1 Perpendicular Recording and Patterned Media

The technology based on perpendicular magnetic recording suggested by Prof. Iwasaki in 1975 replaced conventional longitudinal recording, which is no longer made up of randomly oriented grain clusters. Instead, a regular array of nanostructures is to be used as a patterned recording media, and their magnetization states will be perpendicular to the plane of the media. In this new technology, 1 bit is stored in each nanostructure, which functions as one magnetic unit. This enhances the storage density greatly. The comparison between conventional media and bit patterned media is shown in Figure 3.5b. The most

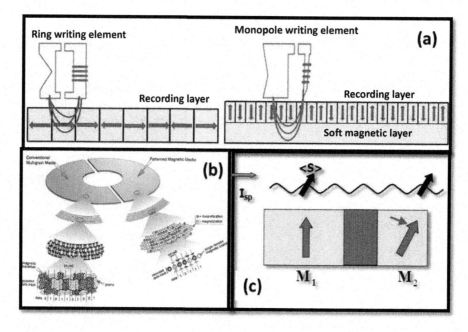

FIGURE 3.5 (a) Schematic diagram for a longitudinal recording medium and perpendicular recording medium, adapted from Wikipedia source [41] under CC BY-SA 3.0. (b) Schematic diagram of storage media showing conventional multigrain media and bit patterned media Adapted from [42] under CC BY 4.0. (c) Schematic showing torque exerted on local magnetization by the injected spin-polarized current. Adapted with permission [2]. Copyright the author (2013), IIT Delhi.

important improvement in this new technology is that the superparamagnetic limit is significantly delayed. Longitudinal and perpendicular recording have their thermal stability limits at about 300 GB/in^2 and 1 Tb/in^2, respectively. In a patterned media, the thermal stability issues occur only at extremely high densities (a few Tb/in^2) [40–41]. Many different kinds of magnetic nanostructures such as rings, disks, and NWs have been proposed to serve as the patterned recording media. NWs have a strong advantage over the other geometry because of their very large H_c when magnetized along the long axis. This is due to its inherent large shape anisotropy. A large H_c is desirable in magnetic recording to prevent the accidental erasure of data by a stray magnetic field. Although this seems to be a great advantage, it does not mean that an NW array would be the ultimate choice for the patterned media. There are many other compatibility issues associated with the servo technology as well as the read/write mechanisms at the read/write head, which need to be resolved.

To use these NW arrays in the recording medium, it is necessary to study their properties, like spin switching mechanisms and damping processes, under external perturbation methods. The external perturbation methods include sources like microwaves, spin-polarized current, magnetic field, etc., which are used for switching the magnetization. The analysis of the magnetic interactions between the NWs and their switching behavior can help to know about the magnetization reversal in these nanostructures and gain insight into cooperative magnetism at the fundamental level.

3.5.1.2 Racetrack Memory

Along with perpendicular bit patterned storage media, racetrack memory is a new design of fast, non-volatile, and high data density storage media, proposed in 2002 by Parkin *et al.* based on voltage/current control of domain wall motion in NWs [43]. A racetrack is a NW having magnetic domains, separated by domain walls, which store bits of data. Electrons have a fundamental quantum property called spin. Sending an electric current along the NW is the spintronic method for moving the domain walls uniformly along the racetrack. Consider the 1–0–1 domain layout. The electrons passing through the initial 1 domain will be spin polarized, with their magnetism aligned in the direction of the 1 domain. Each electron's magnetism will tend to flip to the 0 direction as it crosses the 1–0 domain wall. The magnetism of an electron is determined by its spin, which is a measure of angular momentum. Angular momentum, like energy and linear momentum, is a conserved quantity. Something else must flip from 0 to 1 for the electron to flip from 1 to 0, and that something else is an atom on the 0 sides of the domain wall. The spin-polarized electrons shift the domain wall along with the NW one atom at a time as they flow through it. Thus spin-polarized electric currents are used to manipulate the magnetic domain walls instead of the magnetic field and hence the bits themselves move along their racetrack, passing through a read/write head at a particular location. The magnetized regions are rewritable and non-volatile and require no moving parts, thereby increasing the reliability and speed of the racetrack memory device. Also, a single domain wall can move 150 nm in one nanosecond, leading to very high switching speeds.

3.5.2 Magnonic Crystal Devices

Along with storage media applications, spin-wave excitations (magnons) have also been studied in a 1D magnonic crystal fabricated out of Ni$_{80}$Fe$_{20}$ NWs by Topp *et al.* [44]. These crystals are exceptional candidates for the manufacturing of nanoscale microwave devices. 1D magnonic crystal is a magnetic NW array (Ni-Fe or Co NWs) with bandgap tunability arising from different inter-wire separations. Wang *et al.* designed and fabricated a 1D MC of a periodic array of alternating cobalt and permalloy nanostripes. The spin waves are used as information carriers by the encoding of information into the magnetization amplitude. Different frequencies can be employed as separate information channels, and the information is encoded in the phase of the spin wave, allowing for simultaneous data processing in the same structure. The capability to transmit and process information in a multi-channel magnonic crystal is more efficient than the existing switch-based logic devices. Magnonic crystals on polymer substrates open up the possibility of a strain-controllable multi-channel telecommunication system.

3.5.3 Spintronics Applications

The multilayered NWs made of alternating ferromagnetic and nonmagnetic nanometric layers are of great interest as magnetic sensors and spintronics applications. The current perpendicular to the plane (CPP) giant magnetoresistance (GMR) effect is the basic principle of these devices [45]. A spin-polarized current can be used to change the orientation of a magnetic layer in a magnetic tunnel junction or spin valve, which is known as spin-transfer torque. According to John Slonczewski (1996) and Luc Berger (1996), electron currents in magnetic multilayer systems can transmit angular momentum from one magnetic layer to another, imposing a torque on the local magnetization. They also found that the magnetization may oscillate at microwave frequencies for a long time. (1) The spacer layers between the magnetic layers must be thin (50 nm) so that spins do not depolarize as they pass from one layer to the next; (2) the device must be small (100 nm) (i.e., a NW) so that the amount of spin momentum transported by the electron current is a significant fraction of the angular momentum of the magnetic element; and (3) the configuration must have enough non-linearities to stabilize the precessional orbits. Tsoi *et al.* (1998, 2000) [46] discovered microwave oscillations in magnetic multilayers in a point-contact geometry, and then Kiselev *et al.* (2003) [47] and Rippard *et al.* (2004) [48] reported microwave emission directly in nanoscale device architectures. Spin-torque oscillators, spin-transfer oscillators, and spin-transfer nano-oscillators (STNOs) are all terms used to describe these structures.

3.5.4 Biomedical Applications

3.5.4.1 Biosensors

NWs can be used as medical biosensors since nano-sized sensing wires can be laid down through a microfluidic channel. The NW sensors take up the chemical identifications of particles as they travel through the microfluidic channel and may quickly convey this information to the outside world via a connection of electrodes. These can detect the presence of cancer-related gene mutations and may aid researchers in pinpointing the exact location of such mutations [49].

3.5.4.2 Cancer Cell Destruction

Magnetic hyperthermia is a technique using magnetic nanostructure to destroy cancer cells at a point by producing local heating with the help of an ac magnetic field. Iron-based magnetic materials show higher biocompatibility than Ni- or Co-based materials.

3.5.5 Catalysis

The magnetic NW acts as nano-stirrers in an induced magnetic field and can locally improve the flow of ions/molecules to increase the catalysis process of degradation or hydrogen production. Serra *et al.* employed 25 nm wide Co–Pt mesoporous NWs for the degradation of two pollutants (4-nitrophenol and methylene blue) and obtained the normalized rate constants k_{nor} as high as 20,667 and 21,750 s^{-1} g^{-1} for 4-nitrophenol and methylene blue decline, correspondingly. They also used these NWs for the production of hydrogen from sodium borohydride, and activity values for hydrogen production are as high as 25.0 L H$_2$ g^{-1}/min^{-1} at room temperature [50]. Three distinct organic dye molecules, RB, MB, and RhB, were mineralized in the presence of NaBH4, employing CoO NWs as a catalyst in 130, 81, and 22 minutes, respectively [51]. Sun *et al.* investigated the hydrolysis of ammonia borane utilizing bimetallic Co–Cu NWs as a catalyst, and the activity was calculated to be 6.17 (mol H_2/min/(mol Co–Cu)) in terms of turnover frequency (TOF), and thereafter magnetic NWs can be easily separated with the help of a magnet from the system [52].

3.5.6 Neuromorphic Computation

Neuromorphic computation or non-Boolean computation refers to the complex computation similar to the inherent properties of physical systems. The DW dynamics in magnetic NWs are suitable for implementing this computation. A magnetic nanowire or its array behaves as a DW device. The combination of DW device with a CMOS buffer resembles an artificial neuron. Energy-efficient voltage-controlled nanomagnetic deep neural networks can be realized in real-time by studying the voltage control of domain wall motion in these nanoscale neuromorphic devices [53]. Ababei *et al.* have realized that individual nanoscale DWs have dynamics suitable for creating neuromorphic computing devices with critical dimensions < 1 μm [54]. Neuromorphic spintronics has become an emerging field of research to simulate complex reservoir computation and architectures to mimic brain-like functions for AI devices [55].

3.6 CONCLUSION AND FUTURE SCOPE

The shape geometry and magnetization dynamics of NWs and their arrays have shown promising applications in the fields of spintronics, magnonics, data storage, biosensors, and catalysis, while the new field of artificial neuromorphic computation in these NWs has rekindled the quest to understand domain wall dynamics in these complex nanowire configurations. The synthesis by electrodeposition in templates is successfully and easily done, but the transport measurements are quite challenging. With the advent of powerful and new characterization methods, the field of one-dimensional magnetic NWs can be explored for a large number of spintronics-based devices.

REFERENCES

1. Klabunde, K. J. Introduction to nanotechnology (chapter 1). *Nanoscale Materials in Chemistry* (John Wiley & Sons), 1–13 (2001). doi:10.1002/0471220620.CH1.
2. Kaur, D. Synthesis and study of structurally tailored magnetic nanowires. PhD thesis, IIT Delhi (2013).
3. Prestvik, W. S., Berge, A., Mørk, P. C., Stenstad, P. M. and Ugelstad, J. Preparation and application of monosized magnetic particles in selective cell separation (chapter). *Sci. Clin. Appl. Magn. Carriers (Springer Nature)* 11–35 (1997). doi:10.1007/978-1-4757-6482-6_2.
4. Hergt, R. and Andrä, W. Magnetic hyperthermia and thermoablation. *Magn. Med. A Handb.* Second Ed. (John Wiley & Sons), 550–570 (2007). doi:10.1002/9783527610174.CH4F.
5. Andricacos, P. C., Uzoh, C., Dukovic, J. O., Horkans, J. and Deligianni, H. Damascene copper electroplating for chip interconnections. *IBM J. Res. Dev.* 42 (5), 567–574 (1998).
6. Deligianni, H., Ahmed, S. and Romankiw, L. T. The next frontier: Electrodeposition for solar cell fabrication, electrochem. *Soc. Interface* 20, 47–53 (2011).
7. Ullah, S., Ullah, H., Bouhjar, F., Mollar, M. and Marí, B. Synthesis of in-gap band CuGaS2:Cr absorbers and numerical assessment of their performance in solar cells. *Sol. Energy Mater. Sol. Cells* 180, 322–327 (2018).
8. Kim, S.-S., Na, S.-I., Jo, J., Kim, D.-Y. and Nah, Y.-C. Plasmon enhanced performance of organic solar cells using electrodeposited Ag nanoparticles. *Appl. Phys. Lett.* 93, 073307 (2008).

9. Pauporté, T. and Lincot, D. Electrodeposition of semiconductors for optoelectronic devices: Results on zinc oxide. *Electrochim. Acta* 45, 3345–3353 (2000).
10. Nielsch, K., Müller, F., Li, A.-P. and Gösele, U. Uniform nickel deposition into ordered alumina pores by pulsed electrodeposition. *Adv. Mater.* 12, 582–586 (2000).
11. Blondel, A., Doudin, B. and Ansermet, J. P. Comparative study of the magnetoresistance of electrodeposited Co/Cu multilayered nanowires made by single and dual bath techniques. *J. Magn. Magn. Mater.* 165, 34–37 (1997).
12. Ounadjela, K., Ferré, R. and Louail, L. Magnetization reversal in cobalt and nickel electrodeposited nanowires. *J. Appl. Phys.* 81, 5455–5457 (1997).
13. Piraux, L., George, J. M., Despres, J. F., Leroy, C., Ferain, E. and Legras, R. Giant magnetoresistance in magnetic multilayered nanowires. *Appl. Phys. Lett.* 65, 2484–2486 (1994).
14. Staňo, M. and Fruchart, O. Magnetic nanowires and nanotubes (chapter 3). *Handb. Magn. Mater.* (Elsevier) 27, 155–267 (2018).
15. Córdoba, R., Sharma, N., Kölling, S., Koenraad, P. M. and Koopmans, B. High-purity 3D nano-objects grown by focused-electron-beam induced deposition. Nanotechnology 27, 355301 (2016).
16. Pablo-Navarro, J., Magén, C. and Teresa, J. M. de. Three-dimensional core – shell ferromagnetic nanowires grown by focused electron beam induced deposition. *Nanotechnology* 27, 285302 (2016).
17. Daub, M., Knez, M., Goesele, U. and Nielsch, K. Ferromagnetic nanotubes by atomic layer deposition in anodic alumina membranes. *J. Appl. Phys.* 101, 09J111 (2007).
18. Khalil, A., Lalia, B. S., Hashaikeh, R. and Khraisheh, M. Electrospun metallic nanowires: Synthesis, characterization, and applications. *J. Appl. Phys.* 114, 171301 (2013).
19. Yoon, H. et al. Epitaxially integrating ferromagnetic Fe1.3Ge nanowire arrays on few-layer graphene. *J. Phys. Chem. Lett.* 2, 956–960 (2011).
20. Scott, J. A. et al. Versatile method for template-free synthesis of single crystalline metal and metal alloy nanowires. *Nanoscale* 8, 2804–2810 (2016).
21. Sun, L., Hao, Y., Chien, C. L., Searson, P. C. and Searson, P. C. Tuning the properties of magnetic nanowires. *IBM J. Res. Dev.* 49, 79–102 (2005).
22. Cullity, B. D. and Graham, C. D. Introduction to magnetic materials. *Introd. to Magn. Mater.* (2009) Wiley, United States of America. doi:10.1002/9780470386323.
23. Encinas-Oropesa, A., Demand, M., Piraux, L., Huynen, I. and Ebels, U. Dipolar interactions in arrays of nickel nanowires studied by ferromagnetic resonance. *Phys. Rev. B* 63, 104415 (2001).
24. Brown, W. F. Theory of the approach to magnetic saturation. *Phys. Rev.* 58, 736 (1940).
25. Kronmuller, H. and Fähnle, M. *Micromagnetism and the microstructure of ferromagnetic solids.* Cambridge University Press, 2003.
26. Givord, D., Rossignol, M. F. and Taylor, D. W. Coercivity mechanisms in hard magnetic materials. *J. Phys.* IV 02, C3–95 (1992).
27. Frei, E. H., Shtrikman, S. and Treves, D. Critical size and nucleation field of ideal ferromagnetic particles. *Phys. Rev.* 106, 446 (1957).
28. Aharoni, A. and Shtrikman, S. Magnetization curve of the infinite cylinder. *Phys. Rev.* 109, 1522 (1958).
29. Wernsdorfer, W., Doudin, B., Mailly, D., Hasselbach, K., Benoit, A., Meier, J., Ansermet, J.-Ph. and Barbara, B. Nucleation of magnetization reversal in individual nanosized nickel wires. *Phys. Rev. Lett.* 77, 1873 (1996).
30. O'Barr, R., Lederman, M. and Schultz, S. Preparation and quantitative magnetic studies of single-domain nickel cylinders. *J. Appl. Phys.* 79, 5303 (1998).
31. Ferré, R., Ounadjela, K., George, J. M., Piraux, L. and Dubois, S. Magnetization processes in nickel and cobalt electrodeposited nanowires. *Phys. Rev. B* 56, 14066 (1997).
32. Wegrowe, J.-E., Kelly, D., Franck, A., Gilbert, S. E. and Ansermet, J.-P. Magnetoresistance of ferromagnetic nanowires. *Phys. Rev. Lett.* 82, 3681 (1999).
33. Stoner, E. C. and Wohlfarth, E. P. A mechanism of magnetic hysteresis in heterogeneous alloys. *Philos. Trans. R. Soc. London. Ser. A, Math. Phys. Sci.* 240, 599–642 (1948).
34. Vivas, L. G., Vazquez M., Escrig J., Allende S., Altbir D., Leitao D. C. and Araujo, J. P. Magnetic anisotropy in CoNi nanowire arrays: Analytical calculations and experiments. *Phys. Rev. B* 85, 035439 (2012).
35. Aharoni, A., Angular dependence of nucleation by curling in a prolate spheroid. *J. Appl. Phys.* 82, 1281 (1998).
36. Tannous, C., Ghaddar, A. and Gieraltowski, J. Geometric signature of reversal modes in ferromagnetic nanowires. *EPL (Europhysics Lett.)* 91, 17001 (2010).
37. Lavin, R., Denardin, J. C., Espejo, A. P., Cortés, A. and Gómez, H. Magnetic properties of arrays of nanowires: Anisotropy, interactions, and reversal modes. *J. Appl. Phys.* 107, 09B504 (2010).
38. Vivas, L. G., Escrig, J., Trabada, D. G., Badini-Confalonieri, G. A. and Vázquez, M. Magnetic anisotropy in ordered textured Co nanowires. *Appl. Phys. Lett.* 100, 252405 (2012).

39. Jacobs, I. S. and Bean, C. P. An approach to elongated fine-particle magnets. *Phys. Rev.* 100, 1060 (1955).
40. Nakamura, Y. Perpendicular magnetic recording. *Magn. Storage Syst. Beyond* 2000, 75–102 (2001) Springer, Dordrecht. doi:10.1007/978-94-010-0624-8_4.
41. Perpendicular recording – Wikipedia (webpage). https://en.wikipedia.org/wiki/Perpendicular_recording.
42. Bit Patterned Media (BPM) – Helmholtz-Zentrum Dresden-Rossendorf, HZDR (webpage). www.hzdr.de/db/Cms?pOid=48585&pNid=0.
43. Parkin, S. S. P., Hayashi, M. and Thomas, L. Magnetic domain-wall racetrack memory. *Science* 320, 190–194 (2008).
44. Topp, J., Heitmann, D., Kostylev, M. P. and Grundler, D. Making a reconfigurable artificial crystal by ordering bistable magnetic nanowires. *Phys. Rev. Lett.* 104, 207205 (2010).
45. Blondel, A., Meier, J. P., Doudin, B. and Ansermet, J. -Ph. Giant magnetoresistance of nanowires of multilayers. *Appl. Phys. Lett.* 65, 3019 (1998).
46. Tsoi, M., Jansen, A. G. M., Bass, J., Chiang, W.-C., Seck, M., Tsoi, V. and Wyder, P. Excitation of a magnetic multilayer by an electric current. *Phys. Rev. Lett.* 80, 4281 (1998).
47. Kiselev, S. I., Sankey, J. C., Krivorotov, I. N., Emley, N. C., Schoelkopf, R. J., Buhrman, R. A. and Ralph, D. C. Microwave oscillations of a nanomagnet driven by a spin-polarized current. *Nat.* 425, 380–383 (2003).
48. Rippard, W. H., Pufall, M. R., Kaka, S., Silva, T. J., Russek, S. E. and Katine, J. A. Injection locking and phase control of spin transfer nano-oscillators. *Phys. Rev. Lett.* 95, 067203 (2005).
49. Cui, Y., Wei, Q., Park, H. and Lieber, C. M. Nanowire nanosensors for highly sensitive and selective detection of biological and chemical species. *Science* 80 (293), 1289–1292 (2001).
50. Serrà, A., Grau, S., Suriñach, C.G., Sort, J., Nogués, J. and Vallés, E. Magnetically-actuated mesoporous nanowires for enhanced heterogeneous catalysis. *Appl. Catal. B Environ.* 217, 81–91 (2017).
51. Kundu, S., Mukadam, M. D., Yusuf, S. M. and Jayachandran, M. Formation of shape-selective magnetic cobalt oxide nanowires: Environmental application in catalysis studies. *CrystEngComm* 15, 482–497 (2013).
52. Sun, L., Li, X., Xu, Z., Xie, K. and Liao, L. Synthesis and catalytic application of magnetic Co – Cu nanowires. *Beilstein J. Nanotechnol.* 8, 1769–1773 (2017).
53. Torrejon, J., Neuromorphic computing with nanoscale spintronic oscillators. *Nature* 547, 428–431 (2017).
54. Ababei, R. V., Ellis, M. O. A., Vidamour, I. T., Devadasan, D. S., Allwood, D. A., Vasilaki, E. and Hayward, T. J. Neuromorphic computation with a single magnetic domain wall. *Sci. Reports* 11, 1–13 (2021).
55. Grollier, J., Querlioz, D., Camsari, K. Y., Everschor-Sitte, K., Fukami, S. and Stiles, M. D. Neuromorphic spintronics. *Nat. Electron.* 3, 360–370 (2020).

Synthesis Techniques for Low Dimensional Magnets

4

Kalyani Chordiya[1,2], Gergely Norbert Nagy[1,2], and Mousumi Upadhyay Kahaly[1,2]

1 ELI-ALPS, ELI-HU Non-Profit Ltd., Wolfgang Sandner utca 3., Szeged, Hungary
2 Institute of Physics, University of Szeged, Dóm tér 9, Szeged, Hungary

Contents

4.1	Introduction	60
4.2	Synthesis Techniques	61
	4.2.1 Exfoliation Method	61
	4.2.1.1 Mechanical Exfoliation Method	62
	4.2.1.2 Liquid Exfoliation Method (LEM)	63
	4.2.2 Epitaxy	63
	4.2.2.1 Vapor Phase Epitaxy (VPE)	63
	4.2.2.2 Pulsed Laser Deposition (PLD)	64
	4.2.2.3 Liquid Phase Epitaxy (LPE)	65
	4.2.2.4 Molecular Beam Epitaxy (MBE)	66
	4.2.2.5 One-Pot/Domino/Cascade/Tandem Synthesis	66
	4.2.2.6 Solvothermal Synthesis	66
	4.2.3 Chemical Vapor Deposition (CVD)	66
	4.2.3.1 Routine Growth	67
	4.2.3.2 Molten-Salt-Assisted CVD	67
	4.2.3.3 Thermal CVD	68
	4.2.4 Template-Assisted Synthesis	68
4.3	Summary and Conclusions	69
4.4	Acknowledgments	69
References		70

DOI: 10.1201/9781003197492-4

4.1 INTRODUCTION

For the past few decades, the increased interest in understanding the relationship between the number of spatial dimensions and phase transition has stimulated dedicated research towards synthesizing low dimensional materials. Today, it is a separate branch of material science [1–2]. Reduced dimensions of bulk materials provide a unique opportunity to study exceptional quantum-dimension-related aspects such as electron confinement [3], topologically protected states [4], quantum transport [5], strong planar covalent bonds [6], and atomic thickness [7], which leads to enthralling electronic, magnetic, structural, and optical properties. The quantum confinement changes the (electronic and magnon) density of states [8] and oscillator strength, which consequently leads to the observed differences in the properties of low-dimensional materials with respect to their bulk counterparts. Low-dimensional materials or nanomaterials are materials with at least one dimension smaller than 100 nm and are classified based on their size in x-, y-, and z-dimensions as two-dimensional (2D), one-dimensional (1D), and zero-dimensional (0D) materials [9]. If all three dimensions of a system are confined to the nanoscale (<10 nm) and are composed of only a few (<50) atoms, then such system appears as a dot; for example, clusters, quantum dots (QDs) (in the case of magnets known as magnetic quantum dots, MQDs), and colloids. 1D materials possess linear chain-like structures and are commonly known as nanorods (NR), nanowires (NW), or nanotubes (NT), depending on the aspect ratio (length-to-width ratio). 2D materials show expansion in two dimensions and are known as thin nanofilms, nanosheets, quantum wells, or superlattices [10].

The scoring parameters of low dimensional materials result in their distinctly different properties, such as exotic magnetism, superconductivity, and charge density waves [11] compared to their bulk counterparts. For example, strong ferromagnetic ordering persists at room temperature in VSe_2 2D materials grown through molecular beam epitaxy (MBE) while their bulk counterparts show paramagnetic properties [12]. Layer-dependent ferromagnetism has been observed for CrI_3 and $Cr_2Ge_2Te_6$ nanosheets at low temperatures with an out-of-plane anisotropy [13]. Layer-dependent magnetism was also investigated, leading to the observation of ferromagnetism in four-layer Fe_3GeTe_2 above room temperature [14]. Bilayer CrI_3 was found to show metamagnetic conversion between antiferromagnetic and ferromagnetic states at fixed magnetic fields [15], and such conversion can be controlled by the stacking order in the case of bilayer $CrBr_3$ [16]. For an extensive review on 2D magnetic materials, we recommend the report by Awan et al. [17] on 2D materials, in addition to reports by E. Coronado for molecular magnets [18] and by S. Eom for review on 2D nanomaterials in biomedical application [19]. Methods of synthesis have a massive impact on the material's structure, phase, and magnetic properties. Yue Niu et al. studied the magnetism in bulk and 2D cylindrite [20], and an extensive review was reported on liquid exfoliation by V. Nicolosi et al. [21] and on mechanical exfoliation by E. Gao et al. [22]. A report by N. Poudyal et al. [23] on the controlled synthesis of FePt low dimensional materials using the one-pot synthesis method stands as a concrete example, showing that depending on whether the material forms a nanoparticle, nanowire, or nanorod structure, the magnetic properties appear to be modified (Figure 4.1). A similar study with the same dimension, but with different particle sizes of the $L1_0$-FePt nanoparticles, was reported by C.B. Rong et al. [24] in 2006. This report shows that the magnetic properties, including Curie temperature, magnetization, and coercivity, strongly depend on long-range ordering and particle size. Because the size of the nanomaterials determines the magnetic properties, different nano-architectonics are involved in developing the size and shape of nanomaterials for desired applications. Hence, in this chapter, we attempt to list possible synthesis techniques used in the fabrication of low-dimensional magnetic materials.

The demonstration of mechanical exfoliation of graphite to isolate a single layer of graphene [25] sparked the interest of many researchers to explore the world of 2D materials. Since then, several new 2D materials have been discovered and studied: silicene, germanene, hexagonal boron nitride (h-BN), black phosphorus, transition metal dichalcogenides (TMDs), layered perovskites, oxides, hydroxides, and more. 2D materials have proven to show a wide range of physical properties, such as metallic, semiconductor, insulator, and superconductor, along with the broad scope for tuning the properties through defects and

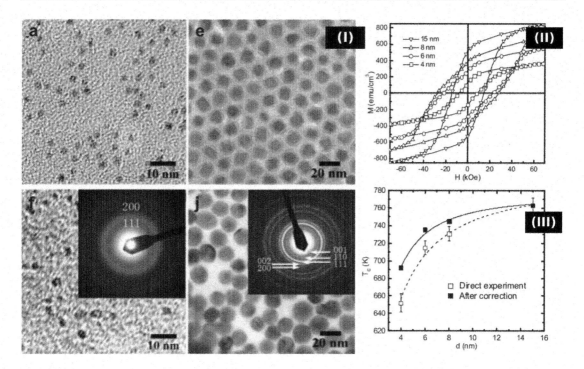

FIGURE 4.1 (I) TEM images of FePt nanoparticles (top two panels) and the SAED patterns (bottom two panels) for salt-annealed particles. (II) Hysteresis loops of salt-annealed nanoparticles. (III) The dependence of the Curie temperature on the FePt particle size. Adapted with permission from [23]. Copyright (2009) AIP Publishing.

functionalization [26] through post-synthetic or pre-synthetic methods. However, instability at ambient conditions and nontrivial detection of magnetism have kept magnetic 2D materials elusive until recently. The development of 2D organic layered polymers, especially in metal-organic frame (MOF) polymers [27], has provided unique examples for magnetic layers with highly versatile functionalization.

4.2 SYNTHESIS TECHNIQUES

The synthesis of low dimensional magnetic materials can be performed via two characteristic methods: top-down and bottom-up. From among the former, in this chapter, we provide a detailed discussion of mechanical and liquid exfoliation, and from among the latter, we describe the method of epitaxy, within which we discuss vapor-, liquid-, and molecular beam epitaxy, chemical vapor deposition, template-assisted synthesis, one-pot synthesis, and solvothermal synthesis. For a clear understanding of the synthesis process, we summarize all the steps involved in the synthesis process for one or more magnetic materials.

4.2.1 Exfoliation Method

Mechanical exfoliation of defect-free, single-layered materials from bulk materials using scotch tape is a well-known method [28]. In general, exfoliation is a process in which layered materials are expanded and separated into single or few-layer sheets by using external factors, such as physical, chemical, electrical, or thermal factor. Depending on the physical factor involved in the separation of non-covalent interactions in between the layers, the exfoliation method can be categorized as micro-mechanical exfoliation, liquid or chemical exfoliation, electrochemical exfoliation, or thermal exfoliation. In this chapter, we discuss the most commonly used exfoliation techniques for synthesizing magnetic materials, such as the mechanical and liquid exfoliation methods.

4.2.1.1 Mechanical Exfoliation Method

Mechanical exfoliation [29] is a top-down technique similar to the method used by Grim and his coworkers to separate single-layer graphene from highly oriented pyrolytic graphite (HOPG) [25]. For the schematics of the stepwise process, please see Figure 4.2(1). In this method, an adhesive tape is pressed against a 2D crystal to attach a few crystal layers to the tape. The tape with layered material is then pressed against a clean surface of choice. Upon peeling, the bottom layer of the crystal is left on the substrate. During this process, one can obtain a single or multilayered flake on the surface due to the cleavage of weak van der Waals forces between two layers. The peeled layered can then be washed with an insoluble solvent such as acetone, isopropanol, and distilled water for 10 minutes in an ultrasonic bath and transferred on the desired substrate. This process can be repeated in order to get undamaged monolayer materials. This method has been employed to obtain 2D magnetic materials such as $CrCl_3$, CrI_3, $EuSn_2As_2$, cylindrite, and

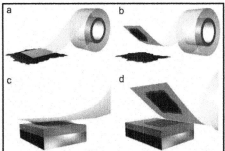
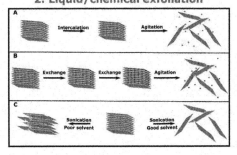
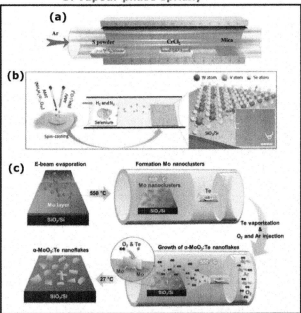

FIGURE 4.2 1. Schematics for the steps involved in micro-mechanical exfoliation of 2D crystals Adapted with permission from [30]. Copyright (2004) American Association for the Advancement of Science. 2. Schematics for the steps involved in liquid exfoliation mechanisms using intercalation, ion exchange, and sonication method. Reproduction from [21]. 3. Summarizing different vapor phase epitaxy techniques used in literature, (a) illustrates the CVD growth for ultrathin Cr_2S_3 flakes on two stacked mica substrates. Adapted with permission from [31]. Copyright (2019) Wiley Online Library. (b) Schematic of the pretreatment of precursors for V-doped WSe_2 by mixing liquid W with V precursors. Adapted with permission from [32]. Copyright (2020) Wiley Online Library. (c) Schematic illustration of the pretreatment of precursors for α-MoO_3:Te nanoflakes from Mo layer deposited by vapor epitaxy using evaporation as a precursor. Adapted with permission from [33]. Copyright (2019) ACS Publications.

some graphene-like materials such as hybrid coordination polymers from bulk materials. This method has several advantages: it is simple, fast, cost-effective, and helps in maintaining structural integrity. On the other hand, there are also certain disadvantages: it is not applicable for large-scale production due to the breaking of monolayers during exfoliation, and the yield of monolayers is low. Nevertheless, many efforts are attributed to making mechanical exfoliation a scalable approach [28].

4.2.1.2 Liquid Exfoliation Method (LEM)

For the large-scale production of monolayered magnetic materials, liquid exfoliation is a facile and accepted approach. Liquid exfoliation is the most common method to delaminate layered coordination polymers, perovskites, oxychalcogenides, oxypnictides [34], and magnetic nanomaterials. Liquid exfoliation can be performed with different approaches; a pictorial summary can be found in Figure 4.2(2), showing ultrasonic cleavage, intercalation, and ion exchange liquid exfoliation. This section discusses the ultrasonic and intercalation-assisted liquid exfoliation methods in detail.

In ultrasonic exfoliation, two components influence the yield of exfoliation: (i) energy input to the bulk materials, which is provided by the sonication process itself; and (ii) the liquid media (such as ionic liquids, aqueous solutions of stabilizers, and organic solvents) to overcome the energy barrier in the interlayers. In typical ultrasonic exfoliation methods, the powdered bulk material is dissolved in liquid media and sonicated with high energy for a longer duration. After this sonication process, the supernatant (containing separated layers) is collected by a pipette or centrifugation, and then it is filtered and dried on the desired substrate. The high-energy ultrasonic waves create cavitation bubbles in the solvent. These bubbles, on sudden bursting, generate high energy via the release of pressure followed by an exfoliation of layers [21].

The intercalation approach uses the diffusion, ion exchange capability of the layered materials, and external forces, such as sonification, to separate the layers. This method has been widely used to functionalize 2D materials with heavy organic groups or metallic groups to enhance their magnetic properties. In this process, the 2D material can be functionalized by the intercalation of ions between the layers. For example, in the synthesis of edge-rich ferromagnetic MoS_2 [35], a solution of pristine MoS_2 powder dissolved in N-Methyl-2-Pyrrolidone (NMP) solvent is sonicated at low power for a few hours, followed by high power sonification for six hours. Finally, the mixture is centrifuged, and the collected supernatant is filtered. The prior-filtered flakes of MoS_2 nanosheets are then dispersed into isopropyl alcohol (IPA) uniformly and dried at a low temperature of 60°C, and the extracted MoS_2 nano-layers containing single, double, and multilayered particles can be stored in a vacuum. The easy way to extract the magnetized MoS_2 is by placing magnets outside its aqueous solution.

4.2.2 Epitaxy

Epitaxy is one of the bottom-up crystal growth approaches, where one or more well-defined orientations with respect to the crystalline seed layer could be achieved. This method is not restricted to 2D nano-fabrication: it can also be used for 1D and 0D crystals or mixed dimension crystals. For an extensive review on the growth of nanostructures of different types and dimensions using the epitaxy growth method, we refer the reader to a review article [36]. This section discusses vapor, liquid, and molecular beam epitaxy growth methods, one-pot synthesis, and solvothermal synthesis in detail, followed by chemical vapor deposition and template-assisted synthesis in separate sections.

4.2.2.1 Vapor Phase Epitaxy (VPE)

In a typical VPE, the precursor metals are vaporized and deposited on the choice substrate. Lee et al. [28] reported growth of ferromagnetic ultrathin α-MoO_3:Te nanoflakes using this method. In this report, a 2 nm thick metal Mo layer was coated on SiO_2/Si substrate using an electron-beam evaporation method.

This assembly of Mo/SiO$_2$/Si substrate was further mounted in a quartz VPE system (in the main reaction zone) along with the Te grains at the extra impurity zone. At a temperature of 550°C, Mo nucleation and vaporization of Te were carried out in their respective zones; with the subsequent addition of O$_2$ reaction gas, α-MoO$_3$:Te was successfully grown.

4.2.2.2 Pulsed Laser Deposition (PLD)

An exciting and promising subtype of VPE is pulsed laser deposition (PLD). The basic idea is the same as for conventional VPE: create a vapor, and deposit it on the substrate. Contrary to that, the vapor is produced by a high-energy ultrashort laser pulse ablating a target in a vacuum close to the substrate. During the ablation, the target material rapidly evaporates and creates an intense high-speed pulse of plasmatic vapor propagating in the general direction of the substrate, see Figure 4.3 for a general layout of PLD. The radiant energy of the reflected laser beam, together with the temperature and kinetic energy of the incoming material, also provides a heating effect on the substrate, which can further influence the result. As the evaporation of the material is based on photon-matter interaction and rapid heating, the created vapor largely reflects the stoichiometric composition of the target, which is often not the case with traditional thermally induced evaporation [37].

As such, the circumstances of deposition are distinctly different from traditional VPE: the deposition rate is very high, and it happens in rapid pulses. Correspondingly, samples prepared with pulsed-laser deposition have disparate structures. P. Ohresser et al. [38] examined this in detail experimentally by creating a Fe film with the width of a few atoms deposited on a Cu (111) substrate. This Cu surface favors face-centered cubic (fcc) crystal growth instead of the natural body-centered cubic (bcc) structure of Fe, and the driving forces behind these two will compete during the formation. Conventional thermal deposition results in several bcc-structured sections, even for a few multilayers (Figure 4.4). Above 3–3.5 layers, the bcc structure becomes very significant, and it grows into being utterly dominant around five multilayers. Using PLD significantly improves the picture: there is practically no bcc structure up until five layers, and they only start appearing as we approach the 6-layer thickness. It also considerably improves the shape of deposition: while thermal growth produces mostly droplet-shaped regions of varying widths, in PLD, the film grows in an almost-perfect layer-by-layer fashion. As a result, we have a close to uniform layer width and structure.

The result of PLD also strongly depends on the target material [39–40]. Nakao et al. examined pulsed laser deposition of FePt alloys, which have promising uses in the medical field as monolayer magnets, and thanks to its biocompatible nature. The Fe content of this composite influences the magnetic

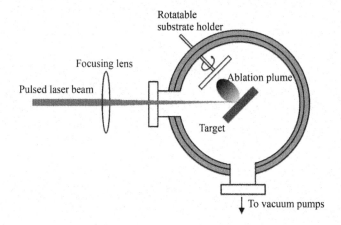

FIGURE 4.3 General schematics of PLD. A pulsed laser beam falls on the target and creates a plasmatic vapor cloud. This ablation plume rapidly expands in the general direction of the sample, where it is deposited. Adapted with permission from [37]. Copyright (2004) Royal Society of Chemistry.

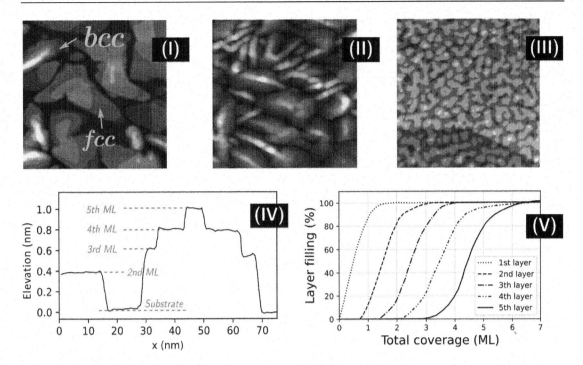

FIGURE 4.4 Properties of a Fe ultrathin layer deposited on a Cu (111) substrate thermally and via pulsed laser. (I): STM topography image of a thermally deposited Fe layer. The bcc structures form a ridge-like structure, while the fcc zones appear as regions of mostly uniform thickness. A cross-section (IV) from the same sample at an arbitrarily chosen fcc zone highlights the layered structure. (II–III): Thermally (II) and pulsed laser deposited (III) Fe layers of 3.5 M. The former is dominated by bcc ridges, while the latter is mostly homogenous in thickness. All three STM images cover an area of roughly 75 nm × 75 nm. (V): The layer filling (as a percentage of total surface area) as a function of monolayer (ML) count, highlighting the almost perfectly sequential nature of the PL-VPE growth. Based on data published in [38].

properties – such as magnetic coercivity and magnetic remnance – significantly. Using an $Fe_{50}Pt_{50}$ target, PLD grown FePt monolayers on Ta substrate resulted in poorer magnetic performance than one created by conventional VPE. Increasing the iron content to $Fe_{70}Pt_{30}$ in the target increased the magnetic remanence from 0.7 T to 0.95 T, even though the thickness of the film was less in the second case. The power of the laser also affected the composition of the fabricated magnetic layer: increasing the power increased the Fe concentration in the deposited material.

4.2.2.3 Liquid Phase Epitaxy (LPE)

MOFs and other crystalline coordination networks (CCNs), due to their high porosity, electronic and magnetic coupling properties between the metal centers (and sometimes the ligands), and distinct capability for functionalization, have posed themselves as potential candidates for low dimensional magnetic materials. LPE of MOFs can be carried out as a stepwise process. For example, in the synthesis of HKUST-1 thin film layers on pretreated nanoparticles [41], thin-film growth is carried out by directly using carboxylic acid terminated magnetic beads. For the growth process, a suspension was prepared by alternately adding $Cu(CH_3COO)_2 \cdot H_2O$ ethanol solution and 1,3,5-benzenetricarboxylic acid in ethanol. The ultrasonic method was used for mixing and fast mass transfer at the solid-liquid interface. The MagPrep silica nanoparticles and the functionalized magnetic beads were then transferred to an Eppendorf tube and washed for residual water. The magnetic particles were separated by placing magnets outside the Eppendorf tube.

4.2.2.4 Molecular Beam Epitaxy (MBE)

MBE is the most powerful technique for making 0D and 2D nanostructures, as it provides reliable control over atom-by-atom growth and composition of NW films, hence providing an advantage for planning and designing of advanced spintronic nanodevices. In the MBE process, effusion cells are used to evaporate the metals used in the composition of the nanostructures, such as Ba and Ti in the growth of $BaTiO_3$ nanowires and are focused onto a heated substrate under ultrahigh vacuum conditions. Several evaporation methods have been reported for the refractory elements (such as Ti), like effusion cells, electron beam gun evaporation, Ti-Ball sublimation source, and metal-organic vapor source. However, the reactivity of metal with ambient oxygen makes the control of the beam fluxes and the composition a non-trivial process. Hence, in situ real-time diagnostic tools and monitoring techniques such as reflection high energy electron diffraction (RHEED) could be employed. For a more thorough review of molecular beam epitaxy of ferroelectric $BaTiO_3$ films on semiconductor substrates, one can refer to a focused review by L. Mazet et al. [42].

4.2.2.5 One-Pot/Domino/Cascade/Tandem Synthesis

Breakthrough in the "one-pot synthesis," which is also known as "domino reaction," "cascade reaction," or "tandem reaction," was marked after Robinson's synthesis of tropinone more than a century ago [43]. Since then, this process has been successfully used to synthesize several organic catalysts, chain systems, hyperbranched polymers, nanoparticles, and single-molecule magnets. As the name suggests, in this method, all the precursors are mixed in a single pot to form homogeneous slurry at room temperature. One-pot synthesis has proven to help cope with two major issues in science: "efficiency" by reducing the number of purification processes and "environmental sustainability" by minimizing waste. In this approach, the reaction pot is heated under inert conditions for the precursors to react with the hybrid nanostructures. One-pot synthesis can be performed as colloidal synthesis or dry synthesis. In the colloidal synthesis method, controlled hybrid nanostructures with excellent uniformity in dispersion, size, and shape, along with high purity, could be achieved. The nanostructures' physical properties can be controlled during the synthesis, and composition could be tuned post-synthesis. During the synthesis, experimental parameters such as temperature, time, solvent, precursors, and the surfactants could be tuned. Post-synthesis methods such as ion exchange reactions can be used to tune the compositions of synthesized epitaxial hybrid nanostructures.

4.2.2.6 Solvothermal Synthesis

Synthesis of magnetic quantum dots can be performed using the solvothermal synthesis method. For example, consider the synthesis of MQDs with Fe_2O_3 magnetic nanoparticles and CdSe quantum dots of varying sizes [44]. To prepare this MQD, CdO and stearic acid were first heated to 150–200°C and then cooled to room temperature. MPs, trioctylphosphine oxide (TOPO), and hexamethylenediamine (HDA) were added to this mixture and then heated to 280–300°C. Se dissolved in trioctylphosphine (TOP) was quickly injected into this mixture within a few seconds, and the resulting CdSe QDs were allowed to grow for different times (1–5 min) to yield different sizes of dots, which are characterized by different colors. The aliquots of the growth solution were quenched by the addition of chloroform. Methanol was then added to the growth solution, and a magnet was applied to the vial containing the sample to separate the magnetic particles.

4.2.3 Chemical Vapor Deposition (CVD)

Another method to synthesize large-area and uniform thickness 2D layered magnetic materials is through a bottom-up approach called the CVD technique (Figure 4.5). An extensive review on the synthesis of

4 • Synthesis Techniques for Low Dimensional Magnets   67

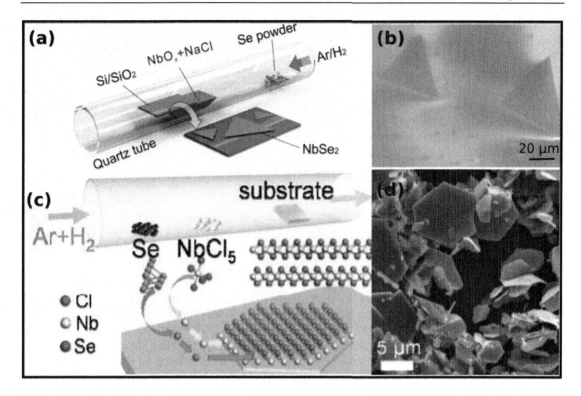

FIGURE 4.5 Synthesis schematics and SEM results for NbSe$_2$ synthesized using two different approaches and substrates (a, b). Adapted with permission from [46]. Copyright (2017) Nature Publishing Group. (a) NbSe$_2$ grown on quartz. (c–d) Synthesis schematics and SEM result. Adapted with permission from [47]. Copyright (2017) Royal Society of Chemistry. (c) Synthesis and SEM image of hexagonal or truncated triangular NbSe$_2$ nanoplates.

two-dimensional magnetic materials by chemical vapor deposition is reported in the literature by H Jiang et al. [45]. CVD synthesis can be characterized by three different methods: (i) routine growth, (ii) molten-salt-assisted method, and (iii) thermal CVD.

4.2.3.1 Routine Growth

In a routine growth method, let us take an example of MoS$_2$ growth, where Mo metal powder is placed on the top of the substrate or in a ceramic boat along with the sulfur powder in a separate boat kept at a short distance from the other boat [48]. Both the boats are placed at the center of the CVD tube. An inert carrier gas such as N$_2$ is then heated and allowed to flow, vaporizing the sulfur and reacting with the metal to form a MoS$_2$ layer on the substrate. However, even though this method provides uniform thickness, it fails to ensure control over the thickness of the layer or crystal structure of the material.

4.2.3.2 Molten-Salt-Assisted CVD

The molten-salt-assisted CVD synthesis method employed by H. Wang et al. [46] (see Figure 4.5a,b for reaction chamber schematics) resulted in a monolayer of NbSe$_2$ crystals of thickness ~1 nm. In the CVD reaction chamber, NbO$_x$, partially oxidized Niobium (Nb), and NaCl powder are placed in an alumina boat located at the center of the furnace. For deposition of the final product, a quartz substrate is placed ~3 mm above the powder mixture. Selenium (Se) powder is placed at upstream of the quartz tube, where the temperature ranges from 300°C to 460°C during the reaction, where a mixture of argon and H$_2$ is used as carrier gas. To synthesize NbSe$_2$ layers, the furnace is heated to 795°C. To signify the impact of the synthesis method for similar materials, we discuss a different synthesis approach for NbSe$_2$. This approach was used by Zou et al. [47] (Figure 4.5c for reaction chamber schematics) to grow well-faceted hexagonal nanoplates NbSe$_2$ nanoplates (~ 80 nm) by CVD in a horizontal tube furnace. Prior to the growth, O$_2$ gas was removed to prevent sample oxidation and purged by Ar gas. Co-flow mixed by Ar and H$_2$ was used as the carrier gas, and niobium chloride (NbCl$_5$) and Se were used as precursors. During the growth, Se powders were located at the upstream region of the furnace (500°C) and NbCl$_5$ powders at the heating center (720°C). A reaction between NbCl$_5$ and gaseous H$_2$ was used as a source of Nb vapor, which is then reacted with Se to synthesize nanostructures of NbSe$_2$. These nanostructures are then deposited to an SiO$_2$ substrate placed at the downstream region of the furnace (~580°C). The SEM images show well-faceted hexagonal NbSe$_2$ nanoplates.

4.2.3.3 Thermal CVD

The thermal CVD process is known to offer very little control over the thickness of the deposited sample, structure, and orientation of nanostructures. For example, typically, CVD synthesis of MoS$_2$, WSe$_2$, or WS$_2$ on Si/SiO$_2$ substrate leads to the synthesis of nanowires-nanoflakes hybrids [49]. Hence, to add an additional handle in controlled deposition, the substrate can be modified or undergo different preparation methods, such as the use of an electron beam [50] to deposit the base metal such as Mo for preparation of MoS$_2$ or by pretreating the substrate, such as coating the Si/SiO$_2$ substrate with graphene oxide (GO) before performing the CVD process. Using such methods, one can fabricate MoS$_2$ by pre-depositing base metal by spin coating precursor solution or depositing base metal using an electron beam on the substrate of choice before the CVD process, Furthermore, sulfur is introduced to react with base metal Mo at 750°C, forming a very thin MoS$_2$ film (from a single layer to a few layers). However, the thickness of the layer is still not reduced to a monolayer and needs more improvisation to make the samples ready for practical applications.

4.2.4 Template-Assisted Synthesis

Template-assisted synthesis is a promising and very effective method for the synthesis of 1D nanostructures. In this method, the pores of a host material can be used as a template to direct the growth of new materials. The templates used in this method are classified as hard, soft, and colloidal templates. Organic templates are commonly known as "soft templates," and templates based on silicon or inorganic materials are referred to as "hard templates." The template synthesis method had been effectively used to synthesize several magnetic nanostructures, such as spinel cobalt ferrite nanowires in mesoporous silica SBA-15 as a host matrix [51] and cylindrical and "pin" or "X" shaped Co nanowires from anodic aluminum oxide (AAO) template and polycarbonate track-etched (PCTE) template grown by the template-assisted electrodeposition technique [52]. For a thorough understanding of template synthesis, we discuss two synthesis processes for spinel cobalt ferrite and Co nanowires. For the synthesis of spinel cobalt ferrite by El-Sheikh et al. [51], first, the hexagonally ordered SBA-15 template was synthesized, followed by the hydrothermal synthesis of CoFe$_2$O$_4$ nanostructures using the iron and cobalt precursor solution in deionized water in a 2:1 molar ratio so that the volume is identical to the volume of the SBA-15 template pores. The precursor solution is added dropwise to a flask containing SBA-15 dispersed in methanol. The resultant mixture is left to stir overnight for a complete impregnation process. Then the flask is left open for the complete removal of the solvent, followed by drying the material in an oven. The prepared samples then undergo a calcination process during which the cobalt ferrite nanowires are released by dissolving the as-prepared nanocomposites in NaOH and collected by repeating centrifugation, washing with water and ethanol, and drying.

During the template synthesis of cylindrical and "pin" or "X" shaped Co nanowires [52], nano-sized pore anodic aluminum oxide (AAO) and polycarbonate track-etched (PCTE) membranes are used as templates. An electrochemical cell is used for electro-deposition, consisting of electrolytes in a container with current passing through working, counter, and reference electrodes. During the deposition process, one side Au coated template provides good contact between the template and Au coated glass slide. A constant potential is maintained through the arranged electrode acting as the working electrode, the Pt metal sheet functions as a counter electrode and the saturated calomel electrode (SCE) as a reference electrode. Fresh sulfate baths/electrolyte solutions are prepared with an acidic pH value of about 4 for $CoSO_4$, H_3BO_3, and sodium lauryl sulfate. An efficient deposition is optimized through several iterations of electro-deposition.

4.3 SUMMARY AND CONCLUSIONS

In summary, this chapter presents possible techniques used to synthesize low-dimensional magnetic materials. Two synthesis approaches presented here are top-down and bottom-up approaches. In top-down approach, exfoliation of layered crystal using mechanical and liquid exfoliation methods is presented. The former approach, even though it is cheap and effective, possesses the disadvantages of small-scale production and the damage of layers. The latter approach is more reliable and also helps in the functionalization of magnetic materials through intercalation techniques, leading to the exchange or adsorption of ions on 2D layered materials. Hence, among the two exfoliation methods, liquid exfoliation is the more reliable and widely used approach.

In the bottom-up approach, one can synthesize 0D materials with the desired orientation and morphology. These 0D materials can be used to grow 1D or 2D nanostructures further. In this approach, we discussed in detail the epitaxy method, such as liquid and vapor phase epitaxy, pulsed laser deposition, molecular beam epitaxy, one-pot synthesis, and solvothermal synthesis; furthermore, template synthesis is discussed. Even though the chemical vapor deposition process is classified as a vapor phase epitaxial method, due to the different approaches, we have discussed it in a separate section. The thickness and orientation of the crystals in the vapor phase epitaxial method, such as CVD, are controlled by pretreatment of the substrate by the deposition of the base metal on the substrate via spin coating or electron beam method. PLD has the potential to prepare high-quality and homogenous monolayers and to reach compositions that are hard to obtain via conventional VPE. Metal-organic frames and crystalline coordination networks are posing as the potential candidates for low-dimensional magnets and liquid phase epitaxial method, solvothermal synthesis, and one-pot synthesis are the most common approaches for the synthesis of these materials. For the atom-by-atom synthesis of 1D materials, molecular beam epitaxy is the most potential candidate and hence provides a special advantage in designing spintronic devices. Another effective method implemented in the synthesis of 1D materials is template-assisted synthesis, where a porous organic, inorganic, or colloidal template is used to guide the growth of low dimensional and bulk materials. We hope that this chapter helps and guides chemists in synthesizing low dimensional materials.

4.4 ACKNOWLEDGMENTS

KC, GN, and MUK thank ELI-ALPS, which is supported by the European Union and co-financed by the European Regional Development Fund (GINOP-2.3.6-15-2015-00001). KC, GN, and MUK acknowledge Project no. 2019–2.1.13-TÉT-IN-2020–00059, which has been implemented with the support provided by the National Research, Development, and Innovation Fund of Hungary, financed under the 2019–2.1.13-TÉT-IN funding scheme, and funding from PaNOSC European project.

REFERENCES

[1] K. Katsumata, "Low-dimensional magnetic materials," *Current Opinion in Solid State and Materials Science*, vol. 2, no. 2, pp. 226–230, 1997.

[2] H. S. Nalwa, *Handbook of nanostructured materials and nanotechnology, five-volume set*. Academic Press, 1999.

[3] A. D. Yoffe, "Low-dimensional systems: Quantum size effects and electronic properties of semiconductor microcrystallites (zero-dimensional systems) and some quasi-two dimensional systems," *Advances in Physics*, vol. 42, no. 2, pp. 173–262, 1993.

[4] Y. Ren, Z. Qiao, and Q. Niu, "Topological phases in two-dimensional materials: A review," *Reports on Progress in Physics*, vol. 79, no. 6, p. 066 501, 2016.

[5] D. Bercioux and P. Lucignano, "Quantum transport in rashba spin – orbit materials: A review," *Reports on Progress in Physics*, vol. 78, no. 10, p. 106 001, 2015.

[6] P. M. Larsen, M. Pandey, M. Strange, and K. W. Jacobsen, "Definition of a scoring parameter to identify low-dimensional materials components," *Physical Review Materials*, vol. 3, no. 3, p. 034 003, 2019.

[7] N. Ilyas, D. Li, Y. Song, H. Zhong, Y. Jiang, and W. Li, "Low-dimensional materials and state-of-the-art architectures for infrared photodetection," *Sensors*, vol. 18, no. 12, p. 4163, 2018.

[8] D. Voiry, H. S. Shin, K. P. Loh, and M. Chhowalla, "Low-dimensional catalysts for hydrogen evolution and CO_2 reduction," *Nature Reviews Chemistry*, vol. 2, no. 1, pp. 1–17, 2018.

[9] M. A. Kebede and T. Imae, "Low-dimensional nanomaterials," in *Advanced Supramolecular Nanoarchitectonics*. Elsevier, 2019, pp. 3–16.

[10] A. N. Vasiliev, O. S. Volkova, E. A. Zvereva, and M. Markina, *Low-Dimensional Magnetism*. CRC Press, 2019.

[11] A. Hirohata, K. Yamada, Y. Nakatani, I. L. Prejbeanu, B. Diény, P. Pirro, and B. Hillebrands, "Review on spintronics: Principles and device applications," *Journal of Magnetism and Magnetic Materials*, vol. 509, p. 166 711, 2020.

[12] C. Gong, L. Li, Z. Li, H. Ji, A. Stern, Y. Xia, T. Cao, W. Bao, C. Wang, Y. Wang, Z. Q. Qiu, R. J. Cava, S. G. Louie, J. Xia, and X. Zhang, "Discovery of intrinsic ferromagnetism in two-dimensional van der Waals crystals," *Nature*, vol. 546, no. 7657, pp. 265–269, 2017.

[13] B. Huang, G. Clark, E. Navarro-Moratalla, D. R. Klein, R. Cheng, K. L. Seyler, D. Zhong, E. Schmidgall, M. A. McGuire, D. H. Cobden, W. Yao, D. Xiao, P. J-Herrero, and X. Xu, "Layer-dependent ferromagnetism in a van der Waals crystal down to the monolayer limit," *Nature*, vol. 546, no. 7657, pp. 270–273, 2017.

[14] Y. Deng, Y. Yu, Y. Song, J. Zhang, N. Z. Wang, Y. Sun, Y. Yi, Y. Z. Wu, S. Wu, J. Zhu, J. Wang, X. H. Chen, and Y. Zhang, "Gate-tunable room-temperature ferromagnetism in two-dimensional Fe_3GeTe_2," *Nature*, vol. 563, no. 7729, pp. 94–99, 2018.

[15] B. Huang, G. Clark, D. R. Klein, D. MacNeill, E. N-Moratalla, K. L. Seyler, N. Wilson, M. A. McGuire, D. H. Cobden, D. Xiao, W. Yao, P. J.-Herrero, and X. Xu, "Electrical control of 2d magnetism in bilayer CrI_3," *Nature Nanotechnology*, vol. 13, no. 7, pp. 544–548, 2018.

[16] W. Chen, Z. Sun, Z. Wang, L. Gu, X. Xu, S. Wu, and C. Gao, "Direct observation of van der Waals stacking – dependent interlayer magnetism," *Science*, vol. 366, no. 6468, pp. 983–987, 2019.

[17] S. U. Awan, S. Zainab, M. D. Khan, S. Rizwan, and M. Z. Iqbal, "2-dimensional magnetic materials for spintronics technology," *RSC Nanoscience*, pp. 91–119, 2020.

[18] E. Coronado, "Molecular magnetism: From chemical design to spin control in molecules, materials and devices," *Nature Reviews Materials*, vol. 5, no. 2, pp. 87–104, 2020.

[19] S. Eom, G. Choi, H. Nakamura, and J.-H. Choy, "2-dimensional nanomaterials with imaging and diagnostic functions for nanomedicine; A review," *Bulletin of the Chemical Society of Japan*, vol. 93, no. 1, pp. 1–12, 2020.

[20] Y. Niu, J. Villalva, R. Frisenda, G. S-Santolino, L. R-González, E. M Pérez, M. G-Hernández, E. Burzurí, and A. C-Gomez, "Mechanical and liquid phase exfoliation of cylindrite: A natural van der Waals superlattice with intrinsic magnetic interactions," *2D Materials*, vol. 6, no. 3, p. 035 023, 2019.

[21] V. Nicolosi, M. Chhowalla, M. G. Kanatzidis, M. S. Strano, and J. N. Coleman, "Liquid exfoliation of layered materials," *Science*, vol. 340, no. 6139, 2013.

[22] E. Gao, S.-Z. Lin, Z. Qin, M. J. Buehler, X.-Q. Feng, and Z. Xu, "Mechanical exfoliation of two-dimensional materials," *Journal of the Mechanics and Physics of Solids*, vol. 115, pp. 248–262, 2018.

[23] N. Poudyal, G. S. Chaubey, C.-B. Rong, and J. P. Liu, "Shape control of FePt nanocrystals," *Journal of Applied Physics*, vol. 105, no. 7, 07A749, 2009.
[24] C.-B. Rong, D. Li, V. Nandwana, N. Poudyal, Y. Ding, Z. L. Wang, H. Zeng, J. P. Liu, "Size-dependent chemical and magnetic ordering in l10-FePt nanoparticles," *Advanced Materials*, vol. 18, no. 22, pp. 2984–2988, 2006.
[25] K. S. Novoselov, A. K. Geim, S. V. Morozov, J. Zhangs, V. Dubonosi, V. Grigorieva and A. A. Firsov, "Electric field effect in atomically thin carbon films," *Science*, vol. 306, no. 5696, pp. 666–669, 2004.
[26] H. Wang, X. Zhang, and Y. Xie, "Recent progress in ultrathin two-dimensional semiconductors for photocatalysis," *Materials Science and Engineering: R: Reports*, vol. 130, pp. 1–39, 2018.
[27] L. León-Alcaide, J. López-Cabrelles, G. M. Espallargas, and E. Coronado, "2d magnetic MOFs with micron-lateral size by liquid exfoliation," *Chemical Communications*, vol. 56, no. 55, pp. 7657–7660, 2020.
[28] M. Yi and Z. Shen, "A review on mechanical exfoliation for the scalable production of graphene," *Journal of Materials Chemistry A*, vol. 3, no. 22, pp. 11 700–11 715, 2015.
[29] J.-G. Park, "Opportunities and challenges of two-dimensional magnetic van der Waals materials: Magnetic graphene?" arXiv preprint arXiv:1604.08833, 2016.
[30] K. Novoselov and A. C. Neto, "Two-dimensional crystals-based heterostructures: Materials with tailored properties," *Physica Scripta*, vol. 2012, no. T146, p. 014 006, 2012.
[31] S. Zhou, R. Wang, J. Han, D. Wang, H. Li, L. Gan, and T. Zhai, "Ultrathin non-van der Waals magnetic rhombohedral Cr_2S_3: Space-confined chemical vapor deposition synthesis and Raman scattering investigation," *Advanced Functional Materials*, vol. 29, no. 3, p. 1 805 880, 2019.
[32] S. J. Yun, D. L. Duong, D. M. Ha, K. Singh, T.L. Phan, W. Choi, Y.M. Kim, Y.H. Lee, "Ferromagnetic order at room temperature in monolayer WSe_2 semiconductor via vanadium dopant," *Advanced Science*, vol. 7, no. 9, p. 1 903 076, 2020.
[33] D. J. Lee, Y. Lee, Y. H. Kwon, S. H. Choi, W. Yang, D. Y. Kim, and S. Lee, "Room-temperature ferromagnetic ultrathin α-MoO_3: Te nanoflakes," *ACS Nano*, vol. 13, no. 8, pp. 8717–8724, 2019.
[34] S. J. Clarke, P. Adamson, S. J. Herkelrath, O. J. Rutt, D. R. Parker, M. J. Pitcher, and C. F. Smura, "Structures, physical properties, and chemistry of layered oxychalcogenides and oxypnictides," *Inorganic Chemistry*, vol. 47, no. 19, pp. 8473–8486, 2008.
[35] G. Gao, C. Chen, X. Xie, Y. Su, S. Kang, G. Zhu, D. Gao, A. Trampert and L. Cai, "Toward edges-rich MoS_2 layers via chemical liquid exfoliation triggering distinctive magnetism," *Materials Research Letters*, vol. 5, no. 4, pp. 267–275, 2017.
[36] C. Tan, J. Chen, X.-J. Wu, and H. Zhang, "Epitaxial growth of hybrid nanostructures," *Nature Reviews Materials*, vol. 3, no. 2, pp. 1–13, 2018.
[37] M. N. R. Ashfold, F. Claeyssens, G. M. Fuge, and S. J. Henley, "Pulsed laser ablation and deposition of thin films," *Chemical Society Reviews*, vol. 33, no. 1, pp. 23–31, 2004. https://doi.org/10.1039/B207644F
[38] P. Ohresser, J. Shen, J. Barthel, M. Zheng, C. V. Mohan, M. Klaua, and J. Kirschner, "Growth, structure, and magnetism of fcc fe ultrathin films on cu(111) by pulsed laser deposition," *Physical Review B – Condensed Matter and Materials Physics*, vol. 59, no. 5, pp. 3696–3706, 1999. https://doi.org/10.1103/PhysRevB.59.3696
[39] M. Nakano, W. Oniki, T. Yanai, and H. Fukunaga, "Magnetic properties of pulsed laser deposition-fabricated isotropic Fe-Pt film magnets," *Journal of Applied Physics*, vol. 109, no. 7, pp. 3–5, 2011. https://doi.org/10.1063/1.3561785
[40] F. Donati, S. Rusponi, S. Stepanow, C. Wäckerlin, A. Singha, L. Persichetti, R. Baltic, K. Diller, F. Patthey, E. Fernandes, J. Dreiser, K. Šljivančanin, K. Kummer, C. Nistor, P. Gambardella, and H. Brune, "Magnetic remanence in single atoms," *Science*, vol. 352, no. 6283, pp. 318–321, 2016. https://doi.org/10.1126/science.aad9898
[41] M. E. Silvestre, M. Franzreb, P. G. Weidler, O. Shekhah, and C. Wöll, "Magnetic cores with porous coatings: Growth of metal-organic frameworks on particles using liquid phase epitaxy," *Advanced Functional Materials*, vol. 23, no. 9, pp. 1210–1213, 2013.
[42] L. Mazet, S. M. Yang, S. V. Kalinin, S. Schamm-Chardon, and C. Dubourdieu, "A review of molecular beam epitaxy of ferroelectric $BaTiO_3$ films on Si, Ge and GaAs substrates and their applications," *Science and Technology of Advanced Materials*, vol. 16, p. 036005, 2015.
[43] R. Robinson, "Lxiii. – a synthesis of tropinone," *Journal of the Chemical Society, Transactions*, vol. 111, pp. 762–768, 1917.
[44] S. T. Selvan, P. K. Patra, C. Y. Ang, and J. Y. Ying, "Synthesis of silica-coated semiconductor and magnetic quantum dots and their use in the imaging of live cells," *Angewandte chemie*, vol. 119, no. 14, pp. 2500–2504, 2007.
[45] H. Jiang, P. Zhang, X. Wang, and Y. Gong, "Synthesis of magnetic two-dimensional materials by chemical vapor deposition," *Nano Research*, vol. 14, no. 6, pp. 1789–1801, 2021.

[46] H. Wang, X. Huang, J. Lin, J. Cui, Y. Chen, C. Zhu, F. Liu, Q. Zeng, J. Zhou, P. Yu, X. Wang, H. He, S. H. Tsang, W. Gao, K. Suenaga, F. Ma, C. Yang, L. Lu, T. Yu, E. H. T. Teo, G. Liu, and Z. Liu, "High-quality monolayer superconductor NbSe$_2$ grown by chemical vapour deposition," *Nature Communications*, vol. 8, no. 1, pp. 1–8, 2017.

[47] Y.-C. Zou, Z.-G. Chen, E. Zhang, F. Xiu, S. Matsumura, L. Yang, M. Hong and J. Zou, "Superconductivity and magnetotransport of single-crystalline NbSe$_2$ nanoplates grown by chemical vapour deposition," *Nanoscale*, vol. 9, no. 43, pp. 16 591–16 595, 2017.

[48] V. Shokhen, Y. Miroshnikov, G. Gershinsky, N. Gotlib, C. Stern, D. Naveh and D. Zitoun, "On the impact of vertical alignment of MoS$_2$ for efficient lithium storage," *Scientific Reports*, vol. 7, no. 1, pp. 1–11, 2017.

[49] T. Järvinen, G. S. Lorite, J. Peräntie, G. Toth, S. Saarakkala, V. K Virtanen, and K. Kordas, "WS$_2$ and MoS$_2$ thin film gas sensors with high response to NH$_3$ in air at low temperature," *Nanotechnology*, vol. 30, no. 40, p. 405 501, 2019.

[50] J. De Teresa, A. Fernández-Pacheco, R. Córdoba, L. Serrano-Ramón, S. Sangiao, and M. R. Ibarra, "Review of magnetic nanostructures grown by focused electron beam induced deposition (FEBID)," *Journal of Physics D: Applied Physics*, vol. 49, no. 24, p. 243 003, 2016.

[51] S. M. El-Sheikh, F. A. Harraz, and M. M. Hessien, "Magnetic behavior of cobalt ferrite nanowires prepared by template-assisted technique," *Materials Chemistry and Physics*, vol. 123, no. 1, pp. 254–259, 2010.

[52] S. Pathak and M. Sharma, "Polar magneto-optical Kerr effect instrument for 1-dimensional magnetic nanostructures," *Journal of Applied Physics*, vol. 115, no. 4, p. 043 906, 2014.

2D Magnetic Systems
Magnetic Properties, Measurement Techniques, and Device Applications

5

Daljit Kaur[a] and Shikha Bansal[b]
a Department of Physics, DAV University, Jalandhar, India
b IIT Guwahati, Guwahati, India

Contents

5.1	Introduction	74
5.2	Prospective Materials	75
	5.2.1 Classification of 2D Magnetic Materials	75
	5.2.1.1 Transition Metal Dichalcogenides	75
	5.2.1.2 Transition Metal Halides	75
	5.2.1.3 Transition Metal Phosphorous Tri-Chalcogenides	76
	5.2.1.4 Metal Oxides	76
	5.2.1.5 Perovskites	76
	5.2.1.6 MXene	76
5.3	Magnetic Properties and Underlying Phenomena of 2D Magnets	76
	5.3.1 Origin of Long-Range Magnetic Interaction	76
	5.3.1.1 Exchange interaction	77
	5.3.1.2 Double exchange interaction	77
	5.3.1.3 Superexchange interaction	77
	5.3.1.4 Ruderman–Kittel–Kasuya–Yosida (RKKY) interaction	77
	5.3.1.5 Dzyaloshinskii-Moriya interaction (DMI)	77
	5.3.1.6 Kitaev interaction	78
	5.3.2 Magnetic Models	78
	5.3.2.1 Heisenberg model	78
	5.3.2.2 Ising model	79
	5.3.2.3 XY model	79
	5.3.3 Magnetic Ordering in Various 2D Magnetic Systems	80
5.4	Measurement Techniques	82
	5.4.1 Optical Probes	82

DOI: 10.1201/9781003197492-5

		5.4.1.1	Polar MOKE	82
		5.4.1.2	MCD	82
		5.4.1.3	Polarization-resolved PL	84
		5.4.1.4	Polarization-resolved Raman scattering	84
		5.4.1.5	Second- harmonic generation (SHG)	84
	5.4.2	Electrical Probes		85
		5.4.2.1	AMR or TMR	85
		5.4.2.2	Anomalous Hall effect (AHE)	85
	5.4.3	Magnetic Imaging Probes		85
		5.4.3.1	MFM	85
		5.4.3.2	Nitrogen vacancy (NV) center magnetometry/scanning single spin magnetometry	86
		5.4.3.3	Spin-polarized scanning tunneling microscope (SP-STM)	87
5.5	Device Applications of 2D Magnetic Materials			87
	5.5.1	Spintronics-Based Devices		87
	5.5.2	Valleytronics-Based Devices		88
5.6	Conclusion – Challenging Issues and Future Scope			89
References				89

5.1 INTRODUCTION

Two-dimensional materials are referred to as a class of layered crystalline solids with a few nanometers of thickness. The single layer of atoms, known as a monolayer, is one atom thick, while a few layers can extend up to a thickness of 20–50 nm but should retain the 2D limit without attaining any 3D bulk properties of the material. Graphene is the first 2D material that was realized in 2004 by peeling off from graphite. Isolation of graphene depicted the possibility to get the atomically stable and thin sheets of crystals which show a variety of excellent properties such as high mobility, high conductivity, and high mechanical strength with long spin diffusion length for spintronics devices compared to the existing materials. The atomic-layer structures and exceptional properties of 2D materials make them potential candidates to be used for a wide variety of applications in optoelectronic devices, sensors, energy storage devices, drug delivery, and DNA sequencing. After this, many novel 2D materials are being realized and studied, like transition metal dichalcogenides (TMDs), MX_2 (M = Mo, W, Nb, Ta; X = S, Se, Te), graphitic carbon nitride (g-C_3N_4), hexagonal boron nitride (h-BN), and 2D oxides including lead, phosphorus, and transition metal oxides, etc. These materials cover a range of properties from metals to semimetals, topological insulators, semiconductors, insulators, and superconductors as well. The doping or heterostructures of these 2D materials with other materials can result in improved and novel properties, achieving high sensitivity as well as high power harvest efficiency.

For a decade, 2D magnetism remained a pipe dream until the discovery of the first 2D ferromagnets, $Cr_2Ge_2Te_6$ and CrI_3, down to the bilayer and monolayer limits in 2017. It was believed that at any nonzero temperature, a one- or two-dimensional isotropic magnet system following the Heisenberg model with finite-range exchange interaction could be neither ferromagnetic nor antiferromagnetic, as stated by the Mermin-Wagner theorem. The experimental realization of 2D magnetic systems was quickly followed by a surge of experimental and analytical work aimed at delving deeper into the realm of 2D magnetism.

Both the 2D ferromagnets, $Cr_2Ge_2Te_6$ and CrI_3, are ferromagnetic (FM) insulators with a low critical temperature T_c for phase transition (45 K for the CrI_3 monolayer and 28 K for the $Cr_2Ge_2Te_6$ bilayer) [1–2]. Soon after the discovery of CrI_3 and $Cr_2Ge_2Te_6$, Fe_3GeTe_2, a conducting FM with nomadic ferromagnetism, was reported with substantially regulated T_c reaching room temperature for trilayer composition [3]. Metallic 1T-phase transition metal dichalcogenides VSe_2 and $MnSe_2$ monolayers also showed ferromagnetism [4–5]. Recent extensive investigations on 2D $MnBi_2Te_4$ have sparked a renewed interest in magnetic topological insulators in the 2D limit [6].

With the help of high-performance computations along with density functional theory, more than 185 intrinsic 2D magnetic materials have been predicted to date [7]. Among them are high T_c FM semiconductors, 100% spin polarization half-metals, and magneto-electrically connected multiferroic crystals that make spintronic applications highly enticing. The combination of orbital, spin, lattice, charge, and valley produces a range of new physical phenomena in 2D magnetic crystals and their architectures. The remarkable characteristic of these materials is their fast response to external stimuli, such as electric doping, pressures, strains, and twisting.

Based on these results, this chapter classifies the newly found 2D magnetic materials by chemical composition and discusses their magnetic properties, types of magnetic ordering, and fit models, followed by a detailed discussion of the current characterization tools and techniques for probing 2D magnetization in these magnetic materials. The future potential of various device applications based on 2D magnetic materials is also being explored.

5.2 PROSPECTIVE MATERIALS

Diverse methods have been used to create many ultrathin 2D materials with various compositions and crystal structures. Layered and non-layered materials are two types of 2D materials. The in-plane atoms in layered materials are linked together by strong chemical bonding within the layer, while these layers are stacked together by a weak vdW force. Hence, it is feasible to exfoliate them into ultrathin 2D sheets down to a monolayer. The magnetic couplings in the materials are found to be extremely responsive to the lattice constant. Thus, modifying the lattice structure is all that is required to regulate the magnetic arrangement. Different types of stacking poly-types exist due to weak coupling between interlayers. Magnetic moments in layered magnetic materials are generated by the orbital and spin angular momenta of the 3D electrons of transition metal ions. Top-down and bottom-up techniques, such as liquid or mechanical exfoliation and chemical vapor deposition (CVD), chemical vapor transport (CVT), or molecular beam epitaxy (MBE) growth, are the most common methods utilized for 2D material creation. The synthetic technique determines the morphology, consistency, quality, cost, and potential uses of any material produced.

5.2.1 Classification of 2D Magnetic Materials

The discovered 2D magnet systems can be classified into the following categories on the basis of their chemical composition:

5.2.1.1 Transition Metal Dichalcogenides

Transition metal dichalcogenides are 2D materials with a layered structure and are represented by the chemical formula AB_2, where A stands for transition metal and B stands for chalcogen (Te, Se, S). Among the several existing TMDs, materials based on Cr and V, such as $CrTe_2$, VTe_2, $CrSe_2$, VSe_2, CrS_2, and VS_2, are found to possess magnetic properties either intrinsically or the magnetism can be induced by adding some impurities. These materials possess semi-metallic characteristics and stability in the 1T phase with A-A stacking [8].

5.2.1.2 Transition Metal Halides

Transition metal halides include transition metal dihalides represented as MX_2 and transition metal trihalides represented as MX_3, where M is the transition metal and X is a halogen. In the dihalides, transition metal cations form a triangular lattice structure in the monolayer, while a honeycomb lattice structure is typical for the monolayer of trihalides. In the Cr trihalide family, the stacking order is very essential. At a temperature below 210 K, CrI_3 in its bulk form shows ferromagnetism and undergoes a structural phase transition from monoclinic to rhombohedral. However, CrI_3 in 2D form has an antiferromagnetic interlayer coupling. Another trihalide, VI_3, a ferromagnetic semiconductor, showed that magnetization can be modulated by stacking orientation and pressure. The monolayers of Fe dihalides – $FeCl_2$, $FeBr_2$, and FeI_2 – are predicted, by theoretical estimations, to transform into spin half metals.

5.2.1.3 Transition Metal Phosphorous Tri-Chalcogenides

Transition metal phosphorous tri-chalcogenides can be represented by the general formula $[M_1M_2]$ $[P_2(X)_6]$ with basal plane sublattice $[P_2(X)_6]$, with M_1 and M_2 as transition metals and X as chalcogen atoms. Corresponding to Cr trihalides, M atoms play a significant role in deciding the magnetic properties of MPX_3. It should be noted that MPX_3 materials display low-temperature antiferromagnetism. $FePS_3$ and $NiPS_3$ possess zigzag-type magnetic structures, whereas $MnPS_3$ has a Neel-type magnetic structure. However, the magnetization direction for $MnPSe_3$ is in-plane, resulting in an XY-type spin model. At a finite temperature, isostructural MPX_3, a bilayer chromium-based chalcogenide, $CrGeTe_3$, is found to possess intrinsic long-range Heisenberg-type ferromagnetic ordering.

5.2.1.4 Metal Oxides

Several theoretical calculations predict that magnetic order can be induced in various metal oxides via applying strain or hydrogenation. Besides this, intrinsic FM has been reported in the MnO_2 monolayer [9]. Nanoribbons of ZnO possess ferromagnetic property, resulting from passivation of edges caused by applied electric field or hydrogenation. Magnetic activity is also reported in some hydroxide materials like nickel hydroxide, $Ni(OH)_2$, which exhibit antiferromagnetic (AFM) behavior. However, applying 4% biaxial compressive strain results in FM ordering. Emerging spintronics is greatly impacted by the strain-induced tunable magnetism in metal hydroxides. Hematene, which is a 2D version of hematite, shows the FM ordering; however, hematite is an AFM material [10].

5.2.1.5 Perovskites

Perovskites and mixed honeycomb oxides are electrically, structurally, magnetically, and thermally adjustable materials. The main characteristics of perovskites that attract their interest in 2D magnetization are spin-gapped ground states, the lack of fixed magnetism at low temperatures, and constrained AFM spin behavior. The potential magnetic ordering in K_2CuF_4, which is an insulating material, is very complex and fascinating. It can be assigned to the hybridization of *p*-states in F and d_{xy} states in Cu.

5.2.1.6 MXene

MXene, which is a family of 2D transition metal carbides, carbonitrides, and nitrides, is shown to have intrinsic magnetization in its pristine form. Typical examples of ferromagnetic MXenes include Cr_2C, Fe_2C and Ti_2C, Cr_2Br, Fe_2Br, and Ti_2Br. The surface-functionalized M_2X with OH, F, or O groups can be represented as M_2XG (G = O, F, or OH). An additional degree of freedom in M_2X facilitates control of magnetic ground states. Theoretically, it is shown that the surface functionalization in Cr_2C leads to a transition from FM to AFM, in addition to the transition from metal to insulator [11].

5.3 MAGNETIC PROPERTIES AND UNDERLYING PHENOMENA OF 2D MAGNETS

5.3.1 Origin of Long-Range Magnetic Interaction

The magnetism of any material compound arises due to the resultant spin and orbital magnetic moments of atoms that constitute the compound molecule. The microstructural/spatial arrangement of magnetic atoms/ions in a compound decides the type of magnetic ordering possible in it. The interaction/overlapping between the first, second, and third neighboring atomic orbitals with the magnetic atomic orbital can result in various exchange energies, either leading to ferromagnetic, ferrimagnetic, or antiferromagnetic long-range ordering. There are different types of interactions that have different energies, which are as follows:

5.3.1.1 Exchange interaction

The electrons are spin-1/2 particles, and a system of two interacting or non-interacting electrons requires its total wavefunction (product of spatial and spin wavefunction) to be anti-symmetric. This leads to the four possible combinations, three of which have a symmetric spin wavefunction known as a triplet state (total spin of the system, S = 1) and one anti-symmetric spin state (S = 0). The difference in the energy of singlet and triplet states can be parametrized by $\vec{S_1}.\vec{S_2}$, and the term in the Hamiltonian of two interacting spin-1/2 particles (electrons here) is as follows:

$$\hat{H}^{spin} = -2J\left(\vec{S_1}.\vec{S_2}\right). \qquad (1)$$

where J is the exchange constant or exchange integral and is a measure of the difference in the energy of a singlet state and a triplet state, $J = \frac{E_s - E_t}{2}$. If $J > 0$ then $E_s > E_t$ and hence triplet state (S = 1) is favored, and if $J < 0$, then $E_t > E_s$ and hence singlet state (S = 0) is favored. This is the case when both the electrons are on the same atom; if the two electrons are on neighboring atoms, then the bonding orbitals of the atoms lead to a singlet state (anti-parallel alignment), while anti-bonding orbitals lead to a triplet state (parallel alignment). If the electrons in neighboring magnetic atoms interact via an exchange interaction, this is known as a "direct exchange" [12].

5.3.1.2 Double exchange interaction

It is a ferromagnetic exchange interaction that occurs owing to the mixed valency property of a magnetic ion, e.g., manganese oxides where the Mn ion can exist in two oxidation states, +3 and +4, particularly in $La_{1-x}Sr_xMnO_3$. The e_g electron from an Mn^{+3} ion can hop to a neighboring Mn^{+4} magnetic ion if there is a vacancy in its e_g orbital, but it requires that spin must remain the same during hopping. Therefore, ferromagnetic alignment of neighboring ions is necessary to hop from the donating ion to the receiving ion.

5.3.1.3 Superexchange interaction

A superexchange is defined as an indirect exchange interaction between non-neighboring magnetic ions which is mediated by a non-magnetic ion, placed in between the magnetic ions, e.g. Mn^{2+} manganese ions connected via O^{2-} oxygen ions in MgO. Superexchange is a second-order process that is derived from the second-order term of perturbation theory.

5.3.1.4 Ruderman–Kittel–Kasuya–Yosida (RKKY) interaction

The indirect exchange interaction in metals mediated by spin-polarized conduction electrons from one magnetic ion to another magnetic ion is called the RKKY interaction. It does not involve the direct coupling of neighboring magnetic moments. It is a long-range interaction and has a characteristic oscillatory coupling whose wavelength is π/k_F, where k_F is the Fermi propagation constant. It can be ferromagnetic or antiferromagnetic.

5.3.1.5 Dzyaloshinskii-Moriya interaction (DMI)

This is an anti-symmetric exchange interaction that takes place between neighboring spins of magnetic atoms. The DMI term in the Hamiltonian of a magnetic system is as follows:

$$H_{DMI} = \vec{D}_{ij}.\left(\vec{S}_i \times \vec{S}_j\right). \qquad (2)$$

where S_i is the spin of the i-th magnetic ion and D_{ij} is the Dzyaloshinskii-Moriya vector. The vector D_{ij} is proportional to the vector product $R \times r_{ij}$ of the vector R that specifies the displacement of the ligand (for example, oxygen) and the unit vector r_{ij} along the axis connecting the magnetic ions i and j. The DMI is generally observed in chiral magnetic compounds (non-collinear structure) due to strong spin-orbit coupling and in antiferromagnets or at an interface in thin magnetic films and multilayers due to broken inversion symmetry. This term induces a relative tilt between neighboring spins or canting of spins, giving rise to a weak FM behavior.

5.3.1.6 Kitaev interaction

The Kitaev interaction is a specific anisotropic exchange coupling, generally observed in honeycomb-lattice materials (α-RuCl$_3$, Na$_2$IrO$_3$) with 90° metal-oxygen-metal bonds and strong spin-orbit interaction. The Kitaev interaction tries to align the spins connected by any given bond along with a single one of the principal axes of the structure, but the threefold symmetry of the honeycomb lattice means that this "bond-dependent magnetic anisotropy" is frustrated since all three possible spin directions are equally probable and no long-range order can be observed. The Hamiltonian of Kitaev's term is:

$$H_{Kitaev} = K S_i^\gamma S_j^\gamma \qquad (3)$$

where K is bond-dependent anisotropic Kitaev constant and S are spins at adjacent sites i and j. Xu *et al.* have theoretically computed and analyzed that significant Kitaev interaction does exist in 2D CrI$_3$ and CrGeTe$_3$ and that the interplay of Kitaev interaction with single-ion anisotropy can explain the observed magnetic anisotropy in these 2D materials [13–14].

5.3.2 Magnetic Models

In general, the magnetism in any 3D material results from spontaneously broken time-reversal symmetry, which is not the case in a continuous 2D system where symmetries cannot break spontaneously at finite temperature and the system remains isotropic, as stated by the Mermin-Wagner theorem. But the presence of magnetic anisotropy in the 2D system plays an important role in the arousal of its magnetic property by introducing an explicitly broken spin rotational symmetry in these systems. Magnetic anisotropy in a 2D system arises from magnetocrystalline anisotropy due to spin-orbit interaction and magnetostatic dipole-dipole interaction due to lattice distortions. The magnetic states in these systems are a result of the interplay of thermal fluctuations, quantum fluctuations, and spin-orbit interactions at the atomic scale. The theory of 2D magnetism is completely new, and established models of 3D magnetism, to find magnetic order and critical temperatures, cannot be applied in these systems.

The long-range magnetic order exists below a certain critical temperature, and above that the thermal energy induces spin fluctuations, thereby destroying the magnetic order and hence the FM or AFM state changes into a paramagnetic state. There are three types of magnetic models or governing equations that can explain the spin ordering behavior and transition (or critical) temperatures that are described in the following. We discuss the models that have been realized in 2D magnetic systems explored experimentally. These are:

5.3.2.1 Heisenberg model

If the spin on a lattice site can orient in any direction, it has a spin dimensionality of 3 as shown in Figure 5.1. The isotropic Heisenberg model was developed by W. Heisenberg in 1928. It is based on the spin interactions of nearest neighboring atoms or ions whose Hamiltonian is given as:

$$H_{Heisenberg} = -\frac{J}{2} \sum_{i,j} \vec{S_i} \cdot \vec{S_j} \qquad (4)$$

where J is the Heisenberg exchange term; when $J < 0$, interaction is called FM, and AFM when $J > 0$. The summation runs over the nearest neighbors (magnetic) atoms. Equation 1 cannot hold for the 2D magnetic order until the terms that explicitly break the spin-rotational symmetry are introduced. Hence,

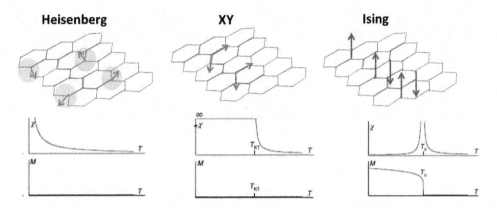

FIGURE 5.1 The three models of magnetic ordering: Heisenberg, XY, and Ising, based on spin dimensionality 3, 2, and 1 respectively, along with the expected temperature variation of their susceptibilities, χ, and magnetization, M. Adapted with permission from [17]. Copyright (2019) Springer Nature.

the Hamiltonian with additional terms related to single-ion anisotropy and anisotropic exchange between neighboring atoms is considered as follows:

$$H = -\frac{J}{2}\sum_{i,j}\vec{S_i}.\vec{S_j} - \frac{\lambda}{2}\sum_{i,j}S_i^z S_j^z - A\sum_i (S_i^z)^2 \qquad (5)$$

where λ and A are the 'on-site' and 'intersite' (or exchange) magnetic anisotropy constants which are evaluated from density functional theory (DFT), including spin-orbit coupling, by considering differences in the energy of in-plane and out-of-plane spin configurations. 2D materials such as $Cr_2Ge_2Te_6$, $MnPS_3$, and $MnPSe_3$ were found to be Heisenberg magnets with a small contribution of magnetic anisotropy in $Cr_2Ge_2Te_6$ [15].

5.3.2.2 Ising model

If the spin on the lattice site can orient only in a direction parallel or anti-parallel to the plane of the lattice, i.e., it has a spin dimensionality of 1, then the model used to express spontaneous magnetization is called a 2D Ising model. The Ising Hamiltonian in the absence of an external magnetic field is given as:

$$H_{Ising} = -\sum_{i,j} J_{ij}\sigma_i\sigma_j \qquad (6)$$

where $J_{i,j}$ is Ising interaction term, J > 0 for ferromagnetism, and the sum is over pairs of nearest neighbor sites. The exact solution by Onsager of the 2D Ising model shows that a phase transition to a magnetically ordered phase occurs at a temperature $T_c > 0$, known as the Ising transition. In this case, the anisotropy of the system, which favors a specific spin component, opens a gap in the spin-wave spectrum, thus suppressing the effect of thermal fluctuations. 2D materials such as a monolayer of CrI_3, $FePS_3$, and $FePSe_3$ are well described by the Ising model [16].

5.3.2.3 XY model

The XY model is applied when the spin dimensionality is 2, i.e. when spin can rotate or orient only in the x-y plane of the lattice. The magnetic susceptibility diverges below a finite temperature called the Berezinskii, Kosterlitz, and Thouless (BKT) transition temperature; T_{BKT} and a quasi-long-range order can be observed as a new type of topologically ordered magnetic phase is evolved by the creation of vortex-antivortex pairs. The Hamiltonian for the XY model is as follows:

$$H_{xy} = -J_{xy}\sum_{i,\lambda} S_i^x S_{i+\lambda}^x + S_i^y S_{i+\lambda}^y \qquad (7)$$

where J_{XY} is the nearest neighbor exchange interaction term, and the sum is over the nearest sites. The magnetic behavior of 2D materials like Rb_2CrCl and $NiPS_3$ can be explained by the XY model. The long-range order depends on both the symmetry of the order parameter and the type of spin-spin interactions, that compete with intrinsic fluctuations of either quantum or thermal nature. Strong fluctuations can easily destroy magnetic ordering in low-dimensional systems.

The generic magnetic Hamiltonian for such systems can be written as follows:

$$H = -J_{xy}\sum_{i,\lambda}(S_i^x S_{i+\lambda}^x + S_i^y S_{i+\lambda}^y) - J_I \sum_{i,\lambda} S_i^z S_{i+\lambda}^z \qquad (8)$$

where J_{xy} and J_I are spin-exchange energies on the basal plane and along the c-axis, respectively; S_i^α is the ($\alpha = x, y,$ or z) component of total spin; and i and λ run through all lattice sites and all nearest neighbors, respectively. All three fundamental models can be realized with the generic Hamiltonian: $J_{xy} = 0$ for the Ising model, $J_I = 0$ for the XY model, and for the Heisenberg model, $J_{xy} = J_I$. There is another model known as the XXZ model which has been taken into account for some 2D materials instead of the Heisenberg model to fully explain the magnetic ordering [17].

5.3.3 Magnetic Ordering in Various 2D Magnetic Systems

The exfoliation of bulk layered materials from 3D form to atomically thin few layers or monolayer of 2D material either preserves the bulk magnetic order or is destroyed by quantum or thermal fluctuations by lattice distortions in monolayers. There are two types of magnetic coupling or ordering observed in these layered vDW materials – intralayer coupling and interlayer coupling. It is generally observed that the spins on atoms in the same layer are aligned parallel or anti-parallel to each other, thus giving rise to FM or AFM character. But some materials like CrI_3, $CrCl_3$, $NiBr_2$, $NiPS_3$, etc. are known as layered antiferromagnets in which the interlayer coupling is AFM, i.e. the spins on the adjacent layer align opposite to that of the first layer. Hence the monolayer of these materials shows FM hysteresis, but the bilayer exhibits zero magnetization (AFM character) and this can be observed from characteristic spin-flip transition at a high field value, in its magnetization reversal behavior [18–19]. The magnetic properties of different 2D materials that have been explored theoretically or experimentally, fitted with different magnetic models, are summarized in Table 5.1 [20–25].

TABLE 5.1 Magnetic Properties (Magnetic Ordering, Transition Temperature [Curie/Neel], Magnetic Moment, Spin Model) of Various 2D Magnetic Materials

MATERIAL	MAGNETIC ORDER	BANDGAP	CRYSTAL STRUCTURE (IN 2D)	T_C/T_N	MOMENT (μ_B/ATOM)	MAGNETIC MODEL FIT
Metal di- and tri-halides						
CrI_3	FM (monolayer), A-type AFM (few layer)	Insulator	C2/m	45 K (1L)	2.98 μ_B/Cr	Ising
$CrCl_3$	A-type AFM (bi-layer)	Insulator	C2/m	16 K (1L)	3.04 μ_B/Cr	XY
$CrBr_3$	FM (monolayer)	Insulator	$R\bar{3}$	27 K (1L)	6 μ_B/Br	Ising & Heisenberg
VI_3	FM	Insulator	$R\bar{3}$	46 K	2 μ_B/V	Ising
$RuCl_3$	Frustrated magnet	Semiconductor	$P3_112$	7 K	–	Kitaev-Heisenberg
$FeCl_2$	A-type AFM	Insulator	$R\bar{3}m$	24 K	–	Ising
$FeBr_2$	A-type AFM	Insulator	$P\bar{3}m1$	17 K	–	Ising

MATERIAL	MAGNETIC ORDER	BANDGAP	CRYSTAL STRUCTURE (IN 2D)	T_C/T_N	MOMENT (μ_B/ATOM)	MAGNETIC MODEL FIT
FeI$_2$	AFM	Insulator	P$\bar{3}$m1	9 K	*	Ising
NiI$_2$	Helimagnet	Semiconductor (1.25 eV)	R$\bar{3}$m	79 K	–	–
NiBr$_2$	FM (intralyer) AFM (interlayer)	Semiconductor	R$\bar{3}$m	46 K (to C AFM), 20K (to IC)	3.25 μ_B/Ni	–
CoBr$_2$	Helimagnet	Insulator	P$\bar{3}$m1	19 K	5.53 μ_B/Co	–
MnI$_2$	Helimagnet	Semiconductor (3.34 eV)	P$\bar{3}$m1	–	5 μ_B/Mn	Mean field Heisenberg
CoI$_2$	Helimagnet	Insulator	P$\bar{3}$m1	–	2.73 μ_B/Co	–
VCl$_2$	Triangular AFM	Semiconductor	P$\bar{3}$m1	35.8 K	3 μ_B/V	Frustrated, Heisenberg with weak easy-axis anisotropy
VBr$_2$	Triangular AFM	Semiconductor	P$\bar{3}$m1	28.6 K	~3 μ_B/V	Frustrated, Heisenberg with weak easy-axis anisotropy
Transition metal phosphorus trichalcogenides						
FePS$_3$	Zig-zag type AFM	Insulator	C2/m	104 K	4 μ_B/Fe	Ising
NiPS$_3$	Zig-zag type AFM	Insulator	C2/m	155 K	1.03 μ_B/Ni	XY
MnPS$_3$	Neel-type AFM	Insulator	C2/m	78 K	-	Heisenberg
FePSe$_3$	Zig-zag type AFM	Insulator	R$\bar{3}$	115 K	4.9 μ_B/Fe in the basal plane	Ising
MnPSe$_3$	Neel-type AFM	Insulator	R$\bar{3}$	74 K	4.7 μ_B/Mn in basal plane	XY
Ternary tellurides						
CrSiTe$_3$	FM	Insulator	R$\bar{3}$	32 K		Ising
CrGeTe$_3$	FM (bilayer)	Insulator	R$\bar{3}$	30 K		Heisenberg
Fe$_3$GeTe$_2$	FM (monolayer)	Metal	P63/mmc	130 K		Ising
Fe$_4$GeTe$_2$	FM (7 layer)	Metal	R$\bar{3}$m	270 K		Ising
Fe$_5$GeTe$_2$	FM (28 nm)	Metal	R$\bar{3}$ &R$\bar{3}$m	310 K		Ising
MnBi$_2$Te$_4$	FM (intralyer) AFM (interlayer)	Topological insulator	R$\bar{3}$m	25 K	~5 μ_B/Mn	Frustrated
Transition metal oxyhalides						
CrOCl	AFM	Insulator	Pmmn	13.5 K	–	–
TiOCl	Spin-Peierls	Insulator	Pmmn	91 K, 45 K	–	–
VOCl	AFM	Insulator	Pmmn	80.5 K	–	–
FeOCl	AFM	Insulator	Pmmn	14 K	–	–
TiOBr	Spin-Peierls	Insulator	Pmmn	67 K, 28 K	*	–

MATERIAL	MAGNETIC ORDER	BANDGAP	CRYSTAL STRUCTURE (IN 2D)	T_c/T_N	MOMENT (μ_B/ATOM)	MAGNETIC MODEL FIT
Transition metal dichalcogenides						
VSe$_2$	FM (monolayer)	Metal	P$\bar{3}$m1	300 K	0.6 μ_B/V	Ising
VTe$_2$	FM (few layers)	Metal	P$\bar{3}$m1	300 K	-	Ising
MnSe$_2$	FM (monolayer)	Metal	P$\bar{3}$m1	300 K	3.0 μ_B/Mn	Ising
CrSe$_2$	AFM (monolayer)	Metal	P$\bar{3}$m1	160 K	2.44 μ_B/Cr	–
1T-CrTe$_2$	FM (monolayer)	Metal	P$\bar{3}$m1	300 K	0.21 μ_B/Cr	–

Table adapted with permission from [8]. Copyright the authors, some rights reserved; exclusive licensee John Wiley & Sons. Distributed under a Creative Commons Attribution License 4.0 (CC BY) https://creativecommons.org/licenses/by/4.0/.

5.4 MEASUREMENT TECHNIQUES

5.4.1 Optical Probes

The magnetization measurements of ultrathin monolayer samples are very challenging as atomically thin layers are prone to degradation in ambient conditions, and the magnetic moment of the monolayer must be carefully measured without any spurious signal from magnetic impurities. Hence, the standard magnetometer techniques like superconducting quantum interference device (SQUID), vibrating sample magnetometer (VSM), or neutron diffraction/scattering were not much preferred. Instead, the non-contact, powerful optical magnetization probes have been used to study 2D magnetic materials. The most common magnetization techniques for layered magnetic materials that are based on the optical probing method are polar MOKE (magneto-optical Kerr effect), magnetic circular dichroism (MCD), polarization-resolved photoluminescence (PL) spectroscopy, and micro-Raman spectroscopy. The working principle of each technique is as follows:

5.4.1.1 Polar MOKE

Polar MOKE is generally used when the magnetization of the sample, M, lies out of the plane and the linearly polarized light is incident on the sample and the component reflected normal to the plane of sample undergoes a rotation of its polarization vector, a phenomenon due to magnetic circular birefringence (MCB). The angle of rotation, θ_K, depends on the magnetization of the sample (material and substrate both) and the excitation wavelength of polarized light. MCB introduces the difference in polarization of right circularly and left circularly polarized light when reflected from the magnetized sample. This is due to the different contributions of magnetization to the imaginary part of optical conductivity, σ of the sample. In the case of linearly polarized light (with electric field E_x and E_y) propagating along the z-direction, this MCB originates from the anti-symmetric off-diagonal elements in the complex optical conductivity tensor $\sigma_{xy} = -\sigma_{yx}$, which are dependent on the magnetization, M [26].

5.4.1.2 MCD

Magnetic circular dichroism measures the difference in the absorption of right circularly and left circularly polarized light, due to the different contributions of magnetization to the real part of optical conductivity, σ of the sample. If the wavelength of polarized light lies in the X-ray range, it is called X-ray magnetic circular dichroism (XMCD). MOKE measures the changes in the polarization by measuring the angle of rotation of polarized light, while XMCD/MCD measure the changes in the ellipticity of the reflected polarized light. The ellipticity of polarized light depends on the real part of optical conductivity, which further depends on the sign of magnetization. The MCD measurement does not require any background subtraction in comparison to the MOKE signal. The extraction of optical conductivity values is difficult as there can be intermixing of real and imaginary parts in both MOKE and MCD signals, but these techniques are very sensitive in detecting even very small magnetic moments of any monolayer. The signal-to-noise ratio can be improved by using wavelengths that are resonant with the optical transitions of the magnetized samples. Both the MOKE and MCD techniques have been mainly used for Ising type and Heisenberg type 2D magnets because they can probe out-of-plane magnetization only. The comparison of MOKE and field-dependent MCD measurements on the CrI$_3$ monolayer and bilayer is shown in Figure 5.2a–c. To probe the magnetism of XY 2D magnet and 2D magnetic semiconductor, inelastic light scattering–based measurements are used. Although the use of X-rays instead of the laser source in MCD can help to measure the in-plane magnetization along with a determination of elements by X-rays, giving the spin and orbital momentum of magnetic atoms in XMCD [27].

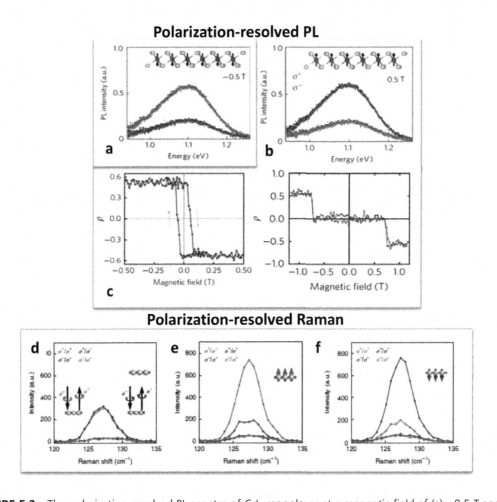

FIGURE 5.2 The polarization-resolved PL spectra of CrI$_3$ monolayer at a magnetic field of (a) −0.5 T and (b) 0.5T for σ+ (polarization component has higher PL intensity) than σ− (component at −0.5 T and vice-versa at 0.5 T) circular polarization components, acquired at 15 K under linearly polarized excitation. (c) Field-dependent circular polarized PL signal for CrI$_3$ monolayer and bilayer. Adapted with permission from [28]. Copyright (2018) Springer Nature. Helicity-resolved Raman spectra of the A$_{1g}$ mode of the CrI$_3$ monolayer in the (d) paramagnetic state at 60 K and the two spin states, (e) spin-up state and (f) spin-down state, at 15 K. Insets (d): Raman scattering channels with identical helicities between the incident and scattered light. Adapted with permission from [30]. Copyright (2018) Springer Nature.

5.4.1.3 Polarization-resolved PL

To probe out-of-plane magnetism in monolayers of 2D semiconductor, field-dependent PL measurements can be used to trace a hysteresis curve for FM ordering observed in a monolayer of CrI_3, while AFM spin-flip feature in magnetization reversal can also be traced in bilayer CrI_3, as shown in Figure 5.2 by Seyler et al. [28]. There are two ways to measure the sample magnetization – the first method is to use linearly polarized incident light and measure the normalized difference in intensity of left circularly and right circularly polarized reflected light that directly depends on the magnetization of the sample, while the second method is to use circularly polarized incident light and measure either left or right circularly polarized luminescence (emitted light), in which the average luminescence intensity of spin-polarized states is served as the reference. Magnetization is the relative difference between the intensity of emitted luminescence and that of the reference.

5.4.1.4 Polarization-resolved Raman scattering

Micro-Raman spectroscopy is based on an inelastic scattering process from the atoms of the given sample, which detects the magnetic order of the vDW magnets by measuring the phonon modes coupled to magnetic excitations above and below the transition temperature in FM and AFM magnetic materials. Above Curie and Neel temperatures, where long-range magnetic order is absent, the Raman spectrum has a broad background signal due to thermal-magnetic fluctuations, whereas below T_c/T_N, the long-range magnetic order results in sharp peaks either due to magnons, spinons (fractional spin excitations in frustrated magnets), or Brillouin zone folding. For example, the monolayer $FePS_3$ magnetic order was first detected by Raman measurement, which showed a broad feature at low energy values, but when it is cooled down to below the T_N, the broad feature splits into four sharp peaks. These four modes are Raman-active zone-boundary modes that resulted from Brillouin zone folding about a zone center due to the presence of the zig-zag AFM order [29]. However, the nature of magnetic order cannot be determined directly, but the precise determination of many layers can be done by Raman spectroscopy [30–31]. Sandilands et al. measured the effective exchange constant J of exfoliated $Bi_2Sr_2Dy_xCa_{1-x}Cu_2O_{8+\delta}$ using atomic force and polarized Raman spectroscopy by measuring its two-magnon joint density of states [32]. The helicity resolved Raman spectra of the CrI_3 monolayer above and below the Curie temperature are shown in Figure 5.2d–f. Figure 5.2d clearly shows equal Raman intensity for two different scattering channels where helicity is preserved and negligible intensity is observed for reversed helicity in the case of the paramagnetic state of CrI_3. The Raman spectra of up-spin and down-spin magnetization for the ferromagnetic state of CrI_3 are given in Figure 5e and 5f, which show that scattering is more in the same helicity channel.

5.4.1.5 Second-harmonic generation (SHG)

Second-harmonic generation is a newly established, highly sensitive optical technique that is based on a nonlinear optical process where two photons of the same frequency convert to a single photon of twice the fundamental frequency. This SHG takes place in material systems where inversion symmetry breaks down due to electric dipole approximation. The SHG can be observed in materials that do not possess lattice inversion symmetry (noncentrosymmetric), this process is called "i-type SHG," as it is time-invariant. When the material is centrosymmetric, then electric dipole allowed SHG can be observed if its magnetism comes into play by breaking both space- and time-reversal symmetries. Hence, reciprocal SHG that does not have time-inversion symmetry, known as "c-type SHG," can be easily measured to detect AFM order in bulk crystals and surface FM in thin films. Sun et al. have measured strong SHG at 5 K in layered antiferromagnetic CrI_3 bilayer (below transition temperature of 40 K) without magnetic field. As the magnetic field of −1 T is applied, the interlayer AFM coupling vanishes due to the alignment of spins in both layers, thereby restoring space- and time-inversion symmetry, and hence no SHG can be observed, thus confirming FM interlayer coupling in CrI_3 bilayers at high fields [8, 33].

5.4.2 Electrical Probes

The techniques that are based on the measurement of spin-dependent electrical conductivity of the magnetic sample come under the category of electrical probes. These include anomalous Hall effect (AHE), magnetoresistance measurements like TMR, GMR, or AMR of ultrathin monolayers, or spin device heterostructures of magnetic materials. The electrical contacts to the material and various geometries of junction heterostructures decide the mechanism of magnetotransport measurements taken on different spin-based device configurations. Each of the electrical measurement methods is explained as follows:

5.4.2.1 AMR or TMR

Anisotropic magnetoresistance (AMR) is generally observed in 2D magnets owing to the shape anisotropy of monolayer thin films and magnetocrystalline anisotropy of the material. Different resistance values are obtained when the magnetic field is applied longitudinally and transverse to the direction of the current in-plane of the monolayer. In-plane and out-of-plane giant or tunnel magnetoresistance (GMR or TMR) measurements of spin valves, ferromagnetic/nonmagnetic ferromagnetic multilayers, or magnetic tunnel junctions (MTJs) based on vDW heterostructures can help to directly probe spin device functionalities. But these transport measurements require good contacts with top and bottom metallic electrodes, along with clean interfaces with layered materials. The MTJs can be of two types: the first type involves 2D magnetic metallic materials as electrodes with oxide or insulator sandwiched between each electrode, e.g., Fe_3GeTe_2/hBN/Fe_3GeTe_2 structure (hBN is hexagonal boron nitride), and the second type is a 2D magnetic insulator as a tunneling barrier for metallic or graphene electrodes, e.g. graphene/CrI_3/graphene acting as a vertical spin filter device [34]. A large TMR of 160% is observed in the first type of MTJ, while the TMR in the second type of spin filter has shown nearly 100% value when a spin-flip transition of AF CrI_3 layers changes to FM order by applying a high magnetic field. The magnetic exchange and anisotropy fields of 2D CrI_3 can be determined from these measurements. As the number of CrI_3 layers is increased, the TMR value of the spin filter increases quickly, with values as large as 10^6 % for 10 layers [35].

5.4.2.2 Anomalous Hall effect (AHE)

The 2D ferromagnets possess Hall resistance R_{xy} in an applied magnetic field, which is contributed by both normal Hall effect (NHE) term that depends on the strength of the magnetic field applied and the AHE term, that depends on the magnetization of the sample. In the case of itinerant 2D ferromagnets, the contribution from the NHE term is negligible compared to the AHE term such that the hysteresis loops can be traced by measuring field-dependent R_{xy} in the case of Fe_5GeTe_2 monolayers at 220 K. The magnetism of 2D FM insulators can be measured by proximity-induced AHE measurements by depositing a metallic conducting Pt film on the insulator material, for example, Pt/$CrGeTe_3$ shows clear AHE and the field-dependent AHE shows hysteresis with abrupt transition points as shown in Figure 5.3. A quantized AHE has been observed in few-layer $MnBi_2Te_4$ showing A-type AF ordering [26, 36].

5.4.3 Magnetic Imaging Probes

5.4.3.1 MFM

MFM is similar to AFM in that the magnetic topography or magnetic domain texture of the sample is scanned by a tip coated with a ferromagnetic material like CoFe, Co, etc. The stray magnetic field gradient is measured through changes in the tapping frequency of the tip. It is difficult to infer the magnetization of the sample from these measurements, but the different magnetic domain states and unique structures like single domain, multi-domain, stripy, vortex, antivortex, skyrmion, etc. can be imaged. Lohmann *et al.* employed MFM to study the sharp transition points obtained in the field-dependent AHE of the

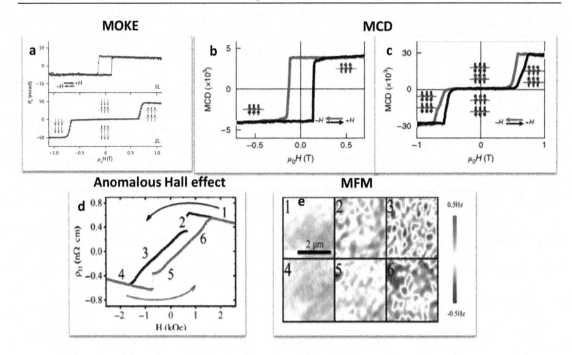

FIGURE 5.3 (a) The MOKE measurements of Kerr rotation angle, θ_K, with applied magnetic field on CrI_3 monolayer (1L), which showed ferromagnetic hysteresis, and bilayer (2L), showed spin-flip reversal at ±0.65 T with zero magnetization at low field. Adapted with permission from [2]. Copyright (2019) Springer Nature. Field-dependent MCD measurements for monolayer (b) and bilayer (c) CrI_3 at 4 K. Adapted with permission from [37]. Copyright (2019) Springer Nature. (d) Field-dependent AHE curve for the $Pt/CrGeTe_3$ heterostructure and (e) magnetic force microscopy (MFM) images of the $Pt/CrGeTe_3$ heterostructure at different values of an applied magnetic field. The nucleation of magnetic domains leads to abrupt variation points in the hysteresis shown in (d). Adapted with permission from [36]. Copyright (2019) Springer Nature

$Pt/CrGeTe_3$ heterostructure [36]. As shown in Figure 5.3e, the nucleation of magnetic domains leads to abrupt variation points in the hysteresis,

5.4.3.2 Nitrogen vacancy (NV) center magnetometry/ scanning single spin magnetometry

The nitrogen-vacancy defect in diamond leads to the determination of quantitative as well as qualitative information on the sample magnetization. The scanning single spin magnetometry, also known as NV center magnetometry, is equipped with a diamond tip having a single spin at the center of diamond NV that acts as an ideal sensitive nanoprobe that scans over the sample surface to detect stray magnetic fields and hence provides clear magnetic images. It can give information about a ferromagnet, a ferrimagnet, or an A-type antiferromagnet, but AFM material samples whose total net magnetization is zero cannot be detected by this method. The spatial resolution of this nanoscale imaging technique is approximately 10 nm, which is useful for studying nanoscopic domains and magnetic textures. The quantitative information is obtained when the stray magnetic fields of the sample split the spin sublevels of the NV center known as Zeeman splitting, and the spin-selective luminescence process of NV center helps to determine the sample magnetization [38]. The magnetization of the CrI_3 monolayer measured by diamond NV magnetometry was 16.1 μ_B nm^{-2}, whereas the theoretical value was found to be 14.7 μ_B nm^{-2} for a single layer of fully polarized Cr^{3+} spins. The bilayer CrI_3 has been shown to exhibit no net out-of-plane magnetization, which confirms the antiferromagnetic interlayer coupling in atomically thin CrI_3 as expected [39].

5.4.3.3 Spin-polarized scanning tunneling microscope (SP-STM)

Spin-polarized STM uses an atomically sharp magnetic tip of Cr or Fe that scans the sample surface and detects the tunneling current from the sample surface. The differential tunneling conductance (dI/dV) is generally measured using a non-magnetic W tip or a magnetic Cr tip. The field-dependent (dI/dV) conductance measurements result in hysteresis loops with different plateaus. Chen *et al.* determined the interlayer FM and AFM coupling in CrBr$_3$ by SP-STM [40]. The H-type stacked bilayer (when the orientation of one layer is rotated by 180° with respect to the second layer) and a monolayer of CrBr$_3$ showed normal hysteresis loops, which confirms the FM coupling, while the R-type stacked bilayer (same orientation of both layers) exhibited irregular hysteresis loops with four plateaus for the fields in the range of −0.5 T to 0.5 T whereas outside this range the normal hysteresis was observed, which suggested AFM interlayer coupling. Unlike MFM and NV center magnetometry, SP-STM can detect both AFM and ferrimagnetic order along with FM domains.

5.5 DEVICE APPLICATIONS OF 2D MAGNETIC MATERIALS

5.5.1 Spintronics-Based Devices

The magnetic properties and good electrical conductivity of 2D materials make them a promising candidate to study spintronic-based device applications like magnetic tunnel junctions, spin valves, spin transistors, current-induced magnetization switching for magnetic memories, spin-torque nano-oscillators, etc. The electric control of magnetization in a 2D material is more effective than in a 3D material owing to an atomically thin layer that is highly sensitive to an electric signal, and large values of electric field ~1 V/nm can be built across these vDW heterostructures [26]. The spin filtering mechanism provided by gate control along with tunneling magnetoresistance in spin filter MTJ made up of graphene/CrI$_3$/graphene shows a high TMR of 100% as shown in Figure 5.4. Table 5.2 provides the reports of MTJs, spin valves,

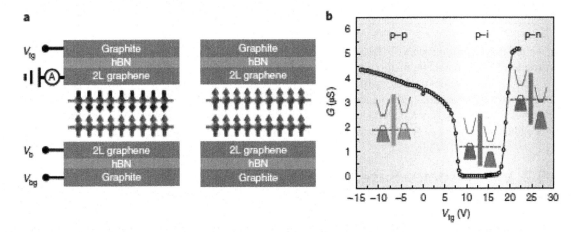

FIGURE 5.4 (a) The schematic of spin tunneling field-effect transistors made of bilayer CrI$_3$ showing low and high tunnel conductance states. (b) Gate control of tunnel conductance as a function of top gate voltage at zero magnetic fields. (Insets show the band alignments of the p–p, p–i, and p–n tunnel junctions. Here p is hole-doped, i is intrinsic, and n is electron-doped graphene.) Adapted with permission from [43]. Copyright (2019) Springer Nature.

TABLE 5.2 Spintronics Device Heterostructures Made Up of 2D Magnetic Materials and Their Properties Studied

S. NO	VDW HETEROSTRUCTURE	PROPERTY STUDIED	REF.
1.	AFM bilayer CrI$_3$ with graphene layers acting as contact electrodes h-boron nitride (hBN): the gate dielectric	Magnetization switching by electric field: FM state of CrI$_3$ bilayer for field value less than 0.2 V/nm^{-1} to an AFM state for field value greater than 0.7 V/nm^{-1}	[41]
2.	Vertical MTJ of hBN/10L CrI$_3$/hBN	Out-of-plane TMR values are as high as 10,000% at temperature < 10 K	[42]
3.	2L graphene/2L CrI$_3$/2L graphene	A dual-gated tunnel field-effect transistor (TFET): On-off (high and low conductance ratio) ratio of 400% is achieved	[43]
4.	Graphene/CrI$_3$/graphene	Spin filter MTJ: spin filtering and large TMR value of 100% when CrI$_3$ layer act as tunneling barrier while TMR value of four-layer CrI$_3$ in spin filter MTJ is 19,000% at temperature 2 K	[34–35]
5.	Fe$_3$GeTe$_2$/hBN/Fe$_3$GeTe$_2$ tunneling spin valve	Spin-dependent tunneling through vDW interfaces with TMR of ~160% at 4.2 K	[44]
6.	Fe$_{1/4}$TaS$_2$/Ta$_2$O$_5$ oxide/Cr$_{1/3}$TaS$_2$	TMR of 13% is observed with native oxide Ta$_2$O$_5$	[45–46]
7.	hBN/graphene/MnPS$_3$/graphene/hBN	MTJ with different layers of MnPS$_3$ as tunneling barrier, 15% TMR observed in 13 L and 6 L devices and established the presence of a spin-flop transition as a distinctive signature of antiferromagnetism	[47]
8.	hBN encapsulated graphite/CrCl$_3$/graphite MTJ	Magneto-conductance measurements reach a few 100% (depending on the applied bias and sample thickness); complete phase diagram of a magnetic system revealing subtle phases that arise due to finite-size effects	[23]
9.	Graphene/CrBr$_3$/graphene	Magnon assisted inelastic tunneling through FM CrBr$_3$ barrier at low temperature	[48]
10.	CrI$_3$/graphene (2L)/CrI$_3$	Spin valve (theoretical) devised and found that in-plane conductance is controlled by a spin proximity effect	[49]

and spin field-effect transistors (FETs) devised theoretically or realized experimentally using vDW heterostructures of 2D magnetic materials.

Apart from these device heterostructures, the spin degree of freedom of electrons in 2D magnetic materials can be modulated by magnetic or non-magnetic dopants, strain, and proximity effects from topological conductors, superconductors, multiferroic materials, piezoelectric materials, semimetals, and semiconductors, which can lead to various novel phenomena and properties.

5.5.2 Valleytronics-Based Devices

Valleytronics is based on the manipulation of the valley degree of freedom of electrons in semiconductors to process and store information. An ideal valleytronic material is a semiconductor whose electronic band structure has more than one degenerate but inequivalent valley state (a valley state is formed by a local minimum in a conduction band and a local maximum in a valence band at the same or different crystal momentum, k) located at or near the X symmetry points in the Brillouin zone. The energy of a valley state can be tuned by strain or magnetic field and is termed valley polarization. 2D valleytronic materials can be polarized by magnetic proximity with 2D materials. This concept was realized recently when a monolayer of WSe_2 was magnetically coupled to 2D CrI_3 below its Curie temperature, which led to the emergence of spontaneous circularly polarized PL at zero applied magnetic field. The spontaneous Zeeman splitting of ~2 meV between right-handed and left-handed PL has been observed due to valley polarization in monolayer WSe_2 by magnetic exchange interaction. The magnetic field changes the magnetization of the CrI_3 layer, which further affects the Zeeman splitting and PL handedness [50].

5.6 CONCLUSION – CHALLENGING ISSUES AND FUTURE SCOPE

To sum up, the 2D magnetic materials family is growing rapidly with the help of the first principle detection of new prospective materials. Its plethora of structural, optical, electronic, and magnetic properties can be explored in addition to its integration in 3D vDW heterostructures for spintronics applications. The main challenges faced are during the growth and design of layered materials and their device heterostructures because of instability and degradation of material in ambient atmospheric conditions. The measurement techniques for various properties of these materials have large noise compared to the signal. Therefore, new detection methods must be devised. Some of the 2D magnetic materials can be topological insulators, semimetals, multiferroics, ferroelectric, or semiconductors, which gives rise to new physical phenomena. Their magnetic properties can be tuned by strain, doping, gating, or proximity-induced effects from other materials and will be useful for further reading or writing purposes for magnetic memory devices.

REFERENCES

1) C. Gong et al., "Discovery of intrinsic ferromagnetism in two-dimensional van der Waals crystals," *Nature*, vol. 546, no. 7657, pp. 265–269, 2017, doi: 10.1038/nature22060.
2) B. Huang et al., "Layer-dependent ferromagnetism in a van der Waals crystal down to the monolayer limit," *Nature*, vol. 546, no. 7657, pp. 270–273, 2017, doi: 10.1038/nature22391.
3) Y. Deng et al., "Gate-tunable room-temperature ferromagnetism in two-dimensional Fe3GeTe2," *Nature*, vol. 563, no. 7729, pp. 94–99, 2018, doi: 10.1038/s41586-018-0626-9.
4) M. Bonilla et al., "Strong room-temperature ferromagnetism in VSe_2 monolayers on van der Waals substrates," *Nat. Nanotechnol.*, vol. 13, no. 4, pp. 289–293, 2018, doi: 10.1038/s41565-018-0063-9.
5) D. J. O'Hara et al., "Room temperature intrinsic ferromagnetism in epitaxial manganese selenide films in the monolayer limit," *Nano Lett.*, vol. 18, no. 5, pp. 3125–3131, 2018, doi: 10.1021/ACS.NANOLETT.8B00683.
6) Y. Deng et al., "Quantum anomalouas hall effect in intrinsic magnetic topological insulator $MnBi_2Te_4$," *Science*, vol. 367, no. 6480, pp. 895–900, 2020, doi: 10.1126/SCIENCE.AAX8156.
7) N. Mounet et al., "Two-dimensional materials from high-throughput computational exfoliation of experimentally known compounds," *Nat. Nanotechnol.*, vol. 13, no. 3, pp. 246–252, 2018, doi: 10.1038/s41565-017-0035-5.
8) S. Yang, T. Zhang, and C. Jiang, "van der Waals magnets: Material family, detection and modulation of magnetism, and perspective in spintronics," *Adv. Sci.*, vol. 8, no. 2, pp. 1–31, 2021, doi: 10.1002/advs.202002488.
9) M. Kan, J. Zhou, Q. Sun, Y. Kawazoe, and P. Jena, "The intrinsic ferromagnetism in a MnO_2 monolayer," *J. Phys. Chem. Lett.*, vol. 4, no. 20, pp. 3382–3386, 2013, doi: 10.1021/JZ4017848.
10) A. Puthirath Balan, S. Radhakrishnan, C. F. Woellner, S. K. Sinha, L. Deng, C. Reyes, B. M. Rao, M. Paulose, R. Neupane, A. Apte, V. Kochat, R. Vajtai, A. R. Harutyunyan, C.-W. Chu, G. Costin, D. S. Galvao, A. A. Martí, P. A. van Aken, O. K. Varghese, C. S. Tiwary, A. M. M. R. Iyer, and P. M. Ajayan, "Exfoliation of a non-van der Waals material from iron ore hematite," *Nat. Nanotechnol.*, vol. 13, no. 7, pp. 602–609, 2018, doi: 10.1038/s41565-018-0134-y.

11) C. Si, J. Zhou, and Z. Sun, "Half-metallic ferromagnetism and surface functionalization-induced metal – insulator transition in graphene-like two-dimensional Cr_2C crystals," *ACS Appl. Mater. Interfaces*, vol. 7, no. 31, pp. 17510–17515, 2015, doi: 10.1021/ACSAMI.5B05401.

12) J. Junquera, "Introduction to Heisenberg model." https://personales.unican.es/junqueraj/JavierJunquera_files/CA/5.Introduction_to_Heisenberg_model.pdf

13) C. Xu, J. Feng, H. Xiang, and L. Bellaiche, "Interplay between Kitaev interaction and single ion anisotropy in ferromagnetic CrI_3 and $CrGeTe_3$ monolayers," *NPJ Comput. Mater.*, vol. 4, no. 1, 2018, doi: 10.1038/s41524-018-0115-6.

14) J. A. Sears, L. E. Chern, S. Kim, P. J. Bereciartua, S. Francoual, Y. B. Kim, and Y.-J. Kim, "Ferromagnetic Kitaev interaction and the origin of large magnetic anisotropy in α-$RuCl_3$," *Nat. Phys.*, vol. 16, no. 8, pp. 837–840, 2020, doi: 10.1038/s41567-020-0874-0.

15) A. Kumar, "Note on magnetism in 2D materials models for 2D magnetism," pp. 4–9, 2019, doi: 10.13140/RG.2.2.24443.87842.

16) D. Torelli and T. Olsen, "First principles Heisenberg models of 2D magnetic materials: The importance of quantum corrections to the exchange coupling," *J. Phys. Condens. Matter*, vol. 32, no. 33, p. 335802, 2020, doi: 10.1088/1361-648X/ab8664.

17) [M. Gibertini, M. Koperski, A. F. Morpurgo, and K. S. Novoselov, "Magnetic 2D materials and heterostructures," *Nat. Nanotechnol.*, vol. 14, no. 5, pp. 408–419, 2019, doi: 10.1038/s41565-019-0438-6.

18) B. Huang, G. Clark, E. N.-Moratalla, D. R. Klein, R. Cheng, K. L Seyler, D. Zhong, E. Schmidgall, M. A. McGuire, D. H. Cobden, W. Yao, D. Xiao, P. J.-Herrero, and X. Xu, "Layer-dependent ferromagnetism in a van der Waals crystal down to the monolayer limit," *Nature*, vol. 546, no. 7657, pp. 270–273, 2017, doi: 10.1038/nature22391.

19) N. Chandrasekharan and S. Vasudevan, "Magnetism and exchange in the layered antiferromagnet $NiPS_3$," *J. Phys. Condens. Matter*, vol. 6, no. 24, p. 4569, 1994, doi: 10.1088/0953-8984/6/24/017.

20) H. Zhang, "High-throughput design of magnetic materials," *Electron. Struct.*, vol. 3, no. 3, p. 033001, 2021, doi: 10.1088/2516-1075/ABBB25.

21) B. K. Rai, A. D. Christianson, D. Mandrus, and A. F. May, "Influence of cobalt substitution on the magnetism of $NiBr_2$," *Phys. Rev. Mater.*, vol. 3, no. 3, pp. 1–7, 2019, doi: 10.1103/PhysRevMaterials.3.034005.

22) J. A. Sears, M. Songvilay, K. W. Plumb, J. P. Clancy, Y. Qiu, Y. Zhao, D. Parshall, and Y.-J. Kim, "Magnetic order in α – $RuCl_3$: A honeycomb-lattice quantum magnet with strong spin-orbit coupling," *Phys. Rev. B – Condens. Matter Mater. Phys.*, vol. 91, no. 14, pp. 1–5, 2015, doi: 10.1103/PhysRevB.91.144420.

23) Z. Wang, M. Gibertini, D. Dumcenco, T. Taniguchi, K. Watanabe, E. Giannini, and A. F Morpurgo, "Determining the phase diagram of atomically thin layered antiferromagnet $CrCl_3$," *Nat. Nanotechnol.*, vol. 14, no. 12, pp. 1116–1122, 2019, doi: 10.1038/s41565-019-0565-0.

24) Y. Li, D. Chen, X. Dong, L. Qiao, Y. He, X. Xiong, J. Li, X. Peng, J. Zheng, and X. Wang, "Magnetic and electric properties of single crystal $MnI2$," *J. Phys. Condens. Matter*, vol. 32, no. 33, p. 335803, 2020, doi: 10.1088/1361-648X/AB8983.

25) K. S. Burch, D. Mandrus, and J. G. Park, "Magnetism in two-dimensional van der Waals materials," *Nature*, vol. 563, no. 7729, pp. 47–52, 2018, doi: 10.1038/s41586-018-0631-z.

26) K. F. Mak, J. Shan, and D. C. Ralph, "Probing and controlling magnetic states in 2D layered magnetic materials," *Nat. Rev. Phys.*, vol. 1, no. 11, pp. 646–661, 2019, doi: 10.1038/s42254-019-0110-y.

27) P. Huang, P. Zhang, S. Xu, H. Wang, X. Zhang, and H. Zhang, "Recent advances in two-dimensional ferromagnetism: Materials synthesis, physical properties and device applications," *Nanoscale*, vol. 12, no. 4, pp. 2309–2327, 2020, doi: 10.1039/c9nr08890c.

28) K. L. Seyler, D. Zhong, D. R. Klein, S. Gao, X. Zhang, B. Huang, Efrén Navarro-Moratalla, L. Yang, D. H. Cobden, M. A. McGuire, W. Yao, D. Xiao, P. Jarillo-Herrero, and X. Xu, "Ligand-field helical luminescence in a 2D ferromagnetic insulator," *Nat. Phys.*, vol. 14, no. 3, pp. 277–281, 2017, doi: 10.1038/s41567-017-0006-7.

29) J.-U. Lee, S. Lee, J. H. Ryoo, S. Kang, T. Y. Kim, P. Kim, C.-H. Park, J.-G. Park, and H. Cheong, "Ising-type magnetic ordering in atomically thin $FePS_3$," *Nano Lett.*, vol. 16, no. 12, pp. 7433–7438, 2016, doi: 10.1021/ACS.NANOLETT.6B03052.

30) B. Huang, J. Cenker, X. Zhang, E. L. Ray, T. Song, T. Taniguchi, K. Watanabe, M. A. McGuire, Di Xiao, and X. Xu, "Tuning inelastic light scattering via symmetry control in the two-dimensional magnet CrI_3," *Nat. Nanotechnol.*, vol. 15, no. 3, pp. 212–216, 2020, doi: 10.1038/s41565-019-0598-4.

31) C.-T. Kuo, M. Neumann, K. Balamurugan, H. J. Park, S. Kang, H. Wei Shiu, J. H. Kang, B. H. Hong, M. Han, T. W. Noh, and J.-G. Park, "Exfoliation and raman spectroscopic fingerprint of few-layer $NiPS_3$ Van der Waals crystals," *Sci. Reports*, vol. 6, no. 1, pp. 1–10, 2016, doi: 10.1038/srep20904.

32) L. J. Sandilands, J. X. Shen, G. M. Chugunov, S. Y. F. Zhao, S. Ono, Y. Ando, and K. S. Burch, "Stability of exfoliated Bi2 Sr2 Dyx Ca1-x Cu2 O8+δ studied by Raman microscopy," *Phys. Rev. B – Condens. Matter Mater. Phys.*, vol. 82, no. 6, 2010, doi: 10.1103/PHYSREVB.82.064503/FIGURES/6/THUMBNAIL.

33) Z. Sun, Y. Yi, T. Song, G. Clark, B. Huang, Y. Shan, S. Wu, D. Huang, C. Gao, Z. Chen, M. McGuire, T. Cao, D. Xiao, W.-T. Liu, W. Yao, X. Xu, and S. Wu, "Giant nonreciprocal second-harmonic generation from antiferromagnetic bilayer CrI$_3$," *Nature*, vol. 572, no. 7770, pp. 497–501, 2019, doi: 10.1038/s41586-019-1445-3.
34) T. Song, X. Cai, M. W.-Y. Tu, X. Zhang, B. Huang, N. P Wilson, K. L. Seyler, L. Zhu, T. Taniguchi, K. Watanabe, M. A. McGuire, D. H. Cobden, D. Xiao, W. Yao, and X. Xu, "Giant tunneling magnetoresistance in spin-filter van der Waals heterostructures," *Science*, vol. 360, no. 6394, pp. 1214–1218, 2018, doi: 10.1126/SCIENCE.AAR4851.
35) "One million percent tunnel magnetoresistance in a magnetic van der Waals Heterostructure," *Nano Lett.*, vol. 18, no. 8, pp. 4885–4890, 2018, https://pubs.acs.org/doi/10.1021/acs.nanolett.8b01552
36) M. Lohmann, T. Su, B. Niu, Y. Hou, M. Alghamdi, M. Aldosary, W. Xing, J. Zhong, S. Jia, W. Han, R. Wu, Y.-T. Cui, and J. Shi, "Probing magnetism in insulating Cr2Ge2Te6 by induced anomalous hall effect in Pt," *Nano Lett.*, vol. 19, no. 4, pp. 2397–2403, 2019, doi: 10.1021/ACS.NANOLETT.8B05121.
37) S. Jiang, L. Li, Z. Wang, K. F. Mak, and J. Shan, "Controlling magnetism in 2D CrI$_3$ by electrostatic doping," 2018, doi: 10.1038/s41565-018-0135-x.
38) B. Huang, M. A. McGuire, A. F. May, D. Xiao, P. Jarillo-Herrero, and X. Xu, "Emergent phenomena and proximity effects in two-dimensional magnets and heterostructures," *Nat. Mater.*, vol. 19, no. 12, pp. 1276–1289, 2020, doi: 10.1038/s41563-020-0791-8.
39) Z. Wang, C. Tang, R. Sachs, Y. Barlas, and J. Shi, "Proximity-induced ferromagnetism in graphene revealed by the anomalous hall effect," *Phys. Rev. Lett.*, vol. 114, no. 1, p. 016603, January 2015, doi: 10.1103/PhysRevLett.114.016603.
40) W. Chen, Z. Sun, Z. Wang, L. Gu, X. Xu, S. Wu, and C. Gao, "Direct observation of van der Waals stacking – dependent interlayer magnetism," *Science*, vol. 366, no. 6468, pp. 983–987, 2019, doi: 10.1126/SCIENCE.AAV1937.
41) S. Jiang, J. Shan, and K. F. Mak, "Electric-field switching of two-dimensional van der Waals magnets," *Nat. Mater.*, vol. 17, no. 5, pp. 406–410, 2018, doi: 10.1038/s41563-018-0040-6.
42) Z. Wang, I. G.-Lezama, N. Ubrig, M. Kroner, M. Gibertini, T. Taniguchi, K. Watanabe, A. Imamoğlu, E. Giannini, and A. F. Morpurgo, "Very large tunneling magnetoresistance in layered magnetic semiconductor CrI$_3$," *Nat. Commun. 2018 91*, vol. 9, no. 1, pp. 1–8, 2018, doi: 10.1038/s41467-018-04953-8.
43) S. Jiang, L. Li, Z. Wang, J. Shan, and K. F. Mak, "Spin tunnel field-effect transistors based on two-dimensional van der Waals heterostructures," *Nat. Electron. 2019 24*, vol. 2, no. 4, pp. 159–163, 2019, doi: 10.1038/s41928-019-0232-3.
44) Z. Wang, D. Sapkota, T. Taniguchi, K. Watanabe, D. Mandrus, and A. F. Morpurgo, "Tunneling spin valves based on Fe3GeTe2/hBN/Fe3GeTe2 van der Waals Heterostructures," *Nano Lett.*, vol. 18, no. 7, pp. 4303–4308, July 2018, doi: 10.1021/ACS.NANOLETT.8B01278.
45) M. Arai, R. Moriya, N. Yabuki, S. Masubuchi, K. Ueno, and T. Machida, "Construction of van der Waals magnetic tunnel junction using ferromagnetic layered dichalcogenide," *Appl. Phys. Lett.*, vol. 107, no. 10, p. 103107, 2015, doi: 10.1063/1.4930311.
46) Y. Yamasaki, R. Moriya, M. Arai, S. Masubuchi, S. Pyon, T. Tamegai, K. Ueno, and T. Machida, "Exfoliation and van der Waals heterostructure assembly of intercalated ferromagnet Cr$_{1/3}$TaS$_2$," *2D Mater.*, vol. 4, no. 4, pp. 1–21, 2017, doi: 10.1088/2053-1583/aa8a2b.
47) G. Long, H. Henck, M. Gibertini, D. Dumcenco, Z. Wang, T. Taniguchi, K. Watanabe, E. Giannini, and A. F. Morpurgo, "Persistence of magnetism in atomically thin MnPS$_3$ crystals," *Nano Lett.*, vol. 20, no. 4, pp. 2452–2459, April 2020, doi: 10.1021/ACS.NANOLETT.9B05165.
48) D. Ghazaryan, M. T. Greenaway, Z. Wang, V. H. Guarochico-Moreira, I. J. Vera-Marun, J. Yin, Y. Liao, S. V. Morozov, O. Kristanovski, A. I. Lichtenstein, M. I. Katsnelson, F. Withers, A. Mishchenko, L. Eaves, A. K. Geim, K. S. Novoselov, and A. Misra, "Magnon-assisted tunnelling in van der Waals heterostructures based on CrBr$_3$," *Nat. Electron.*, vol. 1, no. 6, pp. 344–349, 2018, doi: 10.1038/s41928-018-0087-z.
49) C. Cardoso, D. Soriano, N. A. García-Martínez, and J. Fernández-Rossier, "Van der Waals spin valves," *Phys. Rev. Lett.*, vol. 121, no. 6, p. 67701, 2018, doi: 10.1103/PhysRevLett.121.067701.
50) J. R. Schaibley, H. Yu, G. Clark, P. Rivera, J. S. Ross, K. L. Seyler, W. Yao, and X. Xu, "Valleytronics in 2D materials," *Nat. Rev. Mater.*, vol. 1, no. 11, pp. 1–15, 2016, doi: 10.1038/natrevmats.2016.55.

3D Magnonic Structures as Interconnection Element in Magnonic Networks

6

A. A. Martyshkin, S. A. Nikitov, and A. V. Sadovnikov

Laboratory of Magnetic Metamaterials, Saratov State University, Saratov, Russian Federation

Contents

6.1	Introduction	93
6.2	In-Plane Magnonic Transport	95
6.3	Magnonic Intermodal Coupler	100
6.4	Out-of-Plane Magnonic Transport	102
6.5	Three-Dimensional Magnonic Topology	104
6.6	Conclusion	106
6.7	Acknowledgments	107
References		107

6.1 INTRODUCTION

Low power consumption, easy integration, compatibility with optional metal oxide semiconductor (CMOS) architecture, short wavelength, anisotropic properties, negative group velocity, nonreciprocity, and efficient tuning using various external stimuli are just some of the possibilities that spintronics allows us to use when creating devices based on magnon principles. The key advantages offered by spin waves (SW) [1] for data processing are the scalability down to sub-nanometer dimensions due to SW peculiar features of SW dispersion. The frequency range of SW excitation spreads from GHz to hundreds of THz. At the same time, SWs are used for data processing in the wide temperature range from ultra-low to room

DOI: 10.1201/9781003197492-6

temperatures [1]. Historically, SWs were utilized to develop passive and active RF microwave devices [1]. In the last decade, progress in lithography has made it possible to develop technologies for the fabrication of structures on the top of thin-film magnetic films, which has given rise to a variety of mechanisms and methods for studying spin-wave excitations in micro- and nanostructures. The use of structured ferromagnetic films as waveguide structures opens up new possibilities for creating devices and concepts such as magnon memory, logic, transistors, converters, RF components (filters, diodes, circulators, directional couplers, half-adders), neuromorphic computing, etc.

Similar to the idea of building integrated circuits based on CMOS principles, concepts for creating computational layers based on magnon principles have been proposed during the last decade in the concept of "magnon spintronics" [2]. This part of spintronics which refers to the information transport and processing by spin waves is known as magnonics [3]. The integration of functional blocks based on magnonic principles into multi-element networks is carried out using magnon waveguide devices, in which the spin-wave amplitude and/or phase is encoded the information signal, providing additional advantages of non-volatility and low operating power. In addition, the ability to read the state of magnetization using inductive methods, magnetic tunnel junctions, and fabrication of spin-wave guiding media on semiconductor substrates provides the next step towards the integration of magnonics with CMOS technology [4].

Low packing density is one of the most vulnerable points of criticism of magnonic devices combined in magnonic networks since the attenuation of the spin-wave signal is a serious obstacle to the assembly of a computational unit limited to spatial scales commensurate with the SW propagation length. The use of magnetic films with the lowest decay rate is one of the approaches to overcoming these limitations – in particular, the use of yttrium iron garnet (YIG) films makes it possible to achieve the propagation length of a spin-wave signal up to several millimeters without significant losses. A typical spin-wave guide structure contains a non-magnetic substrate on which a magnetic film is grown. Gallium-gadolinium garnet (GGG) is most commonly used as a substrate material. To form a waveguide, there are a large number of methods for growing a magnetic film: in particular, liquid-phase epitaxy, deposition, and laser ablation. YIG most meets the requirements of low attenuation of the spin-wave signal since it has a narrow absorption line of ferromagnetic resonance (typically 0.5 Oe at 10 GHz). The simplest waveguide structures are confined in lateral direction magnetic films of YIG. High-efficiency control of spin-wave propagation in ultra thin yttrium iron garnet by the spin-orbit torque was considered in [5]. With the help of such structures, it became possible to excite the spin torque with efficient signal transmission over extended regions. Investigation of inhomogeneous tapered waveguide structures showed that, under the width constraint, it is necessary to take into account the multimode propagation of spin waves due to energy redistribution between modes [6, 7]. It should be noted that in magnonics, not only the Gilbert damping [1] but also the damping associated with the spin-wave scattering is important. The fabrication of magnonic waveguides also affects the spin-wave group velocity, which influences the signal attenuation [1, 8, 9].

Nevertheless, the spatial limitation of the scale prompts us to reduce the area of the device when increasing the computational units in the magnon network. Increasing the density of elements in CMOS is a serious problem and requires consideration of Joule heating when designing integrated circuits. The absence of current transfer in spin-wave devices opens up new ways of solving spatial constraints that go beyond the traditionally studied plane geometries. The expansion of magnon networks into the third dimension reduces the area occupied by individual spin-wave devices. Therefore, the utilization of three-dimensional (3D) magnon computational schemes leads to the solution of the problem of routing the spin-wave signal between the individual spin-wave devices inside the magnonic networks. While adjusting the spin-wave scattering losses by performing micromagnetic simulations is a fast alternative to experiments, the optimization of device shape is a complicated task that requires the use of the whole set of parameters to ensure feasibility. The alternative way to interconnect the magnonic units is to use the effects of spin-wave coupling, e.g. by utilization of the concept of spin-wave couplers [10–12]. The effects of spin-wave coupling control based on the engineering of the stray field profile lead to a linear and nonlinear switching regime [13], which manifests itself in the variation of the coupling length with the variation of the spin-wave amplitude. This functional interconnection element can be used as a multiport device, which acts as a signal switch when spin waves of different amplitude and/or frequency are guided to different outputs of the directional coupler [14]. This idea of spin-wave coupling in a lateral direction has opened up new promising prospects for the architecture of spin-wave logic [15–16].

The creation of waveguide structures with a violation of translational symmetry [17–18] has made it possible to create devices based on magnon logic, which are based on the concept of three-dimensional magnon topologies, which will significantly reduce the size of logical elements. One of the options for using the three-dimensional concept of magnonic devices has been demonstrated using magnonic crystals [18–19] and dielectric YIG [20, 21] materials representing a meander structure of ferromagnetic segments located at an angle of about 90 degrees relative to each other. This design was chosen to be the first step towards the demonstration of the spin-wave transport in orthogonal to the film surface direction [22]. Control of the spin-wave spectra in 3D magnonic crystals allows SW to propagate along with orthogonal segments of the film. Using the method of ferromagnetic resonance, strong peaks of excitation of spin waves were revealed in the spectra of spin waves propagating in the structure, as well as with stationary modes. Using Brillouin light scattering (BLS) experiments for a single meander CoFeB structure and a CoFeB/Ta/NiFe magnetic bilayer, the presence of forbidden bands was demonstrated.

The meander structure having vertical segments with certain dynamic magnetization profiles are connected and provide SW signal transmission between layers of horizontal segments that are located at different heights, representing a new degree of freedom for controlling spin-wave excitations. Under such conditions, it is important to study the mechanisms associated with the transfer of a spin signal in a multilevel topology of a magnon network based on three-dimensional structures. Of greatest interest is the possibility of transferring a spin-wave signal between the functional levels of the magnon network.

This chapter describes the different types of structures based on magnetic films of yttrium iron garnet grown on gallium-gadolinium garnet substrates. In the first section, a description is given of waveguide structures with translation symmetry breaking performed in a plane. The second section is devoted to laterally located transversely bounded waveguide structures. In the third section, a combination of translational symmetry-breaking waveguides and lateral transversely bounded waveguides are considered. The fourth section demonstrates possible options for combining planar and vertical structures into a single elementary computing unit for magnon networks. The "Conclusion" section is devoted to the discussion of the possibilities of using three-dimensional waveguide structures to create a new type of logical device based on magnon principles with a three-dimensional topology.

6.2 IN-PLANE MAGNONIC TRANSPORT

In this section, we consider the possibilities of propagation of a spin-wave signal in a planar structure with translation symmetry breaking. A schematic representation of the investigation structure is shown in Figure 6.1a, which is an orthogonal in-plane connection of magnetic stripes.

Using the method of micromagnetic modeling in the MuMax3 software package, based on the numerical solution of the Landau-Hilbert equation, it is possible to describe the precession of the magnetization vector M_s in the effective magnetic field $H_{eff} = H_0 + H_{demag} + H_{ex} + H_a$, where H_0 is the external magnetic field, H_{demag} is the demagnetizing field, H_{ex} is the exchange field, and H_a is the anisotropy field. Taking into account that the magnetocrystalline anisotropy of YIG is negligible in comparison with the shape anisotropy, the value of H_a was assumed to be zero. It is well known that shape anisotropy plays a decisive role in the transfer of spin waves along the magnonic bend [13]. Isotropic YIG films of various thicknesses and widths with the following material parameters, such as saturation magnetization $M_s = 139 G$, Gilbert damping parametr $a = 10^{-3}$ and non-uniform exchange constant $A_{ex} = 3.614 \times 10^{-2} cm^{-2}$ were considered.

To reduce the SW reflections from the boundaries of the computational domain, we added absorbing boundary layers (ABL) with an exponentially decreasing attenuation coefficient alpha edge at the beginning of the input and the end of the output sections of the [23] waveguide structure. The sources of excitation of the spin-wave signal were located directly next to the ABL regions at the ends of the [24]

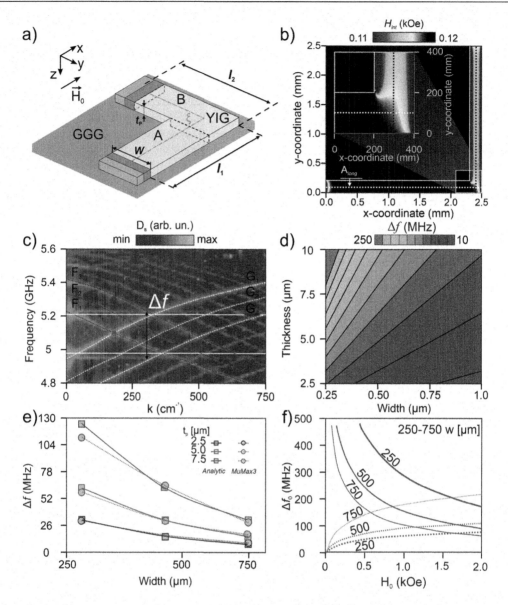

FIGURE 6.1 Numerical characterization of the in-plane magnonic bend with a right-angled junction. (a) Schematic of the two-dimensional in-plane (AB, extending across the XY-plane). (b) The amplitude of the internal magnetic field. (c) Calculated map of SW dispersion characteristics across the A and B waveguide sections for 500 μm-wide YIG stripes of thickness 10 μm. (d) Color-coded map of the function Δf (w, t) for a range of stripe thicknesses and widths as indicated. (e) Dependence of Δf on the stripe's width w and thickness tp obtained from the micromagnetic modeling (solid curve) and analytical calculations (dotted curves). (f) Dependence of Δf_0 on H in waveguides with a width as indicated for BVMSWs (solid curve) and MSSWs (dotted curve), respectively.

waveguides, as shown by the red regions in Figure 6.1a. The structures were placed in an external uniform magnetic field of $H_0 = 1200$ Oe.

From the solution of the static problem, we obtained the spatial distribution of the internal static magnetic field H_{eff} in the waveguide system. Figure 6.1b shows the distribution of the y-component of the field H_{eff} in the plane of the waveguide structure when the field H_0 is oriented along the y-axis. In section

A, the internal field H_{eff} practically coincides in magnitude with H_0, since in this case, magnetization occurs along the length of the side of the section ($l > w$) and the y-component of the demagnetization field H_{demag} is small. In section B, magnetization occurs along the short sides of the sections, the influence of demagnetizing fields is significant, and the internal fields are less than the external ones by an amount of the order of $\Delta H = 50$ Oe.

Investigation of spin-wave excitations in an unbounded ferrite film has shown that BVMSWs [25] exist in different frequency bandwidths, which are separated by the ferromagnetic resonance frequency of the YIG film $f_0 = \gamma \sqrt{H_{int}(H_{int} + 4\pi M_s)}$ [26].

The breaking of translational symmetry in a transversely bounded waveguide leads to a change in the internal magnetic fields, as a result of which a general MSW overlap bandwidth is formed since a decrease in the value of the internal magnetic field in a bounded waveguide leads to a downward shift of the frequency f_0. The frequency overlap of the dispersion branches [27] of the BVMSW and MSSW in the orthogonal connection of the magnetic stripes provides the SW type transformation effect in the structures with bending [26].

To calculate the effective dispersion characteristics, a broadband dynamic magnetic field $b_z(t) = b_0 sinc(2\pi f_c t)$ was used to excite spin waves with a cutoff frequency $f_c = 6$GHz and an amplitude $b_0 = 10$ m Oe. Making the assumption that due to the small size of the investigation structure, the alternating magnetic field of the source has a uniform distribution over the YIG film thickness. Along the center of the waveguide section, the time-varying behavior of the magnetization $m_{perp}(i) = \int^V m_{perp}(x,y,z)dV$, where V is the volume of the regional sections, the discretized time step $\Delta t = 75$ns, and the duration $T = 600$ns. Using the technique of two-dimensional Fourier transform, it is possible to extract a two-dimensional map of the distribution of the quantity $D(k,f) = \frac{1}{N}\sum_{i=1}^{N}\theta_2 \mid m_{perp}(x,y,z,y)\mid^2$, where k is the wavenumber, θ_2 is the 2D Fourier transform operator, i is the cell number, and $N = 256$ is the number of cells along the central part of the waveguide. The mapping $D(k,f)$ is the square of the modulus of the magnetization amplitude and allows one to reconstruct the effective dispersion characteristics for spin-wave modes in waveguides. In the (k,f) plane, one can also identify local maxima of $D(k,f)$, which physically correspond to modes of width n-th order [1] with transverse wavenumbers $\kappa = \frac{n\pi}{w}$.

On the obtained map of dispersion characteristics (Figure 6.1c) for YIG-thickness $10\,\mu m$ and width $500\,\mu m$, the curves F_n and G_n ($n = 1,2,3$.) indicate SW width modes propagating through sections A and B along the x-axis and y-axis, respectively. Here, the indices n denote the number of anti-nodal points characteristic of the limited SW phase across the magnon band [28]. The frequency overlap region Δf is shown in Figure 6.1c, where the modes of the first width ($n = 1$) BVMSW (F_1) and MSSW (G_1) coexist. It should be noted that a change in the width and thickness of the stripes, along with a change in the strength of the external magnetic field, can adjust the frequency band Δf of the magnon coupling. The change in Δf in the YIG strip can be controlled by changing the internal magnetic field as a result of the shape anisotropy and/or a shift of the lower/upper cutoff frequency for spin waves of MSSW/BVMSW types, as a consequence of the dispersion transformation for spin waves, propagating along a limited magnon waveguide [1]. The influence of the spatially varying magnitude of the internal magnetic field on the value of $\Delta f(w,t)$ can be estimated using the inhomogeneous profile H_{int} inside each waveguide for the case of tangential magnetization [29]. Using this approach, the bandwidth can be estimated as a function of t and w. In Figure 6.1d the value of the function $\Delta f(w,t)$ is for a range of thicknesses ($2.5 < t < 10$)μm and widths ($250 < w < 1000$μm). Obviously, as the bandwidth decreases and the thickness increases, the frequency bandwidth increases. Using the dispersion characteristics from [1], it is possible to estimate the cutoff frequency shift for BVMSW and MSSW. The results of micromagnetic modeling and analytical evaluation clearly show good agreement for the considered YIG film (Figure 6.1e) with increasing thickness and decreasing waveguide width structure leading to an increase in the area of coexistence of BVMSW and MSSW in the Δf frequency band. Thus, the influence of the geometric parameters of the waveguide structure becomes obvious when creating an in-plane bend structure. Next, we analyze the effect of the uniform magnetic field strength on the value of Δf_0. Note that Δf_0 is defined for both MSSW and BVMSW as the difference in f_0 that results from the confinement of a continuous magnetic film. Figure 6.1f shows Δf_0 versus H_0 in different waveguide widths, with the solid and dotted curves representing BVMSW and MSSW, respectively. These results show that an increase in H_0 results in an increase in Δf_0 for MSSW and a decrease for BVMSW.

The ease of fabrication of planar bend magnon waveguides using chemical or ion-beam etching or laser ablation makes it possible to consider various types of waveguide transitions: rounded bend (R_b), sharp corner (S_b), and diagonal connection (D_b). Fabricated crystalline YIG films with a thickness of 10 μm [$Y_3Fe_2(FeO_4)_3$ (111)] with saturation magnetization $4\pi M_s = 1750$ G, epitaxially grown on substrates from gadolinium gallium garnet GGG [$Gd_3Ga_5O_{12}$ (111)] was considered. The ferromagnetic resonance line width for YIG was 0.5 Oe. The films were formed using a local laser ablation system (LLAS) based on a Nd:YAG fiber laser with a 2D galvanometric scanning module (Cambridge Technology 6240H) operating in a pulsed mode with a pulse duration of 50 ns and an energy of $50\,mJ$. The width of the YIG waveguides was $w = 500$ μm bandwidth. Three different types of in-plane bend were fabricated: rounded bend (R_b), sharp corner (S_b), and diagonal connection (D_b). BLS spectroscopy of magnetic materials has been used to detect the propagation of spin waves along plane magnon bends with various R_b, S_b, D_b constructions. The method is based on the effect of inelastic light scattering by coherently excited magnons [28]. Light with a wavelength $532\,nm$ generated by a single-frequency laser (Spectra Physics Excelsior EXLSR-532–200-CDRH) was focused into a spot $25\,\mu m$ and power 1 mW on the surface of the structures under study. The microwave signal from the Anritsu MG3692C signal generator was fed to an input microstrip antenna $30\,\mu m$ to excite the SW. The structure was magnetized with the uniform magnetic field $\mu_0 H_0 = 1200$ Oe generated by a GMW 3472-70 electromagnet and directed along the x-axis to turn on the BVMSW excitation.

Using BLS microscopy, we track the propagation of spin waves through the AB sections of the microwave with different types of connections (right-angle R_b, sharp angle S_b and diagonal connection at $45°D_b$). A precision positioning system was used to obtain maps of the spatial distribution of dynamic magnetization using the BLS method. The experiment was carried out in a quasi-backscattering configuration, with the intensity of the reflected optical signal proportional to the square of the dynamic magnetization $I_{BLS} = \sqrt{(m_x^2 + m_y^2)}$ into the optically probed region. Then the stationary spatial distribution I_{BLS} was obtained for various values of the input frequency of spin waves. Changing the frequency of the input signal allows one to control the mode composition of the spin-wave signal in the B section. For the case of the waveguide junction consisting of a right-angled bend (second column of images in Figure 6.2b and e, the mode composition of the MSSW with increasing frequency propagating in the B section changes [28–30]. At the frequency $f = 5.020$ GHz, for example, the BVMSW excited in the A section propagates through the bend and converts to a MSSW while maintaining its modal composition. By increasing the frequency to $f = 5.050$ GHz, it is possible to instead obtain a two-mode regime of propagation mode of the MSSW [31] (Figure 6.2e).

Spatial distribution maps of the I_{BLS} intensity for the values of the input signal frequency $f_1 = 5.020$ GHz, $f_2 = 5.050$ GHz are shown in Figure 6.2a–f. The structures were always acted upon by an external magnetic field $H_0 = 1200$ Oe directed along the x-axis for effective excitation of the BVMSW in the input section A [26]. The input power was $P_0 = -10$ dBm, ensuring that the SWs were always in a linear excitation and propagation mode. Spatial scanning was carried out over an area of 2.5×5 μm^2.

The design of magnon bends in the form of 45° connections of images in Fig.2c, f are provided for the formation of a superposition of symmetric and antisymmetric modes in section B, which manifests itself in the form of a "snake-like" or "zigzag" pattern of the intensity of spin waves inside B. As shown in Figure 6.1c, the spin-wave spectra inside B consist of the $G_{1,2,3}$ modes. The characteristic snake-like pattern is formed mainly due to the contribution and interference of the co-propagating modes G_1 and G_2 [1, 32–33]. The efficiency of the propagation of spin waves through the bend is mainly determined by two factors, as has already been clarified due to the inhomogeneous profile of the internal magnetic field in the bending region which is shown in [13] for a rounded magnon bend when the orientation of the displacement magnetic field in the plane changes. Another factor is the breaking of translational symmetry, which

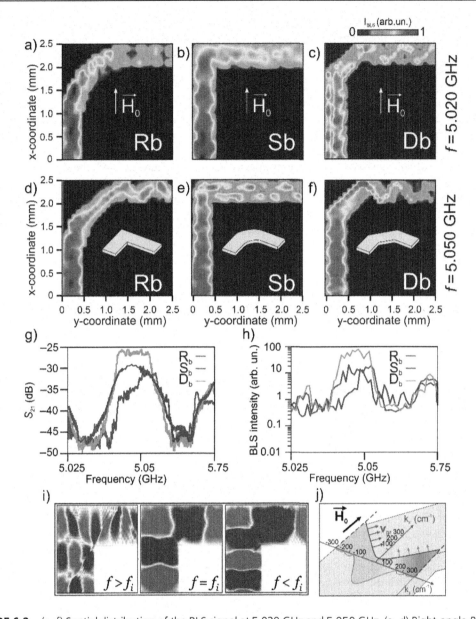

FIGURE 6.2 (a–f) Spatial distribution of the BLS signal at 5.020 GHz and 5.050 GHz. (a–d) Right-angle R_b, (b–e) sharp angle S_b, and (c–f) diagonal connection at 45° D_b. (g) Microwave measurement of the absolute value of frequency transmission coefficient S_{21} of the signal in the output sections B by VNA for bends with different types of connections R_b, S_b, and D_b. (h) Intensity transmission measured by BLS in the B section for bends R_b, S_b, and D_b. (i) Wavelength conversion in AB structure for the cases of $f > f_i$, $f = f_i$ and $f < f_i$. (j) Schematic visualization of isofrequency curves characteristic of SWs propagating across the diagonal junction. The magnetic field H_0 is directed along the x-axis.

leads to the formation of SW width modes with finite transverse wavenumbers. The latter, in particular, determines the conditions for the excitation of higher modes of spin waves in cross-section B.

To understand the influence of these factors on the transmission of a spin wave through a bend, we then concentrate on the transmission spectra obtained using microwave measurements, BLS microscopy, and micromagnetic techniques. Along with BLS, microwave spectroscopy was used to obtain a matrix of complex S-coefficients using a Keysight PNA-X Vector Network Analyzer (VNA). The amplitude-frequency and effective dispersion characteristics were extracted through the frequency dependence of the modulus and argument of the coefficient S_{21} in the case when the signal was excited and detected using a microstrip antenna.

Figure 6.2g shows the results of a microwave measurement of the absolute value of the frequency gain S_{21} of the signal in the output sections B for various bends R_b, S_b, D_b. Here we use a microstrip microwave line to excite and detect the precession of spin waves in the region of short-circuited microstrip transducers. It can be seen that the BVMSW propagating along the input section is transformed into MSSW, and the frequency spectra of the signal power at the output in the R_b and S_b structures are practically equal. In a magnon bend, which consists of two sections interconnected at an angle of 45°, the signal transmission is higher. The BLS method is also suitable for measuring the frequency dependence of the intensity of spin waves in the output region inside B [28, 34]. To plot the frequency dependences for each bend design, we integrate the BLS signal collected in the area of the output transducer that was previously used for BLS measurements (Figure 6.2h). Scanning results using the BLS technique show the best propagation of a spin-wave signal in a structure with a diagonal connection.

Changing the SW wavelength after turning can also affect the propagation length – firstly, due to the different group velocities of the SW wave in sections A and B, and secondly, due to the decrease in the propagation length with decreasing wavelength, since the propagation length of the spin wave is determined by the internal damping of the YIG along with the wavelength (damping per unit wavelength). The situation with a change in the slope of the dispersion is not considered here, since this factor becomes significant only when the width of the YIG waveguide approaches $100\,\mu m$ for a strip thickness of $10\,\mu m$. The second situation with a change in wavelength after SW rotation is manifested in distinct lobes of the frequency dependence S_{21}. Spin-wave signal undergoes a wavelength transformation in accordance with the spectra of width mode dispersion. This is confirmed by numerical simulations shown in Figure 6.2i, which considers three situations: conversion to a longer wavelength region ($f < f_i$), preservation of the wavelength ($f = f_i$), and conversion to a shorter wavelength ($f > f_i$).

One of the interpretations of the change in the efficiency of the transfer of spin waves can be associated with the propagation of the SW beam, which undergoes multiple reflections from the geometric edges of the waveguide, in accordance with the (usually anisotropic) laws of spin-wave optics. Bending of the inhomogeneous profile of the internal magnetic film was considered in [35, 36]. This is accompanied by the propagation of a caustic beam through the junction at an angle of 45° degrees, which is close to the cutoff angle of the magnetostatic spin wave in a tangentially magnetized YIG film [37–39].

Figure 6.2j schematically shows the isofrequency curve for SW propagating in a flat magnetized ferromagnetic YIG film [38]. On the Fig.2j schematically show the spectra of transverse wavenumbers k_y in the case of diffraction of spin waves from the end of the first section of the waveguide [36–38]. This sketch demonstrates that the group velocity of the spin waves points in the same direction as the D_b junction cross-section for wavenumbers satisfying $k_y < -50\,cm^{-1}$ [38].

6.3 MAGNONIC INTERMODAL COUPLER

Theoretical and experimental study of magnetostatic waves propagating through two magnetized channels formed by two rectangular magnetic films of YIG demonstrated a strong influence of spin-wave modes on the transfer of a spin-wave signal from one channel to another. A schematic representation of the investigated structure is shown in Figure 6.3a. In particular, using BLS, it is possible to obtain an image of a two-dimensional map of spin waves propagating in each magnetic stripe to study the intermodal interaction of magnetic stripes (Figure 6.3b). A single-crystal film of YIG with saturation magnetization $4\pi M_s = 1750\,G$, epitaxially grown on a GGG substrate $500\,\mu m$ thick, was used to fabricate waveguides using ablation by local heating by focused laser light. Effective coupling between adjacent waveguides

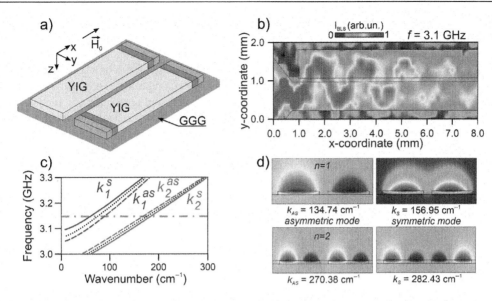

FIGURE 6.3 (a) Schematic of the side-coupled magnetic stripes. (b) Spatial distribution of the BLS signal at 3.134 GHz (c) Dispersion characteristics for the coupled magnetic stripes of width w = 200 μm for first three symmetric (solid lines) and antisymmetric (dashed lines) modes. With the dotted lines, the dispersion of 1st and 2nd modes of a single magnetic stripe of width w = 200 μm are plotted. The dashed-dotted line marks the frequency f = 3.134 GHz, which corresponds to the frequency used in the BLS experiment. (d) Profiles for first (n = 1) and second (n = 2) transverse symmetric (left column) and antisymmetric (right column) modes calculated with the finite element method (FEM) at the frequency of f_1 = 3.134 GHz. All profiles were calculated for H_0 = 600 Oe. The wavenumber of each mode is shown inside each frame.

can be achieved if the propagating spin waves in each strip have the same propagation constants that satisfy the phase-matching condition. Therefore, identical strips were made, separated from each other by a gap d = 40 μm with a width of w = 200 μm and a thickness of t = 10 μm. Laterally located magnetic microwave guides were placed in a uniform magnetic field of H_0 = 1200 Oe, created by an electromagnetic GMW 3472-70 in the plane of the waveguides along the y-axis for efficient excitation of MSSWs using microwave transducers, the arrangement of which is schematically shown in Figure 6.2a. Light with a wavelength of 532 nm, generated by a single-frequency laser (Spectra Physics Excelsior EXLSR-532-200-CDRH), was focused into a spot with a diameter of 25 m and a power of 1 mw on the surface of the structures under study. The microwave signal from an Anritsu MG3692C signal generator was fed to an input microstrip antenna 30 μm wide.

Numerical simulation reveals mode coupling properties and can be used for the geometric design of a flat directional multimode spin-wave coupler for magnetic applications. The dispersion of symmetric and antisymmetric modes and the profiles of transverse modes were calculated by the finite element method (FEM). The symmetric mode in the x-direction corresponds to the case when the amplitudes of the magnetic potentials in two YIG films have the same phase values (along the abscissa), and in the antisymmetric mode, they are phase-shifted by 180°. Dispersion characteristics shown in Figure 6.3c were calculated using the FEM, taking into account the inhomogeneous profile of the internal magnetic field, since the lateral limitation of each YIG band leads to a decrease in the internal magnetic field in the MSSW configuration. We consider only the first two transverse modes of coupled bands: symmetric $k_1^s; k_2^s$ and antisymmetric $k_1^{as}; k_2^{as}$. It is seen that the lower cutoff frequency for symmetric lower-order modes is higher than for other modes. This cutoff frequency determines the beginning of the frequency range of the effective dipole interaction of spin waves. The dispersion for two transverse modes of a single magnetic stripe with a width of w = 200 μm is shown by dashed curves in Figure 6.3c.

It can be seen from the obtained dispersion characteristics that the bond length is longer when operating in a multimode mode than in a single-mode one. Note that the bond length becomes shorter the smaller the distance d between the waveguides. At $d = 0\,\mu m$, the dispersion characteristic of the first symmetric modes reaches the dispersions of the first and second transverse modes of a band with a width of $w = 400\,\mu m$. First, antisymmetric modes in this case pass into the second transverse mode in a waveguide with a width of $w = 400\,\mu m$. In case $d > 0$, both the first symmetric and antisymmetric modes are converted to the first transverse mode of the $200\,\mu m$ wide strip. To show this, we analyzed the mode profiles of the first three modes corresponding to the eigenmodes of coupled magnetic waveguides (Figure 6.3d). The coupling length is longer for high-order modes due to a decrease in the overlapping area of the mode profiles for higher-order modes. In particular, using this configuration of the mode profiles, we can perform the expansion of the modes in terms of the intensity of the spin waves (BLS map). In addition, the propagation constant of the symmetric mode is less than that of the antisymmetric mode for modes of odd order (n_1). On the contrary, for even modes (n_2), the propagation constant of the symmetric mode is greater than that of the antisymmetric mode.

6.4 OUT-OF-PLANE MAGNONIC TRANSPORT

SW signal transmission in the out-of-plane direction is one of the important ways to increase the number of computational elements in the MN. The use of orthogonally connected magnetic stripes provides a simple means for fabricating non-planar waveguides in a three-dimensional structure – for example, meander waveguide [1, 39–40]. Thus, to find out the mechanism of signal transmission through the BC structure (Figure 6.4a) with broken translational symmetry across the yz-plane, we use micromagnetic calculations. The structure is always acted upon by an external magnetic field $H_0 = 1200\,Oe$ directed along the x-axis for effective excitation of the MSSW in the B section. Since the concept of magnon logic [41] is based on the use of spin-wave interference effects, both the amplitude and the phase of the SW in the output section of the [42–43] structure are important.

Compared to the in-plane transition bandwidth, the BC structure is expected to have a larger bandwidth since the SW retains its dispersion type in both sections. Here, we note that for a single waveguide with a length $l_1 + l_2$, the bandwidth of the excited SW can be limited either by the dispersion of the magnetostatic wave or by the antenna width. The first factor gives $\Delta f(w,t) = \gamma \left[H_0 + 2\pi M_s \sqrt{H_0(H_0 + 4\pi M_s}\right] = 542$ MHz for $H_0 = 1200\,Oe$, while the last one gives $\Delta f = 0.5\,GHz$ for an antenna width of $30\,\mu m$ power density P_{out} just before the transition in section B (solid black curve) and in the output section C (red dashed curve). The signal transmission bandwidth is defined as $\Delta f = f_2 - f_1 = 0.63\,GHz$.

To conduct an experimental study of the transfer of spin waves through a vertical junction of magnetic stripes, it was made by the orthogonal connection of two short-circuited microstrip lines with the same grounding shielding electrode (the inset of Figure 6.4a). The magnon waveguides had a width and thickness of $200\,\mu m$ and $10\,\mu m$, respectively. S-parameters were measured by microwave spectroscopy using a VNA connected to a microstrip line. Figure 6.4b shows the signal detected by the microstrip transducer located at the output of section C, while the MSSW is driven in the input section B. The bandwidth estimated from our experimental measurements is on the order of $250\,MHz$, which is less than Δf obtained from micromagnetic calculations shown in Figure 6.4c. This discrepancy between the passbands is due to the presence of a mechanical gap at the junction of two YIG waveguides in the experimental setup. We emphasize, however, that this experimental result serves as proof of the concept of spin-wave transfer at the orthogonal junction of two sections. A more physically real and applicable situation arises, for example, in waveguides in the form of a meander [4] or in studies of BLS and/or microwave radiation [1, 39–40]. The dip in the transmission frequency, shown in Figure 6.4c, can also be associated with an integer number of half waves between the input antenna and the transition region B and C. In particular,

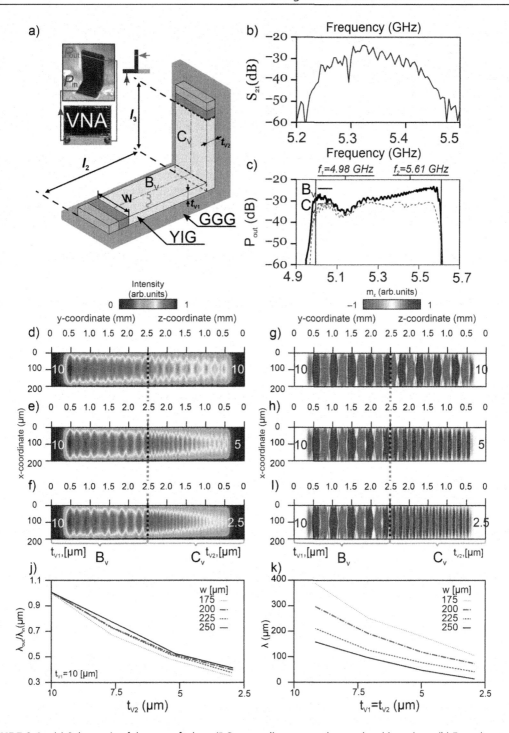

FIGURE 6.4 (a) Schematic of the out-of-plane (BC, extending across the yz-plane) junctions. (b) Experimentally measured SW transmission S_{21}. (c) Numerical results of micromagnetic calculations of the spectral power density P_{out}. (d–f) Spatial distribution of $I(x,y)$ with width $w = 200\,\mu m$. (g–i) Spatial distribution of the $|m_x|^2$ component of dynamic magnetization at a frequency of 5.1 GHz at $w = 200\,\mu m$. (j) Dependence of the wavelength in the C section on the thickness t_{v2} of the out-of-plane waveguide with varying width as indicated. (k) Dependence of the wavelength in the C section on the thickness $t_{v1} = t_{v2}$ with varying width as indicated. All data shown is obtained at 5.1 GHz and $H_0 = 1200\,Oe$.

the dip at 5.3 GHz is caused by the presence of an air gap, which inevitably arises behind the mechanical connection between sections B and C.

Micromagnetic modeling also shows that SW propagates along with a vertical transition without a magnetostatic wave-type transformation since the frequency bandwidth is the same frequency for measuring spectral characteristics at the end of sections B and C. Based on the absence of a spin wave–type transformation when passing through the vertical connection, it can be argued that the change in the design of the bend of the magnon transition outside the plane is not critical, in contrast to the spin-wave transport in planar waveguides with translation symmetry breaking.

At the same time, it is known that the specific features of the structure of the three-dimensional magnon meander associated with SW arise due to the unequal thickness of the vertical and horizontal segments [39]. Let us consider the propagation of a spin wave through a bend BC with a constant thickness of the first segment $t_{V1} = 10\,\mu m$ and a variable thickness t_{V2} (in the range $2.5 < t < 10\,\mu m$) of the second segment C. For this, a continuous (CW) signal $b_0(t) = b_0 sin(2\pi f t)$, where $b_0 = 0.1 Oe$ with a predetermined frequency $f = 5.1 GHz$, is applied along the z-axis in the region in Figure 6.4a. The resulting dynamic magnetization was recorded over a total duration of 300 ns throughout the structure BC. The spatial distribution of the intensity of the dynamic magnetization $I(x,y)$ in section B and $I(z,y)$ in section C, as well as the distribution of the dynamic component of magnetization $|m_z|$, are shown in Figure 6.4d–f and Figure 6.4g–i, respectively. The wavelength of SW propagating along C is shorter than in B. Here we should note that the magnetostatic spin wave type is conserved in BC since the SW wave vector is permanently orthogonal to the static magnetization.

A change in the thickness t_{V2} section C leads to significant changes in the contribution of volume magnetic charges to the dispersion of the propagating MSSWs [4]. At the same time, the fact that the demagnetizing field H_{demag} decreases for thinner waveguides leads to an increase in H_{eff} in section C as t_{V2} decreases, which allows the excitation of SWs with shorter wavelengths at a fixed frequency. This is confirmed by the results of micromagnetic calculations shown in Figure 6.4j, which also shows for narrower bands the conversion factor $\kappa = \dfrac{\lambda_{out}}{\lambda_{in}}$ is lower for higher values t_{V2}. In addition, if the width of both sections B and C increases simultaneously at $t_{V1} = t_{V2}$, the SW wavelength decreases, while increasing the thickness $t_{V1} = t_{V2}$ leads to an increase in the SW wavelength (Figure 6.4k).

It should be noted that a step in thickness in a single YIG waveguide leads to the formation of a region with an inhomogeneous profile of the internal magnetic field, which serves as a source for the generation of short dipolar exchange SWs [44, 45]. A vertically articulated BC waveguide structure with an unequal thickness of each section can also be used to generate shortwave SWs since it can be considered as a scan of the presented waveguide [45]. Thus, the out-of-plane connection of two YIG strips can also be used for the design of predefined characteristics of signal transmission. Our results show that the SW amplitude can be easily estimated using the fact that the SW type is conserved in such a structure.

6.5 THREE-DIMENSIONAL MAGNONIC TOPOLOGY

Here we focus on the structure ABC depicted in Figure 6.5a. Micromagnetic simulations and experimental measurements was used to analyze the signal propagation of the in-plane AB and out-of-plane BC junctions.

Experimental setup of the three-dimensional magnonic junction of three YIG stripes shown on the inserted panel in Figure 6.5a is made of two mechanically connected YIG waveguides. Waveguide sections AB are made in the form of a single L-shaped waveguide formed using the laser ablation method with the width $w = 200\,\mu m$ and the length $l_{1,2} = 2\,mm$ from double-sided YIG/GGG/YIG film. A vertical section C of the same width is mechanically attached to the end of section

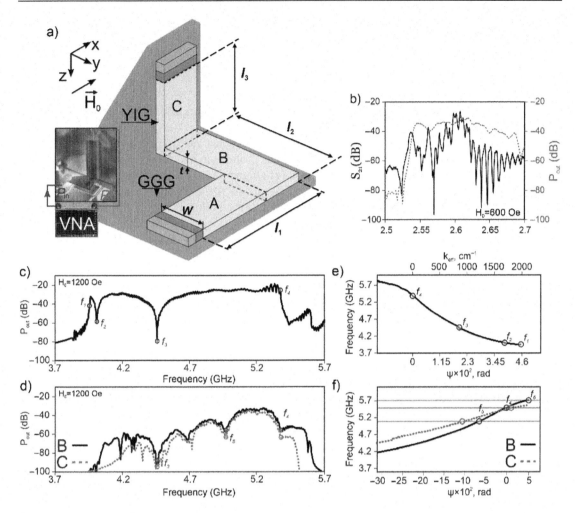

FIGURE 6.5 (a) Schematic of the three-dimensional ABC waveguide structure. (b) The absolute value of the frequency transmission coefficient S_{21} of the signal in the output sections C measured by VNA. Results of micromagnetic calculations of the three-dimensional magnonic junction. (c) Spectral power density $P_{out}(f)$ evaluated at the region indicated in (a) and (e) the corresponding phase $\psi(f)$. (d–f) The same type of result as shown in panels (c) and (e) obtained at different positions shown by a dotted line at the exit of sections A, B, and C (a). The vertical axis corresponds to the frequency and horizontal to the effective wavenumber $k_{eff}(f)$. Data are plotted only for section A where the BVMSW is excited and propagated. The uniform magnetic field of $H_0 = 1200\,Oe$ is directed along the long axis of section A throughout.

B. The experimental transmission coefficient S_{21} (Figure 6.5b, solid curve) using a vector analyzer was obtained with external magnetization by the field $H_0 = 600\,Oe$ demonstrates bandwidths up to $100\,MHz$. Micromagnetic modeling with the same external magnetization field demonstrates a similar form of the transmission coefficient (Figure 6.5b, dotted curve) in the same frequency range. The observed dips in the experimental data were associated with the air gap at the connection with the out-of-plane section. This reveals itself with the spin-wave reflection from the interconnection area and should be investigated additionally.

Next, the spectral characteristics of the SW transport across the three-dimensional junction ABC were obtained only in micromagnetic calculations. Throughout, the structure was subject to a uniform magnetic field $H_0 = 1200\,Oe$ directed along the x-axis to efficiently excite BVMSWs along section A.

The power spectral density of the output signal is extracted from the time-resolved calculations, obtained across the cross-sections at the outputs of each of the A, B, and C sections using the Fourier transform. The effective wavenumbers were also calculated in the frequency range in which BVMSW and MSSW can exist, as $k_{eff} = \psi_f - \psi_0(f)/l_s$ where $\psi(f) = \int_0^{l_s} k(o) do$, $(o = x, y, z)$ is the SW phase shift in the corresponding section relative to the initial phase at the source.

Figure 6.5c shows the power spectral density BVMSW $P_{out}(f)$ obtained at the output of section A. The bandwidth is $f_4 - f_1 = 1.424$ GHz. High-frequency oscillations P_{out}, close to f_4, arise due to the presence of an integer number of half waves between the input antenna and the junction of sections A and B. The dips at frequencies f_2 and f_3 and the sharp drop at f_1 are related to the antenna width used in the simulation. The frequency range above f_4 corresponds to higher-order BVMSW modes. As Figure 6.5e shows in section A, the $k_{eff}(f)$ dependence corresponds to the BVMSW dependence and qualitatively coincides with the dispersion dependence for F_1. The power spectral density of the MSSW obtained in sections B and C (Figure 6.5f) shows the presence of a region is $f_5 < f < f_4$, where the effective propagation of MSSW and BVMSW occurs. It should be noted that the bandwidth of the signal through the ABC is determined mainly by the frequency range in which the MSSW and BVMSW coexist. The dependence $k_{eff}(f)$ cannot be extracted at the end of segment B, since the variance in each of the sections has a different gradient. At the same time, the phase shift for the signal found at the end of section B can be directly extracted from the simulation (Figure 6.5f). Finally, the frequency dependence of the phase $f(\psi)$ has a positive slope, as in the MSSW modes. Note that the shape $f(\psi)$ here corresponds to the sum of the phase shift in both A and B.

In the region below the f_5 frequency, modes with a wider MSSW can be distinguished in the form of lobes in the spectra, as in [13]. The phase versus frequency dependence at the end of section C also has a positive slope, as for the MSSW. The 3D magnon junction design, based on two bends, enables frequency selective wave propagation with intrinsic attenuation originating primarily from the material (for YIG, 20 dBm path loss from section A to the end of section C). At the same time, the phase shift for this structure can be easily calculated, and the predetermined phase-frequency characteristics also allow for construction, for example, for a Mach-Zehnder interferometer for a three-dimensional magnon scheme.

6.6 CONCLUSION

In conclusion, the next steps towards the assembling of the functional units into magnonic networks could be performed with the simultaneous use of spin-wave couplers and structures with broken translational symmetry. This approach provides additional advantages of non-volatility and low operating power functional spin-wave devices. The expansion of magnonic networks into the third dimension using the versatile interconnection elements reduces the area occupied by individual magnonic blocks. Microwave spectroscopy along with space-resolved Brillouin light scattering helps elucidate the mechanisms of spin-wave transport in the interconnection elements. Micromagnetic simulation is a fast alternative to experiments. A simulation was performed to adjust the spin-wave scattering losses and provide the optimization of device shape to ensure feasibility. Experimental methods and micromagnetic simulations are in good agreement with each other.

The mechanisms associated with the transfer of a spin-wave signal in a multilevel topology of a magnonic network are based on the conversion of the type of magnetostatic spin wave from surface to backward volume and vice versa. At the same time, the lateral L-shape unit performs the conversion of spin-wave wavelength. The directional coupler acts as the spatial-frequency selective device. The most prominent design of magnonic networks based on three-dimensional magnonic structures could be realized with a combination of these principles (Figure 6.6).

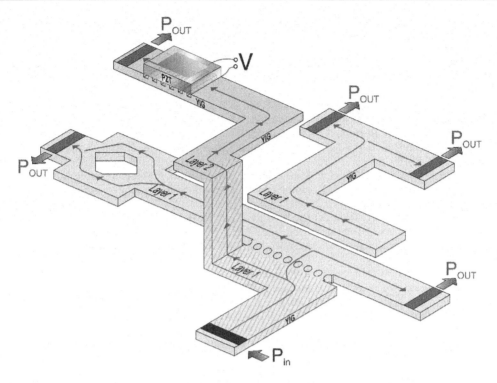

FIGURE 6.6 The concept of 3D magnonic networks with the interconnection of magnonic units in the form of magnonic couplers and structures with broken translational symmetry.

6.7 ACKNOWLEDGMENTS

This work was supported by the Russian Science Foundation (No. 20–79–10191).

REFERENCES

1. D. Stancil and A. Prabhakar, *Spin Waves: Theory and Applications*, Springer, New York, 2009.
2. A. V. Chumak, V. I. Vasyuchka, A. A. Serga, and B. Hillebrands, "Magnon spintronics", *Nat. Phys.*, vol. 11, pp. 453–461, 2015.
3. S. A. Nikitov, D. V. Kalyabin, I. V. Lisenkov, A. N. Slavin, Yu. N. Barabanenkov, S. A. Osokin, A. V. Sadovnikov, E. N. Beginin, M. A. Morozova, Yu. P. Sharaevsky, Yu. A. Filimonov, Yu. V. Khivintsev, S. L. Vysotsky, V. K. Sakharov, and E. S. Pavlov, "Magnonics: A new research area in spintronics and spin wave electronics", *Phys. Usp.* vol. 58, no. 10, 2015.
4. A. V. Sadovnikov, S. A. Nikitov, E. N. Beginin, S. E. Sheshukova, Yu. P. Sharaevskii, A. I. Stognij, N. N. Novitski, V. K. Sakharov, and Yu. V. Khivintsev, "Route toward semiconductor magnonics: Light-induced spin-wave nonreciprocity in a YIG/GaAs structure", *Phys. Rev. B* vol. 99, p. 054424, 2019.
5. M. Evelt, V. E Demidov, V. Bessonov, S. O. Demokritov, J. L. Prieto, M. Muñoz, J. Ben Youssef, V. V. Naletov, G. de Loubens, O. Klein, M. Collet, K. Garcia-Hernandez, P. Bortolotti, V. Cros, and A. Anane, "High-efficiency control of spin-wave propagation in ultra-thin yttrium iron garnet by the spin-orbit torque", *Appl. Phys. Lett.*, vol. 108, p. 172406, 2016.

6. D. V. Kalyabin, A. V. Sadovnikov, E. N. Beginin, and S. A. Nikitov, "Surface spin waves propagation in tapered magnetic stripe", *Journal of Applied Physics*, vol. 126, p. 173907, 2019.
7. T. W. O'Keeffe and R. W. Patterson, "Magnetostatic surface-wave propagation in finite samples", *J. Appl. Phys.*, vol. 49, pp. 4886–4895, 1978.
8. S.N. Bajpai, "Excitation of magnetostatic surface waves: Effect of finite sample width", *J. Appl. Phys.*, vol. 58, p. 910, 1985.
9. E. N. Beginin, A. V. Sadovnikov, Yu. P. Sharaevskii, and S. A. Nikitov, "Multimode surface magnetostatic wave propagation in irregular planar YIG waveguide", *Solid State Phenomena*, vol. 215, pp. 389–393, 2014.
10. A. V. Sadovnikov, E. N. Beginin, S. E. Sheshukova, D. V. Romanenko, Y. P. Sharaevskii, and S. A. Nikitov, "Directional multimode coupler for planar magnonics: Side-coupled magnetic stripes", *Appl. Phys. Lett.*, vol. 107, p. 202405, 2015.
11. A. V. Sadovnikov, A. A. Grachev, S. E. Sheshukova, Y. P. Sharaevskii, A. A. Serdobintsev, D. M. Mitin, and S. A. Nikitov, "Magnon straintronics: Reconfigurable spin-wave routing in strain-controlled bilateral magnetic stripes", *Phys. Rev. Lett.*, vol. 120, p. 257203, 2018.
12. Q. Wang, M. Kewenig, M. Schneider, R. Verba, F. Kohl, B. Heinz, M. Geilen, M. Mohseni, B. Lägel, F. Ciubotaru, C. Adelmann, C. Dubs, S. D. Cotofana, O. V. Dobrovolskiy, T. Brächer, P. Pirro, and A. V. Chumak, "A magnonic directional coupler for integrated magnonic half-adders", *Nat. Electron.*, vol. 3, pp. 765–774, Dec. 2020.
13. A. V. Sadovnikov, S. A. Odintsov, E. N. Beginin, S. E. Sheshukova, Yu. P. Sharaevskii, and S. A. Nikitov, "Toward nonlinear magnonics: Intensity-dependent spin-wave switching in insulating side-coupled magnetic stripes", *Phys. Rev. B* vol. 96, p. 144428, 2017.
14. Yu. P. Sharaevsky, A. V. Sadovnikov, E. N. Beginin, M. A. Morozova, S. E. Sheshukova, A. Yu. Sharaevskaya, S. V. Grishin, D. V. Romanenko, and S. A. Nikitov, "Coupled spin waves in magnonic waveguides", Chapter 2 of the *Book "Spin Wave Confinement: Propagating Waves* (2nd edition) (Edited by Sergej Demokritov). Pan Stanford Publishing Pte. Ltd, 2017.
15. A. V. Sadovnikov, S. A. Odintsov, E. N. Beginin, S. E. Sheshukova, Yu. P. Sharaevskii, and S. A. Nikitov, "Spin-wave switching in the side-coupled magnonic stripes", *IEEE Transactions on Magnetics*, vol. 53, Issue 11, pp. 1–4, 2017.
16. A. B. Ustinov, E. Lähderanta, M. Inoue, and B. A. Kalinikos, "Nonlinear spin-wave logic gates", *IEEE Magnetics Lett.*, vol. 10, p. 5508204, Oct. 2019.
17. A. V. Sadovnikov, C. S. Davies, S. V. Grishin, V. V. Kruglyak, D. V. Romanenko, Y. P. Sharaevskii, and S. A. Nikitov, "Magnonic beam splitter: The building block of parallel magnonic circuitry", *Appl. Phys. Lett.*, vol. 106, pp. 192406, May 2015.
18. A. V. Sadovnikov, C. S. Davies, V. V. Kruglyak, D. V. Romanenko, S. V. Grishin, E. N. Beginin, Y. P. Sharaevskii, and S. A. Nikitov, "Spin wave propagation in a uniformly biased curved magnonic waveguide", *Phys. Rev. B*, vol. 96, p. 060401, Aug. 2017.
19. G. Gubbiotti, A. Sadovnikov, E. Beginin, S. Nikitov, D. Wan, A. Gupta, S. Kundu, G. Talmelli, R. Carpenter, I. Asselberghs, I. P. Radu, C. Adelmann, and F. Ciubotaru, "Magnonic band structure in vertical meander-shaped Co40Fe40B20 thin films", *Phys. Rev. Appl.*, vol.15, p. 014061, 2021.
20. G. Gubbiotti, A. Sadovnikov, E. Beginin, S. Sheshukova, S. Nikitov, G. Talmelli, I. Asselberghs, I. P. Radu, C. Adelmann, and F. Ciubotaru, "Magnonic band structure in CoFeB/Ta/NiFe meander-shaped magnetic bilayers", *Appl. Phys. Lett.*, vol. 118, p. 162405, 2021.
21. V. K. Sakharov, E. N. Beginin, Y. V. Khivintsev, A. V. Sadovnikov, A. I. Stognij, Y. A. Filimonov, and S. A. Nikitov, "Spin waves in meander shaped YIG film: Toward 3D magnonics", *Appl. Phys. Lett.*, vol. 117, p. 022403, 2020.
22. G. Gubbiotti and Sergej A. Nikitov, "Three-dimensional magnonics, the 2021 Magnonics Roadmap", *Journal of Physics: Condensed Matter*, vol. 33, pp. 413001, 2021.
23. G. Venkat, H. Fangohr, and A. Prabhakar, "Absorbing boundary layers for spin wave micromagnetics", *Journal of Magnetism and Magnetic Materials*, vol. 450, p. 34–39, 2018.
24. M. Dvornik, A. N. Kuchko, and V. V. Kruglyak, "Micromagnetic method of s-parameter characterization of magnonic devices", *Journal of Applied Physics*, vol. 109(7), 2011.
25. J. R. Eshbach and R. W. Damon, "Surface magnetostatic modes and surface spin waves", *Phys. Rev.*, vol. 118, p. 1208, 1960.
26. A. G. Gurevich and G. A. Melkov, *Magnetization Oscillations and Waves*. CRC Press, 2020.
27. K. Y. Guslienko and A. N. Slavin, "Boundary conditions for magnetization in magnetic nanoelements", *Physical Review B*, vol. 72(1), p. 014463, 2005.
28. S. Demokritov, "Brillouin light scattering studies of con-fined spin waves: Linear and nonlinear confinement", *Physics Reports*, vol. 348, p. 441, 2001.

29. O. Karlqvist, "Calculation of the magnetic field in the ferromagnetic layer of a magnetic drum", *Trans. Roy. Inst. Techno*, vol. 86(3), 1954.
30. J. Jorzick, C. Krämer, S. O. Demokritov, B. Hillebrands, B. Bartenlian, C. Chappert, D. Decanini, F. Rousseaux, E. Cambril, E. Sondergard, M. Bailleul, C. Fermon, and A. N. Slavin, "Spin wave quantization in laterally confined magnetic structures", *Journal of Applied Physics*, vol. 89, p. 7091, 2001.
31. V. E. Demidov, S. O. Demokritov, K. Rott, P. Krzys-teczko, and G. Reiss, "Linear and nonlinear spin-wavedynamics in macro- and microscopic magnetic confined structures", *Journal of Physics D: Applied Physics*, vol. 41, p. 164012, 2008.
32. O. Biittner, M. Bauer, C. Mathieu, S.O. Demokritov, B. Hillebrands, P.A. Kolodin, M.P. Kostylev, S. Sure, H. Dotsch, V. Grimalsky, Yu. Rapoport, and A. N. Slavin, "Mode beating of spin wave beams in ferrimagnetic Lu/sub 2.04/Bi/sub 0.96/Fe/sub 5/O/sub 12/films", *IEEE Transactions on Magnetics*, vol. 34(4), pp. 1381–1383, 1998.
33. A. V. Sadovnikov, A. A. Grachev, S. E. Sheshukova, Y. P. Sharaevskii, A. A. Serdobintsev, D. M. Mitin, and S. A. Nikitov, "Magnon straintronics: Reconfigurable spin-wave routing in strain-controlled bilateral magnetic stripes", *Physical Review Letters*, vol. 120(25), p. 257203, 2018.
34. T. Sebastian, K. Schultheiss, B. Obry, B. Hillebrands, and H. Schultheiss, "Micro-focused Brillouin light scattering: Imaging spin waves at the nanoscale", *Frontiers in Physics*, vol. 3, p. 35, 2015.
35. K. Vogt, F. Fradin, J. Pearson, T. Sebastian, S. Bader, B. Hillebrands, A. Hoffmann, and H. Schultheiss, "Realization of a spin-wave multiplexer", *Nature Communications*, vol. 5, 2014.
36. T. Schneider, A. A. Serga, A. V. Chumak, C. W. Sandweg, S. Trudel, S. Wolff, M. P. Kostylev, V. S. Tiberkevich, A. N. Slavin, and B. Hillebrands, "Non-diffractive subwavelength wave beams in a medium with externally controlled anisotropy", *Physical Review Letters*, vol. 104, 2010.
37. A. V. Vashkovsky and E. H. Lock, "Properties of backward electromagnetic waves and negative reflection in ferrite films", *Physics-Uspekhi*, vol. 49, p. 389, 2006.
38. E. H. Lock, "The properties of isofrequency dependencesand the laws of geometrical optics", *Physics-Uspekhi*, vol. 51, p. 375, 2008.
39. V. K. Sakharov, E. N. Beginin, Y. V. Khivintsev, A. V. Sadovnikov, A. I. Stognij, Y. A. Filimonov, and S. A. Nikitov, "Spin waves in meander shaped YIG film: To-ward 3D magnonics", *Applied Physics Letters*, vol. 117, p. 022403, 2020.
40. I. Turčan, L. Flajšman, O. Wojewoda, V. Roučka, O. Man, and M. Urbánek, "Spin wave propagation in corrugated waveguides", *Applied Physics Letters*, vol. 118(9), p. 092405, 2021.
41. A. Khitun, M. Bao, and K. L. Wang, "Magnonic logic circuits", *J. Phys. D: Appl. Phys*, vol. 43, p. 264005, 2010.
42. M. P. Kostylev, A. A. Serga, T. Schneider, B. Leven, and B. Hillebrands, "Spin-wave logical gates", *Applied PhysicsLetters*, vol. 87, p. 153501, 2005.
43. K.-S. Lee and S.-K. Kim, "Conceptual design of spin wavelogic gates based on a Mach – Zehnder-type spin wave interferometer for universal logic functions", *Journal of Ap-plied Physics*, vol. 104, p. 053909, 2008.
44. E. N. Beginin, A. V. Sadovnikov, A. Y. Sharaevskaya, A. I. Stognij, and S. A. Nikitov, "Spin wave steering in three-dimensional magnonic networks", *Applied PhysicsLetters*, vol. 112, p. 122404, 2018.
45. S. L. Vysotskii, A. V. Sadovnikov, G. M. Dudko, A. V. Kozhevnikov, Y. V. Khivintsev, V. K. Sakharov, N. N. Novitskii, A. I. Stognij, and Y. A. Filimonov, "Spin-wavegeneration at the thickness step of yttrium iron garnet film", *Applied Physics Letters*, vol. 117, p. 102403, 2020.

Nanostructured Hybrid Magnetic Materials

7

Sha Yang and Wei Liu
School of Materials Science and Engineering, Nanjing University of Science and Technology, Nanjing, China

Contents

7.1	Introduction	111
7.2	Ferromagnetic Film With Semiconductors	112
	7.2.1 Ferromagnetic Metal Film on a Semiconductor	112
	7.2.2 Half-Metallic Magnet on Semiconductors	113
7.3	2D Material With Ferromagnetic Material	115
	7.3.1 Graphene-Based Hybrid Magnetic Material	115
	7.3.2 TMDs-Based Hybrid Magnetic Material	116
7.4	Organic With Ferromagnetic Metal	118
	7.4.1 Nonmagnetic Organic Semiconductors–Based Spinterface	118
	7.4.2 Molecular Magnet–Based Spinterface	120
7.5	Conclusions and Perspectives	121
References		122

7.1 INTRODUCTION

The development in spintronic devices has promoted high requirements for small-size and more versatile magnetic materials. The hybrid spintronics architectures, formed by integrating magnetic materials and nonmagnetic spin hosting materials, have achieved much attention in developing novel spin transistors and logic units. The nonmagnetic materials used in such a hybrid system involve bulk, two-dimensional (2D), and molecular semiconductors (SC), while the magnetic materials include metallic or half-metallic magnets. Compared to traditional magnets, this hybrid architecture combines the advantages of both sides, such as the high Curie temperature and high spin polarization of ferromagnetic (FM) metals and the controllable charge carriers and more tunability of SCs. This combination may provide new types of control over the spins and novel spintronic properties that cannot be achieved in traditional magnetic materials.

In this chapter, we explore the fabrication, structure, interfacial hybridization, and spintronic properties for the hybrid magnetic materials, including FM/bulk SCs, FM/2D material, and FM/organic systems. We briefly introduce the history and recent progress for these hybrid systems, highlighting the key parameters for spintronic devices, such as the interfacial spin polarization, spin injection efficiency, spin relaxation, Curie temperature, and spin-orbital coupling effect. Finally, we summarize some now-existing challenges. Limited by the space and references, this chapter covers some major and general aspects. Fortunately, some excellent reviews [1–3] for various types of hybrid magnetic materials can be referred.

7.2 FERROMAGNETIC FILM WITH SEMICONDUCTORS

Benefiting from the mature molecular beam epitaxy (MBE) technique, the thin-film growth of FM on SCs can be realized. This has promoted extensive investigations on the hybrid FM/SC systems since its first fabrication in 1981 [4]. The SCs exhibit long spin relaxation time and coherence length due to the weaker spin-orbit coupling than the traditional metal used in spintronic devices [5]. Notably, the spin signals through the SCs can be actively controlled by an external field, e.g., a well-defined gate voltage. These advantages provide the feasibility of SCs as an excellent spin hosting medium. The FM/SC hybrid system can present high Curie temperature and easy integration with current magnetic technologies. To date, the MBE and sputtering techniques are the most commonly used growth method to synthesize high-quality hybrid FM/SC structures. Here, we briefly introduce the fabrication, interfacial structure, and spintronic properties of several typical FM/SC systems.

7.2.1 Ferromagnetic Metal Film on a Semiconductor

Ferromagnetic thin film on single-crystal SC substrates has the longest history in hybrid magnetic systems. During the last decades, hybrid systems, such as Fe/GaAs, Fe/InAs, Co/GaAs, Ni/GaAs, have been developed. Among them, Fe/GaAs(001) is one of the most promising hybrid magnetic systems due to a small mismatch in lattice parameters between Fe (2.866 Å) and GaAs (5.654 Å). The Fe/GaAs(001) system was first fabricated in 1987 by the MBE technique [6]. Before epitaxial growth of Fe film, the bulk GaAs substrate was preprocessed by chemo-mechanical polishing to obtain a clean and high-quality surface. Auger electron spectroscopy (AES) and reflection high-energy electron diffraction (RHEED) were used to manage the interfacial structure and film thickness during the growth process. It was observed that the Fe film grows as single-crystal α-Fe with the (100) surface and with the in-plane <100> axes of the Fe and GaAs. When the thickness of Fe film was larger than 30 nm, the continuous streaks in RHEED patterns disappeared. This indicates that the flatter structure of the Fe film can be destroyed when the thickness is too large.

The atomic interfacial structure can be observed by high-angle annular dark-field scanning transmission electron microscopy (HAADF-STEM) [7]. As shown in Figure 7.1a, two typical atomic structures gave been observed at the Fe/GaAs(001) interface. Region I and II denote an abrupt structure and a partially intermixed structure, respectively. The detailed spacing structure can be seen in the left panel of Figure 7.1b. Different from structure I, a Fe atom is inserted to the middle position between the two As atoms in structure II. This can be evidenced by the intensity profiles in Figure 7.1b, which follows a $Z^{1.5}$ to Z^2 dependency for structure II [8]. The formation of interface structures I and II could arise from the different surface reconstructions before Fe epitaxial growth. The distinct interfacial spacing would result in varied spin transport. Ab initio calculations show that the injection efficiency of minority carriers can be enhanced in the partially mixing regions. This indicates a positive relationship between the interface

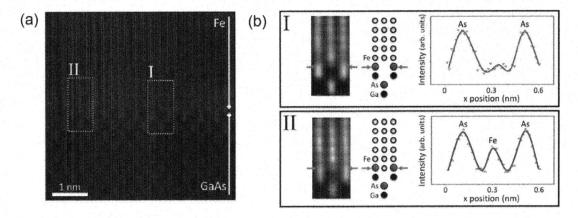

FIGURE 7.1 Interfacial structure at the Fe/GaAs(001) interface. Adapted with permission from [7]. Copyright (2013) American Physical Society.

abruptness and the spin injection efficiency. The contact between Fe and GaAs presents a significant influence on the magnetic anisotropy from the <100> easy axes of bulk body-centered cubic (bcc) Fe to an in-plane uniaxial magnetic anisotropy (UMA) along the [110] direction. This variation [9] is caused by an interface anisotropy.

The magnetization manipulation by electric current, termed the interfacial spin-orbit torque (SOT), has been a promising way in developing high-density nonvolatile memory and logic units with high thermal stability. The hybrid interfaces formed by the FM film and spin hosting materials with strong spin-orbit coupling (SOC) are the basic structure for spin-to-charge conversion. In early times, the SOT phenomenon has been mainly studied for the interface formed by ferromagnetic and nonmagnetic metals due to their strong spin-orbit interaction. Interestingly, a strong spin-orbital field was also obtained at the Fe/GaAs interface. Chen et al. [10] observed a robust SOT effect for the Fe/GaAs interface at room temperature. Both ferromagnetic exchange and spin-orbital interactions exist at the interface. In particular, the ferromagnetic exchange contributes to the generation of pure spin currents, while the SOI effect contributes to the spin-to-charge conversion. By spin-orbit ferromagnetic resonance and spin pumping, the spin current and charge can be mutually converted at room temperature. Notably, the magnitudes of such an interfacial SOT effect can be comparable to that at FM/nonmagnetic metal interfaces. Besides, within the Fe/GaAs(001) interface, the interplay of Rashba and Dresselhaus spin-orbit interaction leads to anisotropic magnetic damping with a twofold C_{2v} symmetry [11], which has been demonstrated by the anisotropic density of states at the hybrid interface.

7.2.2 Half-Metallic Magnet on Semiconductors

Despite huge advances on the FM/SC interfaces, the spin injection efficiency from the FM to SC is still low. This substantially arises from the huge conductivity mismatch between the spin-dependent and low resistance of FM and the dramatically larger spin-independent resistance of the SC. As a result, the spin polarization of the injected current into the SCs is usually smaller than 1%. To overcome this problem, the half-metallic magnets come into view – i.e., half-metallic oxides and Heusler alloys. The spin polarization at the Fermi level of a pure FM usually ranges from 0 to 1. When it comes to the half metals, either the spin-up or the spin-down states are empty at the Fermi level, which leads to a spin polarization up to 1. Varying with the direction of the spins, the half-metallic magnets can be conductive and insulate. In addition to larger spin polarization, the half-metallic magnets usually present high Curie temperature. This makes the half-metallic magnets excellent alternatives in designing hybrid architectures.

Fe_3O_4 is a typical example of a half-metallic oxide magnet, which has high polarization (approaching 100%) at the Fermi level and high Curie temperatures (up to 858 K). Additionally, Fe_3O_4 is one of the most stable iron oxides. These advantages enable Fe_3O_4 to be an excellent magnet to inject highly spin-polarized carriers into the SCs. Up to date, the Fe_3O_4 film has been successfully fabricated on many types of SCs, such as GaAs, InAs, Si, and GaN. Among them, the most investigated system is the Fe_3O_4/GaAs structure. The lattice constant of GaAs and Fe_3O_4 are 5.654 and 8.396 Å, respectively. In contrast to the aforementioned Fe/GaAs interfaces, when the Fe_3O_4 film epitaxially grows on the GaAs(100) substrate, there is a rotation of the Fe_3O_4 cell by 45° relative to GaAs(100). The is due to the fact that the (100)<011> directions of Fe_3O_4 present a better lattice match with GaAs(100)<010> directions than (100)<010> directions [12].

The magneto-optical Kerr effect (MOKE) loops of Fe_3O_4/GaAs(100) demonstrated a dependence of the uniaxial magnetic anisotropy on the epitaxial thickness. The Fe_3O_4 film with thickness less than 1.0 nm presents a nonmagnetic characteristic. When the thickness ranges from 1.0 to 2.0 nm, the films are superparamagnetic. When Fe_3O_4 film thickness is in the interval [2.0, 6.0 nm], a UMA can be observed. Increasing the film thickness up to 6 nm, the easy axis of the Fe_3O_4 film gradually rotates to [1] direction from [0–11], which is perpendicular to that in Au/Fe/GaAs systems.

Another type of commonly used half-metallic magnets, Heusler alloys, are intermetallic compounds with a formula of X_2YZ with $L2_1$ structure or XYZ with Cl_b structure, where X and Y are transition metals and Z is the main group elements such as Si, Ge, or a nonmagnetic metal. The discovery of half-metallic magnets originates from the early studies of Heusler alloys. The NiMnSb was the first verified half-metallic Heusler alloy by Groot and coworkers in 1983 [13]. The spin polarization for bulk NiMnSb can be up to 100%. However, the surface spin polarization significantly varies with the specific structure and composition, which arise from the different oxidation or repeated sputter-anneal cycles. For example, the enrichment of the Mn element at the surface dramatically decreases the spin polarization. In this context, a high-quality NiMnSb surface must be used for forming a hybrid FM/SC interface. Borghs et al. [14] found that the flux of the Sb elements has a significant influence on the surface structure during the epitaxial growth of NiMnSb on GaAs. When the Sb beam-equivalent-pressure flux $\varphi_{Sb} = 10\ \varphi_{Ni}$, high-quality NiMnSb films were achieved with magnetization near the bulk value.

The NiMnSb/GaAs interface has achieved much attention due to a relatively small lattice mismatch (the bulk lattice constants of NiMnSb and GaAs are 5.903 and 5.653 Å, respectively) [15]. The epitaxial growth of NiMnSb film on GaAs shows a cube-on-cube orientation relationship, where the (001) and [100] of NiMnSb are parallel to that of GaAs. Using the sputtering method, Zhao et al. [16] studied the interfacial structures and spintronic properties of NiMnSb/GaAs hybrid magnets. The NiMnSb film shows a tetragonal distortion for the epitaxial growth. However, the sign of the distortion is opposite to the lattice mismatch between NiMnSb and GaAs. For the distorted NiMnSb, the lattice constants a = b = 5.937 ± 0.002 Å and c = 5.921 ± 0.002 Å. This may result from the periodical misfit dislocations near the NiMnSb/GaAs interface. Notably, the tetragonal distortion of the NiMnSb file induces spin-orbit interaction within the crystal. Interestingly, the SOT is five times larger than those structures by MBE. In particular, the sputtered NiMnSb presents both field-like and damping-like torque, while the MBE-grown NiMnSb only shows field-like torque. Notably, the effective field generation efficiency of sputtered NiMnSb is comparable to those reported from heavy metal/ferromagnet systems [17].

In contrast to XYZ Heusler alloys, the X_2YZ form presents two magnetic sublattices, where the two X atoms occupied the tetrahedral lattice sites and induce magnetic interactions between these two X atoms. Due to the two different magnetic sublattices, the X_2YZ Heusler compounds can show ferromagnetism, ferrimagnetism, and half-metallic ferromagnetism. When the full Heusler compounds become half metallic, a fully spin-polarized density of states at the Fermi level can be obtained. Fe_2MnSi is a typical X_2YZ half-metallic Heusler compound. However, epitaxial Fe_2MnSi thin film has a Curie temperature of 210 K, which is lower than the room temperature. Miyao and coworkers [18] found that both the metallicity and Curie temperature can be rationally tailored by controlling the composition of Fe and Mn. The $Fe_{3-x}Mn_xSi$ alloys with compositions in the range $0.75 \leq x \leq 1.5$ exhibit a half-metallic feature. When x is in the interval [0.5, 0.75], a high spin polarization larger than 0.9 can be obtained. However, the increase of Fe composition leads to significantly decreased Curie temperature. For example, Fe_3Si has a high Curie temperature of more than 800 K with a spin polarization of 45±5%. Therefore, to obtain a half-metallic magnet with both considerable spin polarization and Curie temperature, the key is to control the composition of metal X and Y. For example, $Fe_{0.6}Mn_{2.4}Si$ shows a Curie temperature above 300 K and a spin polarization larger than 0.9.

7.3 2D MATERIAL WITH FERROMAGNETIC MATERIAL

Compared with the traditional bulk SCs, the 2D materials present exclusive advantages, such as well-controlled atomic layer thickness and near-perfect surface without dangling bonds. The combination of 2D layers and FMs can help establish the hybrid architecture and manipulate the spintronic properties at an atomic level, which provides new opportunities in spintronics. Such an interface can be prepared by the chemical vapor deposition (CVD) or mechanical exfoliation technique of 2D layers on the FM substrate. Here, we discuss the structural and spintronic properties arising from the hybridization for several typical 2D materials on FM surface – i.e., graphene and transition metal dichalcogenides (TMDs).

7.3.1 Graphene-Based Hybrid Magnetic Material

Graphene has achieved much attention in spintronics since it was fabricated in 2004 [19]. The graphene exhibits well-controlled atomic-level thickness, long spin diffusion length, and weak spin-orbital coupling. This motivates the search for graphene-based spintronic phenomena, e.g. interfacial spin injection, perpendicular magnetic anisotropy (PMA), quantum spin Hall effect, and tunneling magnetoresistance [20–21].

The graphene spin valve was first fabricated by Hill et al., where the NiFe electrode was used for injecting the spin-polarized carriers [22]. The graphene monolayer was prepared by exfoliation of highly oriented pyrolytic graphite. A 10% magnetoresistance was achieved as the magnetization of the electrodes was switched from the parallel to the antiparallel state at room temperature. The obtained magnetoresistance is significantly larger than that in the NiFe electrodes. This demonstrates the potential of graphene as the nonmagnetic conductor in the spin valve.

The integration of graphene with a Ni and Co surface are more commonly studied systems due to a lattice mismatch of less than 2% and direct epitaxial growth of graphene on the Ni and Co surface. The conductance for parallel and antiparallel orientations of Ni magnetizations are shown in Figure 7.2. The magnetoresistance of such an interface can be up to 100%. Similar magnetoresistance values can also be obtained for the $Ni/Gr_n/Co$ and $Co/Gr_n/Co$ hybrid systems. The magnetoresistance of such a hybrid system significantly depends on the number of the graphene layer below five layers [23]. The interfacial bonding between graphene and Ni and the number of graphene layers has a significant influence on the spin injection. An ideal $Ni/Gr_n/Ni$ junction (n denotes the number of graphene layers) can provide 100% spin injection. However, in practice, the interface structure would be rough, which would decrease spin injection efficiency from 100% to 70%.

Similar to the FM/SC hybrid systems, there is also a conductivity mismatch problem that would decrease the spin injection efficiency into graphene. This can be solved by inserting Al_2O_3 layers into graphene/FM to increase the spin-dependent resistance. As exemplified by graphene/Al_2O_3/Co four-terminal contact devices [24], the spin polarization of 0.1 can be obtained at room temperature. Given the low intrinsic spin-orbit interaction in graphene layers, the spin transport for graphene/Al_2O_3/Co can be up to micrometer-scale distances. By utilizing a magnetic field, switched spin signals were achieved, which are insensitive to the temperature. By improving the fabrication techniques, such as decreasing the impurity and defect, the spin relaxation length and relaxation times can be further modified.

Although the graphene shows a weak SOC effect, the graphene on Co films can significantly improve the PMA up to twice that of pristine Co films [25]. The scale of the PMA linearly increases as a function

116 Fundamentals of Low Dimensional Magnets

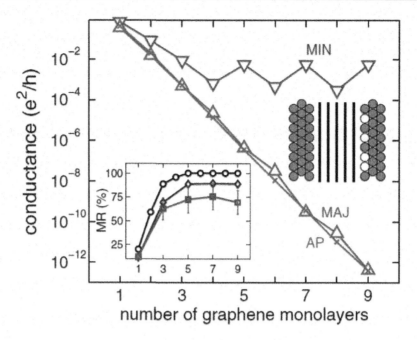

FIGURE 7.2 Conductance and magnetoresistance at the Ni/Gr$_n$/Ni interface. Adapted with permission from [23]. Copyright (2007) American Physical Society.

of heterostructure thickness. This provides the possibility for graphene/FM in novel spintronic devices that simultaneously satisfy longer relaxation length and larger PMA than FM metal/oxide or FM metal/NM metal. Recently, chiral magnetism has been observed for graphene/ferromagnetic metal interfaces. This is a shocked and exciting phenomenon for graphene-based spintronics. The chiral magnetism is usually induced by the interfacial Dzyaloshinskii-Moriya interaction (DMI). However, the DMI at the graphene/FM interface has been thought to be weak since the DMI scales with the SOC and the graphene has weak SOC. Significant DMI was achieved at the graphene-ferromagnetic metal interfaces [26]. Notably, the DMI can be significantly improved with more NiCoGr layers.

7.3.2 TMDs-Based Hybrid Magnetic Material

TMDs have the chemical formula MX$_2$, where M is a transition metal (e.g., Mo, W, and Te) and X is a chalcogen element (e.g., S and Se). Compared with graphene, TMDs exhibit more diversity, varying with the specific combinations of chalcogen and transition metal elements [27]. This offers more tunability and functionality when coupled to FMs. Notably, the giant spin-orbit splitting in the valence band edge of TMDs can significantly suppress the spin relaxation, which consequently leads to a long spin lifetime. For example, the spin lifetime of n-type MoS$_2$ and WS$_2$ monolayer can be up to 3 ns at 5 K, which is two to three orders of magnitude longer than typical excitation recombination times. In this context, the TMDs could enable both strong spin-orbit coupling and long spin lifetimes within a single FM/SC hybrid system

Shao et al. [28] demonstrated a high charge-spin conversion efficiency in an MoS$_2$/CoFeB hybrid system. The MoS2 flakes were obtained by the CVD method where the transition metal trioxides and S were vaporized in a chamber under a controlled temperature and gas environment. After that, the CoFeB was deposited on the top of MoS$_2$ by magnetron sputtering technique [28]. Figure 7.3a shows anomalous Hall resistance as a function of the out-of-plane magnetic field, which indicates an in-plane easy plane with an effective anisotropy field of −1 T. Figure 7.3b shows a sin2φ relation well for the planar Hall resistance

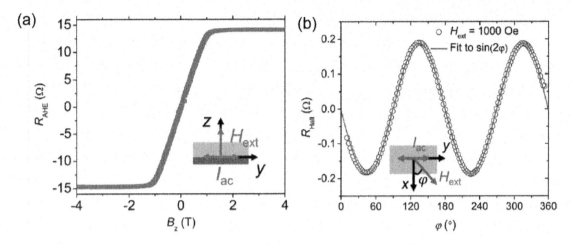

FIGURE 7.3 Hall resistance as a function of (a) magnetic field and (b) in-plane azimuthal angle at the MoS2/CoFeB interface. Adapted with permission from [28]. Copyright (2016) American Chemical Society.

as a function of in-plane azimuthal angle under an in-plane external magnetic field. Note that similar saturation anomalous Hall resistance, planar Hall resistance, and effective anisotropy field have also been observed in the CoFeB/SiO$_2$ case. This indicates that the contact between MoS$_2$ and CoFeB presents no influence on the magnetic properties of CoFeB. The current through the MoS$_2$/CoFeB interface would induce a net spin accumulation transverse to the current direction in the film plane. The current-induced spin polarization generates a strong field-like torque and a negligible damping-like torque. Notably, the relaxation time of spin polarization could be longer than 1 ns due to the coupled valley and spin in monolayer MoS$_2$.

When the ferromagnetic metal induces nonzero local magnetization in the TMDs, the magnetic proximity effect emerges, which decays exponentially away from the interface. Dolui et al. [29] systematically investigated the proximity effect within MoSe$_2$/Co, WSe$_2$/Co, and TaSe$_2$/Co. The hybridization between the ferromagnetic metal and monolayer TMD induces dramatic transmutation of the electronic and spin structure of Co, which is not only limited to the first layer in direct contact with the TMD. The spin-orbit torque is observed even at the eighth Co layer away from the interface. The injected spin-polarized current induces nonequilibrium spin density, leading to spin textures in each Co layer due to the spin-orbit coupling.

To achieve high spin injection efficiency in the MXene/FM interface, Cheng et al. [30] proposed a novel solution using sub-picosecond spin-current pulses with a strongly out-of-equilibrium nature. As exemplified by the MoS$_2$/Co system, the spin current was observed to be orders of magnitude larger than that obtained by traditional techniques, when an ultrashort spin-polarized current pulse was injected from Co into a monolayer MoS$_2$. The strong spin-orbit interaction in MoS$_2$ leads to high spin-to-charge conversion efficiency. This enables the monolayer MoS$_2$ not only to receive a spin injection but also to selectively convert the spin current into a charge current. In this context, the ultrashort spin-current pulses can be used as a technologically viable information carrier for terahertz spintronics.

Shi et al. [31] demonstrate high-efficiency charge-to-spin conversion in WTe$_2$/Ni$_{81}$Fe$_{19}$ heterostructures, which can be realized with low power consumption. The WTe$_2$ is a Weyl semimetal and possesses strong spin-orbit coupling and non-trivial band structures for both surface and bulk states. The charge-to-spin conversion efficiency at the interface increases with the thickness of WTe$_2$, which suggests a bulk contribution to the spin current generation. For the heterostructures with thick WTe$_2$, a strong frequency-dependent charge-to-spin conversion efficiency was observed, which could be related to a long spin diffusion length. Furthermore, a chiral domain wall tilting in WTe$_2$/Ni$_{81}$Fe$_{19}$ was induced by an effective field induced by the interfacial DMI. Further efforts can be made to explore more Weyl semimetals used as SOC medium with the FMs in SOT applications.

7.4 ORGANIC WITH FERROMAGNETIC METAL

Organic materials are promising for constituting spintronic devices due to their expected long spin lifetime, extremely decreased size, and abundant diversity. The combination of the ferromagnetic metal with the molecule electronic would provide a novel possibility in manipulating the spins within in a sub-nanometer scale. However, transferring spins between the organic molecules and FM magnets remains a huge challenge due to the distinct conductivities. Within such an interface, the electronic hybridization between the organic and FM sides would dominate the spintronic properties. This promotes the field of the spinterface science that was first proposed in 2010 [32]. In what follows, we introduce recent progress in organic/FM-based spinterfaces, with a special focus on electronic hybridization and magnetic coupling at the interface. Here, we classify these spinterfaces into two types according to the magnetism of the molecules, i.e., nonmagnetic molecules and magnetic molecules on FM.

7.4.1 Nonmagnetic Organic Semiconductors–Based Spinterface

Synthetic organic chemistry allows the fabrication of π-conjugated organic semiconductors (OSE) structures with high freedom that cannot yet be obtained in inorganic SCs. The 8-hydroxy-quinoline aluminum (Alq_3) has been the most attractive organic SC to serve as a spacer in FM/OSE/FM spin valves due to its easy deposition as thin film and integration with a variety of FM electrodes. The pioneering work has been done by Xiong et al. [33], where the $La_{0.67}Sr_{0.33}MnO_3$ (LSMO)/Alq_3/Co hybrid architecture was used as the magnetic tunneling junction. The LSMO shows a half-metallic ferromagnetic characteristic and has a 100% spin polarization. Within such a spin valve, a magnetoresistance of 40% was obtained at 11 K, which is comparable to that obtained in metallic giant magnetoresistance spin valves.

Since then, the unique spintronic phenomena of the sandwich FM/OSC/FM hybrid structures have attracted many efforts on uncovering the underlying mechanism of the spin injection from FM to the OSCs. The efficiency of the organic-based spin valve largely depends on the spin transport throughout and the spin injection into the nonmagnetic molecule. However, it is difficult to study the corresponding local spin transport of the OSCs due to an average effect of the architecture of traditional spin-valve devices. In contrast to previous established LSMO/Alq3/Co magnetic tunnel junctions (MTJs), Barraud et al. [34] fabricated spin-valve devices from LSMO/Alq3 bilayers using a conductive-tip atomic force microscope nanolithography process. The thickness of Alq3 can be controlled by manipulating the atomic force microscope tip penetration into the layer. Then, the Co was inserted into the holes. This technique produces MTJ with a cross-section of a few tens of nm^2, which can realize the characterizing of the local magnetic properties of the sandwich structure. The hybridization between the FM and organic molecules induces new polarized states in the first molecular layer that have direct contact with the FM, which can increase the effective spin polarization of the electrodes. After applying a magnetic field, huge magnetoresistance up to 300% was achieved at 2 K. Using a real-time spin-resolved pump probe, Steil et al. [35] studied the spin-dependent dynamics of the Co/Alq3 hybrid electronic structure. It was observed that the excited electrons under external stimuli are trapped at the Co/Alq3 interface by spin-dependent confining potentials. This further demonstrates that interfacial hybridization plays a critical role in determining the spintronic properties at the spinterface. Hence, by chemical functionalization of the organic molecule, the spin transport could be rationally controlled.

Raman et al. [36] demonstrate that spin-dependent resistance arises from the interface magnetoresistance effect. In their experiments, zinc methyl phenalenyl (ZMP) was grown on a Co surface. The other side of the electrode can be either Cu or permalloy. This device shows an unexpected interfacial magnetoresistance of more than 20% at room temperature. When an ultrathin layer of the Al_2O_3 insulator was inserted into ZMP and Co, the resistance changes until the permalloy magnetization was switched. These results indicate that the magnetoresistance arises from the interface magnetoresistance effect rather than conventional spin-conserved tunneling transport through the organic barrier between two FM electrodes. The Co surface layer couples to a bilayer ZMP molecule. The ZMP layer with direct contact to the Co surface is magnetic, while the second layer acts as a spin filter. The molecular orbitals of the first layer ZMP molecule are significantly broadened, showing a metallic characteristic. Due to the hybridization with the Co layer, the molecule shows a weak magnetism that is antiparallel to the Co substrate. In addition, for the second layer of the ZMP, the large distance between the molecule and the magnetic surface suggests a weak chemical bonding, which keeps the molecular-type discrete energy levels near the Fermi level. In particular, a spin split of 0.14 eV was obtained for the LUMO of the second-layer ZMP molecule, which leads to different barrier heights for the two spin channels. The switching of the interfacial magnetization at room temperature indicates a large surface magnetic anisotropy energy. Given the fact that the ZMP molecule shows weak spin-orbit coupling due to its composition of light elements, the large magnetic anisotropy energy could arise from the interfacial hybridization.

The hybridization between the adsorbate and the FM largely depends on the interfacial bonding state, either physisorption or chemisorption. The chemisorption of the aromatic molecules can significantly tailor the local spin polarization [37] of the FM substrate. As exemplified by benzene on the Fe surface, the hybridization between the d states of Fe and the atomic orbitals of C atoms in benzene significantly modify the spin polarization at the Fermi level. Due to the spin splitting of the Fe d states, the adsorbed benzene molecule exhibits a small net magnetic moment. In contrast to chemisorption, the physisorbed molecule usually leads to a negligible influence on the electronic properties of the substrate and molecule [38]. Interestingly, by delicate design, the physisorption and chemisorption of an organic molecule on an FM surface can coexist [39], which can lead to switchable spintronic properties in a single hybrid spinterface. As exemplified by trafluoropyrazine ($C_4N_2F_4$) on Ni(111), both chemisorption and physisorption states can be obtained. There is a small barrier between the two states, which can be observed from the binding energy curves in Figure 7.4c. The reversible bistable adsorption states lead to switchable interface magnetic properties (Figure 7.4a–b). Compared to the physisorbed spinterface, the chemisorbed molecules significantly lower the magnetic moments of the metal atoms below the adsorbate (0.64 vs 0.27 μ_B). This can be evidenced from the magnetic moments (μ) of Ni atoms as a function of the adsorption distance, where μ is positively related to the adsorption distance. Furthermore, the chemisorption significantly decreases the spin polarization of the Ni surface around the Fermi level. Similar switchable

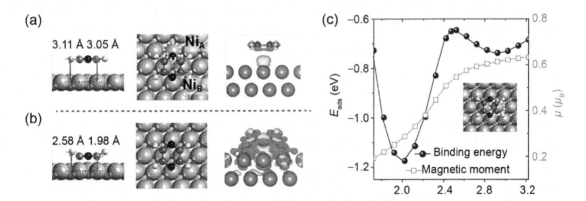

FIGURE 7.4 (a) Adsorption structure, charge transfer, (b) binding energies, and magnetic moment as a function of the adsorption distance. Adapted with permission from [39]. Copyright (2020) American Chemical Society.

spinterfaces based on the bistable adsorption state have also been observed for a series of halogenated aromatic hydrocarbons on ferromagnetic Ni and Co surfaces.

7.4.2 Molecular Magnet–Based Spinterface

Single-molecule magnets represent the smallest bistable magnetic systems, which makes them ideal candidates for quantum computation. However, the thermal spin fluctuations usually lead to random magnetization at room temperature. The magnetic coupling between the molecule and FMs has a significant influence on the magnetization ordering. The magnetic exchange by coupling the SMM with FM could significantly affect the magnetization within the interface. Therefore, the deposition of molecular magnets on FMs may provide an effective solution to manipulate the magnetization in a controllable way. Wende et al. [40] observed ordered molecular structure and magnetization for octaethylporphyrin Fe(III) chloride (OEP) on 5 ML Ni/Cu(100) surface. The Fe and Ni atoms show identical hysteresis loops. These results demonstrated that a ferromagnetic exchange coupling exists between the Fe moments in the porphyrin molecules and the magnetic substrate.

Such a ferromagnetic exchange coupling has also been demonstrated in Tb double-decker complexes (TbPc$_2$) on Ni films, in which the molecule and substrate easy axes are collinear [41]. Figure 7.5 presents the x-ray absorption spectra (XAS) and x-ray magnetic circular dichroism (XMCD, I$^-$ – I$^+$) spectra of the Ni and Tb with out-of-plane and in-plane Ni substrate at 8 K. Both the Ni substrate and Tb atom in the molecule show strong remanent XMCD intensity, which remains stable during the whole measurement process. However, the sign of the XMCD of the Tb atom is opposite to that of Ni, indicating an antiferromagnetic coupling. Furthermore, the magnetization of TbPc2 depends on the easy axis of the ferromagnetic substrate. When the molecule is deposited on the in-plane Ni surface instead of out-of-plane, the XMCD intensity is reduced, leading to a decreased magnetic anisotropy. In this context, the matching between the ferromagnetic substrate and molecular magnetic anisotropy can effectively stabilize the molecular magnetization without external magnetic fields.

As mentioned prior, the monolayer of molecular magnets on FM surface can be controlled from paramagnetic to ferromagnetic. However, the strong interaction between the FM and molecular magnet can induce strong orbital deformation to the stacked molecules, which may destroy their intrinsic properties. In this context, multi-spinterfaces, formed by a multilayer molecular magnet, are a more promising approach to achieve controllable molecular magnetization at the interface. Yoo et al. [42] demonstrated the emergence of a multi-spinterface for paramagnetic cobalt-octaethylporphyrin (CoOEP) molecules on a Fe film. The antiferromagnetic molecular exchange bias was observed in the heterostructure. As the ground state, the neighbor layers are antiferromagnetically coupled. However, the CoOEP film can also show different magnetic coupling with the ferromagnetic CoOEP interfacial layer. The varied energy differences of the spinterface with excited spin configurations demonstrate a strong dependence of the magnetic stability on the spin configurations. This consequently leads to a switchable spin state of the interfacial ferromagnetic CoOEP layer. Therefore, the molecules stacked by van der Waals forces enable more diversity of magnetic states in a single hybrid system with respect to the monolayer molecule deposited spinterfaces.

Besides these molecular magnets with static magnetism, the spin crossover (SCO) complexes are promising candidates in fabricating hybrid magnetic systems. Within such a system, the spin state of a transition metal ion can be controlled by switchable ligands [43] between high- and low-spin states under electrical or light stimuli. While the ferromagnet film contact with the LS-state molecule remains its intrinsic magnetization, the HS-state SCO molecule could significantly alter the magnetic properties, i.e., magnetic anisotropy and spin polarization. Gueddida et al. [44] demonstrated a strong magnetic coupling between the Fe(1,10-phenanthroline)2(NCS)2 (Fe-phen) molecules and Co surfaces. Despite previous encouraging work for the hybrid spinterface by SCO molecules on FM surface, many challenges remain, such as dissipation and missing of switching ability [45] due to the extremely strong interfacial coupling. These problems should be addressed in the future.

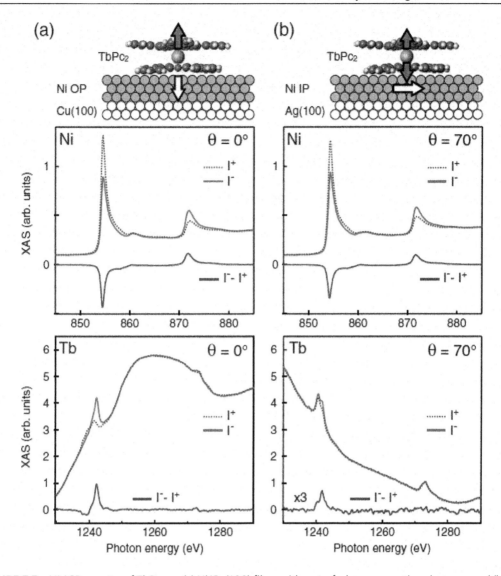

FIGURE 7.5 XMCD spectra of TbPc$_2$ on (a) Ni/Cu(100) films with out-of-plane magnetic anisotropy and (b) Ni/Ag(100) films with in-plane magnetic anisotropy at 8 K. Adapted with permission from [41]. Copyright (2011) American Physical Society.

7.5 CONCLUSIONS AND PERSPECTIVES

In this chapter, we reviewed the recent progress of several hybrid magnetic materials, i.e., the ferromagnetic film with SCs, 2D material with ferromagnet, organic molecules with ferromagnetic metal. The integrated hybrid system combines the advantages of magnetic materials and nonmagnetic materials, which show excellent spintronic properties, such as high spin polarization, high Curie temperature, long spin relaxation distance, and strong spin-orbit torque effect. By delicate consideration of the material resource of both sides, a hybrid system that satisfies all the spintronic parameter could be achieved.

We start by introducing the hybrid systems based on metallic or half-metallic ferromagnet film on bulk SCs, which have the longest history. They usually present high spin polarization and Curie temperature. Notably, a recent study has also shown a strong spin-orbit torque phenomenon within such a hybrid system. Secondly, we briefly introduce the recent progress of the 2D materials-based hybrid systems. The 2D material has a well-controlled atomic layer thickness and a near-perfect surface without dangling bonds. When integrated with the ferromagnets, large spin polarization, robust magnetic anisotropy, and strong SOC effect can be realized in a single hybrid system based on graphene and TMDs. Finally, we pay special attention to the recently proposed spinterface based on organic molecules and FMs. This hybrid system downscales the control of spin-resolved properties to a molecule level, which not only exhibits large magnetoresistance comparable to the conventional metal-based magnets but also shows some novel phenomenon that cannot be achieved for inorganic-based magnets, such as actively controlled spintronic properties under electrical or light stimuli. Unlimited to the aforementioned systems, the recent proposed hybrid systems based on organic-inorganic perovskite [46–47] and hybrid van der Waals heterostructure [48] have achieved more and more attention. These hybrid systems provide new possibilities in future spintronics in applications of high-density data storage and logic.

REFERENCES

[1] Liu, W.; Wong, P. K. J.; Xu, Y. (2019) Hybrid spintronic materials: Growth, structure and properties. *Prog. Mater Sci. 99*: 27.

[2] Dayen, J.-F.; Ray, S. J.; Karis, O.; Vera-Marun, I. J.; Kamalakar, M. V. (2020) Two-dimensional van der Waals spinterfaces and magnetic-interfaces. *Appl. Phys. Rev. 7*: 011303.

[3] Cinchetti, M.; Dediu, V. A.; Hueso, L. E. (2017) Activating the molecular spinterface. *Nat. Mater. 16*: 507.

[4] Prinz, G.; Krebs, J. (1981) Molecular beam epitaxial growth of single-crystal Fe films on GaAs. *Appl. Phys. Lett. 39*: 397.

[5] Kikkawa, J.; Awschalom, D. (1998) Resonant spin amplification in n-type GaAs. *Phys. Rev. Lett. 80*: 4313.

[6] Krebs, J. J.; Jonker, B.; Prinz, G. (1987) Properties of Fe single-crystal films grown on (100) GaAs by molecular-beam epitaxy. *J. Appl. Phys. 61*: 2596.

[7] Fleet, L.; Yoshida, K.; Kobayashi, H.; Kaneko, Y.; Matsuzaka, S.; Ohno, Y.; Ohno, H.; Honda, S.; Inoue, J.; Hirohata, A. (2013) Correlating the interface structure to spin injection in abrupt Fe/GaAs (001) films. *Phys. Rev. B 87*: 024401.

[8] Zega, T. J.; Hanbicki, A. T.; Erwin, S. C.; Žutić, I.; Kioseoglou, G.; Li, C. H.; Jonker, B. T.; Stroud, R. M. (2006) Determination of interface atomic structure and its impact on spin transport using Z-contrast microscopy and density-functional theory. *Phys. Rev. Lett. 96*: 196101.

[9] Thomas, O.; Shen, Q.; Schieffer, P.; Tournerie, N.; Lépine, B. (2003) Interplay between anisotropic strain relaxation and uniaxial interface magnetic anisotropy in epitaxial Fe films on (001) GaAs. *Phys. Rev. Lett. 90*: 017205.

[10] Chen, L.; Decker, M.; Kronseder, M.; Islinger, R.; Gmitra, M.; Schuh, D.; Bougeard, D.; Fabian, J.; Weiss, D.; Back, C. H. (2016) Robust spin-orbit torque and spin-galvanic effect at the Fe/GaAs (001) interface at room temperature. *Nat. Commun. 7*: 1.

[11] Chen, L.; Mankovsky, S.; Wimmer, S.; Schoen, M.; Körner, H.; Kronseder, M.; Schuh, D.; Bougeard, D.; Ebert, H.; Weiss, D. (2018) Emergence of anisotropic Gilbert damping in ultrathin Fe layers on GaAs (001). *Nat. Phys. 14*: 490.

[12] Lu, Y.; Claydon, J.; Xu, Y.; Thompson, S.; Wilson, K.; Van der Laan, G. (2004) Epitaxial growth and magnetic properties of half-metallic Fe3O4 on GaAs (100). *Phys. Rev. B 70*: 233304.

[13] De Groot, R.; Mueller, F.; Van Engen, P.; Buschow, K. (1983) New class of materials: Half-metallic ferromagnets. *Phys. Rev. Lett. 50*: 2024.

[14] Van Roy, W.; De Boeck, J.; Brijs, B.; Borghs, G. (2000) Epitaxial NiMnSb films on GaAs (001). *Appl. Phys. Lett. 77*: 4190.

[15] Ciccarelli, C.; Anderson, L.; Tshitoyan, V.; Ferguson, A.; Gerhard, F.; Gould, C.; Molenkamp, L.; Gayles, J.; Železný, J.; Šmejkal, L. (2016) Room-temperature spin-orbit torque in NiMnSb. *Nat. Phys. 12*: 855.

[16] Zhao, N.; Sud, A.; Sukegawa, H.; Komori, S.; Rogdakis, K.; Yamanoi, K.; Patchett, J.; Robinson, J.; Ciccarelli, C.; Kurebayashi, H. (2021) Growth, strain, and spin-orbit torques in epitaxial Ni-Mn-Sb films sputtered on GaAs. *Phys. Rev. Mat. 5*: 014413.

[17] Fan, X.; Celik, H.; Wu, J.; Ni, C.; Lee, K.-J.; Lorenz, V. O.; Xiao, J. Q. (2014) Quantifying interface and bulk contributions to spin-orbit torque in magnetic bilayers. *Nat. Commun. 5*: 1.

[18] Hamaya, K.; Itoh, H.; Nakatsuka, O.; Ueda, K.; Yamamoto, K.; Itakura, M.; Taniyama, T.; Ono, T.; Miyao, M. (2009) Ferromagnetism and electronic structures of nonstoichiometric Heusler-alloy Fe3−xMnxSi epilayers grown on Ge (111). *Phys. Rev. Lett. 102*: 137204.

[19] Novoselov, K. S.; Geim, A. K.; Morozov, S. V.; Jiang, D.-e.; Zhang, Y.; Dubonos, S. V.; Grigorieva, I. V.; Firsov, A. A. (2004) Electric field effect in atomically thin carbon films. *Science 306*: 666.

[20] Han, W.; Kawakami, R. K.; Gmitra, M.; Fabian, J. (2014) Graphene spintronics. *Nat. Nanotechnol. 9*: 794.

[21] Li, S.; Larionov, K. V.; Popov, Z. I.; Watanabe, T.; Amemiya, K.; Entani, S.; Avramov, P. V.; Sakuraba, Y.; Naramoto, H.; Sorokin, P. B. (2020) Graphene/Half-Metallic Heusler Alloy: A Novel Heterostructure toward High-Performance Graphene Spintronic Devices. *Adv. Mater. 32*: 1905734.

[22] Hill, E. W.; Geim, A. K.; Novoselov, K.; Schedin, F.; Blake, P. (2006) Graphene spin valve devices. *IEEE Trans. Magn. 42*: 2694.

[23] Karpan, V.; Giovannetti, G.; Khomyakov, P.; Talanana, M.; Starikov, A.; Zwierzycki, M.; Van Den Brink, J.; Brocks, G.; Kelly, P. J. (2007) Graphite and graphene as perfect spin filters. *Phys. Rev. Lett. 99*: 176602.

[24] Tombros, N.; Jozsa, C.; Popinciuc, M.; Jonkman, H. T.; Van Wees, B. J. (2007) Electronic spin transport and spin precession in single graphene layers at room temperature. *Nature 448*: 571.

[25] Yang, H.; Vu, A. D.; Hallal, A.; Rougemaille, N.; Coraux, J.; Chen, G.; Schmid, A. K.; Chshiev, M. (2016) Anatomy and giant enhancement of the perpendicular magnetic anisotropy of cobalt-graphene heterostructures. *Nano Lett. 16*: 145.

[26] Yang, H.; Chen, G.; Cotta, A. A.; N'Diaye, A. T.; Nikolaev, S. A.; Soares, E. A.; Macedo, W. A.; Liu, K.; Schmid, A. K.; Fert, A. (2018) Significant Dzyaloshinskii-Moriya interaction at graphene-ferromagnet interfaces due to the Rashba effect. *Nat. Mater. 17*: 605.

[27] Manzeli, S.; Ovchinnikov, D.; Pasquier, D.; Yazyev, O. V.; Kis, A. (2017) 2D transition metal dichalcogenides. *Nat. Rev. Mater. 2*: 1.

[28] Shao, Q.; Yu, G.; Lan, Y.-W.; Shi, Y.; Li, M.-Y.; Zheng, C.; Zhu, X.; Li, L.-J.; Amiri, P. K.; Wang, K. L. (2016) Strong Rashba-Edelstein effect-induced spin-orbit torques in monolayer transition metal dichalcogenide/ferromagnet bilayers. *Nano Lett. 16*: 7514.

[29] Dolui, K.; Nikolić, B. K. (2020) Spin-orbit-proximitized ferromagnetic metal by monolayer transition metal dichalcogenide: Atlas of spectral functions, spin textures, and spin-orbit torques in Co/MoSe2, Co/WSe2, and Co/TaSe2 heterostructures. *Phys. Rev. Mat. 4*: 104007.

[30] Cheng, L.; Wang, X.; Yang, W.; Chai, J.; Yang, M.; Chen, M.; Wu, Y.; Chen, X.; Chi, D.; Goh, K. E. J. (2019) Far out-of-equilibrium spin populations trigger giant spin injection into atomically thin MoS2. *Nat. Phys. 15*: 347.

[31] Shi, S.; Liang, S.; Zhu, Z.; Cai, K.; Pollard, S. D.; Wang, Y.; Wang, J.; Wang, Q.; He, P.; Yu, J. (2019) All-electric magnetization switching and Dzyaloshinskii-Moriya interaction in WTe2/ferromagnet heterostructures. *Nat. Nanotechnol. 14*: 945.

[32] Sanvito, S. (2010) Molecular spintronics: The rise of spinterface science. *Nat. Phys. 6*: 562.

[33] Xiong, Z.; Wu, D.; Vardeny, Z. V.; Shi, J. (2004) Giant magnetoresistance in organic spin-valves. *Nature 427*: 821.

[34] Barraud, C.; Seneor, P.; Mattana, R.; Fusil, S.; Bouzehouane, K.; Deranlot, C.; Graziosi, P.; Hueso, L.; Bergenti, I.; Dediu, V.; Petroff, F.; Fert, A. (2010) Unravelling the role of the interface for spin injection into organic semiconductors. *Nat. Phys. 6*: 615.

[35] Steil, S.; Grossmann, N.; Laux, M.; Ruffing, A.; Steil, D.; Wiesenmayer, M.; Mathias, S.; Monti, O. L. A.; Cinchetti, M.; Aeschlimann, M. (2013) Spin-dependent trapping of electrons at spinterfaces. *Nat. Phys. 9*: 242.

[36] Raman, K. V.; Kamerbeek, A. M.; Mukherjee, A.; Atodiresei, N.; Sen, T. K.; Lazic, P.; Caciuc, V.; Michel, R.; Stalke, D.; Mandal, S. K.; Bluegel, S.; Muenzenberg, M.; Moodera, J. S. (2013) Interface-engineered templates for molecular spin memory devices. *Nature 493*: 509.

[37] Atodiresei, N.; Brede, J.; Lazic, P.; Caciuc, V.; Hoffmann, G.; Wiesendanger, R.; Bluegel, S. (2010) Design of the local spin polarization at the organic-ferromagnetic interface. *Phys. Rev. Lett. 105*: 066601.

[38] Lach, S.; Altenhof, A.; Tarafder, K.; Schmitt, F.; Ali, M. E.; Vogel, M.; Sauther, J.; Oppeneer, P. M.; Ziegler, C. (2012) Metal-organic hybrid interface states of a ferromagnet/organic semiconductor hybrid junction as basis for engineering spin injection in organic spintronics. *Adv. Funct. Mater. 22*: 989.

[39] Yang, S.; Li, S.; Ren, J.-C.; Butch, C. J.; Liu, W. (2020) Reversible control of spintronic properties of ferromagnetic metal/organic interfaces through selective molecular switching. *Chem. Mater. 32*: 9609.

[40] Wende, H.; Bernien, M.; Luo, J.; Sorg, C.; Ponpandian, N.; Kurde, J.; Miguel, J.; Piantek, M.; Xu, X.; Eckhold, P. (2007) Substrate-induced magnetic ordering and switching of iron porphyrin molecules. *Nat. Mater. 6*: 516.
[41] Rizzini, A. L.; Krull, C.; Balashov, T.; Kavich, J.; Mugarza, A.; Miedema, P. S.; Thakur, P. K.; Sessi, V.; Klyatskaya, S.; Ruben, M. (2011) Coupling single molecule magnets to ferromagnetic substrates. *Phys. Rev. Lett. 107*: 177205.
[42] Jo, J.; Byun, J.; Lee, J.; Choe, D.; Oh, I.; Park, J.; Jin, M. J.; Lee, J.; Yoo, J. W. (2020) Emergence of multispinterface and antiferromagnetic molecular exchange bias via molecular stacking on a ferromagnetic film. *Adv. Funct. Mater. 30*: 1908499.
[43] Kumar, K. S.; Ruben, M. (2021) Sublimable spin-crossover complexes: From spin-state switching to molecular devices. *Angew. Chem. Int. Ed. 60*: 7502.
[44] Knaak, T.; González, C.; Dappe, Y. J.; Harzmann, G. D.; Brandl, T.; Mayor, M.; Berndt, R.; Gruber, M. (2019) Fragmentation and distortion of terpyridine-based spin-crossover complexes on Au (111). *J. Phys. Chem. C 123*: 4178.
[45] Ossinger, S.; Naggert, H.; Kipgen, L.; Jasper-Toennies, T.; Rai, A.; Rudnik, J.; Nickel, F.; Arruda, L. M.; Bernien, M.; Kuch, W. (2017) Vacuum-evaporable spin-crossover complexes in direct contact with a solid surface: Bismuth versus gold. *J. Phys. Chem. C 121*: 1210.
[46] Odenthal, P.; Talmadge, W.; Gundlach, N.; Wang, R.; Zhang, C.; Sun, D.; Yu, Z.-G.; Vardeny, Z. V.; Li, Y. S. (2017) Spin-polarized exciton quantum beating in hybrid organic – inorganic perovskites. *Nat. Phys. 13*: 894.
[47] Wang, J.; Zhang, C.; Liu, H.; Liu, X.; Guo, H.; Sun, D.; Vardeny, Z. V. (2019) Tunable spin characteristic properties in spin valve devices based on hybrid organic-inorganic perovskites. *Adv. Mater. 31*: 1904059.
[48] Sierra, J. F.; Fabian, J.; Kawakami, R. K.; Roche, S.; Valenzuela, S. O. (2021) Van der Waals heterostructures for spintronics and opto-spintronics. *Nat. Nanotechnol. 16*: 856.

8 Methods for the Syntheses of Perovskite Magnetic Nanomagnets

Xinhua Zhu
National Laboratory of Solid State of Microstructures, School of Physics, Nanjing University, Nanjing, China

Contents

8.1	Introduction	126
8.2	Classification of Perovskite Magnetic Nanomagnets	126
8.3	Synthesis Methods of Perovskite Magnetic Nanomagnets	128
	8.3.1 Zero-Dimensional (0D) Perovskite Magnetic Nanomagnets	128
	8.3.1.1 Solid-State Reaction Method	128
	8.3.1.2 Reactive Grinding Process	130
	8.3.1.3 Combustion Method	131
	8.3.1.4 Sol-Gel Method	133
	8.3.1.5 Hydrothermal/Solvothermal/Microwave-Hydrothermal Method	134
	8.3.1.6 Microemulsion Route	139
	8.3.2 One-Dimensional (1D) Perovskite Magnetic Nanomagnets	140
	8.3.2.1 Top-Down Approach	141
	8.3.2.2 Bottom-Up Approach	141
	8.3.3 Two-Dimensional (2D) Perovskite Magnetic Nanomagnets	144
	8.3.4 Three-Dimensional (3D) Perovskite Magnetic Nanomagnets	147
8.4	Potential Applications of Perovskite Magnetic Nanomagnets	150
	8.4.1 Magnetic Memory Devices	150
	8.4.2 Spintronic Devices	151
	8.4.3 Magnetic Sensors	151
	8.4.4 Magnetic Refrigeration	152
	8.4.5 Biomedical Applications	152
	8.4.6 Catalysis	154
	8.4.7 Environmental Purification	154
8.5	Conclusions	155
References		156

DOI: 10.1201/9781003197492-8

8.1 INTRODUCTION

Magnetic nanomagnets (MNMs) exhibit many intriguing magnetic properties and have important applications in data recording, magnetic sensors, and biomedicine [1–2]. These potential benefits of MNMs are ascribed to their intrinsic small geometrical sizes and intriguing magnetic properties originating from their finite size, surface effects, collective reactions, and unique transport properties [3]. To understand the excellent magnetic properties of MNMs and implement new functionalities, much effort has been made to synthesize MNMs with desired physical properties and microstructures [4–5]. It is found that the magnetic properties of MNMs are closely related to their geometric sizes and shapes via two distinct phenomena [3]. One is shape anisotropy, which is related to the demagnetizing field. By choosing the shapes of the MNMs, their magnetic properties can be adjusted. The other one is configurational anisotropy, which contributed to the energy competition between magnetostatic energy and exchange energy. It not only determines whether the MNMs have a single domain state but also the non-uniformities of magnetization. Understanding the property variation within the geometry of MNMs can open up a corridor for designing new MNMs where their magnetic properties can be modulated for a special application with high precision. Along with the fast development of computer science and cloud computing technology, miniaturization of electronic devices is highly requested, which pushes the dimensions of new electronic devices into the deep nanometer region. However, in conventional electronic devices based on two-dimensional (2D) planar nanostructures (with single or multilayered magnetic layers), their thickness is comparable with some characteristic magnetic length scale, and single magnetic domain states appear in the vertical direction of such simple geometry, thus restricting the functionality to the substrate plane [6]. Consequently, complex and useful magnetic behaviors are not allowed to be exploited via interfacial effects between the layers. Recently, these fundamental bottlenecks in the normal 2D planar structure stimulate the development of nanosized objects with non-planar structures, enabling nanomagnetism to extend into three-dimensional (3D) MNMs, where 3D magnetic spin textures can be utilized not only in-plane but also out-of-plane. Therefore, many novel magnetic effects associated with geometry, topology, and chirality in 3D MNMs can be observed [7–8]. During the past several decades, many different MNMs have been prepared. They include single component nanostructures (e.g. Fe, Co, and Ni metal nanoparticles and nanowires), metal alloys (e.g. FePd, $CoPt_3$, and FePt nanoparticles), metallic oxide nanoparticles such as Fe_3O_4 nanoparticles, and metal carbides (e.g. Fe_2C, Fe_3C, and Fe_5C_2) as well as spinel-type ferromagnets such as MFe_2O_4 (M = Mg, Fe, Co, Mn) and multicomponent nanostructures such as heterostructures (e.g. Fe_3O_4@Au@Ag, FePt@Au) and exchange-coupled nanomagnets ($Nd_2Fe_{14}B$/α-Fe, $SmCo_5$@Co, and $SmCo_5$@Sm_2Co_{17} nanoparticles) [4–5]. In addition, MNMs with a perovskite structure have also received much attention because of their excellent magnetic properties and unique structural and compositional flexibilities [9–10]. In this chapter, we focus on the synthesis methods for MNMs with a perovskite structure, which is considered to be the next generation of advanced nanomagnets. Special attention is given to the new development of 3D MNMs with unexpected magnetic properties. Finally, some potential applications of MNMs in magnetic memory devices, spintronic devices, magnetic sensors, magnetic refrigeration, biomedical applications, catalysis, and environmental purification are also introduced and discussed.

8.2 CLASSIFICATION OF PEROVSKITE MAGNETIC NANOMAGNETS

Due to extremely small geometrical sizes, perovskite magnetic nanomagnets exhibit many different magnetic properties from their counterparts, such as parent bulk materials, which form a rich area and grow rapidly in condensed matter physics. According to the dimensional numbers at the nanometer length scale, perovskite oxide nanomagnets are classified as (i) zero-dimensional (0D) magnetic nanoparticles, where all three dimensions are constructed in the nanometer length scale, (ii) one-dimensional (1D) magnetic nanomagnets (e.g. magnetic nanowires and nanorods, nanotubes, nanofibers, and nanoribbons), (iii) 2D magnetic nanomagnets, such as magnetic thin films, magnetic nanodot arrays, nanosheets, nanoplates, and nanowalls), and (iv) 3D magnetic nanostructures (i.e., 3D nanohelices, 3D nanoactuators, and 3D nanoellipsoids). Some representative examples of perovskite magnetic nanomagnets with different dimensions are demonstrated in Figure 8.1.

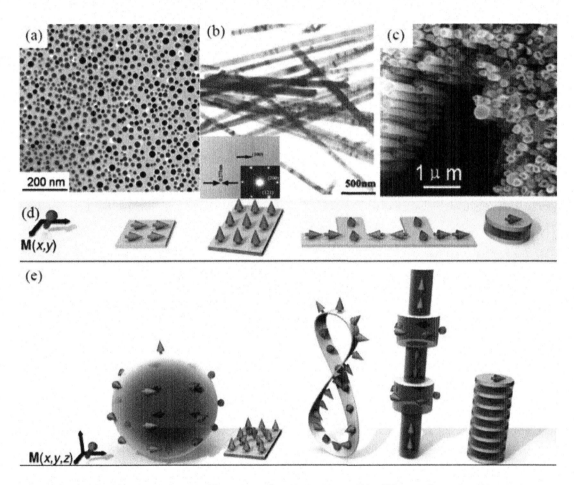

FIGURE 8.1 Classifications of perovskite magnetic nanomagnets with different dimensions into nanometer scale. (a) 0D magnetic BiFeO$_3$ nanoparticles, where all three dimensions are in the nanometer length scale. Adapted with permission from [35]. Copyright (2011) American Ceramic Society. (b) 1D La$_{0.5}$Ca$_{0.5}$MnO$_3$ single-crystalline nanowires revealed by transmission electron microscopy (TEM) image (insets are a selected area diffraction pattern and an HRTEM image). Adapted with permission from [54]. Copyright (2004) Royal Society of Chemistry. (c) 1D BiFeO$_3$ nanotubes revealed by scanning electron microscope (SEM) image. Adapted with permission from [61]. Copyright (2005) American Institute of Physics. (d) Some representative 2D nanomagnets such as single-domain magnets, multilayered magnetic elements with vertical anisotropy, regular arrays of nanostrip with protrusions for domain wall trapping, and two-layered magnets with antiferromagnetic coupling due to indirect exchange by an interfacial effect between the two layers. (e) 3D nanomagnets such as a magnetic sphere exhibiting vortex domain configuration, a magnetic thin film element with a skyrmion, a Möbius strip with perpendicular magnetization, and cylindrical magnetic nanowires with a modulated diameter and magnetic configurations dependent upon the diameter. Adapted with permission from [6]. Copyright (2017) Springer Nature. The article was printed under a CC-BY license.

8.3 SYNTHESIS METHODS OF PEROVSKITE MAGNETIC NANOMAGNETS

8.3.1 Zero-Dimensional (0D) Perovskite Magnetic Nanomagnets

Zero-dimensional perovskite magnetic nanomagnets are classified as one kind of important magnetic nanomaterials and their magnetic properties can be well modulated by controlling sizes and shapes and compositions during the synthesis process. To date, they have become a research hotspot in condensed matter physics. So far, many of methods have been used to synthesize 0D perovskite magnetic nanomagnets. A summary of typical synthetic strategies, such as the solid-state reaction method, reactive grinding, the combustion method, the sol-gel process, the hydrothermal/solvothermal process, and the microemulsion route, is presented as follows.

8.3.1.1 Solid-State Reaction Method

The solid-state reaction method is a conventional route for the synthesis of perovskite oxide nanomagnets, in which different precursors are weighted, mixed, and grinded followed by sintering at high temperatures to get magnetic nanoparticles. The final products obtained by this method usually have coarse particles, poor size distribution, and impure phases [11]. As a typical representative of the perovskite manganite family, $La_{1-x}Sr_xMnO_3$ (LSMO, $x \sim 0.3$) nanoparticles have been widely investigated due to their large magnetization and high Curie temperature (T_c). Navin and Kurchania employed a solid-state reaction method to synthesize $La_{0.7}Sr_{0.3}MnO_3$ nanoparticles [12]. They found that the obtained powders crystallized in a rhombohedral phase structure ($R\bar{3}c$ space group) without any impurity. Their average crystallize size was about 23 nm deduced from the x-ray diffraction (XRD) data. Magnetic property measurements revealed that the magnetization (M) versus magnetic field (H) loops saturated at 500 mT. The saturation magnetization (M_s) and coercive field (H_c) were determined to be 91.04 emu/g and 9.8 mT, respectively. The magnetic behavior of the LSMO nanoparticles can be interpreted by the double-exchange (DE) mechanism. Magnetizations of the LSMO nanoparticles as a function temperature measured under zero-field cooling (ZFC) and field cooling (FC) modes exhibits a large bifurcation between the ZFC and FC curves near 300 K, which suggests the presence of both spin-glass-like behavior and superparamagnetism (SPM) in the LSMO nanoparticles.

To improve the particle size dispersion of the LSMO nanoparticle, Ortiz-Quinonez et al., [13] synthesized well-dispersed LSMO nanoparticles by using salt-assisted solid-sate synthesis, where metal acetylacetonate precursors were grinded with NaCl as dispersing medium and subsequently the products were air-annealed and washed out. The representative scanning electron microscopy (SEM) images of the $La_{1-x}Sr_xMnO_3$ nanoparticles are shown in Figure 8.2, and the particle size distribution histograms are presented as insets. It was found that the sizes of the $La_{1-x}Sr_xMnO_3$ particles with quasi-spherical morphology varied from 30 nm to 225 nm, exhibiting partially fused and interconnected states. For the $SrMnO_3$ sample, the particle size was distributed in the range of 100–550 nm. In addition, the average particle size of $La_{1-x}Sr_xMnO_3$ increased from 90 nm to 129 nm as the x value increased from 0.0 to 0.5, but a larger average size (255 nm) was found in the $SrMnO_3$ ($x = 1.0$) sample. The average crystallite sizes of the $La_{1-x}Sr_xMnO_3$ nanoparticles were estimated to be 111 ($x = 0.0$), 110 ($x = 0.3$), and 124 ($x = 0.5$), respectively, from the XRD patterns, matching well with the data obtained from the SEM images. The magnetic properties of these LSMO nanoparticles are demonstrated in Figure 8.3. A paramagnetic (PM) behavior is observed in the nonstoichiometric $LaMnO_{3+\delta}$ nanoparticles at 300 K (Figure 8.3a) whereas a ferromagnetic (FM) phase appears at 1.8 K with M_s of 59.54 emu/g (see Figure 8.3b). That is owing to the formation of La^{3+} vacancies in the $LaMnO_{3+\delta}$ nanoparticles, which makes some Mn^{3+} ions change as Mn^{4+} cations to maintain the lattice electric neutrality within the $LaMnO_{3+\delta}$ particles. The observed FM

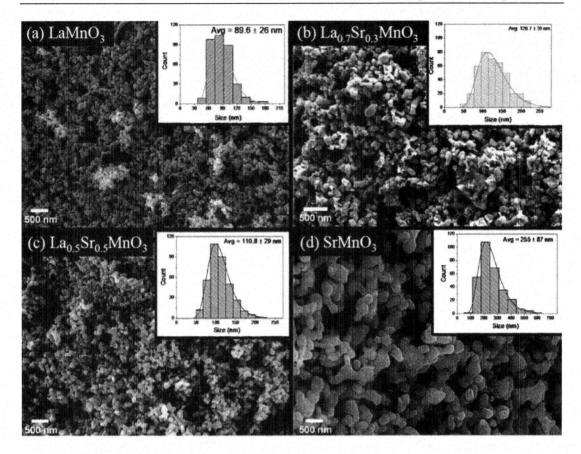

FIGURE 8.2 Representative SEM images taken from the as-synthesized $La_{1-x}Sr_xMnO_3$ nanoparticles. (a) $x = 0.0$, (b) $x = 0.3$, (c) $x = 0.5$, and (d) $x = 1.0$. Insets are the corresponding particle size distribution histograms along with their log-normal fits. Adapted with permission from [13]. Copyright (2020) Elsevier.

state at low temperature is ascribed to the strong DE interactions in the $LaMnO_{3+\delta}$ particles. At 300 K the $La_{0.7}Sr_{0.3}MnO_3$ nanoparticles exhibit a mixed (SPM–PM) behavior with M_s of ~ 26 emu/g (Figure 8.3b), but at 1.8 K M_s increased up to 56.90 emu/g. That was similar to that reported for the $La_{0.7}Sr_{0.3}MnO_3$ nanoparticles prepared by other methods such as mechanical ball milling with high energy [14] and facile molten salt synthetic route [15]. Similarly, the SPM ordering phenomenon was also observed in the $La_{0.5}Sr_{0.5}MnO_3$ nanoparticles at 300 K with M_s of ~ 29.59 emu/g, whereas the FM ordering appears at 200 K with M_s = 41.87 emu/g and H_c = 31.0 Oe, respectively (see Figure 8.3a–b). Based on the different M_s values of the $La_{1-x}Sr_xMnO_3$ nanoparticles at different temperatures (Figure 8.3b), the $La_{0.5}Sr_{0.5}MnO_3$ nanoparticles were found to be the best candidate for room-temperature magnetic hyperthermia. It was also noticed that antiferromagnetic (AFM) behavior appeared in the $SrMnO_3$ nanoparticles at all the temperatures (Figure 8.3a–b), which was attributed to the interactions of Mn^{4+}-O-Mn^{4+} bonds in $SrMnO_3$ nanoparticles. Figure 8.3c shows the magnetizations of the $La_{1-x}Sr_xMnO_3$ (x = 0.0, 0.3, 0.5, and 1.0) nanoparticles as a function of temperature measured under ZFC and FC modes with applied magnetic field of 100 Oe. In the $LaMnO_{3+\delta}$ nanoparticles, a magnetic transition from paramagnetic (PM) to ferromagnetic (FM) phase was observed around 300 K. The irreversibility temperature (T_{irr}), below which a large bifurcation appeared between the ZFC and FC curves, was found to be 130 K. The T_{irr} values for the $La_{0.5}Sr_{0.5}MnO_3$ and $La_{0.7}Sr_{0.3}MnO_3$ nanoparticles were determined to be 318 and 325 K, respectively (Figure 8.3c). Notably, in the $La_{0.7}Sr_{0.3}MnO_3$ nanoparticles, their FC and ZFC curves cross at the temperature T_{irr} rather than merge, as marked by a circle. Such a crossing/inversion between the ZFC and FC data

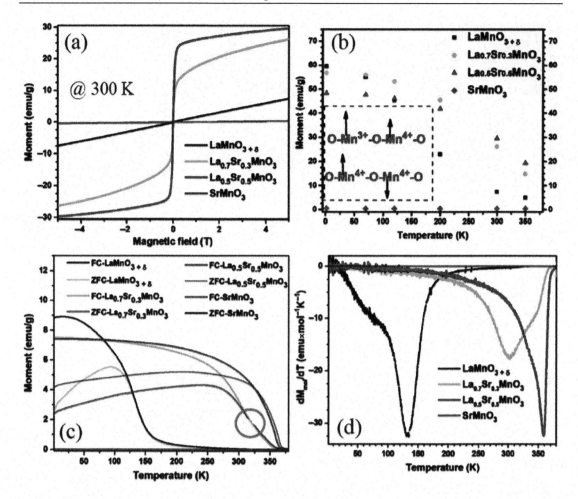

FIGURE 8.3 Magnetic properties of the as-synthesized La$_{1-x}$Sr$_x$MnO$_3$ nanoparticles. (a) *M-H* hysteresis loops measured at 300 K, (b) saturation magnetization (*M$_s$*) measured at 5 T as a function of temperature (inset demonstrates the relative magnitude and alignment of the atomic magnetic moments in Mn^{3+} and Mn^{4+} cations, (c) *M-T* curves measured under ZFC and FC modes at 100 Oe. (d) The first derivative of molar magnetization (*M$_{mol}$*) for the La$_{1-x}$Sr$_x$MnO$_3$ nanoparticles versus temperature, where the *T$_c$* values can be determined. Adapted with permission from [13]. Copyright (2020) Elsevier.

was reported in the La$_{0.275}$Pr$_{0.35}$Ca$_{0.375}$MnO$_3$ nanoparticles [16] and (La$_{1-x}$Ln$_x$)$_{0.67}$Ca$_{0.33}$MnO$_3$ (Ln = Nd and Sm; x = 0.0–0.5) nanoparticles [17]. This abnormal magnetic behavior can be ascribed to the existence of a frustrated spin system or a spin-glass-like surface disorder. The magnetic Curie temperature, T_c, is determined by the minimum of the first derivative of molar magnetization (M_{mol}) with respect to temperature, as depicted in Figure 8.3d. The T_c values of the La$_{1-x}$Sr$_x$MnO$_3$ nanoparticles (x = 0, 0.3, and 0.5) were determined to be 133, 303, and 358 K, respectively. The T_c value increases gradually with increasing the strontium mole fraction (x), which can be interpreted by the interionic spin interactions with the nearest neighbors.

8.3.1.2 Reactive Grinding Process

The reactive grinding process is a noble method for fabricating perovskite magnetic nanoparticles (with a high surface area) at room temperature. This process has the main feature of chemical reaction and phase transformation taking place simultaneously due to high-intensity mechanical milling by high-energy balls. The induced mechano-chemical activation by high-energy ball milling is the key step during the reactive grinding process, which influences the physicochemical properties of the initial reactants and the synthesized mechanism. Recently, cobalt-based perovskite oxide nanoparticles such as LaCoFeO$_3$ [18] and LaCo$_{1-x}$Fe$_x$O$_3$ [19] have been prepared by the reactive grinding method, and these nanoparticles exhibit high catalytic performance for the oxidation of hydrocarbons. That is ascribed to their high specific surface areas and high defects induced during the reactive grinding process.

Despite the fact that the particle sizes of the final products synthesized by the reactive grinding process at room temperature are on the nanometer scale, small batch sizes and much longer processing times restrict its commercial application in large-scale industrial production. Furthermore, some unfavorable contaminants coming from the milling media are the by-products of the high-energy ball-milling process and are harmful to the quality of the synthesized perovskite nanoparticles.

8.3.1.3 Combustion Method

The combustion method involves a quick and auto-continued exothermic reaction between the selected metal precursors, and the organic fuels, which act as a igniting agent and the complexing agents. Based on the different combustion rates, the combustion reaction is divided into three categories: decomposition, deflagration, or detonation. In this method, the gelation is formed after evaporating the solvents because of the poly-condensation of complexing agents. And then, in the drying process, the cationic precipitation is hindered by the immobile metal citrate complexes, favoring the formation of chemical homogeneity in the precursor solution. After drying, further annealing of the gel at a high temperature is required to obtain the crystalline nanoparticles with a pure phase. In general, ethylenediaminetetraacetic acid (EDTA) and citric acid (CA) are chosen as chelating agents in the combustion method. In comparison to citric acid, EDTA has a stronger chelating power, allowing it to chelate a greater number of cations. The latter, on the contrary, has a significantly stronger gelating ability and hence serves as an organic fuel to start the combustion reaction [20]. However, a strong exothermic and non-uniform combustion reaction takes place as citric acid is used as a chelating agent. That is detrimental to the morphology control in the final products. Therefore, the mixed citric acid and EDTA are usually utilized as a hybrid chelating agent. In addition, other organic fuels such as urea and glycine are also used in the combustion synthesis of perovskite oxide nanoparticles. Since a long time is required for the combustion process in the case of urea, the involved reactants can be fully combusted. Therefore, the obtained final products exhibit good crystallinity and high porosity. However, in the case of glycine as organic fuel, its ignition process of the reactant mixture is faster; thus, the flame temperature becomes much higher because of its higher melting point. Due to the short combustion reaction time, the involved reactants cannot be completely combusted. Shinde *et al.* [21] synthesized the La$_{1-x}$Sr$_x$MnO$_3$ nanoparticles by simple solution combustion technique and then annealed them at 600°C. High purity metal nitrate salts of La, Sr, and Mn were used as initial materials, and polyvinyl alcohol (PVA) was utilized as a fuel. Transmission electron microscopy (TEM) images revealed the crystalline nature of these as-synthesized La$_{1-x}$Sr$_x$MnO$_3$ nanoparticles. The average particle size varied from 30 nm to 40 nm. The S-shaped M-H loops were observed in all the La$_{1-x}$Sr$_x$MnO$_3$ (x = 0.1–0.3) nanoparticles, which also exhibit a sharp magnetic transition from the FM to PM phase, as shown in Figure 8.4a–c. The T_c value is increased from 275 to 285 K as the molar fraction of Sr (x) is increased from 0.1 to 0.3. Silva *et al.* [22] also synthesized LaNi$_{1-x}$Co$_x$O$_3$ perovskite nanoparticles by the same method. They found that the substitution of Ni for Co had apparent effects on the structural, textural, and reductive properties of the as-synthesized perovskite nanoparticles, which crystallized in a rhombohedral perovskite structure. Asefi *et al.* [23] synthesized nearly pure BiFeO$_3$ nanoparticles by solution combustion method with different fuel to oxidant ratios (φ). The crystallite size of the as-combusted BiFeO$_3$ nanopowders decreased from 32 nm to 21 nm as the φ value increased from 0.5 to 2.0, and the corresponding magnetization of the BiFeO$_3$ nanopowders also increased from 0.5 to 11.1 emu/g. Recently, ReFeO$_3$ (Re = La, Pr, Nd, Sm, and Gd) nanoparticles are also synthesized by the sol-gel combustion method [24]. XRD patterns revealed that all the ReFeO$_3$ nanoparticles had an orthorhombic perovskite

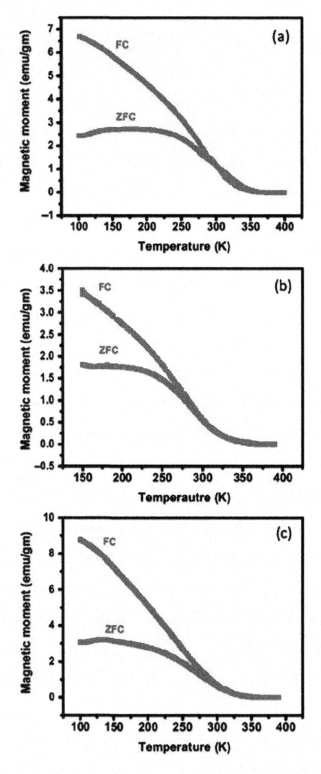

FIGURE 8.4 M-T curves of the as-synthesized La$_{1-x}$Sr$_x$MnO$_3$ (x = 0.1–0.3) nanoparticles measured under ZFC and FC modes with applied field H = 50 Oe. (a) x = 0.1, (b) x = 0.2, and (c) x = 0.3. Adapted with permission from [21]. Copyright (2012) Elsevier.

structure with the Pbnm space group. All nanoparticles exhibited weak FM behavior except for SmFeO$_3$. The magnetic behaviors of ReFeO$_3$ nanoparticles are originally from the magnetizations of the Re and Fe sublattices and a core-shell model.

8.3.1.4 Sol-Gel Method

The sol-gel method is a widely used method to synthesize perovskite magnetic nanoparticles from the liquid phase, where different metal precursors are mixed together in a solvent to get a solution. This method involves hydrolysis and condensation of alkoxide-based precursor solutions, conversion of the sol to gel, and then transformation of the gel into crystalline powders by heat treatment of organic matter. The main features of the sol-gel method are versatility, low processing temperature, low cost, environmentally friendly, and easy fabrication in the laboratory. It offers opportunities for access to organic-inorganic materials. Its processing parameters, such as types of reactants, the concentration, the pH value, and the annealing temperature, have great influences on the structural control, stoichiometry, and physical properties of perovskite oxide nanoparticles. For instance, Shlapa et al. [25] reported the influence of synthesis temperature on the structural and magnetic properties of the sol-gel-derived Nd-doped La$_{0.7}$Sr$_{0.3}$MnO$_3$ nanoparticles with crystallite sizes in the range of 30–39 nm. XRD patterns demonstrated that the synthesized La$_{0.7-x}$Nd$_x$Sr$_{0.3}$MnO$_3$ nanoparticles crystallized in the structurally distorted perovskite structure with space group $R\bar{3}c$. It was found that both crystallographic and magnetic parameters were not monotonously dependent upon the Nd content. In the past decade, the aqueous sol-gel method has been widely used for the synthesis of perovskite manganite nanoparticles. However, its major drawback of a high hydrolyzed rate usually leads to large particle size, high particle growth rate, less controllable morphology, and anionic impurities at low annealed temperatures. In contrast, a non-aqueous sol-gel method can stabilize many small particles via breaking the C-O linkages and creating a metal-oxygen bond without hydrolysis [26]. Sadhu and Bhattacharyya utilized a non-aqueous sol-gel method to synthesize La$_{0.71}$Sr$_{0.29}$MnO$_3$ nanoparticles with a small crystallite size of ~ 20 nm [27]. Figure 8.5 shows the microstructural characterization of these La$_{0.71}$Sr$_{0.29}$MnO$_3$ nanoparticles. Figure 8.5c reveals the lattice fringes with a spacing of 0.282 nm are resolved in the HRTEM image, corresponding to the interplanar spacing of the (110) plane. Low-field magnetoresistance was enhanced in these La$_{0.71}$Sr$_{0.29}$MnO$_3$ nanoparticles, approaching 29.8% at 30 K with 50 mT, and the high-field magnetoresistance reached 56% under a magnetic field of 5T.

Xia et al. [28] also reported the structural characterization and magnetic properties of Ln-doped (La$_{1-x}$Ln$_x$)$_{0.67}$Ca$_{0.33}$MnO$_3$ (Ln = Nd and Sm; x = 0.0–0.5) perovskite nanoparticles synthesized via the sol-gel process. Figure 8.6 and Figure 8.7 demonstrate the XRD patterns of the Nd-doped and Sm-doped (La$_{1-x}$Ln$_x$)$_{0.67}$Ca$_{0.33}$MnO$_3$ nanoparticles, respectively. They found that all these nanoparticles crystallized into an orthorhombic structure with a Pnma space group. With increasing the Ln-doping concentration, the particle unit cell volume decreased at a faster rate in the Sm^{3+} doping than in the Nd^{3+} doping due to the steric effect. In addition, nanoparticles became much more agglomerated, and their morphology changed to an irregular shape from a spherical one. Smaller M_s and higher H_c were observed in the Ln-doped nanoparticles compared to the pristine ones. This was attributed to the presence of AFM interactions. The temperature dependence of the dc magnetizations of the Nd-doped (La$_{1-x}$Nd$_x$)$_{0.67}$Ca$_{0.33}$MnO$_3$ (x = 0.0–0.5) nanoparticles measured under ZFC and FC modes with a magnetic field of 0.05 T is shown in Figure 8.8. All the nanoparticles underwent a PM-FM phase transition with a reduction of T_c by increasing the Nd-doping concentration. Similarly, the freezing temperature (T_f) and irreversibility temperature (T_{irr}) also shifted towards lower temperatures.

The sol-gel method is also used to synthesize perovskite BiFeO$_3$ (BFO) nanoparticles. Wu et al. [29] carried out this work and characterized the microstructures of high-purity BFO nanoparticles synthesized via a sol-gel process. Structural characterizations by XRD, Raman, and IR spectra confirmed the BFO nanoparticles crystallized into a distorted rhombohedral perovskite structure with a R3c space group. The particle size of BFO nanoparticles varied from 30 to 200 nm. A weak ferromagnetic behavior was observed at 2 K with M_r and H_c equal to 0.117 emu/g and 1408 Oe, respectively. Modified sol-gel processes

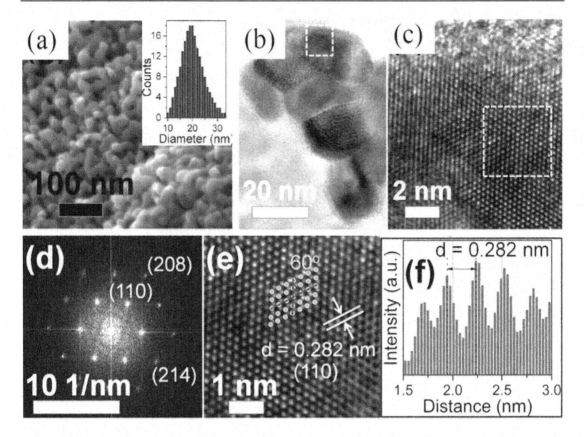

FIGURE 8.5 Microstructural characterizations of the $La_{0.71}Sr_{0.29}MnO_3$ nanoparticles synthesized by non-aqueous sol-gel method. (a) Field-emission SEM image. Inset is the corresponding diameter histogram. (b) TEM image and (c) high-resolution TEM image of the selected area in panel b. (d) Fast Fourier transformation (FFT) of the selected area in panel c. (e) Inverse FFT and (f) height profile to identify the crystallographic planes and interplanar distance in the nanoparticles. Adapted with permission from [27]. Copyright (2014) American Chemical Society.

(e.g., Pechini method, polymer complex solution, and glycol-gel reaction) are also used to synthesize BFO nanoparticles.

8.3.1.5 Hydrothermal/Solvothermal/Microwave-Hydrothermal Method

Hydrothermal and solvothermal methods are often used to synthesize perovskite nanomaterials. The only difference between them is that the hydrothermal process is usually carried out in an aqueous medium, whereas the solvothermal process is performed in an organic medium such as NH_3, methanol, ethanol, benzyl alcohol, and n-propanol. In the hydrothermal process, the processing meters, such as the types of starting materials, reaction temperature, pH value, concentration, and type of mineralizers, play an important role in the compositions and morphologies of the synthesized products [30]. For instance, Remya et al. [31] demonstrated a tunable morphology and size of the perovskite BFO nanostructures, where BFO nanoparticles, nanorods, nanoflakes, and microflowers were produced during the surfactant-assisted hydrothermal process. A schematic diagram illustrating the tuning of the morphologies of the BFO nanostructures is demonstrated in Figure 8.9, where the corresponding morphology revealed by SEM images is also presented. It was found that the sphere-like BFO nanoparticles were obtained as urea was used as surfactant agent plus with potassium hydroxide (KOH), while BFO nanorods were obtained

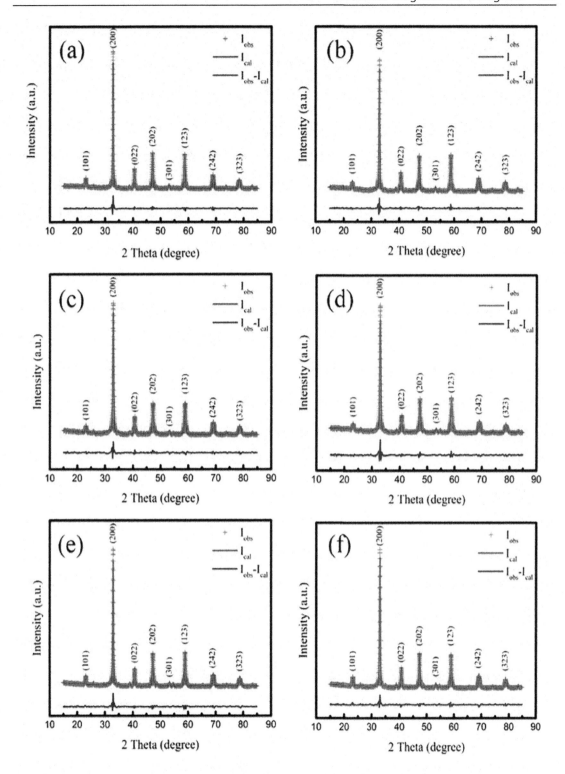

FIGURE 8.6 Experimental (grey cross), calculated (red line), and the difference between them (blue line) of the x-ray powder diffraction patterns of the Nd-doped $(La_{1-x}Nd_x)_{0.67}Ca_{0.33}MnO_3$ ($x = 0.0–0.5$) nanoparticles measured at room temperature. (a) $x = 0.0$, (b) $x = 0.1$, (c) $x = 0.2$, (d) $x = 0.3$, (e) $x = 0.4$, and (f) $x = 0.5$. Adapted with permission from [28]. Copyright (2021) Elsevier.

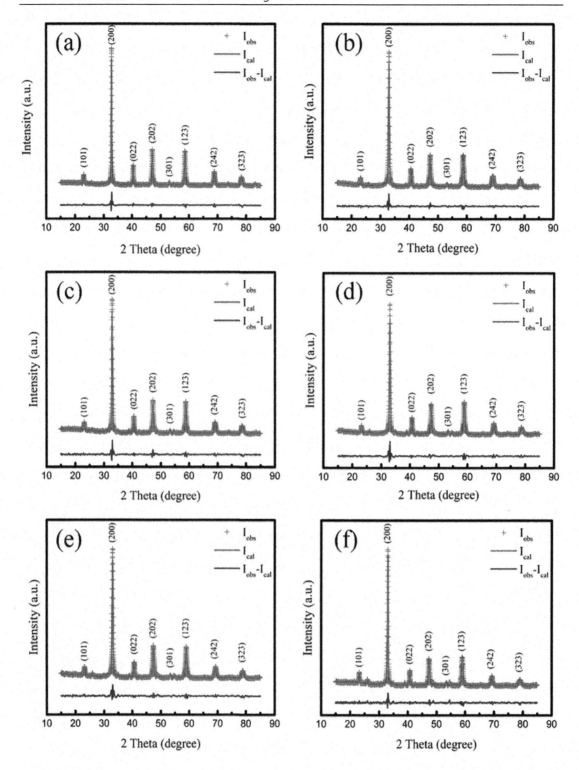

FIGURE 8.7 Experimental (grey cross), calculated (red line), and the difference between them (blue line) of the x-ray powder diffraction patterns of the Sm-doped $(La_{1-x}Sm_x)_{0.67}Ca_{0.33}MnO_3$ (x = 0.0–0.5) nanoparticles measured at room temperature. (a) x = 0.0, (b) x = 0.1, (c) x = 0.2, (d) x = 0.3, (e) x = 0.4, and (f) x = 0.5. Adapted with permission from [28]. Copyright (2021) Elsevier.

8 • Perovskite Magnetic Nanomagnets 137

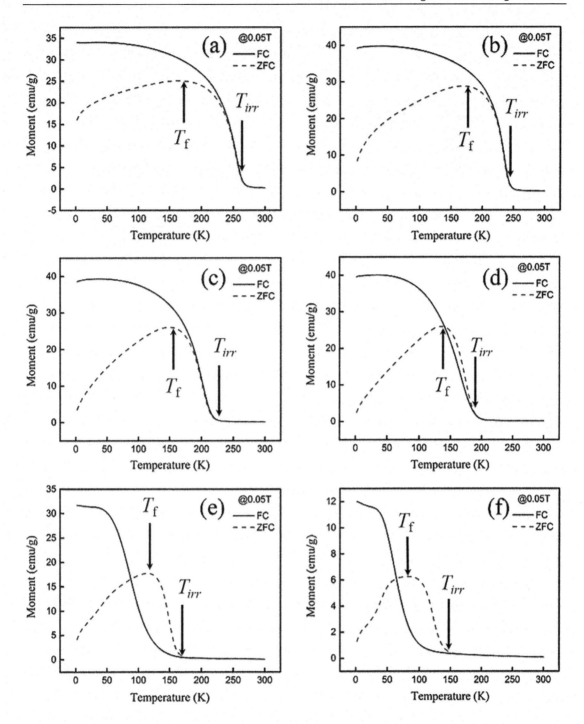

FIGURE 8.8 Temperature dependence of dc magnetizations of the Nd-doped $(La_{1-x}Nd_x)_{0.67}Ca_{0.33}MnO_3$ (x = 0.0–0.5) nanoparticles measured under ZFC and FC modes and with the magnetic field of 0.05 T. (a) x = 0.0, (b) x = 0.1, (c) x = 0.2, (d) x = 0.3, (e) x = 0.4, and (f) x = 0.5. Adapted with permission from [28]. Copyright (2021) Elsevier.

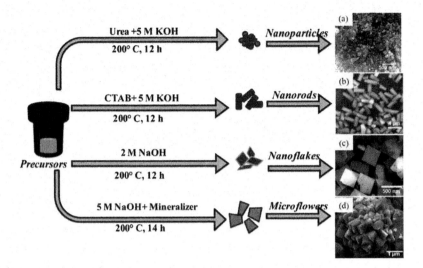

FIGURE 8.9 Schematic diagram illustrating the tunable morphologies of BFO nanostructures under different hydrothermal conditions. Adapted with permission from [31]. Copyright (2020) Elsevier.

when the surfactant agent was changed to cetyltrimethylammonium bromide (CTAB). BFO nanoflakes were obtained via the normal hydrothermal process under 2 M NaOH without any surfactant agent. These nanoflakes had sharp edges and smooth surfaces, as shown in Figure 8.9c. It is found that the concentration of NaOH in the precursor solution is a key parameter for the growth of nanoflakes, which is controlled by the Ostwald ripening mechanism.

The room temperature magnetic properties of BFO nanostructures with different morphologies grown by the hydrothermal process are shown in Figure 8.10. The M_s values of the nanoparticles, nanoflakes, nanorods, and microflowers were determined to be 0.36, 0.27, 0.55, and 0.67 emu/g, respectively. The higher M_s value observed in the microflowers is attributed to their larger shape anisotropy. It is also found that the added mineralizers and reaction conditions play an important role in the phase structure and morphology of the hydrothermal final products. Han et al. [32] reported that during the hydrothermal synthesis of BFO nanostructures, the α-Bi_2O_3 phase was the main hydrothermal product at lower temperatures (e.g. 150°C). However, its dissolution rate was enhanced as increasing the temperature from 150°C to 175°C, and a pure BFO phase was obtained. Despite the temperature being increased from 175°C to 225°C, the phase does not change, only leading to larger particle sizes. In addition, the increases of pH value, reaction time, and temperature could enhance the solubility of α-Bi_2O_3 effectively, promoting the formation of pure BFO phases. When replacing KOH with NaOH in the hydrothermal process, a stronger alkaline precursor solution and a higher reactive temperature lead to stoichiometric products. For example, the addition of a small amount of H_2O_2 into the stronger alkaline conditions (with a pH value of 14) favors the formation of pure phase BFO [33], while the nonstoichiometric product of $Bi_{12}Fe_{0.63}O_{18.945}$ is formed under moderate alkaline precursor solution (pH = 8–12) and low reaction temperature. Besides the hydrothermal reaction conditions, the alkali metal ions also strongly affect the chemistry of the synthesized phase. BFO nanoparticles with pure phase structure were formed at 200°C with KOH concentrations of 7 mol/L and 12 mol/L, whereas NaOH and $LiNO_3$ solutions stabilized the final products of $Bi_2Fe_4O_9$ and $Bi_{12}(Bi_{0.5}Fe_{0.5})O_{19.5}$, respectively, disregarding their concentrations [34].

Microwave-hydrothermal process is only a modified hydrothermal process. The precursors used in the microwave-hydrothermal process are similar to those used in the hydrothermal process, but the autoclave in the microwave-hydrothermal process is heated via microwave radiation. Such a heating method offers some advantages over conventional heating during the hydrothermal process, such as no direct contact between the heating source and the heated objects with energy and time savings due to rapid internal

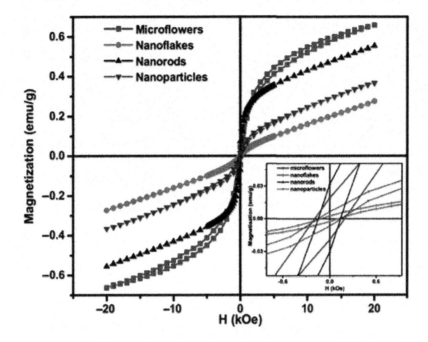

FIGURE 8.10 Room temperature magnetization (*M*)-magnetic field (*H*) hysteresis loops of the BFO nanoparticles, nanorods, nanoflakes, and microflowers grown by surfactant-assisted hydrothermal method. Inset is enlarged hysteresis loops. Adapted with permission from [31]. Copyright (2020) Elsevier.

heating. Recently, perovskite BFO nanoparticles with a particle size of 10–50 nm were synthesized by the microwave-hydrothermal process, where Na_2CO_3 was used with KOH as the mineralizer [35]. In addition, BFO nanocubes with an average size of 50–200 nm were also synthesized by this method [36]. Recent reviews provide more details about the reaction mechanism of the microwave-hydrothermal process for perovskite oxide nanomaterials [37–38].

8.3.1.6 Microemulsion Route

The microemulsion route is the preferred method, as it can synthesize nanoscale particles with excellent surface morphology. The microemulsions are characterized as thermodynamically stable droplets (with an average size of 5–50 nm) formed by mechanical stirring. They offer some characteristic features, such as the exploration of novel crystallization pathways and the possibility to crystallize the targeted particles at room temperature [39]. In a water-oil microemulsion system, the nanosized aqueous micelles (water phase) with spherical morphology are dispersed within an oil matrix, acting as a nanoreactor cavity and templates for the synthesis of perovskite oxide nanoparticles. Lim et al. [40] synthesized the $Nd_{0.67}Sr_{0.33}CoO_{3-\delta}$ (NSC) nanoparticles with a particle size of 20–50 nm and a specific surface area of ~ 13 m²/g via an inverse microemulsion method, which is schematically depicted in Figure 8.11. These NSC nanoparticles exhibit good electrocatalytic activity, especially in the oxygen evolution reaction. Such good catalytic performance is attributed to the formation of the Co^{3+}/Co^{4+} redox couple and the large surfaces. By this method, BFO nanoparticles were synthesized, where the microemulsion system consisted of CTAB/water/isooctane/butanol [41]. In addition, the Ni and Co co-doped $BiFe_{1-x}Ni_xCo_xO_3$ ($x = 0$–0.2) nanoparticles were also synthesized by the microemulsion method, where the CTAB solution was used as a surfactant agent [42]. It is found that the $BiFe_{0.85}Ni_{0.15}Co_{0.15}O_3$ nanoparticles exhibit ferromagnetic behavior with a maximum M_s value of 10.92 emu/g. Such enhanced magnetism is ascribed to the doped elements with a different magnetic moment as well as the magnetic Fe^{3+} ions.

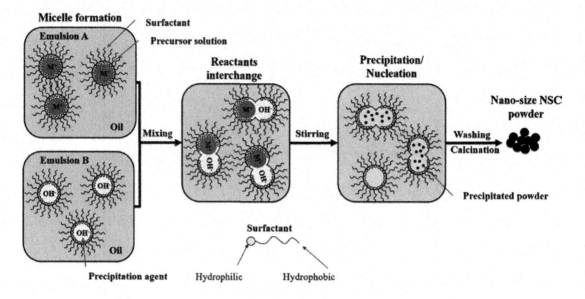

FIGURE 8.11 Schematic diagram illustrating the synthesis of $Nd_{0.67}Sr_{0.33}CoO_{3-\delta}$ (NSC) nanoparticles via an inverse microemulsion method (water-in-oil system). Adapted with permission from [40]. Copyright (2018) Elsevier.

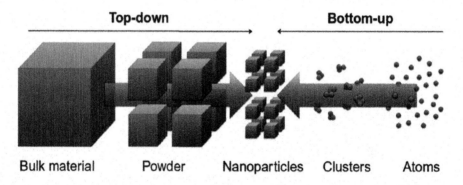

FIGURE 8.12 Schematic diagram illustrating the top-down and bottom-up nanofabrication approaches. Adapted with permission from [44]. Copyright (2015) Elsevier.

8.3.2 One-Dimensional (1D) Perovskite Magnetic Nanomagnets

One-dimensional perovskite magnetic nanomagnets (e.g., nanowires, nanofibers, nanotubes, and nanoribbons) have been given much attention due to their fascinating size-, shape-, and material-dependent physical properties. Their intriguing electronic, magnetic, and optical properties have enabled them to be widely used in the fields of high-density memories, logic devices, barcodes, catalysis, and solar cells [43]. Up to now, two rational approaches have been developed for preparing these 1D perovskite magnetic nanostructures, which are called top-down and bottom-up approaches, respectively, as illustrated in Figure 8.12 [44]. In the top-down approach, one begins with bulk material and then cuts it into smaller pieces until the desired nanosize is achieved, while in the bottom-up approach, one starts with atoms or molecules, which are used as building units to assemble the desired complex structures.

8.3.2.1 Top-Down Approach

The top-down approach works principally on nanolithographic techniques such as e-beam, x-ray, and focused ion beam (FIB) [45]. Lithographic techniques are standard examples of top-down fabrication techniques, that construct diverse nanostructures, viz. nanowires, nanostripes, and patterned magnetic thin films with high repeatability. However, in a top-down approach, internal stresses, surface effects, structural defects, and contamination are usually introduced to the final nanostructures. Recently, Lorena et al. [46] fabricated perovskite $La_{2/3}Ca_{1/3}MnO_3$ nanowires with a large aspect ratio (length-to-width over 300) from a single crystal by the FIB method. Details are shown in Figure 8.13. An enhanced magnetoresistance behavior was observed in the $La_{2/3}Ca_{1/3}MnO_3$ nanowire with a diameter of 150 nm at 0.1 T magnetic field. This is due to the strain release at the edges and the destabilization of the insulating regions. The resistivity and the metal-insulator transition temperature (T_{MI}) are found to be greatly sensitive to the width of the nanowires. By reducing the nanowire diameter, the activation energy and residual resistivity were decreased. Remarkably, an enhanced magnetoresistance observed at T_{MI} with a low magnetic field enables these nanowires to be implemented in the spintronic devices. Besides the perovskite oxide magnetic nanowires, the FIB-based top-down approach is also used to fabricate metal nanowires.

8.3.2.2 Bottom-Up Approach

The bottom-up approach works principally on constructing nanostructures from the bottom up via self-assembly and self-organization of atoms, molecules, or clusters. During the whole process, the Gibbs energy of the system is reduced, which enables the self-organization process to take place spontaneously; therefore, the energy state of the formed nanostructures approaches the thermodynamic equilibrium. The bottom-up approach is dependent upon the chemical synthesis and the formed mesoscopic patterns at the initial stage of the reaction. Therefore, via a bottom-up approach, perovskite magnetic nanowires/nanotubes can be fabricated with suitable templates (e.g., anodic alumina oxide [AAO], silicon, and

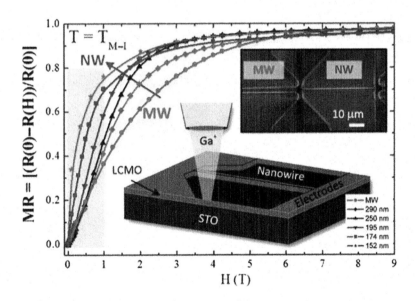

FIGURE 8.13 Magnetoresistance (MR) (defined as (R(0)-R(H))/R(0)) of the $La_{2/3}Ca_{1/3}MnO_3$ nanowires with different widths (w = 290, 250, 195, 174, and 152 nm) compared with that of the micro-wire (width = 5 μm) at T_{MI}. Insets are schematic diagrams illustrating the fabrication of $La_{2/3}Ca_{1/3}MnO_3$ nanowires by FIB and SEM images of the $La_{2/3}Ca_{1/3}MnO_3$ nanowire and micro-wire. Adapted with permission from [46]. Copyright (2014) American Chemistry Society.

polycarbonate membranes with ordered cylindrical porosity) [47]. Up to date, several methods following this approach have been developed, such as the sol-gel process, chemical vapor deposition, molecular beam epitaxy, and molecular self-assembly. Among them, the sol-gel template method is the most popular one, and it involves the template of AAO membranes and filling them up with the desired polymeric precursors via the sol-gel process. After heat treatment, the formed perovskite magnetic materials are subsequently separated from the template by etching the latter away. Finally, perovskite oxide nanowires/nanotubes can be prepared. The sizes, shapes, and structural features of the assembly are well controlled by the used templates. For example, perovskite $La_{1-x}Ca_xMnO_3$ (LCMO, $x = 0.20$) nanowires and an ordered array of $La_{0.67}Sr_{0.33}MnO_3$ nanowires were fabricated by the sol-gel template method, where AAO membranes were used as templates [48]. Similarly, both $La_{0.6}Sr_{0.4}CoO_3$ nanowires and nanotubes were fabricated by injecting the $La_{0.6}Sr_{0.4}CoO_3$ sol into the cylinder pores of the AAO template [49]. In addition, multiferroic BFO nanowires and nanotubes are also synthesized by the sol-gel template method [50–52]. Some representative 1D BFO nanostructures are shown in Figure 8.14. Figure 8.14a displays the TEM image of BFO nanowires fully separated from the AAO template, and Figure 8.14b shows the TEM image of a single BFO nanowire [50]. A SEM image of the BFO nanotubes is shown in Figure 8.14c [51], and Figure 8.14d is the TEM image of a single BFO nanotube [52]. As demonstrated in these SEM/TEM images, the sizes and shapes of perovskite oxide nanowires and nanotubes can be well controlled by the templates; however, these perovskite oxide nanowires/nanotubes generally have a polycrystalline

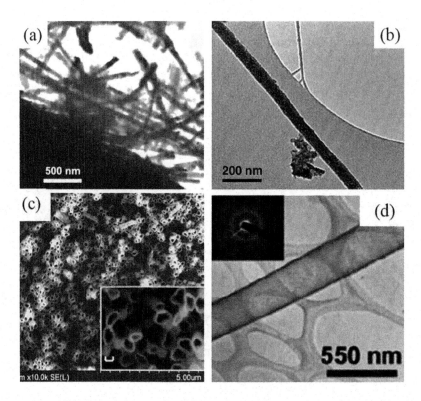

FIGURE 8.14 Structural characterizations of BFO nanowires and nanotubes synthesized by the sol-gel template method. (a) TEM image of the BFO nanowires completely separated from AAO template, (b) TEM image of a single BFO nanowire. Adapted with permission from [50]. Copyright (2005) Elsevier. (c) SEM image of BFO nanotubes. Inset shows a vertical BFO nanotube array at higher magnification (scale bar = 100 nm). Adapted with permission from [51]. Copyright (2008) Elsevier. (d) TEM image of the BFO nanotubes (inset shows the selected area electron diffraction [SAED] patterns of an individual nanotube). Adapted with permission from [52]. Copyright (2004) Royal Society of Chemistry.

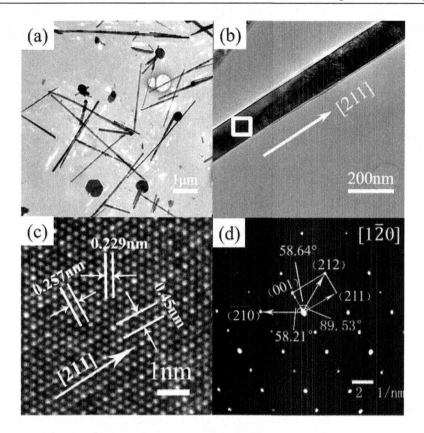

FIGURE 8.15 Microstructural characterizations of the BiFeO$_3$ nanowires grown by a modified hydrothermal method. (a) TEM image of BiFeO$_3$ nanowires, (b) TEM image of a single BiFeO$_3$ nanowire, (c) HRTEM image taken from the rectangular area in (b), and (d) SAED pattern taken from a single BiFeO$_3$ nanowire. Adapted with permission from [53]. Copyright (2011) Royal Society of Chemistry.

structure rather than a single-crystalline nature. That is attributed to the heterogeneous nucleation on the pore walls of the used templates.

Besides the template-assisted methods, 1D perovskite oxide nanostructures (e.g., nanowires, nanorods, nanotubes) have also been synthesized by template-free methods (e.g. hydro- and solvothermal synthesis, molten-salt method, electrophoretic deposition method, and electrospinning process). Single-crystalline BFO nanowires were also synthesized by an improved hydrothermal method [53]. Figure 8.15 shows their microstructural characterizations. Both the HRTEM image and the SAED pattern reveal the single-crystalline nature of the BFO nanowires. Figure 8.16 shows the magnetization measurements under ZFC and FC modes and *M-H* hysteresis loops at different temperatures. A spin-glass transition below the freezing temperature, T_f (= 55 K), was observed in these BFO nanowires. Single-crystalline perovskite La$_{0.5}$Ca$_{0.5}$MnO$_3$ nanowires were also synthesized by the hydrothermal method [54]. These nanowires grew along the [100] direction with a diameter of ~ 80 nm and lengths up to several tens of micrometers.

Recently, Xia et al. [55] reported on the hydrothermal growth of single-crystalline (La$_{0.6}$Pr$_{0.4}$)$_{0.67}$Ca$_{0.33}$MnO$_3$ manganite nanowires. These nanowires had an orthorhombic structure (Pnma space group). Their lattice parameters (*a*, *b*, and *c*) satisfy a relationship of $a \approx c \approx b/\sqrt{2}$. Structural characterizations of these nanowires by TEM and HRTEM are shown in Figure 8.17. TEM images (Figure 8.17a and b) reveal that these nanowires have a clean and smooth surface with diameters of 60–120 nm and an average length of ~ 2.0 μm. HRTEM images (Figure 8.17c and d) confirm the single-crystalline nature of the nanowire with [100] growth direction. At a low temperature of 2 K, an *M-H* hysteresis loop was

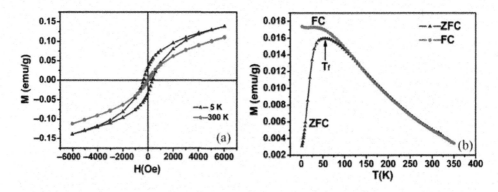

FIGURE 8.16 Magnetic characterizations of the BiFeO$_3$ nanowires grown by a modified hydrothermal method. (a) M-H hysteresis loops of the BiFeO$_3$ nanowires measured at different temperatures. (b) Temperature dependence of the magnetizations of the BiFeO$_3$ nanowires measured under ZFC and FC modes with a magnetic field of 100 Oe. Adapted with permission from [53]. Copyright (2011) Royal Society of Chemistry.

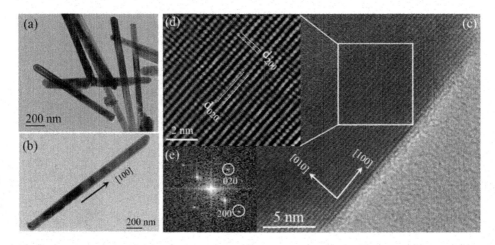

FIGURE 8.17 Structural characterizations of the (La$_{0.6}$Pr$_{0.4}$)$_{0.67}$Ca$_{0.33}$MnO$_3$ nanowires grown by hydrothermal method. (a) Typical TEM image of the nanowires. (b) TEM image taken from a single nanowire. (c) HRTEM image of a single nanowire shown in (b). (d) Fourier-filtered image of the position marked by the square in (c), where two sets of (200) and (020) planes are resolved. (e) FFT pattern of the local HRTEM image marked by a square in (c), where sharp reflections are observed. Adapted with permission from [55]. Copyright (2021) American Ceramic Society.

observed in these nanowires. The magnetic phase transition from PM to FM was observed at 224 K, and a bifurcation between the $M_{ZFC}(T)$ and $M_{FC}(T)$ curves appeared at 273 K. Datta et al. [56] also performed detailed studies on the hydrothermal growth mechanism of perovskite manganite La$_{1-x}$A$_x$MnO$_3$ (A = Sr, Ca; x = 0.3 and 0.5) nanowires and proposed a phase diagram of hydrothermal synthesis of perovskite manganite La$_{1-x}$A$_x$MnO$_3$ (A = Sr, Ca; x = 0.3 and 0.5) nanostructures.

8.3.3 Two-Dimensional (2D) Perovskite Magnetic Nanomagnets

The magnetic properties of ordered arrays of perovskite magnetic nanoparticles integrate the magnetic properties of the building blocks with the collective properties arising from the particle interactions. That

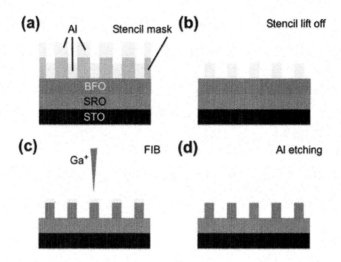

FIGURE 8.18 Schematic diagram illustrating the fabricated process of periodic arrays of BFO nanoislands. (a) Al evaporation through stencil mask over the epitaxial BFO thin film, (b) mask lifting out and resulting in an array of Al dots on the BFO film, (c) ion beam milling, yielding BFO nanoislands capped with Al, and (d) chemical etching to completely remove the remaining Al, leading to the resulting BFO nanoislands. Adapted with permission from [59]. Copyright (2013) American Institute of Physics.

enables the novel possibilities of material design, where both the magnetic properties of the nanoparticles and their interactions can be modulated. This is beneficial for improving the stored density of magnetic memory devices and designing new spin sensors and spintronic devices. Recently, theoretical work has been done to investigate the magnetic properties of 2D arrays of magnetic nanoparticles with different sizes, shapes, and array structures [57]. The results show that without magnetic anisotropy, vortex and flux closure structures are formed in 2D arrays of magnetic nanoparticles with infinite numbers, which are also confirmed by experimental observations [58]. To better understand the collective magnetic properties of 2D arrays of perovskite oxide nanoparticles, high-quality samples of 2D perovskite magnetic nanomagnets should be fabricated by top-down or bottom-up approaches.

Morelli et al. [59] fabricated a periodic array of BFO nanoislands by a template-assisted FIB technique. First, epitaxial BFO thin films were deposited on the vicinal $SrTiO_3(100)$ substrate (coated with a $SrRuO_3$ [SRO] thin layer as a bottom electrode) by the pulsed laser deposition (PLD) method. Then, periodic arrays of 45 nm thick Al dots were fabricated on the BFO thin film via a stencil mask (with an aperture diameter of 400 nm) (Figure 8.18a and b). A focused Ga^+ ion beam with a large beam current was employed to remove the parts of the BFO thin film coated by Al dots (Figure 8.18c) to yield BFO islands covered by Al. Finally, the remaining Al was completely etched away by using a 10% KOH aqueous solution. Finally, periodic arrays of epitaxial BFO nanoislands with a diameter ~ 250 nm could be fabricated, as schematically shown in Figure 8.18d. By using the template-assisted FIB method, the precise positions, shapes, and sizes of the BFO nanostructures can be well-tuned. However, the consumed time and low throughput restrain the application of the top-down approach to large-scale, inexpensive production of 2D perovskite oxide magnetic nanostructures.

In the past decade, bottom-up approaches have been used to fabricate large-scale production of 2D magnetic nanostructures. Among them, the template-based method is the most widely used technique for the growth of nanowire arrays. For instance, $La_{0.67}Sr_{0.33}MnO_3$ nanowire arrays were fabricated via the sol-gel template method, where AAO membranes were used as templates [60]. SEM and TEM images demonstrated that a vast quantity of $La_{0.67}Sr_{0.33}MnO_3$ nanowires were synthesized. These nanowire arrays exhibited good ferromagnetic properties at 10 K and 300 K respectively, and their Curie temperature was about 350 K. Zhang et al. [61] also prepared ordered multiferroic BFO nanotube arrays by a sol-gel

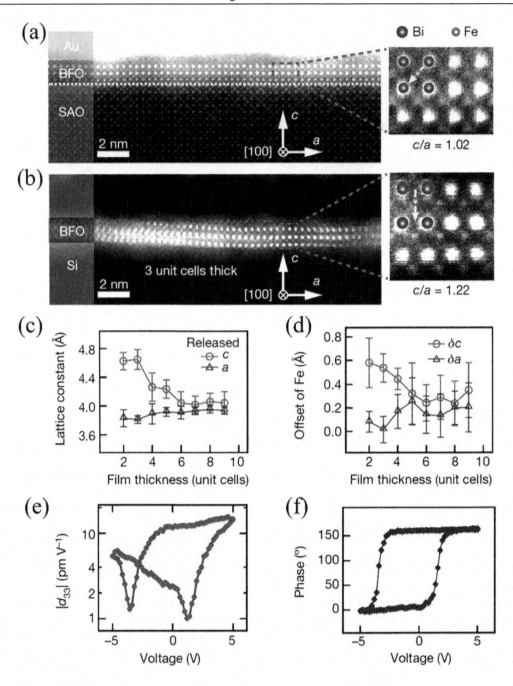

FIGURE 8.19 High polarization and structural lattice distortions in the ultrathin freestanding BFO films. (a, b) Cross-sectional high-angle annular dark-field images of a three-unit-cell BFO film before and after releasing the film, showing an R-like phase with polarization along with the <111> directions and a T-like phase with polarization along the <001> direction, respectively. (c, d) The c/a ratio and the offset (δ_c and δ_a) of Fe ions from the centers of four neighboring Bi ions as a function of the thickness of freestanding BFO films, showing the evolution from an R-like to a T-like phase transition as the thickness of the film decreases. The error bars in c and d represent the fitting error of the lattice constants. (e, f) PFM amplitude-voltage butterfly loop and phase-voltage hysteresis loop of a four-unit-cell freestanding BFO film on a conductive silicon substrate, demonstrating the polarization are switchable. d_{33} is the out-of-plane piezoelectric coefficient. Adapted with permission from [62]. Copyright (2019) Springer Nature.

method utilizing AAO membrane as templates. As confirmed by the XRD pattern and SAED pattern, after post-annealing at 700°C, these BFO nanotubes had a polycrystalline microstructure. Recently, the freestanding ultrathin BFO films with a thickness down to a monolayer (2D BFO nanostructures) were also prepared by a bottom-up approach, where the water-soluble $Sr_3Al_2O_6$ was used as a sacrificial buffer layer [62]. As the thickness of the BFO membrane approaches the 2D limit, the freestanding BFO films exhibit giant tetragonality and large polarization, as demonstrated in Figure 8.19. With this ability, it is expected that 2D perovskite oxide membranes could have promising applications in 2D correlated quantum phases. However, at present, the fabrications of perovskite oxide 2D nanostructures are still in their initial stages, and much work remains to be done in future applications in the field of novel microelectronic devices.

8.3.4 Three-Dimensional (3D) Perovskite Magnetic Nanomagnets

Three-dimensional magnetic nanostructures provide a wide range of exciting physical phenomena in nanomagnetism, where curved geometries provide new types of exotic magnetic configurations, such as topologically protected and chiral magnetic textures [6, 63]. The 3D magnetic memories allow multiple data bits to be stored magnetically above a single electronic memory cell on an integrated circuit, improving the performance of magnetic random access memory (MRAM) by increasing the number of data stored in magnetic layers within the stack [64]. Therefore, the growth of high-purity, narrow 3D ferromagnetic structures is the gateway to a broad range of opportunities for both fundamental and technological applications. Up to date, three main technique routes have been developed for the synthesis of 3D nanomagnets, which are physical, chemical, and 3D nano-printing methods (Figure 8.20) [6].

The physical vapor deposition (PVD) method can be applied to fabricate 3D magnetic nanostructures as the deposition process is performed onto pre-patterned 3D scaffolds rather than flat substrates. However, conformal deposition and material shadowing are the main challenges of this method. The anisotropic feature of the PVD method makes it impossible to fabricate ultra-small nanohelices with a strong magneto-chiral dichroic effect. Instead, another complementary approach for fabricating 3D nanostructures is rolled-up nanotechnology, which makes full use of the internal strain gradients to produce tubular magnetic geometries with controllable diameter, curvature, and thickness [65]. 3D nanostructures are often fabricated by chemical deposition methods. For example, electrochemical deposition in

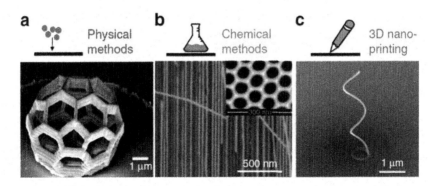

FIGURE 8.20 Schematic diagram illustrating the three main technique routes for fabricating 3D nanomagnets. (a) A buckyball made by two-photon optical lithography, where cobalt is sputtered on top. (b) Cylindrical nanowires electrodeposited from a $CoSO_4$ electrolyte on alumina templates (see inset). (c) A nanospiral fabricated by 3D nano-printing using $Co_2(CO)_8$. Adapted with permission from [6]. Copyright (2017) Springer Nature. The article was printed under a CC-BY license.

combination with porous alumina or gyroid polymer templates enables the synthesis of nanohelices and branched 3D networks with variable diameters (< 50 nm) and length scales (~1 μm) [66]. The remarkable interfacial control between materials achieved during the chemical deposition process allows investigation of the possible spin-transport effects at the interface between layers. However, atomic-scale control in the thickness direction is not available, which hinders the exploiting of spin-transport effects in the perpendicular direction. In recent years, 3D nano-printing realized via focused electron beam induced deposition (FEBID) has been developed, and much progress has been made in FEBID computational methods [67]. This allows the FEBID technique to fabricate individual complex prototype 3D nanostructures with a spatial resolution of a few tens of nanometers. Thus, FEBID is a powerful tool for fabricating 3D nanomagnetism, which allows one to study the emerging physics involved in the complex 3D geometries (e.g. 3D spin textures, new types of magnetic domain walls, and their dynamics) as well as nano-prototyping. Up to date, 3D geometries such as 3D actuators, nanohelices, nanoellipsoids, and nanobridges have been created by the FEBID method [6]. As an additive nanofabrication technique, FEBID offers great versatility to fabricate complex motifs and architectural designs at the nanoscale based on numerous materials – and, in particular, ferromagnetic materials for high-density, low-power applications, such as memories, sensors, and actuators. It is expected that the combination of the FEBID technique with other techniques, such as atomic layer deposition, having the ability to control the deposition at an atomic scale and to deposit conformally on very high aspect ratio structures, could provide exciting opportunities in the emerging field of 3D nanomagnetism [6].

Recently, a 3D nano-template in combination with the PLD method has been used to fabricate a $(La_{0.275}Pr_{0.35}Ca_{0.375})MnO_3$ (LPCMO) nanobox (3D magnetic nanostructure) [68]. In this method, a 3D nano-patterned substrate is developed and used as a 3D nano-template. Its fabricated flowchart by the PLD method is schematically shown in Figure 8.21. Figure 8.22 demonstrates the fabrication process of the LPCMO nanobox, its structural characterization by SEM, and the well width of the LPCMO nanobox

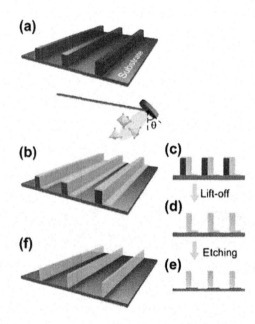

FIGURE 8.21 A flowchart of the 3D nano-template PLD method for fabricating perovskite oxide nanostructure. (a) First, template wall structures are patterned onto a substrate by nano-imprint lithography using an organic resist (blue region). (b) Perovskite oxide is then deposited onto the side surface of the template patterns by PLD. (c) Cross-sectional image of (b). Cross-sectional images for the nanowall-wire structure after (d) lift-off and (e) etching. (f) Finally, self-standing perovskite oxide nanowall-wire arrays are obtained. Adapted with permission from [68]. Copyright (2012) IOP Publishing.

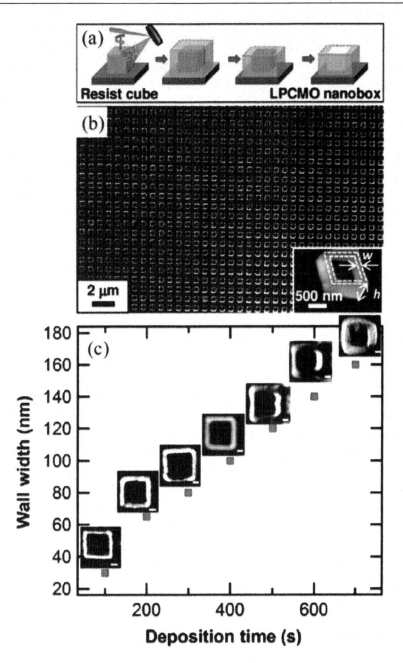

FIGURE 8.22 (a) Schematic diagram illustrating the fabrication procedure of the $(La_{0.275}Pr_{0.35}Ca_{0.375})MnO_3$ nanoboxes and (b) typical SEM image. (c) Wall width of the $(La_{0.275}Pr_{0.35}Ca_{0.375})MnO_3$ nanoboxes as a function of the deposition time. Adapted with permission from [69]. Copyright (2013) American Institute of Physics.

as a function of the deposited time [69]. By changing the deposition time, the wall width of the LPCMO nanoboxes was controlled from 30 nm to 160 nm. An insulator-metal transition was observed in these LPCMO nanoboxes at a higher temperature than that in the corresponding film. This indicates that the LPCMO nanoboxes have potential applications in the field of oxide spintronics. The 3D nano-template in conjunction with the PLD technique offers a new path to fabricating 3D perovskite oxide nanostructures, which have promising applications in 3D nanomagnet logic.

8.4 POTENTIAL APPLICATIONS OF PEROVSKITE MAGNETIC NANOMAGNETS

8.4.1 Magnetic Memory Devices

As a new kind of nonvolatile memory, MRAM can eventually replace the silicon-based DRAM that is based on CMOS technology. MRAM can retain the stored data even if the power is switched off. Therefore, the operating system stored on a computer with MRAM does not require time to boot up from the hard disk. The operation of magnetic memory cells relies on the two stable magnetic states of nanomagnets with opposite magnetizations, which are used as storage elements for binary data. The data in MRAM is read by measuring the electrical resistance of the cell to distinguish between the two states. Thanks to giant magnetoresistance (GMR), this signal is large enough to be useful for magnetic devices. To improve the storage density of magnetic data in MRAM, self-assembled arrays of high anisotropy magnetic nanoparticles as both patterned bit media and as a substitute for thin-film media will be highly required during the next several years. Normally, the elaboration process of high-sensitive cells in MRAM passes through the following two steps that are often used in the semiconductor industry. The first step is the pattern fabrication by photon or electron sensitive polymer (resist) via lithography technique, and the second step is transferring these nanostructures onto the perovskite manganite films via dry etching by UV lithography, scanning electron beam lithography, or x-ray lithography technique. After nanolithography, the pattern transferring can be accomplished by using direct etching assisted with a resist as a mask or by using a metallic lift-off process followed by etching. This lift-off process is the preferred method for manganite etching because these colossal magnetoresistance (CMR) oxides are very hard materials to be etched in comparison with metals. Normally, by such a lift-off process, only micrometer-sized $La_{0.7}Sr_{0.3}MnO_3$ dots with smooth edges and surfaces can be etched into the corresponding films. Recently, regular arrays of epitaxial $La_{2/3}Sr_{1/3}MnO_3$ magnetic oxide nanodots were fabricated by the PLD method as well as electron beam lithography and argon ion exposure [70]. The atomic force microscopy (AFM) image of these perovskite magnetic nanodots is displayed in Figure 8.23. These magnetic oxide nanodots had a diameter smaller than 100 nm with a height of about 37 nm while retaining their crystallinity, epitaxial structure, and ferromagnetic behavior after the fabrication process. That enables the perovskite magnetic nanodots to have potential applications in magnetic massive memory devices and oxide spintronics.

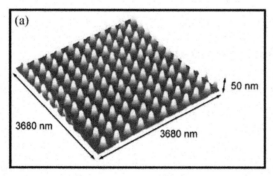 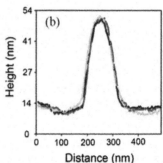

FIGURE 8.23 AFM image of perovskite $La_{2/3}Sr_{1/3}MnO_3$ oxide nanodot arrays. (a) 3D AFM image of a 3680 nm × 3680 nm section of the array. (b) AFM profiles of three representative dots from the AFM image in (a). Dot diameters are less than 100 nm with a height of 37 nm. Adapted with permission from [70]. Copyright (2005) Wiley-VCH Verlag GmbH & Co. KGaA, Weinheim.

8.4.2 Spintronic Devices

Spintronic devices work on the spin of an electron (the intrinsic angular momentum of an electron) to carry, process, and encode information. These new devices have the advantages of nonvolatility, high data-processing speed, low electric power consumption, and increased integration densities in comparison with conventional semiconductor devices that work on the electronic charge. It is envisioned that adding the spin degree of freedom into the conventional semiconductor devices or the use of a spin degree of freedom alone will add much more capability and enhanced performance to the electronic products. The merging of spintronics with electronics, photonics, and magnetics will ultimately lead to new spin-based multifunctional devices such as spin transistors, spin FETs (field-effect transistors), spin LEDs (light-emitting diodes), spin RTDs (resonant tunneling devices), and spin modulators.

As one of the most important spintronic devices, spin valves have been widely used in the magnetic data storage industry. They work principally on the GMR effect, which is described as the resistance of an FM/NM (non-magnetic)/FM multilayer highly dependent upon the relative alignment between the two FM layers. Despite the GMR research being mainly focused on transition metals in the past decade, however, recently spin valves have turned to perovskite oxides, especially the FM perovskite manganites [71]. The rapid development of computer science and cloud computing technology require high-density hard-disk drives, which stimulate intensive research on magnetic tunnel junctions (MTJs) with a large tunneling magnetoresistance (TMR) effect at room temperature [72]. In the past several decades, numerous efforts have been made to develop $La_{0.67}Sr_{0.33}MnO_3$-based MTJ devices due to their 100% spin polarization (half-metallic character). The first $La_{0.67}Sr_{0.33}MnO_3$-based MTJ device was fabricated by an IBM research group, where the $La_{0.67}Sr_{0.33}MnO_3$ layer was used as electrodes and the $SrTiO_3$ layer was used as a tunnel barrier with a thickness of 3–6 nm [73]. The TMR of these MJT devices was reported to be 83% at 4.2 K. In the following years, much larger TMR values were achieved in the $La_{0.67}Sr_{0.33}MnO_3$-based MTJ devices. For example, a low-temperature TMR of 1850% was achieved in the MTJs with a stacked structure of LSMO/STO/LSMO [74]. The corresponding spin polarization rate of 95% was extracted from these experimental data, which confirmed the half-metallicity of $La_{0.67}Sr_{0.33}MnO_3$. Despite these encouraging results, the $La_{0.67}Sr_{0.33}MnO_3$-based MTJ devices are still in their infancy stage and face some detrimental problems such as lower operating temperature than room temperature, irregular switching behaviors, and bias voltage-dependence of the TMR effect [75].

8.4.3 Magnetic Sensors

As passive devices, magnetic sensors can be used to detect the presence of a magnetic field and process this information. Magnetic sensors based on the GMR effect have important applications in mechanical and navigational systems, and as read heads for hard disks. The CMR effect is also observed in nanostructured perovskite manganite films, which can be used for the development of CMR-B-scalar sensors at room temperature [76]. For example, magnetic sensors based on $La_{0.83}Sr_{0.17}MnO_3$ films have been used to measure the magnetic field distributions in railguns [77] and nondestructive pulsed-field magnets [78]. To develop magnetic sensors that can be operated at cryogenic temperatures for measuring high magnetic fields, nanostructured $La_{1-x}Ca_xMnO_3$ films are highly preferred due to their higher sensitivity and lower memory effects in comparison with $La_{1-x}Sr_xMnO_3$ films [78]. For the manganite cobaltite films used for magnetic sensors, their signal sensitivity response to the magnetic field change is a very important factor, which can be increased by Co substitution for Mn in the $La_{1-x}Sr_xMnO_3$ films. The Co substitution for Mn with the amount of Co/(La + Sr) = 0.12 enables the $La_{0.79}Sr_{0.21}Mn_{1.05}Co_{0.12}O_3$ films to have a high sensitivity of 2.5 mV/T, allowing them to fabricate high pulsed magnetic field sensors used at cryogenic temperatures [79].

8.4.4 Magnetic Refrigeration

The magnetocaloric effect (MCE) and its direct application in magnetic refrigeration are popular topics due to their potential improvements in the energy efficiency of cooling and temperature control systems. That enables this technology to have the merits of high security, low energy consumption, low pollution, and low capital cost. Recently, perovskite oxide magnetic nanoparticles have been used as potential candidates for magnetic refrigeration based on an enhanced magnetocaloric effect. Mahato et al. [80] reported a large magnetic entropy change (ΔS = 12.5 J kg^{-1} K^{-1}) in the $La_{0.7}Te_{0.3}MnO_3$ nanoparticles near T_C with a magnetic field change of 50 kOe. Yang et al. [81] also reported a maximum value of ΔS of 1.01 and 1.20 J kg^{-1}K^{-1} for the $La_{0.7}Ca_{0.3}MnO_3$ nanoparticles with an average size of 30 and 50 nm and under a magnetic field of 15 kOe, respectively. That indicates the $La_{0.7}Ca_{0.3}MnO_3$ nanoparticles have promising applications in magnetic refrigeration at room temperature.

As compared with perovskite oxide magnetic nanoparticles, perovskite magnetic manganite thin films are more desirable for magnetic refrigeration since the particle size distribution and their interactions in the films can broaden ΔS over a wide temperature range, thus enhancing the relative cooling power (RCP). In addition, it is easy to integrate thin films and heterostructures into electronic devices with a planar structure, and the large surface areas and interfaces are beneficial to heat exchange, improving the performance of the devices by maximizing the entropy change, reducing hysteresis losses, modulating the operation temperature, and reducing the magnetic field required for an enhanced MCE [82]. It was reported that the $La_{0.67}Ca_{0.33}MnO_3$ film with a thickness of 200 nm had a larger $|\Delta S|$ ($\approx$ 6 J kg^{-1}K^{-1} at $\mu_{0\Delta}$ = 5 T) while the corresponding $|\Delta S|$ for the $La_{0.67}Ca_{0.33}MnO_3$ nanoparticles (with a particle size of 33 nm) was only about 4.9 J kg^{-1}K^{-1} [83]. It is reported that the strains formed in the epitaxial $La_{0.67}Ca_{0.33}MnO_3$ films grown on various substrates play a vital role in their magnetocaloric properties. A representative example is the $La_{0.7}Ca_{0.3}MnO_3/BaTiO_3$ heterostructured films, where the structural phase transition of $BaTiO_3$ film is used to manipulate the lattice strain of $La_{0.7}Ca_{0.3}MnO_3$ thin films, leading to a giant reversible extrinsic MCE in the $La_{0.7}Ca_{0.3}MnO_3$ film. That is ascribed to the strong spin-lattice coupling between $La_{0.7}Ca_{0.3}MnO_3$ and $BaTiO_3$ films [84]. These encouraging results promote further studies of MCE in magnetic manganite thin films. Figure 8.24a shows the isothermal entropy change $|\Delta S|$ vs. isothermal heat $|Q|$ measured near room temperature for some selected magnetocaloric perovskite oxide bulks and thin films, and the corresponding RCP values vs. applied magnetic field ($\mu_o H$) is shown in Figure 8.24b [82]. As can be seen from Figure 8.24, the magnetocaloric thin films exhibit higher $|\Delta S|$ values for higher $|Q|$, and the strain-mediated magnetocaloric mechanism appears to be a novel route to optimizing the magnetocaloric response of perovskite manganite films, indicating their important applications in magnetic refrigeration.

8.4.5 Biomedical Applications

Perovskite magnetic nanoparticles exhibit some unique features, such as the SPM state, which is a result of the effect of thermal energy on a ferromagnetic nanoparticle. At the SPM state, the magnetic moment of each nanoparticle is disorderedly oriented due to thermal fluctuation, thus leading to no net magnetization. However, under an external magnetic field, they tend to be aligned along the direction of the applied field and exhibit a very large magnetization. Therefore, the perovskite magnetic nanoparticle can be targeted by an external magnetic field in an on-off fashion. Based on this characteristic, perovskite magnetic nanoparticles have promising applications in biomedicine, such as magnetic hyperthermia therapy (MHT), drug delivery, and magnetic resonance imaging (MRI) for cancer therapy. In the MHT application, Kuznetsov et al. [85] first reported the self-controlled heating effects in the LSMO nanoparticles under an alternating magnetic field (AMF) with an amplitude of 90 Oe and a frequency of 800 kHz. Similarly, self-controlled hyperthermia effects were also reported in the $La_{0.75}Sr_{0.25}MnO_3$ nanoparticles with high heating efficacy [86]. In the following research, the effect of magnetic field amplitude (H in the range of 83.8–502.8 Oe) on the temperature kinetics and the specific absorption rate (SAR) value of oleic acid (OA)/polyethylene

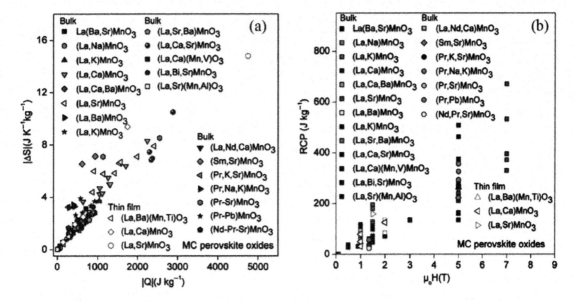

FIGURE 8.24 (a) Isothermal entropy change |ΔS| vs. isothermal heat |Q| for some selected magnetocaloric perovskite oxides. (b) Relative cooling power (RCP) calculated at various applied magnetic fields for some magnetocaloric perovskite oxides. The solid and open symbols indicate magnetocaloric parameters obtained for bulk and thin films, respectively. Adapted with permission from [82]. Copyright (2019) Wiley-VCH Verlag GmbH & Co. KGaA, Weinheim.

glycol (PEG)-coated/capped LSMO-based magnetic nanoparticles is systematically investigated under the fixed magnetic field frequency of 265 kHz [87]. Analogously, the influence of the magnetic frequency on the heating efficiency of $La_{0.75}Sr_{0.25}MnO_3$ nanoparticles is also investigated, and the results show a super-linear relationship between the SAR value (i.e. from 15–35 W/g) and the frequency (i.e. f = 150 to 300 kHz) while maintaining a 400 W power supply [86]. In addition, the influence of the crystallite size on the heat-induced properties of the $La_{0.77}Sr_{0.23}MnO_3$ nanoparticles in MHT application is also reported, and the optimal crystallite size is determined under AMF with an amplitude of 22.2 Oe and a frequency of 370 kHz [88]. In addition, the self-controlled heating effects of LSMO nanoparticles can be greatly improved by encapsulating them with suitable coatings (e.g. silica, OA, PEG, glycine) [87].

Since the sizes of perovskite magnetic nanoparticles can be controlled from a few nanometers to tens of nanometers, therefore, they have potential applications in the field of biology. As they are coated with biological molecules, a controllable method of "tagging" or "addressing" can be realized by interacting with or binding them to a biological entity. Based on this principle, a magnetic nanoparticle-based drug delivery system can be developed, which transports medical drugs to specific sites and subsequently controls the drug/gene release remotely under real-time monitoring.

As a powerful imaging method, MRI is widely used in clinical practice. Its working principle is based on perovskite magnetic nanoparticles acting as contrast agents, which lead to the relaxation time change in the nuclear magnetic resonance imaging process, providing high-resolution images and monitoring the tissue morphology and anatomical details. Because the perovskite magnetic nanomagnets have very large magnetic moments under a static magnetic field, the dipolar interactions between the SPM cores and surrounding solvent protons lead to the reductions of relaxation times, which decrease both in longitudinal and transverse modes. Such changes in relaxation times result from several factors, while size and composition represent the essential ones [89]. Recently, Haghniaz et al. [90] reported that the LSMO nanoparticles coated with dextran exhibited enhanced properties of both positive (longitudinal) and negative (transverse) contrast. Later, Veverka et al. [91] combined $La_{0.65}Sr_{0.35}MnO_3$ nanoparticles with fluorescent moieties to develop two dual-mode diagnostics, such as MRI and fluorescence imaging methods. Perovskite manganite $La_{0.65}Sr_{0.35}MnO_3$ nanoparticles coated with silica shell layer demonstrated relatively high transverse relaxivity at 20°C under the applied magnetic fields of 0.5 T.

8.4.6 Catalysis

Besides their magnetic properties, perovskite MNMs can also be used as efficient catalysts for water splitting due to their high chemical stability, large surface-to-volume ratio, and fast response to the applied magnetic field. Recently, Cheng et al. [92] performed systematic studies on the oxygen evolution reaction (OER) of perovskite $La_{1-x}Sr_xCoO_3$ (with x = 0.0–1.0) both from experiment and theory. They found that Sr substitutions could enhance the OER activity of $La_{1-x}Sr_xCoO_3$ powders synthesized by a modified sol-gel process. Density-functional theory calculations show that the OER activity comes from the electronic band structure; the increased band overlapping between the occupied O 2p valence band and the empty Co 3d conduction band could enhance the OER activity. In another work, they investigated the relationships between the OER activity and several physicochemical properties such as conductivity, oxygen vacancy content, and flat-band potential for perovskite oxides with different compositions (e.g. $La_{1-x}Sr_xCoO_{3-\delta}$ series, $LaMO_{3-\delta}$ series [M = Cr, Mn, Fe, Co, Ni], $Ba_{0.5}Sr_{0.5}Co_{0.8}Fe_{0.2}O_{3-\delta}$, and $PrBaCo_2O_{6-\delta}$) [93]. They found that the proposed activity descriptors are not isolated from each other; instead, they are mutually correlated with the OER activity, allowing ones to build up an OER/multidescriptor correlation. However, Mefford et al. [94] demonstrated that oxygen vacancy defects played an important role in improving the electrocatalysis of oxygens on the surface of $La_{1-x}Sr_xCoO_{3-\delta}$ oxide, whereby they may control the physical parameters of ionic diffusion rates and reflect the underlying electronic structure of the catalyst. They proposed a vacancy-mediated mechanism, providing a deep understanding of the high activity for OER catalysts and allowing for the rational design of the electrolysis of water by using surface chemistry parameters, as demonstrated by *ab initio* modeling. Thus, the role of oxygen vacancy defects should be paid much attention to, which is a critical factor in the benchmarking of metal oxide oxygen electrocatalysts and the advancement of the mechanistic theory underlying the OER. Besides perovskite oxide nanoparticles, perovskite $SrNb_{0.1}Co_{0.7}Fe_{0.2}O_{3-\delta}$ nanorods fabricated by an electrospinning process exhibit much high OER and HER (hydrogen evolution reaction) activities in conjunction with good durability in alkaline solutions [95]. Such enhanced OER and HER activities in the $SrNb_{0.1}Co_{0.7}Fe_{0.2}O_{3-\delta}$ perovskite nanorods as compared with their bulk counterpart is attributed to their larger electrochemical surface areas, faster charge transferring rate, and a larger number of oxygen species associated with surface oxygen vacancies. The good performance of these bifunctional OER and HER catalysts opens up corridors for the development of low-cost and earth-abundant catalysts for overall water splitting. Recently, 2D $La_{1-x}Sr_xCoO_3$ thin films as a representative electrode system have also been used to investigate the intrinsic mechanism of OER [96]. It is found that the 2D $La_{1-x}Sr_xCoO_3$ films exhibit different OER activities in different pH electrolyte solutions, and a linear relationship between the OER activity of the LSCO electrode and the solution pH value is observed. That indicates the importance of surface deprotonation on the OER reaction mechanism. The presence of non-protonated lattice oxygen atoms determines the catalytic activity of the LSCO thin film electrode.

8.4.7 Environmental Purification

The water crisis has become a major problem in the 21st century due to the vast amount of wastewater produced by industry and agriculture sectors. Under these circumstances, different techniques have been developed for water purification. Among these techniques, coloring dyes are one of the most challenging materials to be treated. One of the representative coloring agents, rhodamine-B (RhB), is very hard to treat with conventional techniques. Searching for efficient visible-light photocatalysts for water purification has become quite a challenging task. Recently, Dhiman and Singh reported an enhanced catalytic and photocatalytic degradation of RhB by using LaMnO$_3$ nanoparticles synthesized via the sol-gel process [97]. They found that the degradation rate of RhB by LaMnO$_3$ nanoparticles reached 90%–92% under dark conditions, and then it increased above 99% under visible light irradiation. Such excellent catalytic (under dark) and photocatalytic (under visible-light irradiation) properties of LaMnO$_3$ nanoparticles make them a promising candidate for large-scale wastewater treatment. In other studies, LaFeO$_3$ nanoparticles synthesized by different methods are also used as photocatalysts for the photodegradation of RhB under visible light irradiation. The photocatalytic degradation of the pollutants was completed mainly via the direct reaction of the photo-generated holes but to some extent also via superoxide radical formation, especially at the beginning of the degradation experiments. Overall, LaFeO$_3$ nanoparticles can be regarded as a potential photocatalytic material to degrade organic pollutants under visible-light irradiation.

Recently, BFO has gained much attention due to its multiferroic behaviors. Besides its multiferroic property, current studies also demonstrate that BFO is a promising visible-light photocatalyst due to its small bandgap (2.2–2.7 eV), good chemical stability, and good response to visible-light irradiation. In addition, the magnetic properties of BiFeO$_3$ nanoparticles are also beneficial for reprocessing the photocatalyst after the reaction is performed under an external magnetic field. This prevents the damage of the catalyst from the rotations and makes the catalyst cost-effective [98]. To date, numerous studies have reported the enhanced photocatalytic performance of BFO nanoparticles, which are used to degrade RhB and other organic pollutants [99–101]. Furthermore, BFO-graphene nanohybrids were reported to exhibit an enhanced photocatalytic performance [98]. To tailor the inherent structure of BFO, La and Se co-doped BFO nanosheets with a much lower bandgap (~ 1.76 eV) were synthesized by the wet-chemical method [102]. It is found that the La and Se co-doped BFO nanosheets exhibit a much enhanced synergistic response and high visible-light photocatalytic activities. This is attributed to their lower bandgap and sheet-type morphology, which makes easy electron production upon exposure to incident radiation and to a larger contacting area between the catalyst and the adsorbing species. The enhanced visible-light photocatalytic performance of the BFO-based nanoparticles enables them to be potential candidates for water purification.

8.5 CONCLUSIONS

In summary, the synthesis of perovskite magnetic nanomagnets with different dimensions into nanometer scale has been reviewed. By using physical methods such as the solid-state reaction, reactive grinding process, and combustion method, perovskite magnetic nanoparticles can be synthesized. Similarly, several chemical routes can be used to synthesize perovskite magnetic nanoparticles, including the sol-gel, hydro-/solvothermal, microwave-hydrothermal, and microemulsion processes. 1D perovskite magnetic nanomagnets such as La$_{2/3}$Ca$_{1/3}$MnO$_3$ and BiFeO$_3$ nanowires/nanotubes can be nanofabricated via top-down and bottom-up approaches, respectively. Similarly, 2D perovskite magnetic nanomagnets such as ordered arrays of perovskite magnetic nanoparticles are also fabricated by the same approaches. Freestanding 2D perovskite oxide membranes with a thickness down to a monolayer have also been prepared by the bottom-up approach associated with the sacrificial buffer layer. 3D perovskite magnetic nanostructures exhibit many exciting physical phenomena in nanomagnetism, as their curved geometries offer topologically protected and chiral magnetic configurations, which benefit the construction of 3D magnetic memories and improve their performance. Three main routes, including physical, chemical, and 3D nanoprinting methods, have been used to fabricate complex 3D nanomagnet structures. Perovskite magnetic nanomagnets have promising applications in magnetic memory devices, spintronic devices, magnetic sensors, magnetic refrigeration, biomedical applications, catalysis, and environmental purification. The scientific and technical potential of perovskite magnetic nanomagnets is great, and their future research fields are very bright.

REFERENCES

1. Hao R, Xing R, Xu Z, Hou Y, Gao S, Sun S. Synthesis, functionalization, and biomedical applications of multifunctional magnetic nanoparticles. *Adv Mater* 22 (2010) 2729–2742.
2. Kirk KJ. Nanomagnets for sensors and data storage. *Contemp Phys* 41 (2000) 61–78.
3. Cowburn RP. Property variation with shape in magnetic nanoelements. *J Phys D: Appl Phys* 33 (2000) R1–R16.
4. Zhu K, Ju YM, Xu JJ, Yang ZY, Gao S, Hou YL. Magnetic nanomaterials: Chemical design, synthesis, and potential applications. *Acc Chem Res* 51 (2018) 404–413.
5. Lu AH, Salabas EL, Schüth F. Magnetic nanoparticles: Synthesis, protection, functionalization, and application. *Angew Chem Int Ed* 46 (2007)1222–1244.
6. Fernandez-Pacheco A, Streubel R, Fruchart O, Hertel R, Fischer P, Russell P. Cowburn RP. Three-dimensional nanomagnetism. *Nat Commun* 8 (2017) 15756.
7. Streubel R, Fischer P, Kronast F, Kravchuk VP, Sheka DD, Gaididei Y, Schmidt OG, Makarov D. Magnetism in curved geometries. *J Phys D Appl Phys* 49 (2016) 63001.
8. Fernández-Pacheco A, Skoric L, De Teresa JM, Pablo-Navarro J, Michael Huth M, Oleksandr V. Dobrovolskiy OV. (2020) Writing 3D nanomagnets using focused electron beams. *Materials* 13 (2020) 3774.
9. Zeng ZC, Xu YS, Zhang ZS, Gao ZS, Luo M, Yin ZY, Zhang C, Xu J, Huang BL, Luo F, Du YP, Yan CH. Rare-earth-containing perovskite nanomaterials: Design, synthesis, properties and applications. *Chem Soc Rev* 49 (20220) 1109–1143.
10. Xia WR, Pei ZP, Leng K, Zhu XH. Research progress in rare earth-doped perovskite manganite oxide nanostructures. *Nanoscale Res Lett* 15 (2020) 9.
11. Reshi HA, Pillai S, Shelke V. Comparative study on multifunctional behaviour of rare earth manganites with micro and nano grain size. *J Mater Sci Mater Electron* 25 (2014) 3795–3800.
12. Navin K, Kurchania R. A comparative study of the structural, magnetic transport and electrochemical properties of $La_{0.7}Sr_{0.3}MnO_3$ synthesized by different chemical routes. *Appl Phys A* 126 (2020) 100.
13. Ortiz-Quinonez JL, Garcia-Gonzialez L, Cancino-Gordillo FE, Umapada PU. Particle dispersion and lattice distortion induced magnetic behavior of $La_{1-x}Sr_xMnO_3$ perovskite nanoparticles grown by salt-assisted solid-state synthesis. *Mater Chem Phys* 246 (2020) 122834.
14. Phong PT, Manh DH, Nguyen LH, Tung DK, Phuc NX, Lee IJ. Studies of superspin glass state and AC-losses in $La_{0.7}Sr_{0.3}MnO_3$ nanoparticles obtained by high-energy ball-milling. *J Magn Magn Mater* 368 (2014) 240–245.
15. Tian Y, Chen D, Jiao X. $La_{1-x}Sr_xMnO_3$ (x = 0, 0.3, 0.5, 0.7) nanoparticles nearly freestanding in water: Preparation and magnetic properties. *Chem Mater* 18 (2006) 6088–6090
16. Zhao BC, Ma YQ, Song WH, Sun YP. Magnetization steps in the phase separated manganite $La_{0.275}Pr_{0.35}Ca_{0.375}MnO_3$. *Phys Lett* 354 (2006) 472–476.
17. Xia WR, Leng K, Tang QK, Yang L, Xie YT, Wu ZW, Yi K, Zhu XH. Structural characterization, magnetic and optical properties of perovskite $(La_{1-x}Ln_x)_{0.67}Ca_{0.33}MnO_3$ (Ln = Nd and Sm; x = 0.0–0.5) nanoparticles synthesized via the sol-gel process. *J Alloy Compd* 867 (2021) 158808.
18. Kaliaguine S, Neste AV, Szabo V, Gallot JE, Bassir M, Muzychuk R. Perovskite-type oxides synthesized by reactive grinding Part I. preparation and characterization. *Appl Catal A* 209 (2001) 345–358.
19. Szabo V, Bassir M, Neste AV, Kaliaguine S. Perovskite-type oxides synthesised by reactive grinding Part IV. Catalytic properties of $LaCo_{1-x}Fe_xO_3$ in methane oxidation. *Appl Catal B* 43 (2003) 81–92.
20. Deganello F, Marcı G, Deganello G. Citrate nitrate auto-combustion synthesis of perovskite-type nanopowders: A systematic approach. *J Eur Ceram Soc* 29 (2009) 439–450.
21. Shinde KP, Pawar SS, Shirage PM, Pawar SH. Studies on morphological and magnetic properties of $La_{1-x}Sr_xMnO_3$. *Appl Surf Sci* 258 (2012) 7417–7420.
22. Silva GRO, Santos JC, Martinelli DMH, Pedrosa AMG, Souza MJB, Melo DMA. Synthesis and characterization of $LaNi_xCo_{1-x}O_3$ perovskites via complex precursor methods. *Mater Sci Appl* 1 (2010) 39–45.
23. Asefi N, Hasheminiasari M, Masoudpanah SM. Solution combustion synthesis of $BiFeO_3$ powders using CTAB as fuel. *J Electron Mater* 48 (2019) 409–415.
24. Mehrnoush N, Sanavi KD. Study on structural, magnetic and electrical properties of $ReFeO_3$ (Re= La, Pr, Nd, Sm & Gd) orthoferrites. *Physica B-Condens Matter* 612 (2021) 412899.
25. Shlapa Y, Solopan S, Bodnaruk A, Kulyk M, Kalita V, Tykhonenko-Polishchuk Y, Alexandr TA, Anatolii BA. Effect of synthesis temperature on structure and magnetic properties of $(La,Nd)_{0.7}Sr_{0.3}MnO_3$ nanoparticles. *Nanoscale Res Lett* 12 (2017) 100.

26. Niederberger M. Nonaqueous sol – gel routes to metal oxide nanoparticles. *Acc Chem Res* 40 (2007) 793–800.
27. Sadhu A, Bhattacharyya S. Enhanced low-field magnetoresistance in $La_{0.71}Sr_{0.29}MnO_3$ nanoparticles synthesized by the nonaqueous sol-gel route. *Chem Mater* 26 (2014) 1702–1710.
28. Xia WR, Leng K, Tang QK, Yang L, Xie YT, Wu ZW, Yi K, Zhu XH. Structural characterization, magnetic and optical properties of perovskite $(La_{1-x}Ln_x)_{0.67}Ca_{0.33}MnO_3$ (Ln = Nd and Sm; x = 0.0–0.5) nanoparticles synthesized via the sol-gel process. *J Alloy Compd* 867 (2021) 158808.
29. Wu H, Xue PJ, Lu Y, Zhu XH. Microstructural, optical and magnetic characterizations of $BiFeO_3$ multiferroic nanoparticles synthesized via a sol-gel process. *J Alloy Compd* 731 (2018) 471–477.
30. Zhu XH, Zhu JM, Zhou SH, Liu ZG, Ming NB, Hesse D. $BaTiO_3$ nanocrystals: Hydrothermal synthesis and structural characterization. *J Cryst Growth* 283 (2005) 553–562.
31. Remya KP, Prabhu D, Joseyphus RJ, Bose AC, Viswanathan C, Ponpandian N. Tailoring the morphology and size of perovskite $BiFeO_3$ nanostructures for enhanced magnetic and electrical properties. *Mater and Des* 192 (2020) 108694.
32. Han SH, Kim KS, Kim HG, Lee HG, Kang HW, Kim JS, Il Cheon C. Synthesis, and characterization of multifermic $BiFeO_3$ powders fabricated by hydrothermal method. *Ceram Int* 36 (2010) 1365–1372.
33. Han JT, Huang YH, Wu XJ, Wu CL, Wei W, Peng B, Huang W, Goodenough JB. Tunable synthesis of bismuth ferrites with various morphologies. *Adv Mater* 18 (2006) 2145–2148.
34. Wang YG, Xu G, Yang LL, Ren ZH, Wei X, Weng WJ, Du PY, Shen G, Han GR. Alkali metal ions-assisted controllable synthesis of bismuth ferrites by a hydrothermal method. *J Am Ceram Soc* 90 (2007) 3673–3675.
35. Zhu XH, Hang QM, Xing ZB, Yang Y, Zhu JM, Liu ZG, Ming NB, Zhou P, Song Y, Li ZS, Yu T, Zou ZG. Microwave hydrothermal synthesis, structural characterization, and visible-light photocatalytic activities of single-crystalline bismuth ferric nanocrystals. *J Am Ceram Soc* 94 (2011) 2688–2693.
36. Joshi UA, Jang JS, Borse PH, Lee JS. Microwave synthesis of single-crystalline perovskite $BiFeO_3$ nanocubes for photoelectrode and photocatalytic applications. *Appl Phys Lett* 92 (2008) 242106.
37. Xia WR, Lu Y, Zhu XH. Microwave-hydrothermal synthesis of perovskite oxide nanomaterials. In: *Design Strategies for Synthesis and Fabrication*. Ed. K.D. Sattler, 2020, Taylor & Francis and CRC Press, London.
38. Zhu XH, Hang QM. Microscopical and physical characterization of microwave and microwave-hydrothermal synthesis products. *Micron* 44 (2013) 21–44.
39. Sanchez-Dominguez M, Pemartin K, Boutonnet M. Preparation of inorganic nanoparticles in oil-in-water microemulsions: A soft and versatile approach. *Curr Opin Colloid Interface Sci* 17 (2012) 297–305.
40. Lim C, Kim C, Gwon O, Jeong HY, Song HK, Ju YW, Shin J, Kim G. Nano-perovskite oxide prepared via inverse microemulsion mediated synthesis for catalyst of lithium-air batteries. *Electrochimi Acta* 275 (2018) 248–255.
41. Das N, Majumdar R, Sen A, Maiti HS. Nanosized bismuth ferrite powder prepared through sonochemical and microemulsion techniques. *Mater Lett* 61 (2007) 2100–2104.
42. Asif M, Nadeem M, Imran M, Ahmad S, Musaddiq S, Abbas W, Gilani ZA, Sharif MK, Warsi MF, Khan MA. Structural, magnetic and dielectric properties of Ni-Co doped $BiFeO_3$ multiferroics synthesized via microemulsion route. *Physica B-Condens. Matter* 552 (2019) 11–18.
43. Li L, Liang LZ, Wu H, Zhu XH. One-dimensional perovskite manganite oxide nanostructures: Recent developments in synthesis, characterization, transport properties, and applications. *Nanoscale Res Lett* 11 (2016) 121.
44. Proenca MP, Sousa CT, Ventura J, Araujo JP. Electrochemical synthesis and magnetism of magnetic nanotubes. In: *Magnetic Nano- and Microwires: Design, Synthesis, Properties and Applications*. Ed. M. Vazquez, 2015, Woodhead Publishing, Duxford, UK.
45. Biswas A, Bayer IS, Biris AS, Wang T, Dervishi E, Faupel F. Advances in top-down and bottom-up surface nanofabrication: Techniques, applications & future prospects. *Adv Colloid Interf Sci* 170 (2012) 2–27.
46. Lorena M, Luis M, Algarabel PA, Luis AR, Rodríguez CM, José MDT, Ibarra MR. Enhanced magnetotransport in nanopatterned manganite nanowires. *Nano Lett* 14 (2014) 423–428.
47. Shimomura M, Sawadaishi T. Bottom-up strategy of materials fabrication: A new trend in nanotechnology of soft materials. *Curr Opin Colloid Interface Sci* 6 (2001) 11–16.
48. Shankar K, Raychaudhuri A. Growth of an ordered array of oriented manganite nanowires in alumina templates. *Nanotechnology* 15 (2004) 1312–1316.
49. Wang J, Manivannan A, Wu NQ. Sol-gel derived $La_{0.6}Sr_{0.4}CoO_3$ nanoparticles, nanotubes, nanowires and thin films. *Thin Solid Films* 517 (2008) 582–587.
50. Zhang XY, Dai JY, Lai CW. Synthesis and characterization of highly ordered $BiFeO_3$ multiferroic nanowire arrays. *Prog. Solid State Chem* 33 (2005) 147–151.
51. Wei J, Xue D, Xu Y. Photoabsorption characterization and magnetic property of multiferroic $BiFeO_3$ nanotubes synthesized by a facile sol-gel template process. *Scr Mater* 58 (2008) 45–48.

52. Park TJ, Mao YB, Wong SS. Synthesis and characterization of multiferroic BiFeO$_3$ nanotubes. *Chem Commun* (2004) 2708–2709.
53. Liu B, Hu BB, Du ZL. Hydrothermal synthesis and magnetic properties of single-crystalline BiFeO$_3$ nanowires. *Chem Commun* 47 (2011) 8166–8168.
54. Zhang T, Jin CG, Qian T, Lu XL, Bai JM, Li XG. Hydrothermal synthesis of single-crystalline La$_{0.5}$Ca$_{0.5}$MnO$_3$ nanowires at low temperature. *J Mater Chem* 14 (2004) 2787–2789.
55. Xia WR, Leng K, Tang QK, Yang L, Xie YT, Wu ZW, Yi K, Zhu XH. Structural characterization and magnetic properties of single-crystalline (La$_{0.6}$Pr$_{0.4}$)$_{0.67}$Ca$_{0.33}$MnO$_3$ nanowires. *J Am Ceram Soc* 104 (2021) 5402–5410.
56. Datta S, Ghatak A, Ghosh B. Manganite (La$_{1-x}$A$_x$MnO$_3$; A = Sr, Ca) nanowires with adaptable stoichiometry grown by hydrothermal method: Understanding of growth mechanism using spatially resolved techniques. *J Mater Sci* 51 (2016) 9679–9695.
57. Russier V. Calculated magnetic properties of two-dimensional arrays of nanoparticles at vanishing temperature. *J Appl Phys* 89 (2001) 1287–1294.
58. Georgescu M, Klokkenburg M, Erne BH, Liljeroth P, Vanmaekelbergh D, Zeijlmans van Emmichoven PA. Flux closure in two-dimensional magnetite nanoparticle assemblies. *Phys Rev B* 73 (2006) 184415.
59. Morelli A, Johann F, Schammelt N, McGrouther D, Vrejoiu I. Mask assisted fabrication of nanoislands of BiFeO$_3$ by ion beam milling. *J Appl Phys* 113 (2013) 154101.
60. Han DQ, Zhang XW, Wu ZF, Hua ZH, Wang ZH, Yang SG. Synthesis and magnetic properties of complex oxides La$_{0.67}$Sr$_{0.33}$MnO$_3$ nanowire arrays. *Ceram Int* 42 (2016) 16992–16996.
61. Zhang XY, Lai CW, Zhao X, Wang DY, Dai JY. Synthesis and ferroelectric properties of multiferroic nanotube arrays. *Appl Phys Lett* 87 (2005) 143102.
62. Ji DX, Cai SH, Paudel TR, Sun HY, Zhang CC, Han L, Wei YF, Zang YP, Gu M, Zhang Y, Gao WP, Huyan HX, Guo W, Wu D, Gu ZB, Tsymbal EY, Wang P, Nie YF, Pan XQ. Freestanding crystalline oxide perovskites down to the monolayer limit. *Nature* 570 (2019) 87–90.
63. Streubel R, Fischer P, Kronast F, Kravchuk VP, Sheka DD, Gaididei Y. Schmidt OG, Makarov D. Magnetism in curved geometries. *J Phys D Appl Phys* 49 (2016) 363001.
64. Ishigaki T, Kawahara T, Takemura R, Ono K, Ito K, Matsuoka H, Ohno H. A multi-level-cell spin-transfer torque memory with series-stacked magnetotunnel junctions. In: *2010 Symposium on VLSI Technology*, 2010, pp. 47–48, IEEE, Honolulu.
65. Schmidt OG, Eberl K. Nanotechnology: Thin solid films roll up into nanotubes. *Nature* 410 (2001) 168–168.
66. Ruiz-Clavijo A, Ruiz-Gomez S, Caballero-Calero O, Perez L, Martin-Gonzalez M. Tailoring magnetic anisotropy at will in 3D interconnected nanowire networks. *Phys Status Solidi R* 13 (2019) 1900263.
67. Utke I, Hoffmann P, Melngailis J. Gas-assisted focused electron beam and ion beam processing and fabrication. *J Vac Sci Technol B* 26 (2008) 1197–1276.
68. Kushizaki T, Fujiwara K, Hattori AN, Kanki T, Tanaka H. Controlled fabrication of artificial ferromagnetic (Fe,Mn)$_3$O$_4$ nanowall-wires by a three-dimensional nanotemplate pulsed laser deposition method. *Nanotechnology* 23 (2012) 485308.
69. Nguyen TVA, Hattori AN, Fujiwara Y, Ueda S, Tanaka H. Colossal magnetoresistive (La,Pr,Ca)MnO$_3$ nanobox array structures constructed by the three-dimensional nanotemplate pulsed laser deposition technique. *Appl Phys Lett* 103 (2013) 223105.
70. Ruzmetov D, Seo Y, Belenky LJ, Kim DM, Ke X, Sun H, Chandrasekhar V, Eom CB, Rzchowski MS, Pan XQ. Epitaxial magnetic perovskite nanostructures. *Adv Mater* 17 (2005) 2869–2872.
71. Moritomo Y, Asamitsu A, Kuwahara H, Tokura Y. Giant magnetoresistance of manganese oxides with a layered perovskite structure. *Nature* 380 (1996) 141–144.
72. Moodera JS, Kinder LR, Wong TM, Meservey R. Large magnetoresistance at room temperature in ferromagnetic thin film tunnel junctions. *Phys Rev Lett* 74 (1995) 3273–3276.
73. Lu Y, Li XW, Gong GQ, Xiao G, Gupta A, Lecoeur P, Sun JZ, Wang YY, Dravid VP. Large magnetotunneling effect at low magnetic fields in micrometer-scale epitaxial La$_{0.67}$Sr$_{0.33}$MnO$_3$ tunnel junctions. *Phys Rev B* 54 (1996) R8357–R8360.
74. Bowen M, Bibes M, Barthélémy A, Contour JP, Anane A, Lemaître Y, Fert A. Nearly total spin polarization in La$_{2/3}$Sr$_{1/3}$MnO$_3$ from tunneling experiments. *Appl Phys Lett* 82 (2003) 233–235.
75. Majumdar S, van Dijken S. Pulsed laser deposition of La$_{1-x}$Sr$_x$MnO$_3$:thin-film properties and spintronic applications. *J Phys D: Appl Phys* 47 (2014) 034010.
76. Zurauskiene N, Balevicius S, Stankevic V, Kersulis S, Schneider M, Liebfried O, Plausinaitiene V, Abrutis A. B-scalar sensor using CMR effect in thin polycrystalline manganite films. *IEEE Trans Plasma Sci* 39 (2011) 411–416.
77. Haran TL, Hoffman RB, Lane SE. Diagnostic capabilities for electromagnetic railguns. *IEEE Trans Plasma Sci* 41 (2013) 1526–1532.

78. Balevicius S, Zurauskiene N, Stankevic V, Kersulis S, Plausinaitiene V, Abrutis A, Zherlitsyn S, Herrmannsdorfer T, Wosnitza J, Wolff-Fabris F. Nanostructured thin manganite films in megagauss magnetic field. *Appl Phys Lett* 101 (2012) 092407.
79. Zurauskiene N, Rudokas V, Balevicius S, Kersulis S, Stankevic V, Vasiliauskas R, Plausinaitiene V, Vagner M, Lukose R, Skapas M, Juskenas R. Nanostructured La-Sr-Mn-Co-O films for room-temperature pulsed magnetic field sensors. *IEEE Trans Magn* 53 (2017) 4002605.
80. Mahato RN, Sethupathi K, Sankaranarayanan V, Nirmala R. Co-existence of giant magnetoresistance and large magnetocaloric effect near room temperature in nanocrystalline $La_{0.7}Te_{0.3}MnO_3$. *J Magn Magn Mater* 322 (2010) 2537–2540.
81. Yang H, Zhu YH, Xian T, Jiang JL. Synthesis and magnetocaloric properties of $La_{0.7}Ca_{0.3}MnO_3$ nanoparticles with different sizes. *J Alloy Compd* 555 (2013) 150–155.
82. Barman A, Kar-Narayan S, Mukherjee D. Caloric effects in perovskite oxides. *Adv Mater Interfaces* 6 (2019) 1900291.
83. Xiong CM, Sun JR, Chen YF, Shen BG, Du J, Li YX. Relation between magnetic entropy and resistivity in $La_{0.67}Ca_{0.33}MnO_3$. *IEEE Trans Magn* 41 (2005) 122–124.
84. Moya X, Stern-Taulats E, Crossley S, González-Alonso D, Kar-Narayan S, Planes A, Mañosa L, Mathur ND. Giant electrocaloric strength in single-crystal $BaTiO_3$. *Adv Mater* 25 (2013) 1360–1365.
85. Kuznetsov AA, Shlyakhtin OA, Brusentsov NA, Kuznetsov OA. 'Smart' mediators for self-controlled inductive heating. *Eur Cells Mater* 3 (Suppl. 2) (2002) S75–S77.
86. Salakhova RT, Pyatakov AP, Zverev VI, Pimentel B, Vivas RJC, Makarova LA, Perov NS, Tishin AM, Shtil AA, Reis MS. The frequency dependence of magnetic heating for $La_{0.75}Sr_{0.25}MnO_3$ nanoparticles. *J Magn Magn Mater* 470 (2019) 38–40.
87. Thorat ND, Bohara RA, Malgras V, Tofail SAM, Ahamad T, Alshehri SM, Wu KCW, Yamauchi Y. Multimodal superparamagnetic nanoparticles with unusually enhanced specific absorption rate for synergetic cancer therapeutics and magnetic resonance imaging. *ACS Appl Mater Interfaces* 8 (2016) 14656–14664.
88. Das H, Inukai A, Debnath N, Kawaguchi T, Sakamoto N, Hoque SM, Aono H, Shinozaki K, Suzuki H, Wakiya N. Influence of crystallite size on the magnetic and heat generation properties of $La_{0.77}Sr_{0.23}MnO_3$ nanoparticles for hyperthermia applications. *J Phys Chem Solids* 112 (2018) 179–184.
89. Bautista MC, Bomati-Miguel O, Zhao X, Morales MP, González-Carreno T, Perez de Alejo R, Ruiz-Cabello J, Veintemillas-Verdaguer S. Comparative study of ferrofluids based on dextran-coated iron oxide and metal nanoparticles for contrast agents in magnetic resonance imaging. *Nanotechnology* 15 (2004) S154–S159.
90. Haghniaz R, Bhayani KR, Umrani RD, Paknikar KM. Dextran stabilized lanthanum strontium manganese oxide nanoparticles for magnetic resonance imaging. *RSC Adv* 3 (2013) 18489–18497.
91. Veverka P, Kaman O, Kacenka M, Herynek V, Veverka M, Santava E, Lukes I, Jirak Z. Magnetic $La_{1-x}Sr_xMnO_3$ nanoparticles as contrast agents for MRI: The parameters affecting H-1 transverse relaxation. *J Nanoparticle Res* 17 (2015) 33.
92. Cheng X, Fabbri E, Nachtegaal M, Castelli IE, Kazzi ME, Haumont R, Marzari N, Schmidt TJ. Oxygen evolution reaction on $La_{1-x}Sr_xCoO_3$ perovskites: A combined experimental and theoretical study of their structural, electronic, and electrochemical properties. *Chem Mater* 27 (2015) 7662–7672.
93. Cheng X, Fabbri E, Yamashita Y, Castelli IE, Kim B, Uchida M, Haumont R, Puente-Orench I, Thomas J. Schmidt TJ. Oxygen evolution reaction on perovskites: A multieffect descriptor study combining experimental and theoretical methods. *ACS Catal* 8 (2018) 9567–9578.
94. Mefford JT, Rong X, Abakumov AM, Hardin WG, Dai S, Kolpak AM, Johnston KP, Stevenson KJ. Water electrolysis on $La_{1-x}Sr_xCoO_{3-\delta}$ perovskite electrocatalysts. *Nat Commun* 7 (2016) 11053.
95. Zhu YL, Zhou W, Zhong YJ, Bu YF, Chen XY, Zhong Q, Liu ML, Shao ZP. A perovskite nanorod as bifunctional electrocatalyst for overall water splitting. *Adv Energy Mater* 7 (2017) 1602122.
96. Wang XS, Zhou L, Li MX, Luo Y, Yang TY, Wu TL, Li XL, Jin KJ, Guo EJ, Wang LF, XueDong Bai XD, Zhang WF, Guo HZ. Surface protonation and oxygen evolution activity of epitaxial $La_{1-x}Sr_xCoO_3$ thin films. *Sci China-Phys Mechan Astron* 63 (2020) 297011.
97. Dhiman TK, Singh S. Enhanced catalytic and photocatalytic degradation of organic pollutant Rhodamine-B by $LaMnO_3$ nanoparticles synthesized by non-aqueous sol-gel route. *Phys Status Solidi A* 216 (2019) 1900012.
98. Li Z, Shen Y, Yang C, Lei Y, Guan Y, Lin Y, Liu D, Nan CW. Significant enhancement in the visible light photocatalytic properties of $BiFeO_3$-graphene nanohybrids. *J Mater Chem A* 1 (2013) 823–829.
99. Gao F, Chen X, Yin K, Dong S, Ren ZF, Yuan F, Yu T, Zou ZG, Liu JM. Visible-light photocatalytic properties of weak magnetic $BiFeO_3$ nanoparticles. *Adv Mater* 19 (2007) 2889–2892.
100. Bhushan B, Basumallick A, Bandopadhyay SK, Vasanthacharya NY, Das D. Effect of alkaline earth metal doping on thermal, optical, magnetic and dielectric properties of $BiFeO_3$ nanoparticles. *J Phys D: Appl Phys* 42 (2009) 065004.

101. Hengky C, Moya X, Mathur ND, Dunn S. Evidence of high rate visible light photochemical decolourisation of Rhodamine B with $BiFeO_3$ nanoparticles associated with $BiFeO_3$ photocorrosion. *RSC Adv* 2 (2012) 11843–11849.
102. Umar M, Mahmood N, Awan SU, Fatima S, Mahmood A, Rizwan S. Rationally designed La and Se co-doped bismuth ferrites with controlled bandgap for visible light photocatalysis. *RSC Adv* 9 (2019) 17148–17156.

Design of Room Temperature d⁰ Ferromagnetism for Spintronics Application
Theoretical Perspectives

9

Ravi Trivedi[1] and Brahmananda Chakroborty[1,2]

1 High Pressure and Synchrotron Radiation Physics Division, Bhabha Atomic Research Centre, Trombay, Mumbai, India

2 Homi Bhabha National Institute, Anushaktinagar, Mumbai, Maharashtra

Contents

9.1	Motivation and Background	162
9.2	Spintronics Device – Fundamentals and Applications	162
	9.2.1 Applications	163
9.3	Design of Dilute Magnetic Semiconductors (DMS)	164
9.4	What Is the Room Temperature of d⁰ Ferromagnetism?	168
9.5	Mechanism for d⁰ Ferromagnetism	168
	9.5.1 Mean-Field Zener Model	168
	9.5.2 Superexchange Mechanism	168
	9.5.3 Stoner Criteria	169
	9.5.4 RKKY Oscillation	169

9.6	Different Way to Produce d⁰ Ferromagnetism	170
	9.6.1 Substitutional Doping	170
	9.6.2 Defect Induced	171
	9.6.3 Vacancy Induced	174
	9.6.3.1 Cation Vacancy	174
	9.6.3.2 Anion Vacancy	176
9.7	Factors to Control Curie Temperature	177
9.8	Conclusion and Future Directions	178
Reference		180

9.1 MOTIVATION AND BACKGROUND

The advancements of the most recent twenty years have consolidated the unmistakable fields of spintronics and attraction into incredible power. Spintronics (or spin electronics) is a consistently exhausting space of innovative work that converges among magnetism and electronic devices. It targets exploiting the quantum normal for the electrons that are its spin to make new functionalities and new gadgets. Spintronics with semiconductors enjoys the huge benefit of consolidating the capability of the magnetic materials, like the control of current by spin control, non-volatility, and so on, with the capability of the semiconductors (control of current by the gate, coupling with optics, and so on). Spintronics is the name related with innovation that uses both the inherent spin or charge of an electron in transport devices. It is considered that spintronics began in 1988 with the revelation of the GMR impact, which relates to a huge variety of the opposition of a magnetic multifactor under the use of a magnetic field [1]. Spintronics is the catchphrase to propel investigations of spin energized electronic states in solids and at surfaces or interfaces. The utilization of the electron spin with the electron charge as data transfer is the objective for spintronics gadgets [2]. A spintronic device has several extra benefits contrasted with other arising regions:

(1) It contributes similarly to electronic and optoelectronic gadgets.
(2) It can be coordinated with other semiconductor gadgets for differing functionalities.

The reconciliation of magnetic, optoelectronic, and electronic gadgets is accomplished by the way that the manufacture ventures for semiconductor spintronic devices closely resemble the ordinary miniature creation procedures. The target of this chapter is to give a quantitative establishment to this structure block approach, with the goal that novices can comprehend the ideas immediately dissected and fundamentally assessed.

9.2 SPINTRONICS DEVICE – FUNDAMENTALS AND APPLICATIONS

Spintronics is an arising innovation for assembling electronic gadgets that exploit electron spin and its related magnetic properties, rather than utilizing the electrical charge of an electron, to convey data. Spin electronics is additionally called Spintronics, where the twist of an electron is constrained by an outer magnetic field and energizes the electrons. These captivated electrons are utilized to control the electric flow. The objective of spintronics is to foster a semiconductor that can control the magnetism of an electron. When we add the twist level of opportunity to hardware, it will give huge adaptability and usefulness

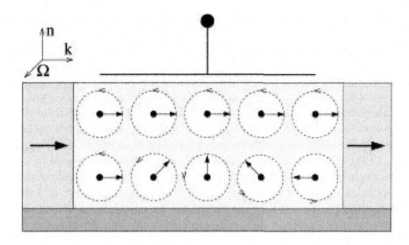

FIGURE 9.1 Scheme of the Datta-Das spin field-effect transistor (SFET). Adapted with permission from **[3]**. Copyright (2004) American Physical Society.

to future electronic items. Magnetic twist properties of electrons are utilized in numerous applications like magnetic memory and magnetic recording.

In electronic devices, data is put away and sent by the progression of power as contrarily charged subatomic particles called electrons, as shown in Figure 9.1. The zeroes and ones of computer paired code are addressed by the presence or nonattendance of electrons inside a semiconductor or other material. In spintronics, data is put away and sent utilizing one more property of electrons called spin. Spin is the characteristic precise energy of an electron; every electron behaves like a minuscule bar magnet, similar to a compass needle, that focuses either up or down to address the spin of an electron. Electrons traveling through a nonmagnetic material regularly have arbitrary twists, so the net impact is zero. Outer magnetic fields can be applied so the spins are adjusted (all up or all down), permitting a better approach to store parallel information as ones (all spins up) and zeroes (all spins down). The impact was first found in a gadget made of different layers of electrically conducting materials. The principal plan of the spintronics device dependent on the metal oxide semiconductor innovation was the primary field impact spin semiconductor proposed in 1989 by Suprio Datta and Biswajit Das of Purdue University. In their gadget, a design produced using indium-aluminum-arsenide and indium-gallium-arsenide gives a channel to two-dimensional electron transports between two ferromagnetic cathodes. One terminal goes about as an emitter and the other as a collector **[3]**. Spintronics is considered as quite possibly the main arising research area, with tremendous potential to give rapid, low-force, and high density and memory electronic gadgets and loweredge ebb and flow and high-power lasers (optoelectronic gadgets) as a hotspot for circularly energized light. While all-metal spintronic gadgets (magneto-electronic) are predominantly centered on memory gadgets, semiconductor spintronics is key for rationale and optoelectronic devices.

9.2.1 Applications

Spintronics is the most encouraging innovation to give multi-utilitarian, high-velocity, low-energy electronic devices with the disclosure and utilization of the giant magnetoresistance impact (GMR) **[4–5]**. Spintronics has, in practically no time, been formed into an alluring field, planning to utilize the spin degree opportunity of electrons as a data transporter to accomplish information stockpiling and intelligent activities. Contrasted with traditional microelectronic gadgets dependent on the charge, spintronic gadgets require less energy to switch a spin state, which can bring about quicker activity speed and lower energy utilization. The spintronic gadget heterostructures are normally developed by subatomic shaft epitaxial on a semi-protecting semiconductor substrate. The Magneto Optoelectronic Integrated Circuit (MOEIC), Figure 9.2a,b, has a horizontal semiconductor spin valve, a cascaded High Electron Mobility Transistor (HEMT) enhancer, and a LED solidly incorporated to show a magneto-electronic switch. A spin valve has two ferromagnetic contacts going about as polarizer and analyzer [6].

The obstruction of the spin valve is low if both polarizer and analyzer are charged a similar way and high in case they are polarized in inverse ways. The cascaded HEMT intensifier in the incorporated circuit enhances the magnetoresistance [6]. An adjustment of the door-to-source voltage of the HEMT changes the channel-to-source current, and consequently the gadget goes about as a trans-conductance enhancer. The light power of the LED changes to the adjustment of the channel-to-source current of the second HEMT. The MOEIC accordingly changes over the twist polarization in the channel locale of the horizontal twist valve to a comparable change in the light force of the LED. The circuit works as a magneto-electronic switch, which adjusts the light force of the LED. Other than the standard uses of hard plate drives in work areas, PCs, or servers, the decrease in their structure factor (one-inch drive) has permitted presenting magnetic circle drives in new customer gadgets – items like camcorders, MP3 players, film recorders, etc. In 1995, advancement was accomplished with the disclosure of room temperature tunneling magnetoresistance (TMR) in magnetic tunnel junctions (MTJs) [7–8]. MTJs are made up of a number of two ferromagnetic layers isolated by a flimsy protecting layer. This stack is sandwiched between a base and a top terminal. An enormous TMR for bimolecular burrow intersections dependent on ferritins immobilized among Ni and E_{GaIn} anodes was announced by Karuppannan et al. [9], as displayed in Figure 9.3.

By applying a bias voltage between these anodes, a current moves through the stack opposite to the interfaces. MTJs are presently carried out in different applications in which their variable opposition is utilized to store data (concerning example in non-unpredictable magnetic recollections, MRAM), to deal with it (magnetic rationale entryways), or to characterize the capacity of CMOS rationale circuits (reprogrammable logic). Similarly, molecular magnets containing lanthanide particles have been getting an expanding level of consideration during the last years. The relevance of atomic magnets on predictable gadgets and innovations – for example, spintronic parts, turn valves, or quantum bits – depends on the accomplishment of magnetic bi-stability with extended unwinding times to such an extent that the particle permits the planned activity. The utilization of lanthanides as fixings in atomic attraction works with the accomplishment of a high energy boundary for charge inversion, a boundary identified with bi-stability, and extended unwinding times, on account of their unquenched orbital energy which improves the magnetic anisotropy of the particle. Similarly, thermoelectric and spintronics applications can be seen in Mn_2MgGe also investigated by Patel et al. [10]. The calculations show that Mn_2MgGe compound has great potential applications as thermoelectric devices. The chemical potential-dependent thermoelectric calculation shows P-type and N-type doping of charge carriers, which can increase the thermoelectric response at high temperatures. The available number of energy states for Mn_2MgGe d orbital electrons is finite at the Fermi level, causing the transition between spin-up electrons from the conduction band to the valence band (Figure 9.4). Thus, the d orbitals are responsible for half metallicity in the Hg_2CuTi type structure of Mn_2MgGe. Spin-polarized band structure and total DOS (density of states) predict the half-metallic ferrimagnetic (HMF) nature of the Hg_2CuTi type structure of Mn_2MgGe.

9.3 DESIGN OF DILUTE MAGNETIC SEMICONDUCTORS (DMS)

The fast progression of microelectronic innovation has provoked interest in reduced, elite gadgets that offer various functionalities. The improvement of usual electronic devices that utilizes the charge of an electron for the handling and putting away of data is nearly immerse because of its low productivity. The improvement of customary electronic gadgets that utilize the charge of an electron for the handling and

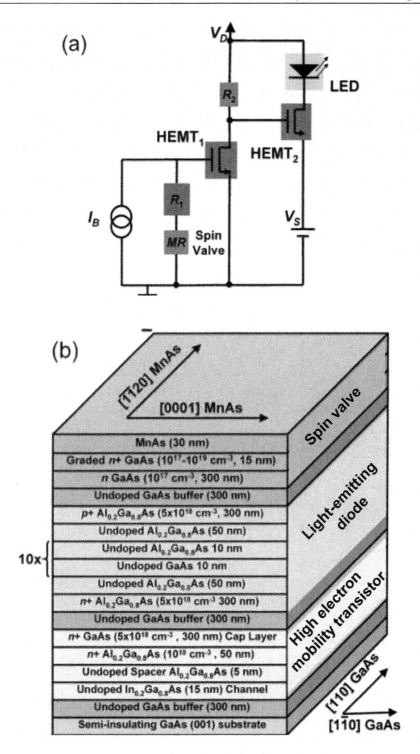

FIGURE 9.2 (a) A schematic of the MOEIC. (b) The heterostructure of the MOEIC consists of epitaxial grown layers. Adapted with permission from **[6]**. Copyright (2008) American Institute of Physics.

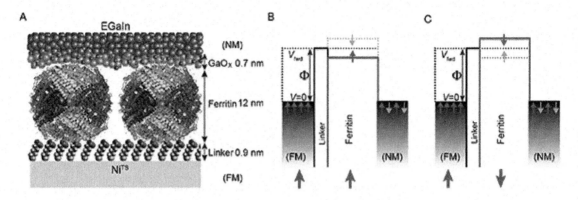

FIGURE 9.3 Schematic illustration of the Ni^{TS}/linker/ferritin//GaO$_x$/EGaIn junction for room temperature TMR. Adapted with permission from **[9]** exclusive licensee [IOP]. Distributed under a Creative Commons Attribution License 4.0 (CC BY).

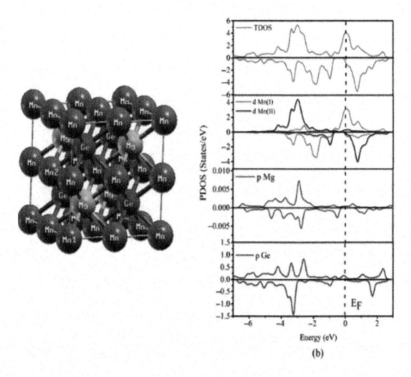

FIGURE 9.4 Ground state minimum energy structure of Mn2MgGe and spin-polarized partial DOS. Adapted with permission from **[10]**. Copyright (2020) Elsevier.

putting away of data is nearly immersed because of its low productivity; Meanwhile, the combination of dilute magnetic semiconducting (DMS) material that utilizes the spin level of opportunity of an electron alongside its charge has made them featured for many utilizations (Figure 9.5). Nonmagnetic semiconductors can be transformed into magnetic semiconductors by doping them with magnetic particles – for example, transition metals (TM, for example, Mn, Fe, Ni, Co, and so forth) or rare earth components (RE, for example, Gd, Eu, Er, and so forth) [11].

TM and RE have unpaired twists that associate with the host semiconductor through trade components and make the semiconductor magnetic [10]. Such semiconductors are called dilute magnetic

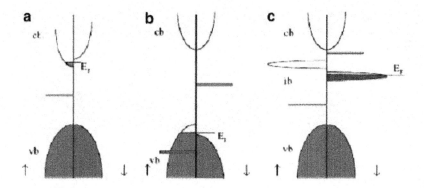

FIGURE 9.5 Some possible schemes for a magnetic semiconductor. Adapted with permission from **[11]**. Copyright (2006) Elsevier.

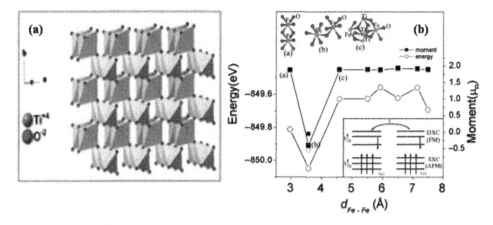

FIGURE 9.6 (a) Crystal structure of TiO$_2$ (rutile). (b) Magnetic moment per iron as a function of the bond distance of Fe-Fe. Adapted with permission from **[14]**. Copyright (2006) American Physical Society.

semiconductors. Practical utilization of 2D attraction will probably require room temperature activity, air steadiness, and (for magnetic semiconductors) the capacity to accomplish ideal doping levels without dopant collection. To conquer these issues, Zhang Fu et al. **[12]** proposed a DMS by vanadium-doped tungsten disulfide monolayer utilizing a dependable single-step film sulfidation technique. Interest in DMSs – for the most part, dilute magnetic oxides (DMOs) or dilute magnetic semiconductor oxides (DMSOs) – has been quickly expanding because of their possible application in spintronics gadgets. DMS can be shaped by adding magnetic pollutions at exceptionally low fixations to the host cross-section, without changing the grid that arose out of the subsequent materials, for many current gadgets, e.g., cutting-edge spintronics-based multifunctional gadgets, due to the presence of ferromagnetism above room temperature **[11–13]**. Double oxide semiconductors, SnO$_2$, show a rutile structure, while TiO$_2$ has both a rutile and anatase structure. Hexagonal ZnO has a wurtzite design, and all are inherently diamagnetic.

Chen et al. **[14]** showed that oxygen opportunities increment the ferromagnetism (FM) in the rutile-organized Fe-doped TiO$_2$ through density functional theory (DFT) computations. More significant imperfection are due to oxygen opportunities or catching of electrons by Fe molecule beginning of DMSs in the system. Zero magnetic minutes were created by Sc and Ni particles while the greatest attraction was seen in Fe and Mn. Rutile-staged Fe-doped TiO$_2$ was investigated by Mallia et al. utilizing DFT for DMS **[15]**. It was noticed that oxygen stoichiometry and the presence of Fe in the TiO$_2$ grid predominantly influences ferromagnetism (Figure 9.6). The opening improves the ferromagnetic coupling J between two Fe dopants, either utilizing the development of a shallow contamination state if the opportunity is near one of the Fe atoms or by an upgrade of the ferromagnetic twofold trade in case they are far away.

9.4 WHAT IS THE ROOM TEMPERATURE OF D⁰ FERROMAGNETISM?

In a magnetic semiconductor, electrical and optical properties might be unpredictably interlinked. Ferromagnetism is clarified by the idea that a few types of particles have a magnetic second. Customarily there have been two ways to deal with attraction in solids, one dependent on nearby minutes, the other on electrons delocalized in tight energy groups. The principal approach is substantial for ionic protecting mixtures, where the magnetic minutes are borne by cations with to some extent filled d or f shells. The quantity of electrons per particle is typically a number, and the cation has a conventional charge state identified with its electron count. In the subsequent methodology, the second approach is conveyed all through the strong in spin split energy groups. For the most part, the Fermi level meets both ↑ and ↓ groups, and the indispensable twist second rule is out of commission. Attraction is a grounded marvel which emerges because of the trade association between the electrons into some degree filled d or f orbital's display this marvel **[16–18]**. Aside from this, magnetic conduct of specific materials with totally filled or zero-filled d or f orbital's is likewise noticed and has stayed under banter among specialists during the last decade **[19–25]**. The magnetism which emerges due to filled or unfilled d or f orbital's is known as d^0 ferromagnetism, and it is a wide field of study for established researchers according to a phenomenological perspective **[19–23]**.

9.5 MECHANISM FOR D⁰ FERROMAGNETISM

Somewhat recently, d^0 attraction related to inherent imperfections or pollutants (without the contribution of regular 3d or 4f magnetic particles) has been seen in an assortment of materials **[26–28]**. Be that as it may, the center instruments overseeing attraction are as yet open to discussion. Here we attempt to clarify the potential components which are identified with instigating d^0 ferromagnetism.

9.5.1 Mean-Field Zener Model

The most broadly utilized hypothesis of the ferromagnetism in FMSs, the "mean-field Zener model" **[29]**, accepts that the Fermi level lies in the Conduction Band (CB) or Valence Band (VB) of the host semiconductors and anticipated that the s,p-d trade cooperations would incite turn parting in these groups. Zener first proposed the model of ferromagnetism driven by the trade collaboration among transporters and confined twists. In any case, this model was subsequently abandoned, as neither the degree of the magnetic electrons nor the quantum motions of the electron twirl polarization around the restricted twists were considered; both of these are presently determined to be basic elements for the hypothesis of magnetic metals.

9.5.2 Superexchange Mechanism

Countless models have been proposed with an end goal to comprehend the beginning of deformity-initiated magnetism. Superexchange **[30–31]** is an aftereffect of the electrons having come from a similar contributor molecule and being combined with the accepting particles' twists. The superexchange wonder is a sort of aberrant magnetic communication. It happens between two cations with unpaired twists limited by an anion without unpaired twists. The coupling can be either ferromagnetic or antiferromagnetic. The supposed Anderson-Goodenough-Kanamori rules figure out which of the two kinds of coupling is available. Superexchange is little bit in contrast with different sorts of magnetic interaction dependent on direct trade of electrons.

9.5.3 Stoner Criteria

A few metals have tight groups and a huge density of states at the Fermi level; this prompts a huge Pauli vulnerability. When Pauli is adequately enormous, it is workable for the band to part unexpectedly, and ferromagnetism shows up. Ferromagnetic trade in metals doesn't generally prompt unconstrained ferromagnetic requests. The Pauli weakness should surpass a specific edge. There should be a particularly huge thickness of states at the Fermi level $N(E_F)$. Stoner applied Pierre Weiss's atomic field thought to the free electron model. In an exceptionally straightforward way, the Stoner standard clarifies why iron is ferromagnetic but manganese, for instance, isn't, although the two components have an unfilled 3d shell and are nearby in the intermittent table: as indicated by this measure, the result of the thickness of states and the trade fundamental should be more noteworthy than solidarity for unconstrained twist requesting to arise [32].

9.5.4 RKKY Oscillation

RKKY stands for Ruderman–Kittel–Kasuya–Yosida [33]. It refers to a coupling mechanism of nuclear magnetic moments or localized inner d- or f-shell electron spins in metal utilizing an interaction through the conduction electrons. The mechanism of the exchange interaction among magnetic ad-atoms on graphene like Fe–graphene indicating that the RKKY (Ruderman–Kittel–Kasuya–Yosida) mechanism plays the dominant role in governing the magnetic order in these systems.

The instrument of the trade association among magnetic adatoms on graphene, like Fe on graphene system, shows that the RKKY component assumes the prevailing part in administering the magnetic request in these frameworks. To comprehend the ferromagnetic requesting in the Fe/Gra with one Fe particle, RKKY assumes a significant part. Here in Figure 9.7, we comprehend that the RKKY trade communication between Fe molecules is intervened by the π electrons of graphene. The times of motions rely upon k_F ($\sqrt{2\pi\sigma_j}$), where σ_j is the viable electron number thickness around the Fermi energy.

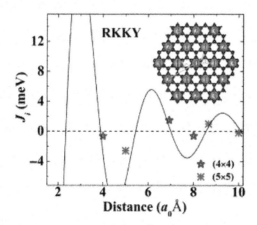

FIGURE 9.7 Exchange interaction parameters between Fe⁰ and the ith neighbors in different Fe/Gra structures. Adapted with permission from [33]. Copyright (2019) American Chemical Society.

9.6 DIFFERENT WAY TO PRODUCE D⁰ FERROMAGNETISM

In the field of spintronics, ferromagnetic semiconductors enjoy uncommon benefits on account of their simple combination into semiconductor gadgets [34]. The greater part of the past examinations has been centered around DMS, in which the magnetic minutes are presented by doping magnetic 3dn TM particles into semiconductors [35–37]. As of late, in any case, startling high-temperature ferromagnetism has been seen in a progression of materials, which don't contain particles with to some extent filled d or f groups [38–41]. This sort of "d⁰ ferromagnetism" [42] gives another chance to look through high-temperature spintronic materials. Nonetheless, it likewise gives a test to comprehend the beginning of the attraction in these materials, i.e., regardless of whether the noticed magnetism is an inborn property of the host material or an outward property relying delicately upon the kind of dopant used to incite the charge. It implies d⁰ ferromagnetism can be delivered by presenting turn minutes in nonmagnetic semiconductors by doping of non magnetic atoms:

9.6.1 Substitutional Doping

Presenting FM in conventional nonmagnetic semiconductors brings about weakened magnetic semiconductors. There are not many elements that assume a fundamental part in the actuation of d⁰ FM, to be specific: (i) pollutant fixation, (ii) exchange coupling between various twists, and (iii) exploratory plausibility.

At extremely low debasement fixation, there may not be any cooperation between the pollution particles. Enlistment of room temperature ferromagnetism from first-standards recreations by nitrogen doping in nonmagnetic yttrium oxide has been concentrated by Chakraborty et al. [43] to investigate d⁰ ferromagnetism. The framework holds its security, as displayed in Figure 9.8a at room temperature, and qualifies the Stoner rules of ferromagnetism. The halfway thickness of states along with the turn thickness plot uncovers that it is the N_{2p} orbital alongside the closest O_{2p} orbital which adds to initiated magnetism. Unblemished yttrium (Y_2O_3), which is nonmagnetic, has a place with space group Ia3̄-(Th_7). The unit cell contains two equivalent cations (Y^{3+}) and one anion (O^{2-}) destination. To examine the reason for ferromagnetic acceptance in N-doped Y_2O_3, the doping of N presents a single opening in the framework. This causes the presence of thin exceptionally restricted pollution groups which display parting of twist adjoining Fermi level, as shown in Figure 9.8b. It is likewise seen that spin parting makes two up-spin levels just as three down-spin levels. Two up-spin channels just as one down-turn channel are involved completely while one more down-spin channel is involved mostly and the excess one is vacant. It is additionally conceivable that the p orbital of the N molecule places the most extreme commitment in creating a magnetic second while there is a considerable gift from the p and d orbitals of O and Y particles separately. Additionally, using In_2O_3, displayed in Figure 9.8c, as a host network, broad estimations dependent on density functional theory have been completed to comprehend the electronic and magnetic properties [44]. For V-, Li-, Na-, and K-doped In_2O_3, the initiated magnetic minutes are essentially confined on the main shell of O molecules around X_{In} locales. The fundamental element of the DOS spectra as delineated in Figure 9.8d is that the E_F crosses the minority twist of pollutant levels. There is solid p-p hybridization between 1NN O atoms, which actuates limited conveyed openings in O 2p states. This load of discoveries emphatically proposes that the limited pp openings of O atoms instigated by salt metal doping assume an indispensable part of the magnetic trade association among O particles. Half-metallic magnetism in Cu-doped ZnO is anticipated by precise maximum capacity linearized expanded plane-wave and DMol3 computations dependent on density functional hypothesis [45]. A net magnetic moment of 1μB is found per Cu. At a Cu convergence of 12.5%, absolute energy computations give the ferromagnetic state as 43 meV lower than the antiferromagnetic state, and it is subsequently anticipated to be the ground state

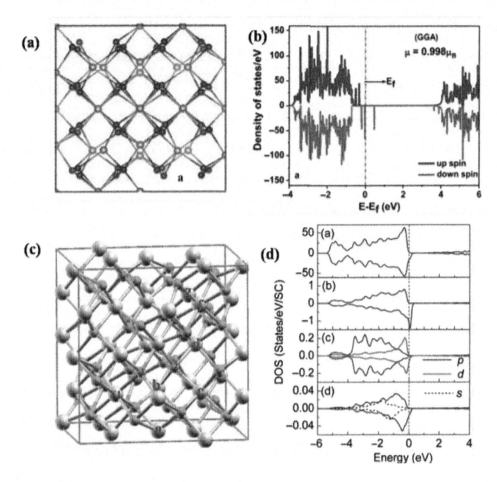

FIGURE 9.8 (a) Schematic structure of yttrium oxide. (b) DOS calculated at generalized gradient approximation (GGA) level. (c) Geometry of In_2O_3, with a, b, c, d, f, and g for different configurations. (d) DOS of different doping in In_2O_3. Adapted with permission from **[43]**. Copyright (2021) Elsevier.

with a T_c assessed. The complete magnetic snapshot of 2μB per supercell, or 1μB per Cu particle, is found because of the half metallicity: Cu is enraptured with a magnetic snapshot of 0.58μB, and the four encompassing O atoms in the CuO_4 tetrahedron are captivated with a magnetic snapshot of 0.04 μB for the top-site O and 0.08μB for the other three O in the basal plane. The more modest magnetic second for the top-site O is because of the twisting of the tetrahedron; the Cu-O bond is longer for the top-side O than the three lying in the basal plane to be around 380 K. The magnetic minutes are confined inside the CuO_4 tetrahedron with ferromagnetic coupling among Cu and O. The electronic states close to EF are overwhelmed by solid hybridization between O 2p and Cu 3d, which suggests that the Cu-O bond is very covalent rather than ionic.

9.6.2 Defect Induced

Attributable to both the intriguing material science and likely applications in spintronics, exceptional endeavors have been devoted to acquiring a full comprehension of room temperature (RT) d^0 magnetism. Notwithstanding, the center instruments administering inherent deformities are as yet open to discussion **[46–47]**. Utilizing hybrid density functional theory, conceivable component for d^0 ferromagnetism intervened by inherent imperfections has investigated by Zhang Z et al. **[48]**. By considering the electronic connections, they tracked down an extra parting of the imperfection states in the Zn opening and, in this way, the chance of acquiring energy by particular filling of opening states, building up ferromagnetism between turn captivated S 3p openings. Somewhat recently, d^0 ferromagnetism related to natural deformities or pollutants (without the inclusion of ordinary 3d or 4f magnetic particles) has been seen in an assortment of oxides, sulfides, and different semiconductors, especially in nanoscale frameworks **[47–49]**. Point surrenders have been recommended a few times as a system for limited minutes and ferromagnetism in oxides. Inborn point imperfection-driven ferromagnetism in case of nonmagnetic mixtures has been recently concentrated on utilizing first-standards techniques in a few frameworks. Striking cases incorporate CaB_6 **[50]**, CaO **[51]**, and SiC **[52]**. The fundamental attribute of the opportunities in this load of frameworks is the high balance, either octahedral or tetrahedral, around the opening site. This constantly prompts a profoundly degenerate single-molecule range, which may then present high-turn states. Also, Kisan et al. **[53]** explored the electronic and magnetic properties of NiO nanocrystals methodically inside the system of density functional theory.

Their outcomes recommend that the opening is made by oxygen or may be by nickel or oxygen are generally liable for ferromagnetism in nanoscale NiO. Point surrenders have been proposed a few times as an instrument for restricted minutes and ferromagnetism in oxides **[54]**. The single molecule density of states and magnetic moment as displayed in figure 9.9, for fixed U = 0:4 W, n_h = 0:12, x = 4%, and three upsides of V = 0:2, 0.45, 0.8 W. The Fermi energy is at nothing. For little V there are no nearby magnetic minutes. As V expands, minutes show up (V = 0:28 W) as the charge and twirl conveyances foster designs

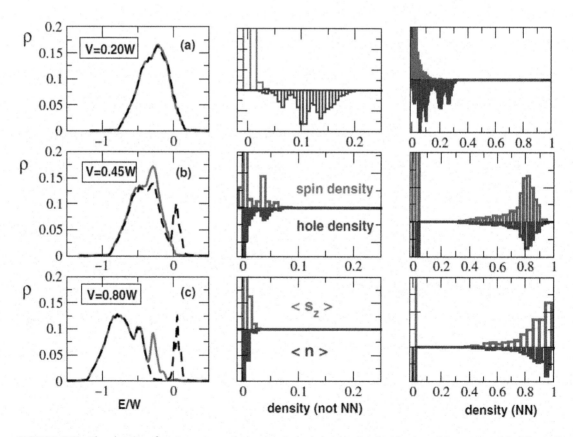

FIGURE 9.9 The density of states, magnetic moment, and charge distribution in oxide compound. Adapted with permission from **[54]**. Copyright (2006) American Physical Society.

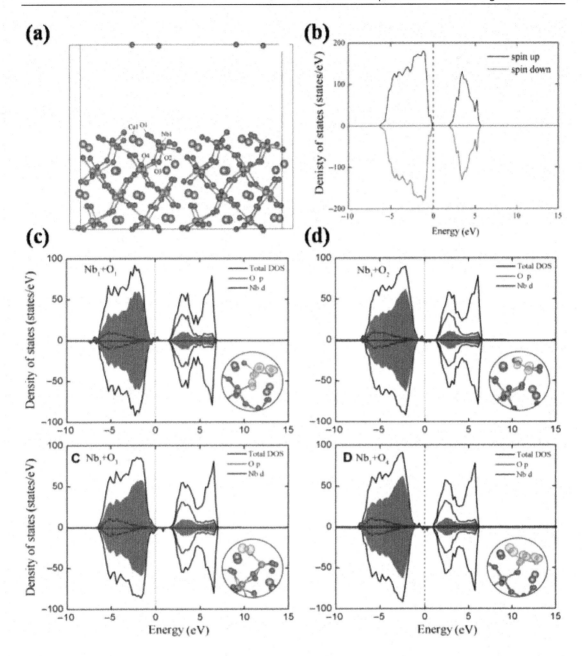

FIGURE 9.10 Table configuration of Ca$_2$Nb$_2$O$_7$ (010) surface with labeled Ca, Nb, and O atoms to be removed for vacancy study. Adapted with permission from [57]. Exclusive licensee Springer. Distributed under a Creative Commons Attribution License 4.0 (CC BY).

limited around the opening (right segment). Afterward (V = 0:45 W), a distinct pollutant band for the minority band parts from the valence band. The first top in quite a while (right section), at low densities, relates to locales with $V_i = V = 3$ and the subsequent wide top to $V_i = V$. As V further expands, they additionally notice a parting of the density of states in the larger part band and openings concentrated around the opportunities, giving immersed magnetic moments.

9.6.3 Vacancy Induced

The proof of opening-incited FM in nonmagnetic frameworks [55–56] suggests that opportunities likewise can assume a significant part in deciding the ferromagnetic conduct of nanoparticles.

To further comprehend d^0 ferromagnetism, hypothetical estimation to investigate the beginning of FM in $Ca_2Nb_{1.9}O_{7-\delta}$ film has been completed by utilizing the DFT. Computations were performed utilizing the plane-wave pseudopotential technique in the Vienna ab initio simulation package (VASP) [57]. In Figure 9.10a, among which O_1 is associated distinctly to one Nb particle, O_2 is facilitated between two adjoined Nb atoms, O_3 interfaces one Nb molecule in the peripheral layer with one more Nb particle in the subsequent layer, while O_4 indicates the one between two Nb atoms in the subsequent layer. Stoichiometric $Ca_2Nb_2O_7$ (010) surface was initially contemplated, and the absolute DOS was displayed in Figure 9.10b. No twirl polarization could be seen around the Fermi level, implying that the stoichiometric $Ca_2Nb_2O_7$ (010) surface is nonmagnetic. There are a few underlying and mathematical elements that control the magnetic conduct of a specific material: the distance between molecules, ligancy (coordination number), and balance. There are a few outcomes to altering the construction of a nanomaterial. The adjustment of the nearby climate of the atoms can be achieved through deformity presentation, like debasements, opening, and opportunity buildings. By decreasing the ligancy and changing the balance, the energy level groups are relied upon to limit. This is the justification for the improvement of attraction in materials with ferromagnetic conduct and could likewise clarify why some nonmagnetic materials display magnetic conduct [58]. One more instrument for the development of FM in weakened magnetic oxide nanoparticles was proposed in [59] and was related to surface or inside situated imperfections in these nanostructures. The creators thought that the FM is shown in a restricted part of the nanosized material, which contains zones wealthy in deserts that can likewise be on a superficial level, zones being portrayed by a limited ("forcefully crested") neighborhood DOS (thickness of states) in the conditions that the Fermi level (situated either beneath or over the versatility edge) will generally not agree with a nearby DOS. One of the as of late famous interests in ZnO is TM-doped ZnO (TM-ZnO) for likely applications in spintronics [60]. After many tests and hypothetical work to comprehend the idea of RT FM in TM-ZnO, primary imperfections are by and large acknowledged to be the reason for the RT FM in TM-ZnO, though transporters (consistently doped magnetic particles) associated with transporter intervened trade are normal side effects of the production of deformities in ZnO.

9.6.3.1 Cation Vacancy

Vacancies defects are one of the most general phenomena in metal oxide materials and are answerable for assorted optoelectronic marvels of both basic and pragmatic importance. Cation vacancies present nearby magnetic minutes just as openings to the host semiconductor and ferromagnetism. The aftereffects of first-standard estimations demonstrate that nonpartisan oxygen opening in ZnO is nonmagnetic [61]; however, Zn vacancy prompts magnetism [62]. The electronic and magnetic properties of cation opportunities in m-HfO_2 are anticipated utilizing thickness practical hypothesis by McKenna et al. [63] as displayed in Figure 9.11a. The hafnium vacancy is found to present a progression of charge change levels in the reach of 0.76–1.67 eV over the valence band most extremely related with openings restricted on adjoining oxygen destinations. The nonpartisan imperfection embraces a S = 2 twist state. Cation opportunities in most metal oxide materials can trap one or various openings. As a rule, these openings are found to confine on individual oxygen particles encompassing the opportunity framing O− particles.

The advanced design of every one of the imperfections conveying a net twist alongside isosurfaces of electron spin density is displayed in Figure 9.11b. For the VX Hf imperfection, three openings are limited on 3C oxygen destinations adjoining the opportunity with a fourth restricted on a 4C oxygen site. The inclination for opening catching on the 3C destinations is reliable with the expanded strength of opening polarons on this sublattice. For the V_{Hf} imperfection, the expansion of an electron takes out the opening on the most unstable 4C oxygen site. The determined electronic density of states for the Hf opportunity

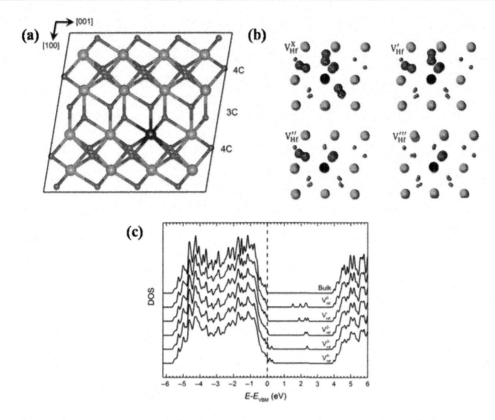

FIGURE 9.11 (a) Schematic ground state structure of HfO$_2$. (b) Isosurface of electron density. (c) Density of states for the Hf vacancy defects. Adapted with permission from **[63]**. Exclusive licensee [APS]. Distributed under a Creative Commons Attribution License 3.0 (CC BY).

surrenders is displayed in Figure 9.11c. The curves are aligned to the bulk valence band maximum using the normal electrostatic potential over particles a long way from the imperfection in various supercells as a typical reference. The V$_{Hf}$ deformity is related to four limited electronic states in the hole somewhere in the range of 1.5 and 2.3 eV over the valence band. To discover the presence of a net magnetic second, the computation of the g tensor for cation opening utilizing the installed group approach recommends that the greatness of the g-tensor parts for cation opportunities are essentially bigger and more anisotropic than some other imperfection. The deformities with the most comparative g tensors are those related with little opening polarons true to form. A lot higher anisotropy is reliable with the much lower evenness of the cation opportunity deserts contrasted with opening polarons. It is additionally striking that there is a huge distinction between the g tensor of the nonpartisan and the contrarily charged opportunity, mirroring their different electronic construction and twist thickness.

Then again, Volnianska et al. **[64]** researched high twist conditions of cation opportunities in GaP, GaN, AlN, BN, ZnO, and BeO utilizing first-standards estimations. The spin energized opening prompted level is situated in the band hole in GaP, ZnO and BeO. In the nitrides, the vacancy states near the valence groups, restricts arrangement of emphatically charged opening in GaN and BN and permits Al opportunity in p-AlN to accept the most noteworthy conceivable S = 2 spin state. The dangling bonds' are high spins conditions of vacancies in semiconductors are very important. Such arrangements occur when the opening prompted multiple level is to some extent involved by electrons with equal twists and contingent upon the genuine occupation; the all-out turn is 1, 3/2, or 2 instead of 0 or 1/2. In this manner, high twist setups relate to the nearby twist polarization of electrons living at the hanging obligations of the opportunity neighbors **[64]**. Here in Figure 9.12, energy levels of

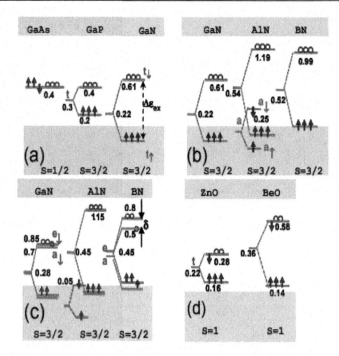

FIGURE 9.12 (a) Energy levels of neutral GaAs, Gap, and GaN, (B) zinc-blende, (c) wurtzite, and (d) w-ZnO or w-BeO. Adapted with permission from [64]. Copyright (2011) American Physical Society.

a neutral V_{cation} (a) in the series of Ga pnictides GaAs, GaP and GaN, and in the nitrides GaN, AlN, and BN with (b) zinc-blende and (c) wurtzite structure, and (d) in w-ZnO and w-BeO, t, e, and a denote the t_2, e_2, and a_1 states, respectively, and the hexagonal splitting δ of t_2 is shown for BN. The numbers give the calculated energies relative to VBT in eV. Energies of the resonance states are shown only schematically. Spins of electrons are indicated by arrows, and empty spheres indicate unoccupied states.

9.6.3.2 Anion Vacancy

Notwithstanding, it has additionally been accounted for that the cation opportunity or anion opening is for the most part connected with enormous arrangement energy. So cation or anion substitution is the best mode to initiate d^0 FM. Yet it is not ensured that the replacement of cations or anions will consistently initiate d^0 FM. The following components assume a pivotal part in the enactment of d^0 FM by presenting opening or electrons in the framework: (i) pollution fixation, (ii) trade coupling between various twists, and (iii) exploratory practicality. Charge through cation-site doping is feasible but not exceptionally proficient because the openings are conveyed into a few adjoining anion destinations. In this manner, it will be more compelling if the opening doping happens straightforwardly at the anion site, for example, C replacement on the O site (CO) [65]. For this situation, the C 2p orbital is limited very much like the O_{2p} orbital, and every CO imperfection makes two openings, similar to V_{Zn}. This shows that for CO, the greater part-turn deformity level is completely involved, though the minority turn level is just to some extent involved, bringing about an all-out magnetic snapshot of 2:00μB. 0:60μB is situated at the CO site, 0:1μB on every one of the four NN Zn destinations, 0:034μB on every one of the 12 hcp NN O locales, and the rest are generally disseminated in the interstitial spaces around these particles. The adjustment of the FM stage noticed for these frameworks can be perceived utilizing the phenomenological band-coupling model [20] (Figure 9.13).

For the disconnected deformity CO, the larger part spin is completely involved, though the minority spin is just too some degree involved. In the FM stage, the 2p states with a similar spin can couple to one another, shaping holding and anti-bonding states. The antiferromagnetic (AFM) stage is balanced out by

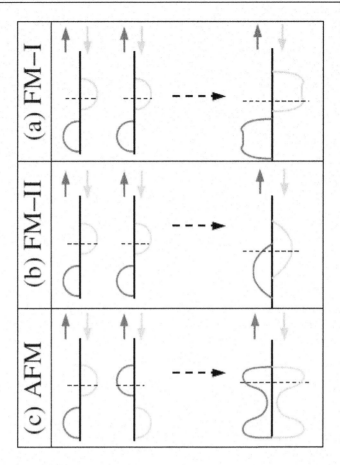

FIGURE 9.13 Schematic illustration of the possible coupling in d⁰ systems. Adapted with permission from [66]. Copyright (2009) American Physical Society.

the superexchange association between the involved larger part spin state on one site and the to some extent involved minority spin state on the other site. The increase in energy is typically bigger in the FM state than in the AFM state because of the second-request nature of the superexchange association in Figure 9.13.

9.7 FACTORS TO CONTROL CURIE TEMPERATURE

Curie temperature relies upon doping fixation as well as on development conditions. Controlling the Curie temperature is undeniably challenging as prompted magnetic moment. Curie temperature is a lot of delicate to development conditions. The Curie temperature relies upon the magnetic minutes and material boundaries of the materials, like susceptibility, dipole minutes, penetrability, permittivity, and so forth. For handily shed van der Waals materials, feeble interlayer magnetic coupling brings about low Curie temperature regardless of whether the material is FM. When peeled from mass materials, more grounded intralayer couplings become predominant and Curie temperature can increase.

Curie temperature additionally relies upon the trade integrals, which is a vital boundary to get a wide working temperature range for microwave gadgets and parts. Crafted by the connection between

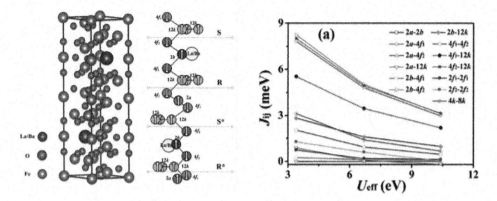

FIGURE 9.14 (a) Schematic representation of La-doped $Ba_{1-x}La_xFe_{12}O_{19}$. (b) The exchange integrals as a function of U_{eff}. Adapted with permission from **[65]**. Copyright (2016) Springer Nature.

the Curie temperature Tc and spin integrals for barium hexaferrites has been principally focused on the undoped tests utilizing the nonlinear fitting strategies: researcher proposed that 12k sublattice of barium hexaferrite is dependent upon a solid trade communication for the connection between R(R*) and S(S*) squares and that Fe_3^+ 2b-O-Fe_3^+ 4f², Fe_3^+ 2a-O-Fe_3^+ 4f¹, Fe_3^+ 4f2-O-Fe_3^+ 12k, and Fe_3^+ 4f¹-O-Fe_3^+ 12k sets of three show the similarly solid trade coupling [66]. As displayed in Fig. 9.14(a), 24 Fe_3^+ particles of M-type hexaferrites are appropriated in five diverse sublattices: 3 equal locales (12k, 2a and 2b) and 2 antiparallel destinations (4f¹ and 4f²) [67]. La-based replacements could add to some Fe3+ changing into Fe2+ at the 2a and 4f² locales [65]. The Curie temperature has expanded with the increment of La substance; the 2a-4f¹, 2a-12k, 2b-4f¹, and 2b-12k communications increment; and the 4f¹-4f², 4f¹-12k, 4f²-12k, 2f¹-2f¹, 2f²-2f², and 4k-8k connections decline, while the 2b-4f² association has a slight change. Figure 9.14b gives the trade integrals as an element of U_{eff} for $Ba_{1-x}La_xFe_{12}O_{19}$ tests. Essentially, the Green function formalism yields the trade pair connections between far-off magnetic particles that are required for quantitative investigations of magnetic excitations including the Curie temperatures. The estimations announced by Kudronovsky et al. [69] exhibit that the basic temperatures are especially relying upon the incorporation of confusion and, specifically, practical distances among magnetic debasements. Basic temperatures of Mn-doped GaAs without and with As antisites show that the T_c increments with the Mn fixation; however, the presence of antisites firmly lessens it, where T_c is straightforwardly identified with trade collaborations, in particular, T_c MFA = $(2x/3) \Sigma R\_=0$ J.0R, where x is the grouping of magnetic molecules. The reliance of Curie temperature on electron connection and electron entropy is likewise settled by Rusz et al. [70] to research the actual properties of the scattered $Ni_{2-x}MnSb$ compounds. The last impact which can impact the worth of T_c is the electronic entropy. The planning is accomplished for T = 0 while fundamental tests show an unimportant impact on T_c for T = 500 K while for T = 1000 K relating changes, which lower assessed T_c, could add up to many K. The strength of these impact increases with temperature. Along these lines, although presenting magnetic marks in nonmagnetic semiconductors is easy, controlling the Curie temperature and prompted magnetic second is extremely troublesome because of a muddled instrument of FM just as different amalgamation conditions affect the Curie temperature.

9.8 CONCLUSION AND FUTURE DIRECTIONS

By effectively utilizing the charge transport of electrons and openings as well as their spins, we can make an assortment of new wonders and useful materials. It is profoundly expected that these new materials

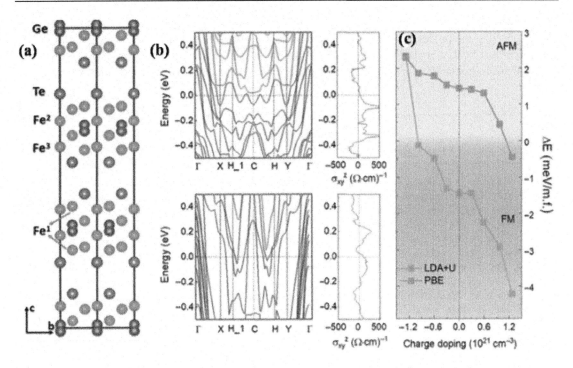

FIGURE 9.15 (a) Schematic crystal structure of Fe$_5$GeTe$_2$. (b) PBE and LDA+U band structure with Hall conductivity. Adapted with permission from [71]. Copyright (2021) American Chemical Society.

and marvels are applied to the advancement of cutting-edge hardware and data innovation. In spintronics applications, metal-based ferromagnetic materials and their multi-facets are generally created for reasonable gadgets. GMR in ferromagnet/nonmagnet metallic multi-facets and TMR in MTJs are utilized for magnetic field sensors in hard-plate drives, adding to the increment of capacity limit. TMR is additionally applied to magnetoresistive irregular access memory (MRAM). These metal-based spintronics gadgets use turn-subordinate vehicle properties in nanometer-thick ferromagnetic/nonmagnetic multi-facets, particularly the magnetoresistance brought about by the distinction in electrical resistivity among equal and antiparallel charge setups. As of late engineers have accomplished record-high electron doping in a layered ferromagnet, causing magnetic stage progress with a huge guarantee for future gadgets. Control of attraction by electric voltage is crucial for creating future low-energy, high-velocity nano-electronic and spintronic gadgets, for example, turn circle force gadgets and twist field-impact semiconductors.

Super high-charge doping-prompted magnetic stage change in a layered ferromagnet permits promising applications in antiferromagnetic spintronic gadgets. Theoretical examinations have additionally arisen [72–74]. F$_5$GT is a recently integrated vdW ferromagnet, which is important because of its high Curie temperature (T$_c$). This T$_c$ approaches room temperature and surpasses the Curie temperature of up to 230 K displayed by the generally researched vagrant vdW FM Fe$_3$GeTe$_2$ (FGT). An orderly examination of the bizarre Hall impact in vdW vagrant FM Fe$_5$GeTe$_2$ (F5GT), as shown in Figure 9.15, is researched as of late by Cheng Tan et al. [74] with both the coercivity and the basic temperature of the ferri-magnetic progress of Fe$_1$ destinations. It relies emphatically upon the density of the F$_5$GT by utilizing DFT estimations with PBE and LDA+U methods. This examination shows that the magnetic ground state in F$_5$GT can be tuned in situ by a door voltage. An electron doping convergence of over 1021 cm^{-3} incited in F$_5$GT by a protonic entryway can trigger a magnetic stage change from FM to AFM.

The chance of electrical charge inversion in magnetic nanostructures at room temperature is turning out to be increasingly reasonable. For instance, a potential way of acquiring it is planning magnetocrystalline anisotropy so that it creates a fastener type movement on cycling the electric field. Another chance is that by utilizing the processional movement achieved when the anisotropy is unexpectedly changed by an electric heartbeat, it ought to be feasible to notice the inversion of charge. Electrical inversion, when acknowledged, should offer an amazingly incredible and productive method of controlling polarization.

REFERENCE

[1] M. N. Baibich, J. M. Broto, A. Fert, F. Nguyen Van Dau, F. Petroff, P. Etienne, G. Creuzet, A. Friederich, and J. Chazelas, Giant magnetoresistance of (001)Fe/(001)Cr magnetic superlattices. *Phys. Rev. Lett.* 61, 1988, 2472.
[2] H. Dery and L. J. Sham, Spin extraction theory and its relevance to spintronics. *Phys. Rev. Lett.* 98, 2007, 046602.
[3] Igor Žutić, Jaroslav Fabian, and S. Das Sarma, Spintronics: Fundamentals and applications. *Rev. Mod. Phys.* 76, 2004, 323.
[4] I. Z. Utic, J. Fabian, and S. D. Sarma, Spintronics: Fundamentals and applications. *Rev Mod Phys.* 76, 2004, 323–386.
[5] J. Sinova and I. Zutic, New moves of the spintronics tango. *Nat Mater* 11(5), 2012, 368–371.
[6] D. Saha, D. Basu, and P. Bhattacharya, A monolithically integrated magneto-optoelectronic circuit. *Appl. Phys. Lett.* 93, 2008, 194104.
[7] T. Miyazaki and N. Tezuka, Giant magnetic tunneling effect in Fe/Al2O3/Fe junction. *J. Magn. Magn. Mater.* 139, 1995, 231.
[8] J. S. Moodera, L. R. Kinder, T. M. Wong, et al., Large magnetoresistance at room temperature in ferromagnetic thin film tunnel junctions. *Phys. Rev. Lett.* 74, 1995, 3273.
[9] S. K. Karuppannan, R. R. Pasula, T. S. Herng, J. Ding, X. Chi, E. D. Barco, S. Roche, X Yu, N. Yakovlev, and S. Lim, Room-temperature tunnel magnetoresistance across biomolecular tunnel junctions based on ferritin. *J. Phys. Mater.* 4, 2021, 035003.
[10] D. Pratik, Patel[a]Jalaja, B. Pandya[a]Satyam M. Shinde[a]Sanjay D. Gupta[b]SomNarayan[c]Prafulla, and K. Jha[c], Investigation of Full-Heusler compound Mn2MgGe for magnetism, spintronics and thermoelectric applications: DFT study. *Comput. Condens. Matter.* 23, 2020, 00472.
[11] F. Zhang, B. Zheng, A. Sebastian, D. H. Olson, M. Liu, K. Fujisawa, Y. T. H. Pham, V. O. Jimenez, V. Kalappattil, L. Miao, T. Zhang, R. Pendurthi, Y. Lei, A. L. Elías, Y. Wang, N. Alem, P. E. Hopkins, S. Das, V. H. Crespi, M. H. Phan, and M. Terrones, Monolayer vanadium-doped tungsten disulfide: A room-temperature dilute magnetic semiconductor. *Adv. Sci.* 7, 2020, 2001174.
[12] J. M. D. Coey, Dilute magnetic oxides. *Curr. Opin. Solid State Mater. Sci.* 10, 2006, 83–92.
[13] J. M. D. Coey, M. Venkatesan, and C. B. Fitzgerald, Donor impurity band exchange in dilute ferromagnetic oxides. *Nat. Matr.* 4, 2005, 173–179.
[14] J. Chen, P. Rulis, L. Ouyang, S. Satpathy, and W. Y. Ching, Vacancy-enhanced ferromagnetism in Fe-doped rutile TiO_2. *Phys. Rev. B* 74, 2006, 235207. [Google Scholar] [CrossRef]
[15] B. Santara, P. Giri, S. Dhara, K. Imakita, and M. Fujii, Oxygen vacancy-mediated enhanced ferromagnetism in undoped and Fe-doped TiO2 nanoribbons. *J. Phys. D: Appl. Phys.* 47, 2014, 235304.
[16] B. D. Cullity, *Introduction to Magnetic Materials*. Reading, MA: Addison-Wesley, 1972.
[17] J. P. J. Qin, J. Nogués, M. Mikhaylova, A. Roig, J. S. Muñoz, and M. Muhammed, Differences in the magnetic properties of Co, Fe, and Ni 250–300 nm wide nanowires electrodeposited in amorphous anodized alumina templates. *Chem. Mater.* 17, 2005, 1829–1834.
[18] J. P. Singh, R. C. Srivastava, H. M. Agrawal, V. R. Reddy, and A. Gupta, Observation of bulk like magnetic ordering below the blocking temperature in nanosized zinc ferrite. *J. Magn. Magn. Mater.* 324, 2012, 2553–2559.
[19] P. Chetri, B. Choudhury, and A. Choudhury, Room temperature ferromagnetism in SnO2 nanoparticles: An experimental and density functional study. *J. Mater. Chem. C* 2, 2014, 9294–9302.
[20] P. Díaz-Gallifa, O. Fabelo, J. Pasán, L. Cañadillas-Delgado, F. Lloret, M. Julve, C. Ruiz-Pérez, Two-dimensional 3d – 4f heterometallic coordination polymers: Syntheses, crystal structures, and magnetic properties of six new Co(II) – Ln(III) compounds. *Inorg. Chem.* 53, 2014, 6299–6308.
[21] M. Venkatesan, C. B. Fitzgerald, and J. M. D. Coey, Thin films: Unexpected magnetism in a dielectric oxide. *Nature* 430, 2004, 630.

[22] M. S. Si, D. Gao, D. Yang, Y. Peng, Z. Y. Zhang, D. Xue, Y. Liu, X. Deng, and G. P. Zhang, Intrinsic ferromagnetism in hexagonal boron nitride nanosheets. *J. Chem. Phys.* 140, 2014, 204701.
[23] Y. Wang, L. Li, S. Prucnal, X. Chen, W. Tong, Z. Yang, F. Munnik, K. Potzger, W. Skorupa, S. Gemming et al. Disentangling defect-induced ferromagnetism in SiC. *Phys. Rev.* 89, 2014, 014417.
[24] A. Sundaresan and C. N. R. Rao, Ferromagnetism as a universal feature of inorganic nanoparticles. *Nano Today* 4, 2009, 96–106.
[25] C. N. Rao, U. T. Nakate, R. J. Choudhary, and S. N. Kale, Defect-induced magneto-optic properties of MgO nanoparticles realized as optical-fiber-based low-field magnetic sensor. *Appl. Phys. Lett.* 103, 2013, 151107.
[26] M. D. Glinchuk, E. A. Eliseev, V. V. Khist, and A. N. Morozovska, Ferromagnetism induced by magnetic vacancies as a size effect in thin films of nonmagnetic oxides. *Thin Solid Films* 534, 2013, 685–692.
[27] D. Gao, J. Zhang, J. Z. Jing, Q. Zhaohui, W. Shi, H. Shi, and D. Xue, Vacancy-mediated magnetism in pure copper oxide nanoparticles. *Nanoscale Res. Lett.* 5, 2010, 769–772.
[28] A. P. Thurber, G. Alanko, G. L. II Beausoleil, K. N. Dodge, C. B. Hanna, and A. Punnoose, Unusual crystallite growth and modification of ferromagnetism due to aging in pure and doped ZnO nanoparticles. *J. Appl. Phys.* 111, 2012, 07C319.
[29] T. Dietl, H. Ohno, F. Matsukura, J. Cibert, and D. Ferrand, Zener model description of ferromagnetism in zincblende magnetic semiconductors. *Science* 287, 2000, 1019–1022.
[30] C. Zener, Interaction between the d Shells in the transition metals phys. *Rev.* 81, 1950, 440.
[31] K. Sato, L. Bergqvist, J. Kudrnovský, P. H. Dederichs, O. Eriksson, I. Turek, B. Sanyal, G. Bouzerar, H. Katayama-Yoshida, V. A. Dinh, T. Fukushima, H. Kizaki, and R. Zelle, *Rev. Mod. Phys.* 82, 2010, 1633
[32] E. C. Stoner, Collective electron ferromagnetism. *Proc. R. Soc. London Ser. A* 165, 1938, 372–414.
[33] Y. Zhu, Y. F. Pan, Z. Q. Yang, X. Y. Wei, J. Hu, Y. P. Feng, H. Zhang, and R. Q. Wu, Ruderman – Kittel – Kasuya – Yosida mechanism for magnetic ordering of sparse fe adatoms on graphene. *J. Phys. Chem. C* 123, 2019, 4441–4445.
[34] S. A. Wolfed, D. Awschalomr, A. Buhrmanj, M. Daughtons, Von Molnárm, L. Roukesa, Y. Chtchelkanovaand, D. M. Treger, Spintronics: A spin-based electronics vision for the future. *Science* 294, 2001, 1488.
[35] H. Ohno, Making nonmagnetic semiconductors ferromagnetic. *Science* 281, 1998, 951.
[36] T. Dietl, H. Ohno, F. Matsukura, J. Cibert, and D. Ferrand, Zener model description of ferromagnetism in zincblende magnetic semiconductors. *Science* 287, 2000, 1019.
[37] T. Jungwirth, J. Sinova, J. Mas̆ek, J. Kus̆era, and A. H. MacDonald, Theory of ferromagnetic (III Mn)V semiconductor. *Rev. Mod. Phys.* 78, 2006, 809.
[38] C. Das Pemmaraju and S. Sanvito, Ferromagnetism driven by intrinsic point defects in HfO(2). *Phys. Rev. Lett.* 94, 2005, 217205.
[39] N. H. Hong, J. Sakai, N. Poirot, and V. Brize, Room temperature ferromagnetism observed in undoped semiconducting and insulating oxide thin films. *Phys. Rev. B* 73, 2006, 132404.
[40] H. Pan, J. B. Yi, L. Shen, R. Q. Wu, J. H. Yang, J. Y. Lin, Y. P. Feng, J. Ding, L. H. Van, and J. H. Yin, Room temperature ferromagnetism in carbon doped ZnO. *Phys. Rev. Lett.* 99, 2007, 127201.
[41] M. A. Garcia, Magnetic properties of ZnO nanoparticles. *Nano Lett.* 7, 2007, 1489.
[42] J. M. D. Coey, d^0 Ferromagnetism. *Solid State Sci.* 7, 2005, 660.
[43] Brahmananda Chakraborty, Prithwish Kumar Nandi, Ajit Kundu, and Yoshiyuki Kawazoe, Nitrogen doping in non-magnetic yttrium oxide: Induction of room temperature ferromagnetism from first principles simulations. *J. Magn. Magn. Mat.* 528, 2021, 167840.
[44] L. X. Guan et al., Nonconventional magnetism in pristine and alkali doped In_2O_3: Density functional study. *J. Appl. Phys.* 108, 2010, 093911.
[45] Lin-Hui Ye, A. J. Freeman, and B. Delley, Half-metallic ferromagnetism in Cu-doped ZnO: Density functional calculations. *Phys. Rev. B* 73, 2006, 033203.
[46] J. A. Chan, S. Lany, and A. Zunger, Electronic correlation in anion p orbitals impedes ferromagnetism due to cation vacancies in Zn chalcogenides. *Phys. Rev. Lett.* 103, 2009, 016404.
[47] A. Zunger, S. Lany, and H. Raebiger, The quest for dilute ferromagnetism in semiconductors. *Physics* 3, 2010, 53.
[48] Zhenkui Zhang, Udo Schwingenschlögl, and Iman S. Roqan, Possible mechanism for d^0 ferromagnetism mediated by intrinsic defects. *RSC Adv.* 4, 2014, 50759–50764.
[49] D. Gao, G. Yang, J. Zhang, Z. Zhu, M. Si, and D. Xue, d0 Ferromagnetism in undoped sphalerite ZnS nanoparticles. *Appl. Phys. Lett.* 99, 2011, 052502.
[50] R. Monnier and B. Delley, Point defects, ferromagnetism, and transport in calcium hexaboride. *Phys. Rev. Lett.* 87, 2001, 157204.
[51] I. S. Elfimov, S. Yunoki, and G. A. Sawatzky, Possible path to a new class of ferromagnetic and half metallic ferromagnetic materials. *Phys. Rev. Lett.* 89, 2002, 216403.

[52] A. Zywietz, J. Furthmüller, and F. Bechstedt, Spin state of cacancies: From magnetic Jahn-Teller distortions to multiplets. *Phys. Rev. B* 62, 2000, 6854.
[53] B. Kisan, J. Kumar, S. Padmanapan, and P. Alagarsamy, Defect induced ferromagnetism in NiO nanocrystals: Insight from experimental and DFT+U study. *Physica B Condensed Matter* 593, 2020, 412319.
[54] Georges Bouzerar and Timothy Ziman, Model for vacancy-induced d0 ferromagnetism in oxide compounds. *PRL* 96, 2006, 207602.
[55] A. Sundaresan and C. N. R. Rao, Ferromagnetism as a universal feature of inorganic nanoparticles. *Nnaotoday* 4, 2009, 96.
[56] L. G. Kong, J. F. Kang, Y. Wang, L. Sun, L. F. Liu, X. Y. Liu, X. Zhang, and R. Q. Han, Oxygen-vacancies-related room-temperature ferromagnetism in polycrystalline bulk co-doped TiO_2, electrochem. *Solid-State Lett.* 9, 2006, G1.
[57] Linije Wu et al., Intrinsic complex vacancy-induced d^0 magnetism in $Ca_2Nb_2O_7$ PLD film. https://doi.org/10.3389/fmats.2021.736011
[58] T. Makarova, Nanomagnetism in otherwise nonmagnetic materials. *arXiv*, 2009; arXiv:0904.1550
[59] J. M. D. Coey, K. Wongsaprom, J. Alaria, M. Venkatesan, Charge-transfer ferromagnetism in oxide nanoparticles. *J. Phys. D Appl. Phys.* 41, 2008, 134012.
[60] F. Pan, C. Song, X. J. Liu, Y. C. Yang, and F. Zeng, Ferromagnetism and possible application in spintronics of transition-metal-doped ZnO films. *Mater Sci Eng, R* 62, 2008, 1.
[61] C. H. Patterson, Role of defects in ferromagnetism in Zn1−xCoxO: A hybrid density-functional study. *Phys. Rev. B* 74, 2006, 144432.
[62] D. Galland and A. Herve, ESR spectra of the zinc vacancy in ZnO. *Phys. Lett. A* 33, 1970, 1.
[63] Keith P. McKenna and David Munoz Ramo, Electronic and magnetic properties of the cation vacancy defect in m-HfO2. *Phys. Rev. B* 92, 2015, 205124.
[64] O. Volnianska, and P. Boguslawski, High spin states of cation vacancies in GaP, GaN, AlN, BN, ZnO and BeO: A first principles study. *Phys. Rev. B. Condens. Matter* 83, 2011, 205205.
[65] Chuanjian Wu, Zhong Yu, Ke Sun, Jinlan Nie, Rongdi Guo, Hai Liu, Xiaona Jiang and Zhongwen Lan, Calculation of exchange integrals and curie temperature for La-substituted barium hexaferrites. *Sci. Rep.* 6, 2016, 36200.
[66] Haowei Peng, H. J. Xiang, Su-Huai Wei, Shu-Shen Li, Jian-Bai Xia, and L. Jingbo, Origin and enhancement of hole-induced ferromagnetism in first-row d^0 Semiconductors. *PRL* 102, 017201, 2009.
[67] Vivek Dixit, Chandani N. Nandadasa, Seong-Gon Kim, Sungho Kim, Jihoon Park, Yang-Ki Hong, Laalitha S. I. Liyanage, and Amitava Moitra, Site occupancy and magnetic properties of Al-substituted M-type strontium hexaferrite. *J. Appl. Phys.* 118, 2015, 203908.
[68] A. M. Van Diepen and F. K. Lotgering, Mössbauer effect in LaFe12O19. *J. Phys. Chem. Solids* 35, 1974, 1641–1643.
[69] J. Kudrnovský, V. Drchal, I. Turek, L. Bergqvist, O. Eriksson, G. Bouzerar, L. Sandratskii, and P Bruno, Exchange interactions and critical temperatures in diluted magnetic semiconductors. *J. Phys.: Condens. Matter* 16 (2004) S5571–S5578.
[70] J. Rusz, L. Bergqvist, J. Kudrnovský, and I. Turek, Exchange interactions and curie temperatures in Ni2−xMnSb alloys: First-principles study. *Phys. Rev. B* 73, 214412, 2006.
[71] Cheng Tan, Wen-Qiang Xie, Guolin Zheng, Nuriyah Aloufi, Sultan Albarakati, Meri Algarni, Junbo Li, James Partridge, Dimitrie Culcer, Xiaolin Wang, Jia Bao Yi, Mingliang Tian, Yimin Xiong, Yu-Jun Zhao, and, Lan Wang, Gate-controlled magnetic phase transition in a van der Waals magnet Fe_5GeTe_2. *Nano Lett*, 21(13), 2021, 5599–5605.
[72] A. A. Serga, M. P. Kostylev, and B. Hillebrands, Formation of guided spin-wave bullets in ferrimagnetic film stripes. *Phys. Rev. Lett.* 101, 2008, 137204.
[73] Kohji Nakamura, Riki Shimabukuro, Yuji Fujiwara, Toru Akiyama, Tomonori Ito, and A. J. Freeman, Giant modification of the magnetocrystalline anisotropy in transition-metal monolayers by an external electric field. *Phys. Rev. Lett.* 102, 2009, 187201.
[74] M. Tsujikawa and T. Oda, Finite electric field effects in the large perpendicular magnetic anisotropy surface Pt/Fe/Pt(001): A first-principles study. *Phys. Rev. Lett.* 102, 247203 (2009).

Crystal Structures and Properties of Nanomagnetic Materials

10

Mirza H. K. Rubel[1] and M. Khalid Hossain[2,3]

1 Department of Materials Science and Engineering, University of Rajshahi, Rajshahi, Bangladesh

2 Advanced Energy Engineering Science, IGSES, Kyushu University, Fukuoka, Japan

3 Atomic Energy Research Establishment, Bangladesh Atomic Energy Commission, Dhaka, Bangladesh

Contents

10.1 Introduction	184
10.2 Crystal and Magnetic Structures	185
10.2.1 Crystal Structure and Morphology of MNPs	185
10.2.2 Magnetic Nanostructures	187
10.2.3 Modeling and Simulations of Nanomagnetic Structures	190
10.3 Basic and Tunned Properties of Nanomagnetic Materials	191
10.3.1 Ferromagnetism	191
10.3.2 Antiferromagnetism	192
10.3.3 Paramagnetism	193
10.3.4 Ferrimagnetism	194
10.3.5 Magnetic Moment and Spin-Orbit Couplings	194
10.3.6 Superparamagnetism	195
10.3.7 Superferromagnetism	196
10.3.8 Superspin Glass and Surface Spin Glass	198
10.4 Application-Oriented Properties	199
10.4.1 Electronic and Optical Properties	199
10.4.2 Thermal Properties	199
10.4.3 Ferroelectric Properties	200

DOI: 10.1201/9781003197492-10

10.4.4 Mechanical Properties 200
10.4.5 Vibrational Properties 201
10.5 Conclusions 202
References 202

10.1 INTRODUCTION

In the present world, nanotechnology and its exploitation have been forecasted and achieved the most significant attention as a distinguished research arena for their unique physio-chemical attributes and biomedical applications [1]. This technology is primarily associated with the fabrication, manipulation, and utilization of promising materials and the development of their properties by taking into account their magnetic and crystal structure, size and dimension, shape, and morphology. The research field of nanoscience covers a wide range of intriguing physics, chemistry, and biomedical sciences of tiny particles in inorganic and organic compounds with novel technological developments as well as challenges. Among various properties of nanomaterials, when magnetism originates owing to the surface and quantum confinement effects on the nanoscale, they are called magnetic nanoparticles (MNPs) of materials. Magnetic materials reveal fundamental properties such as ferromagnetism, antiferromagnetism, ferrimagnetism, and paramagnetism under an externally applied magnetic field. When MNPs achieve a relatively small size, they exhibit a few more unique properties like superferromagnetism, superparamagnetism, superspins, superspin glass, surface spin glass, etc. [2–6]. In these properties, superparamagnetism is the leading property, which is generally shown at an average size of ~2.5 nm, whereas in the case of metal nanoparticles it is down to 4 K [7]. Besides, nanoparticles (NPs) display several other interesting properties – for example, optical, electronic, catalytic, vibrational, optoelectronic, nonmagnetic, diamagnetic, etc. – as well [8]. Extensive research work on MNPs has carried out by numerous researchers in physics, chemistry, and biological sciences to understand and exploit their excellent properties for diverse applications. MNPs have emergent applications in the aforementioned fields as magnetic storage media, nonvolatile memory devices [9], magnetic recording heads [10], spin-torque nano-oscillators (STNOs) [11], spin logic [12], magnetic crystals [2], etc. in the research area of physio-chemical sciences. These magnetic properties in NPs are investigated for slower processes named magnetic vortex dynamics and domain walls in structures while ferromagnetism, superparamagnetism, high magnetic susceptibility, ultrafast demagnetization, and relaxation are accounted for faster processes [4]. MRI, hyperthermic treatment, site-specific drug delivery, tissue engineering, employing cell membranes, cell separation as well as section, gene delivery, magnetorelaximetry, lab-on-a-chip antibacterial agents, biosensing (immunoassays), radiopharmaceuticals, and potential therapeutic applications are just a few of the biomedical applications for nanosized magnetic materials [13–14].

However, morphogenesis of MNPs is technologically very important, as it is associated with phase, shape, and morphology that can be employed for adjusting novel properties of nanomaterials [15]. Nanodots, nanorods, magnetic nanowires, nanotubes, nanostrips, and thin films are all instances of MNPs in various shapes and sizes. The stability of MNPs possessing various shapes is inevitable for applications and essentially relies on the crystal structure of nanomaterials. Therefore, an established relationship among magnetic structures, properties, and crystalline particle sizes is being investigated by researchers [4, 15]. The stable crystal structure of NPs is usually manifested as a cube (Oh), dodecahedron (Ih), icosahedron (Ih), octahedron (Oh), and tetrahedron (Td) versions of different sizes [16]. In these forms of structure, the most durable icosahedral structure exists for particle sizes of < 1.5–2 nm, while 2–5 nm particle size reveals the utmost viable decahedral shape. At times beyond 5 nm, particle size is found for these two phases even in the mesoscale range. Aside from that, face-centered cubic (FCC) NPs with diameters of less than 1.5 nm predominantly exist in the MNPs. In addition, MNPs of materials having Ih and Dh structural shapes have drawn significant interest for their tiny sizes and fixity of various faceting.

Interestingly, magnetic nanostructures with different shapes, sizes, and compositions generate novel properties. It is reported that magnetocrystalline shape anisotropy affects the magnetization direction and behavior of single-domain magnetic NPs (called superspins or macrospins) in Ferro-MNPs [4]. This type of magnetization behavior of NPs is usually governed via antagonism between shape anisotropy and magnetocrystalline anisotropy that subsequently affects the performance of devices under an applied field [4]. In MNPs, formal spins are converted into superspins, and thus atomic crystal motifs are transformed into supracrystals or superlattices of MNPs, and their fascinating characteristics are determined by the existence of superlattices in the materials [4]. Moreover, computer simulations of micromagnetic structures and molecular dynamics have also been carried out to investigate the magnetic structure of NPs [3] and the melting point of various sizes and shapes of nanoparticles [17].

Although MNPs have been successfully fabricated for a long time, they still impose a few drawbacks and adverse influences on the technological field and living cells because of their environmental, hydrophobic, and toxic nature. Furthermore, the nanoscale size of magnetic materials has both good and adverse effects on the properties of MNPs based on their applications [13–14]. For instance, superparamagnetic properties originate in extremely tiny nanoparticles to carry magnetic moments; unfortunately, the NPs are not consolidated into the favored magnetic axis in the recording media for transmitting even a very miniature fragment of knowledge or data. In biomedical applications, the size of MNPs has prolonged effects on the human body, capillary organs, and tissues [13–14]. To resolve these issues, the exploration of new magnetic materials with the perdurable nanostructure of nontoxic metals and their oxides that are presented in this study can be synthesized via improved, modified, and controlled processes to yield nontoxic, environmentally friendly, and compatible products for technological and biomedical purposes. Therefore, the ins and outs of crystal and magnetic structures and relevant properties of MNPs for the development and understanding of the scenario on the nanoscale have been addressed to achieve numerous benefits.

This chapter highlights the crystal and magnetic structures, stability, sizes, shapes, compositions, and absolute properties of metal and metal oxide nanomagnetic materials. In this study, phenomena related to basic and unique magnetic properties, such as ferromagnetism, antiferromagnetism, paramagnetism and superferromagnetism, superparamagnetism, superspin glass, and surface spin-glass states of MNPs, are discussed. The most recent and advanced applications of MNPs in both the technological and biomedical sectors are presented briefly. Further, the incidents involving magnetic and crystal structures and how the behavior of magnetization dynamics impacts the characteristics of MNPs are also pointed out. Finally, the size effects of MNPs on technological applications and biomedical sciences, such as toxicity and possible environmental hazards together with limiting factors and challenges in this research field, are discussed.

10.2 CRYSTAL AND MAGNETIC STRUCTURES

10.2.1 Crystal Structure and Morphology of MNPs

Crystal structural and morphological characterizations are the basic issues to analyze the phase, structure, size, shape, and composition of materials from bulk to nanoscale sizes, and these properties are investigated by X-ray diffraction, transmission electron microscopy (TEM), scanning electron microscopy (SEM), energy dispersive X-ray (EDX), X-ray photoelectron spectroscopy (XPS), Fourier-transform infrared spectroscopy (FTIR), Raman, Brunauer-Emmett-Teller (BET), and Zieta size analyzer techniques. In the midst of them, the X-ray diffraction (XRD) technique is considered one of the best fundamental methods to determine the structure and phase of all dimensions of materials comprising nanomaterials. This characterization technique delivers knowledge on single to multiphase structures, amorphous features, and computed simulation models like Rietveld refinement of nanoscale materials as well. The bulk counterparts of the new phase produce thin films when a phase transition from a bulk structure (3D) to a nanocrystal (2D) occurs as a result of modification of the lattice parameters. Usually, the X-ray diffraction spectra of nanocrystallites display expanded and switched peaks in contrast to bulk ones; these variations are owing to the presence of both size and strain in the materials. The particle size and strain outcomes on the widening of XRD peaks in nanosize particles are calculated and differentiated by the well-known Debye Scherer formula and forming the Williamson-Hall plots, respectively. However, the full width half maximum (FWHM) of nanoparticles, β_{total}, is calculated as the sum of β_{size} and β_{strain} using the following equation (Eq. 1) [18].

$$\beta_{total} = \beta_{size} + \beta_{strain} = \frac{k\lambda}{\beta Cos\theta} + \frac{4\Delta d Sin\theta}{dCos\theta} \tag{1}$$

The first part of Eq. 1, known as the Debye-Scherrer formula, can be used to predict the dimensions of crystallites. In the above equation, β_{size} is the average crystallite size, k is the Scherrer constant (k ~0.9 in the case of spheric particles), which is linked to the shape of the nanocrystals, λ implies the wavelength of the radiation, θ is the Bragg angle, and β FWHM in XRD patterns.

Whereas in the second term of Eq. 1, d is the interplanar spacing and Δd is the difference between those for a specific diffraction peak, which indicates the microstrain, denoting the difference in plane spacing divided by their average magnitude ($\Delta d/d$). The strain [1] of nanocrystals is found by constructing a plot of β_{total} $\cos\theta$ against $4\sin\theta$ that creates a line, where the interception expresses the reverse of crystal size and the corresponding slope generates a certain amount of strain. Furthermore, the lattice contraction that originates from the residual stress can be calculated via Laplace's law as follows (Eq. 2):

$$\Delta P = \frac{4\gamma}{D} \tag{2}$$

In Eq. 2, D indicates liquid droplet diameter, and γ is used to express surface tension. On the other hand, Wulff's theorem [19] is employed for the calculation of deviation in the lattice parameter of solid materials, which can be given as Eq. 3,

$$aDa3\Delta \cong -4\kappa\gamma \tag{3}$$

compressibility κ of bulk phase and Γ. If it is considered that κ does not vary with size reduction, the entity Γ is estimated by utilizing the $\Delta a/a$ versus D-1 plot. The aforementioned strained phenomena were investigated for several bulk nanomaterials, such as $BaTiO_3$ and $PbTiO_3$, and thin films of CeO_2 under high-pressure schemes [20–21]. Sophisticated and important characterization methods such as TEM and SEM are frequently used for the determination of the structure, phase, size, and shape of nanomaterials precisely. The SEM and TEM images in Figure 10.1 show the Au nanoparticle to know the particle size, shape, composition, and selected area electron diffraction (SAED) pattern to support the phase detected by the XRD pattern [22].

XPS is the dominant discerning method for the determination of the appropriate component ratio and existing bonding mechanism of the components, especially in nanomaterials. XPS is a basic spectroscopic method where the binding energy (eV) of electrons in the x-axis vs the number of electrons in the y-axis is plotted in the spectrum for the study of the overall composition and its variation deeply by using surface profiling of materials. Each element bears its own fingerprint binding energy magnitude, which yields a particular set of XPS peaks corresponding to their electronic configuration of shells like 1s, 2s, 2p, and 3s. The EDS technique is utilized to detect the elemental composition from bulk to nanomaterials as semiquantitatively and also qualitatively in atomic and weight percentages.

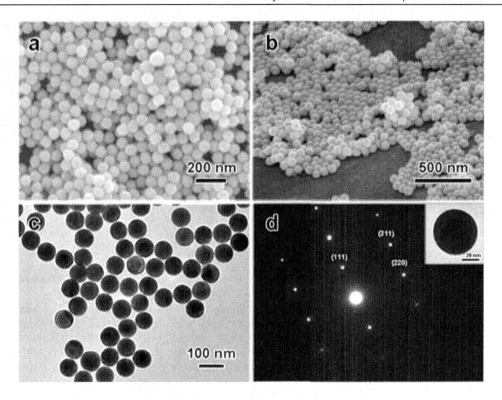

FIGURE 10.1 SEM images of gold (Au) nanosphere particles as (a) top and (b) tilted view, (c) TEM picture of that nanoparticle, and (d) SAED pattern related to the crystal structure (TEM of an Au particle is in inset). Reproduced from [22]. Copyright 2015 Springer Nature.

Vibrational spectroscopy, such as FTIR and Raman, is another important technique to study the local structure of materials that exhibit vibrational modes in spectra. A recent study confirms the vibrational modes of functional Pt NPs (less than 2 nm mean size) have been investigated by FTIR and XPS technique that ensure the functionalization of this nanoparticle. In addition, surface-enhanced Raman spectroscopy (SERS) via a surface plasmon resonance (SPR) scenario was studied to analyze phonon modes in nanostructures and quantum dots of TiO_2, ZnO, and PbS [8].

In recent times, the morphological characteristics of nanoparticles have achieved huge interest because they can be utilized for emerging and impacting important physical properties in NMs. Several characterization tools – for example, polarized optical microscopy (POM), SEM, and TEM – have been used for various morphological studies of the nanostructured phase. Among them, TEM and SEM are considered the best techniques for the characterization of NPs. Figure 10.2 shows TEM micrographs of gold NPs with distinct morphologies, and TEM may also carry information on layered materials at different magnifications [23–28]. Figure 10.3 displays the morphological images from the SEM technique of ZnO nanostructure having a wurtzite structure [15].

10.2.2 Magnetic Nanostructures

Structural analyses such as crystal and magnetic structures of the fabricated nanomaterials are basically and frequently performed by various expletive characterization tools, such as powder XRD (PXRD), high-resolution neutron powder diffraction (NPD), high-resolution transmission electron microscopy (HR-TEM), X-ray absorption near edge structure (XANES), scanning transmission electron microscopy (STEM), and energy-dispersive X-ray spectroscopy (EDS). In the midst of these, the NPD technique is

188 Fundamentals of Low Dimensional Magnets

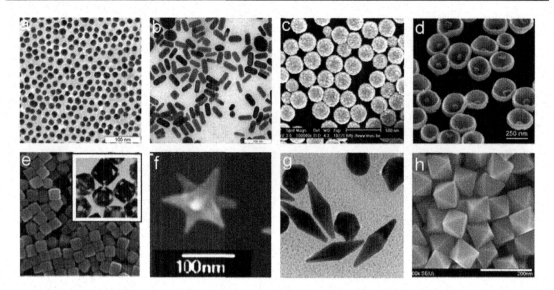

FIGURE 10.2 (a–h) Various shapes of Au nanoparticles are investigated by TEM analysis. (a) and (h) are reproduced from [23]. Copyright (2008) American Ceramic Society. (b) and (c) Reproduced with permission from [24]. Copyright (2010) Elsevier. (d) Reproduced from [25]. Copyright 2009 American Ceramic Society. (e) Reproduced from [26]. Copyright (2006) American Ceramic Society. (f) Reproduced from [27]. Copyright (2006) American Ceramic Society. (g) Reproduced from [28]. Copyright (2005) American Ceramic Society.

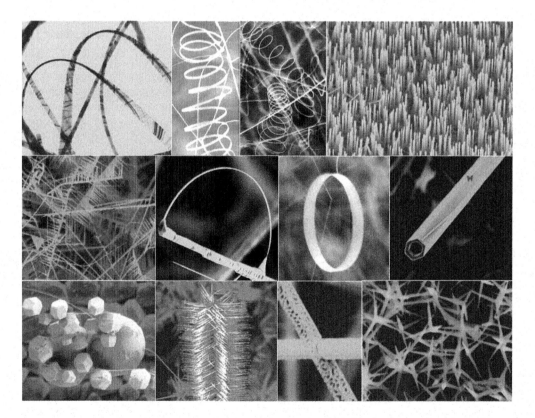

FIGURE 10.3 Amazingly distinguishable morphologies of ZnO nanostructures have been prepared in highly pure form. Reproduced with permission from [15]. Copyright 2010 Elsevier.

mostly employed to determine magnetic structures at room and elevated temperatures as it bears spin moments that scatter out of atomic moments and the magnetic structural model can be obtained from the combined refinement results based on NPD and XRD data. Furthermore, this technique exhibits more beneficial features to determine structures compared to the XRD technique as it can detect lighter elements, lattice distortion, bond length, and even slight elemental deficiency of oxygen in the materials. The variation in the magnetic nanostructures originated due to the ordering of cations in lattices, and related distinguishable magnetic behaviors of ions in nanomaterials deflect the magnetic structures of the crystallites that subsequently affect the magnetic properties.

NPD data indicate long-range magnetic order and also its disappearance with ferromagnetic to antiferromagnetic transformation, and then maintains a ferrimagnetic ordered state in the crystal through superior interchange coupling lattice spins. Skomski et al. show the magnetic structures of four prominent – $MnFe_2O_4$, $CoFe_2O_4$, $NiFe_2O_4$, and $ZnFe_2O_4$ – ferrite magnetic nanocrystals where the magnetic structure is described by a collinear model, using antiparallel moments on the tetrahedral and octahedral sites [29]. According to the theory for a cubic system, the magnetic arbitrary axis is considered towards the <100> direction, which indicates hard magnetization with positive anisotropy, whereas the <111> direction is for soft magnets having negative anisotropy. Notably, the top of nanowires or nanorods has dominant effects of magnetization along the arbitrary convenient axis compared to the magnetization direction. These magnetic behaviors and crystallographic alignment diagrams for the Co nanowire [3] are displayed in Figure 10.4, where the crystal lattice of the nanowire tip influences the magnetic properties. The nanowire of the Co element consists of a hexagonal closed pack (hcp) structure with grains of individual orientations. Figure 10.4a reveals the lattice alignment map of the Co nanowire tip in terms of the z-axis, where dark spots in the crystal represent the inferior durability owing to the superimposition of grains. Such analogous behavior is also observed for the contribution of diffraction peaks at the grain boundaries. Figure 10.4b represents the respective magnetic map by manifesting crystalline area along with the magnetic lines of flux that are pointed along the tip of the nanowire of a given image with respect to the long axis direction.

In addition, the temperature-dependent magnetic nanostructure of Co core and CoO shell [26] is given in Figure 10.5 using neutron powder diffraction patterns [30]. Two basic peaks (111) and (200) in the structure correlate with the bulk CoO structure at 325 K, but a weak (100) peak at $q_2 = 1:475$ Å^{-1} suggests the rock-salt structure of that CoO. Interestingly, the X-ray powder pattern in Figure 10.5c also exhibits these peaks, and the calculation using the FWHM of (111) and (200) peaks by the Scherrer formula represents good agreement with the studied dimensions of the CoO shells. Notably, nanostructural materials

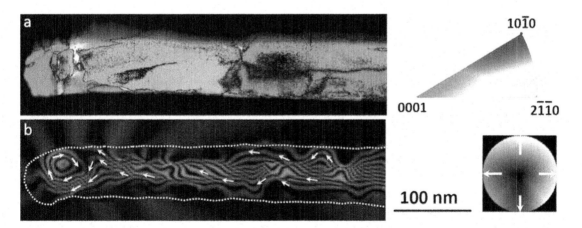

FIGURE 10.4 (a) Reveals the crystallographic orientation along the z-direction to investigate overlapping in crystallites and (b) magnetic structural configuration and flow direction of magnetic lines of Co nanowire, respectively. Reproduced with permission from [3]. Copyright 2015 AIP Publishing.

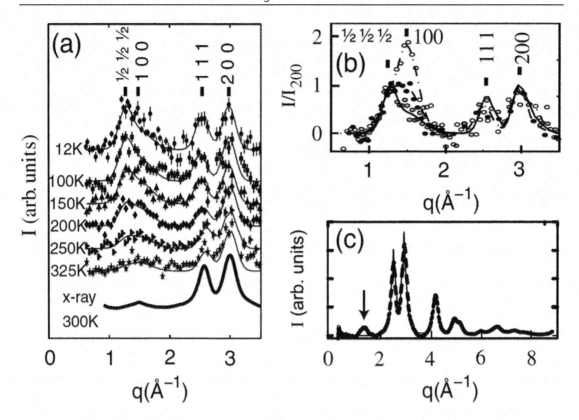

FIGURE 10.5 (a) Indicates ND peaks at different temperatures and an XRD pattern at 300 K of Co core/CoO shell nanoparticles. (b) Shows the ND spectra of various oxidized NPs that are indicated by the circles at 12 K in connection with the (200) nuclear peak. Here solid lines show fitted Gaussian line shapes. (c) The refined structure (black line) and compared XRD pattern (gray points) at 300 K where the arrow indicates a missing (100) peak. Reproduced with permission from [30]. Copyright 2008 American Physical Society.

reveal magnetic domain structures with characteristic magnetization and universal nature, like magnetic vortex cores in materials.

10.2.3 Modeling and Simulations of Nanomagnetic Structures

Numerical simulations which are different from experimental as well as analytical theoretical techniques, are considered a very crucial topic in condensed matter physics. Magnetic nanoparticles are attracting attention as functional materials and are being actively studied. In 1962, Kubo et al. proposed the theory, "In metal, fine particles with a particle size of about several nm, the electronic states become discrete, and their magnetism depends on the number of electrons in the particles".

A simulation model is usually proposed, considering the equilibrium static spin structure and/or the spin dynamics in terms of both time and energy. In the case of MNPs, two classes of numerical simulations are utilized, such as *micromagnetic* [31] and *Monte-Carlo simulations* [32]. Between these, the micromagnetic model deals with the magnetization scenario from the micro and nanometer length scales to express details of the transition behavior between "domains". A micromagnetic simulation is a trustworthy appliance for a theoretical study that utilizes a continuous magnetization vector instead of individual atomic spin to unravel the structure of magnetic domains as well as domain walls [33]. The fundamental theory of micromagnetics was given by Blue et al. by computing the magnetization vector between two domains of a wall. Micromagnetics simulation has important applications in developing innovative permanent magnets and storage as well as sensing technologies, such as magnetization dynamics [34], magnetoresistive random access memory (MRAM) [9], sensors, and magnetic sensor devices [35] by covering present research in spintronic and skyrmion arenas. However, in micromagnetic simulations, the magnetic nature of the studied system is assumed as a continuum approximation [36] without considering localized moments by incorporating a gradual vector field $M(r)$. In the case of applied simulation, the system is separated into cells of typical nanoscale dimension, which must be shorter compared to the exchange distance of the substance. An initial magnetization configuration is chosen at the start. The magnetic moments of every section are then enhanced based on a physical model in the run simulation.

If we consider micromagnetic simulations, the Landau-Lifshitz-Gilbert equation is mostly implemented as an expression of motion of the magnetization vector in an operative field [37], as Eq. 4:

$$d\mathbf{M}_i / dt = -\gamma' \mathbf{M}_i \times \mathbf{H}_{eff} - \alpha\gamma' / M_s \mathbf{M}_i \times (\mathbf{M}_i \times \mathbf{H}_{eff}) \quad (4)$$

where, $\mathbf{H}_{eff} = H + H_{demag} + H_{1K} + H_{ex} + h(t)$ denotes active magnetic field from Zeeman subscription, demagnetizing, anisotropy, and exchange, as well as the time elapsing magnetic fields, and M_i denotes magnetization of cell i, $\gamma' = \gamma/(1 + \alpha^2)$, here γ is the gyromagnetic ratio, M_s the saturation of magnetization, and α implies damping constant for the particular system. It is well that micromagnetic simulations are employed to larger nanomagnets, for $T = 0$. On the other hand, Monte-Carlo simulations are applied [32] for the temperature behavior study of MNPs where the system is used as the background of localized moments. Therefore, every moment is customized based on Monte-Carlo techniques during the simulation process. The Metropolis algorithm [37] has been counted as a familiar approach that computes the statistical probability of a spin-flip or movement along with the Boltzmann factor in the regime of Hamiltonian.

10.3 BASIC AND TUNNED PROPERTIES OF NANOMAGNETIC MATERIALS

Magnetic materials have been developed as an important science and technology branch since ancient times because of their fundamental and unique tunning properties to be designed and processed for targeted applications. The magnetic attributes of materials are illustrated using the basic theory that originated owing to the net magnetic moment (μ_B) of both orbital and spin motions of electrons in atoms of elements, compounds, and molecules. Considering the magnetic states and behaviors, the materials usually exhibit interesting basic properties such as ferromagnetism, ferrimagnetism, antiferromagnetism, paramagnetism, and diamagnetism. Moreover, nanomagnetic materials additionally show several outstanding magnetic properties like superferromagnetism, superparamagnetism, superspin glass, and surface spin-glass nature for their tiny particle sizes. Besides these, nanomagnetic materials display nonmagnetic, electronic, optical, dielectric, mechanical, thermal, vibrational, and catalytic properties that can be tailored for the applications of diverse fields. In this chapter, the aforementioned basic and tuned properties of crystalline nanomagnetic materials are described systematically.

10.3.1 Ferromagnetism

Ferromagnetism, a prominent fundamental property of magnetic materials, originates from the parallel spin alignment and magnetic moments of the constituent atoms of materials referred to as electromagnets or magnetization electrons. Ferromagnetic materials bear permanent dipolar magnetic moments that are spontaneously oriented in a big mob with parallel alignment without an external magnetic field. The ferromagnetic properties of materials under an external field are characterized in detail by a well-known hysteresis loop, where two, primal entities coercivity, and remanence, are importantly considered. Ferromagnetism is a physical property found naturally in iron and magnetite (Fe_3O_4), an oxide of iron, and is often referred to as an intrinsic ferromagnet. It was discovered about 2000 years ago when magnetism attributes of these materials were investigated [38]. Moreover, cobalt, nickel, manganese, gadolinium, Tb, Dy, numerous alloys and compounds, and several rare-earth elements belong to this group. Compared to other classes of materials, ferromagnets are easily and strongly magnetized at saturation levels. At present, ferromagnetic materials are broadly and sophisticatedly employed in various devices essential to our modern life – for example, electric motors and generators, transformers, telephones, loudspeakers, magnetic recording gadgets, floppy discs, cassette tapes, credit card magnetic stripes, etc.

Ferromagnets exhibit a variety of interesting properties, like the creation of a robust magnetic field to attract magnets, magnetization after withdrawal of an applied field, sustained magnetic property at a certain temperature limit called Curie temperature (T_c), strong magnetic field dependence behavior, and outstanding magnetic susceptibility and phase transition both at ambient and elevated temperatures following the Curie-Weiss law. It shows a significant quantity of tiny regions with magnetic dipoles called domains where the magnetization is very strong. A significant number of microscopic regions of magnetic order, known as atomic or ionic magnetic moments, are organized in ferromagnets in a pattern termed domain structure, according to French physicist Pierre-Ernest Weiss's theory. Domain structures present the magnetic structure of materials which can be seen as small particles of magnetites, and their patterns have also been investigated by polarized light and neutrons, electron beams, and X-ray diffractions. Quantum mechanical effects referred to as exchange interactions are also observed in ferromagnetic materials.

10.3.2 Antiferromagnetism

When ordered antiparallel patterns between pairs of near magnetic spins or atomic dipoles are generated via exchange interactions, which is another essential property of magnetic materials, the materials are designated as antiferromagnets. Such magnetic spin alignments for ferromagnets and antiferromagnets are given as follows. At a critical temperature is known as T_c, the spin alignments in antiferromagnetic materials are randomly oriented as well, and the material acts like a paramagnet (Figure 10.6) where the spin arrays in reverse direction cancel the spontaneous magnetizations in ferromagnetic materials [39].

Since the material is paramagnetic at high temperatures, under a definite characteristic temperature, the magnetic dipoles are organized in an ordered antiparallel direction. The transition temperature T_n is also acquainted with the Néel temperature, after the discovery and explanation by French physicist Louis-Eugène-Félix Néel in 1936. Néel temperatures are usually observed for compounds, and the several typical antiferromagnetic kinds of stuff are chromium (311 K), manganese fluoride (67 K), nickel fluoride (73 K), manganese oxide (116 K), and ferrous oxide (198 K). However, the ordered antiferromagnetic phase is a more complex state compared to that of the ferromagnetic phase because a minimum of two sets of dipoles point oppositely. No net spontaneous magnetization is observed on a macroscopic measurement while the identical figure of dipoles of the equivalent dimension on each kit is considered. Most of the insulating chemical compounds exhibit an antiferromagnetic nature via exchanging forces between the magnetic ions. Figure 10.7 shows the spin alignments of a ferromagnet where A is spin alignments for a lattice of one substance and B is that for another. It is seen that the magnetic moments of these components expel each other and produce net a magnetic moment of zero owing to the spin matching of sublattices between A and B. The material is regarded as ferromagnetic while the spin of the lattice is either only A or B type.

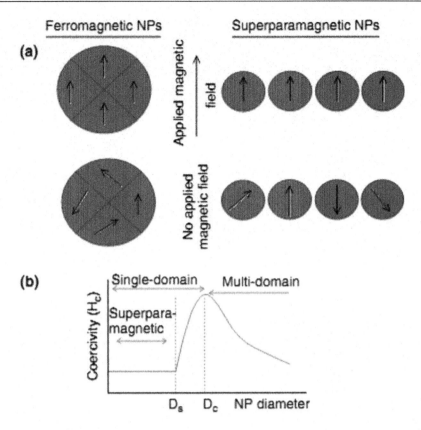

FIGURE 10.6 (a) Magnetic spin alignments of ferromagnetic and superparamagnetic nanomaterials with and without applied magnetic field. (b) The relationship of magnetic domain structures with nano-dimension particles where D_s and D_c are the "superparamagnetism" and "critical" size thresholds, respectively. Reproduced from [39]. Copyright 2012 Springer.

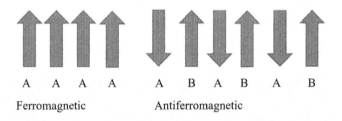

FIGURE 10.7 Simple magnetic spin alignments of ferromagnetic and antiferromagnetic materials.

10.3.3 Paramagnetism

Paramagnetism is the behavior of magnetic materials where the spin moments of unpaired electrons align in either parallel or antiparallel directions that are weakly attracted by a robust magnet. This phenomenon was addressed and investigated by the British scientist Michael Faraday in 1845. Most of the elements, atoms, ions, molecules, and several compounds are paramagnetic, having a permanent magnetic dipole moment. In the midst of them, fierce paramagnetism is observed in compounds containing iron, palladium, platinum, the transition metal series, rare-earth elements, and so on. In paramagnetic materials, electrons of atoms must be partially unfilled in inner shells with unpaired spin quantum numbers.

Therefore, the transition metal series and rare-earth elements with unfilled d and f orbitals respectively exhibit paramagnetic characteristics. Weak paramagnetism is found in many metals in solid forms, such as sodium and other alkali metals, and is temperature independent because antiparallel electron spin is unstable in metals where an external magnetic field affects the spin of loosely bound conduction electrons. Interestingly, the magnetic susceptibility (χ = induced magnetism) of paramagnetic elements always shows a positive value at room temperature.

The temperature-dependent behavior of paramagnetic materials is expressed by a relation known as the Curie-Weiss law and is given by Eq. 5,

$$\chi = \frac{C}{T - Tc} \tag{5}$$

Herein, T = temperature, T_c = Curie temperature (above which magnetization M vanishes), and $C = \mu_0 N_0 \mu_m^2 / 3K$, Curie constant. The paramagnetic state depends on the magnetization of elements and the ratio of the magnetic energy of the individual dipoles to the thermal energy (B/T). However, hard paramagnetism gradually reduces with the rise of temperature as the de-alignment of spin is created for the higher random motion of the atomic magnets.

10.3.4 Ferrimagnetism

Ferrimagnetism originates from the sizes of the magnetization and the collective contributions of the two variant atoms or sublattices in a lattice where a resultant net magnetization is parallel to that of the sublattice with the greater magnetization. This phenomenon was described by Néel and named ferrimagnetism, and the materials that display this behavior are called ferrimagnets. If we consider two ferrimagnetic spin orders of two sublattices A and B with the antiparallel spin alignments of antiferromagnetic and ferrimagnetic lattices, the magnetic moments of sublattices A and B represent a lattice mismatch for the latter. Between them, a larger moment is found for A in contrast to B. Thus, the ferrimagnetic state is truly an antiferromagnetic state having unbalanced magnetic spins in atomic lattices. However, different from the ferromagnetic metals, they show lower electrical conductivity, and in an alternating magnetic field, they greatly reduce the energy due to the loss of eddy currents. Since these losses are for the frequency of the alternating field, these substances have potential in the electronics industry.

10.3.5 Magnetic Moment and Spin-Orbit Couplings

Magnetic moment (μ_B), the core of magnetic properties, is a kind of force that originates from the nonzero spin of electrons in atoms of substances. The magnetization behavior in all sorts of materials, including metals, semiconductors, and insulators, is a consequence of the magnetic moment. The net magnetization in the order of magnitude is expressed by the quantity of Bohr magneton, μ_B, to justify the amount of magnetization suspected in elements. The induced magnetic properties due to magnetic moments are observed in different sizes of particles from micro- to nanoscales and have diverse applications in those fields. The magnetic moments are the vector sum of electronic spin (μ_{spin}) and orbital momentum ($\mu_{orbital}$) in an atom of materials. The basic five magnetic materials are classified as the vector sum of μ_{spin} and $\mu_{orbital}$ following the relation as Eq. 6,

$$\sum (\mu_{spin} + \mu_{orbital}) > 0 \ [\text{Ferromagnetic, Antiferromagnetic, Ferrimagnetic, Paramagnetic}] \tag{6}$$

However, when this relationship holds like $\sum (\mu_{spin} + \mu_{orbital}) = 0$, materials that conform to this form of relationship and exhibit weak expulsion in a magnetic field are called diamagnetic. Based on the electronic configuration, the diamagnetic property is found in the covered electronic subshells as paired electronic spin that yields a net moment of zero. The important diamagnetic elements include Au, Ag, Cu, Be, Bi, and Ge. In an external magnetic field, ZnO·Fe$_2$O$_3$ and SiO$_2$ (quartz) belong to a group of compounds that exhibit a negative susceptibility and weak repulsion. Therefore, diamagnetic materials are non-magnetic and do not have any magnetization effect ideally.

Different types of magnetic moments are observed for the spin-orbit coupling (SOC) phenomenon, which is one of the relativistic effects that originated owing to the nonzero spin of electrons around the orbitals of the atom. It is the electron's spin and its orbital angular momentum that couple the preferential direction of spin with the basic lattice. This relativistic effect increases in magnitude with the nuclear charge (Z) of atoms as well. The spin-orbit coupling phenomenon is also crucially associated with the conversion of charge currents to spin currents and vice versa. This spin-orbit coupling outcome is called the spin Hall effect, or inverse spin Hall effect depending on the direction of the current transformation. Between these two, the earlier one agrees well with the switching of charge currents into spin currents, while the latter one is connected with the turning of spin currents into charge currents. Importantly, the research arena of spin-orbit coupling is called spin-orbitronics [40], mainly related to the spintronics and magnetism scenarios, and its incorporation in material systems is inevitable for the application and computation of numerous interesting physical properties: for instance, the magnetic anisotropy [4], spin relaxation [41], magnetic damping [42], anisotropic magnetoresistance [43], the anomalous Hall effect [44], modeling in semiconductors, Berry curvature, topological insulators, and magnetoelectrics [45].

In the plane-waves and pseudopotential models, the spin-orbit coupling is included using j-dependent pseudopotentials and an electron density with a vector spin at each grid point.

10.3.6 Superparamagnetism

Superparamagnetic properties originate in magnetic products having tiny crystallites – for instance, Fe-containing NPs exhibit superparamagnetism at a dimension of particles lower than 25 nm. The thermal energy in paramagnetic materials can exceed the coupling forces within the neighboring atoms over the Curie point that causes random fluctuations in the magnetization direction to result in a null magnetic moment. However, in a superparamagnetic state, this oscillation affects the direction of magnetization of entire crystallites. The magnetic moments of individual crystalline particles compensate for each other, and the overall magnetic moment becomes zero. Owing to the application of an external field, the behavior is similar to paramagnetism except each atom is independently influenced by an applied field. Ferrite oxide (Fe$_3$O$_4$) is named magnetite, which is an abundant natural ferromagnetic oxide in mining that is extensively employed as superparamagnetic nanoparticles in wide biomedical applications. Superparamagnetism, regarded as the first requirement for biomedical applications (drug delivery) in magnetic NPs, is important because when the applied field is drawn, the magnetization is lost, and thus agglomeration (possible embolization of capillary vessels) has abstained. Besides, superparamagnetic iron oxide nanoparticle (SPION) is used as a biodegradable intact excretion magnetic core with Fe in cells for the application of normal biochemical channels for Fe metabolism.

The examination of the magnetization behavior of NPs in the context of a single domain or monodomain is being performed using a paramagnetic analogy known as the superparamagnetic theory proposed by Bean and Livingston [46]. Domains are the batches of spins that are arranged in the same direction and are separated by domain walls, which have a characteristic width and formation energy. When the particle size is reduced in magnetic particles (multidomain materials), the generation of a single-domain structure is observed, which induces the phenomenon of superparamagnetism as well. Every particle acts like a paramagnetic atom, but each nanoparticle has a very large moment with a definite magnetic state [47]. Weak interparticle magnetic interactions yield superparamagnetic (SPM) behavior also. However, superparamagnetic materials are intrinsically nonmagnetic and are magnetized under an external field. The motion of domain walls is the basic reason for the reversal magnetization in superparamagnetic nanomaterials. The superparamagnetic theory deals with the assumption that all the magnetic moments of the

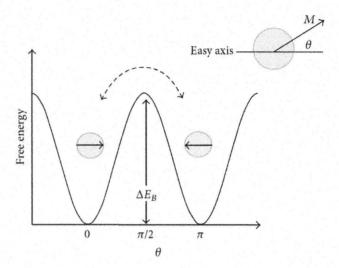

FIGURE 10.8 Schematic presentation of the free energy of an individual domain versus uniaxial anisotropy in the aspect of magnetization direction. In the figure, E_B is used for energy, whereas θ denotes the angle between the magnetization and the convenient axis. Reproduced from [4]. Copyright 2012 Springer.

particles will move coherently, with a magnitude of $\mu = \mu_{at}N$; here μ_{at} indicates the atomic moment and N denotes the number of magnetic atoms that compose such a particle [48].

It is noted that extremely small ferromagnetic (FM) particles have a single domain where the magnetization is frequently in parallel or antiparallel to a specific direction named the convenient axis. The magnetocrystalline structure, shape, strain, and surface anisotropies [49] of materials are all affected by distinct anisotropy contributions.

The energy of a single domain particle with rotation of the easy axis is presented in Figure 10.8 [4]. To explain this phenomenon, an assembly of uniaxial single domain particles having an anisotropy energy density $E = KV \sin2\theta$ is considered where the energy barrier $\Delta E_B = KV$ separates the two energy minima at $\theta = 0$ and $\theta = \pi$ corresponding to the magnetization parallel or antiparallel to the convenient axis. Néel showed [50] that for small enough single-domain particles, KV may become so tiny that energy fluctuations can withdraw the anisotropy energy and spontaneously reverse the magnetization of a particle from one easy direction to the other without an external field. However, when $kBT \gg KV$, the particle fluctuates freely (kB = Boltzmann's constant), and a ferromagnetic NP is considered SPM when the energy obstacle of inverse magnetization is similar to the heat energy (kBT). This scenario has been articulated in different terms, like "apparent paramagnetism" [51], "collective paramagnetism" [52], "quasiparamagnetism" [4], and "subdomain behavior" [53], which implies the isotropic range of superparamagnetism.

10.3.7 Superferromagnetism

The entity of superferromagnetism is used for studying the properties of magnetic nanoparticles. The phenomenon of superferromagnetism was founded at first by Bostanjoglo and Roehkel [54] and then scientist Mørup investigated this phenomenon by observing goethite (FeO(OH)) nanoparticles of 10–100 nm by Mössbauer spectroscopy; after that, it has been broadly employed in analyzing various nanomagnetic systems. However, the magnetic properties of goethite were unable to be explained via existing theories based on the combined excitation of magnetic state and superparamagnetic relaxing state. As a result, Weiss's mean-field model based on interacting stuff has evolved and gives an excellent fit to their outcome, and the mentioned theory has been dubbed superferromagnetism to analyze the characteristics of nanomaterials.

Noteworthy, the superferromagnetic (SFM) state of MNPs has been characterized by ferromagnetic interparticle correlations having a high concentration of nanoparticles with a relatively large size of 12–19 nm. At further increments of concentration, the interparticle interactions become prominent, which can consequently govern some of the FM domain phases between the supermoments of the nanoparticles and the atomic moments within the particles. Therefore, the ensemble of nanoparticles in the FM state is regarded as superferromagnetic. As a result, an SFM domain can be compared with an FM domain; the exception is the substitution of atomic moments by supermoments of the particular nanoparticles. Importantly, the SFM domain state has scarcely been studied for Fe@SiO$_2$ nanoparticles, cobalt and nickel systems, Co-SiO$_2$ granular films, and many nanoparticle systems under the zero-field cooled (ZFC) state. However, the growing nanoparticles are also capable of pushing monodomain particles to multidomain ferromagnets through the leading superspin state of nanoparticles. Thus, it bears significant interest to shift the magnetic behaviors of NPs while conserving the monodomain phase. Figure 10.9 shows

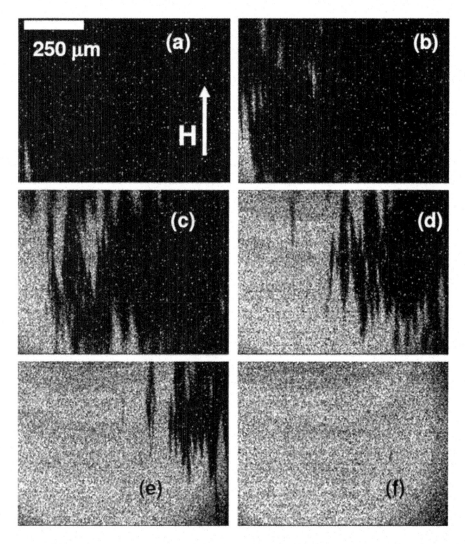

FIGURE 10.9 MOKE microscopic pictures in the longitudinal direction for (Co$_{80}$Fe$_{20}$(1.3 nm)/Al$_2$O$_3$(3 nm))$_{10}$ superferromagnet at ambient temperature using supercoercive fields, $\mu_0 H$ = 0.6 mT, and different time intervals such as t = 2s (a), 3 s (b), 4 s (c), 5 s (d), 6 s (e), and 9s (f). Reproduced with permission from [55]. Copyright 2010 IOP Publishing.

the magneto-optical Kerr effect (MOKE) microscopic images in a longitudinal direction for (Co$_{80}$Fe$_{20}$ (1.3 nm)/Al$_2$O$_3$(3 nm))$_{10}$ superferromagnet [55].

10.3.8 Superspin Glass and Surface Spin Glass

The concentrated form of magnetic nanoparticles can show a transformation scenario from a superparamagnetic state to an amorphous collective phase, which is mentioned as a superspin glass state that has conformity with the orderless and frustrated magnetic phase at lower temperatures in spin glass products [56]. Super spin glass materials exhibit various interesting properties in numerous nanomaterial schemes, such as robust increment of magnetic nonlinearities [57] and dynamic scaling behavior [58] possessing feasible magnitudes of the critical exponents. For truly low interparticle interactions, SPM blocking phenomena have been investigated in assemblages of magnetic nanomaterials, yet this interplay might impact interparticle magnetic natures such as superspin glass (SSG) and superferromagnetism. Spin glass (SG) materials are considered the eminent class of amorphous systems in condensed matter magnetism that has been extensively studied for the long term [59]. The SSG phase is usually characterized by thermoremanent magnetization (TRM), dc, and ac susceptibility measurements under ZFC and field-cooled (FC) conditions in nanomagnetic substances [60]. Moreover, site disorder and magnetic frustrated interaction and SG parameters in superspin glasses yield amazing properties that are delicately comparable to that of ferro- and/or antiferromagnetic materials, which allusively illustrates the arbitrary arrangement of chilled spin alignments [61]. However, the SG state in dilute spins of bulk kinds of stuff gathers the glassy magnetic phase into a single domain of MNPs where the interparticle interaction is trivial. The phases of SSG have been investigated in a frozen condition of ferrofluids [62] or discontinuously organized metal-insulator multilayers (DMIMs) [63] of a fixed dimensional ferromagnet using the ground state of magnetic macromolecules as well. Many researchers have analyzed the important components, like spatial entropy as well as magnetic frustration, related to the dipolar interactivity of superspins in classic dipolar glassy materials [64] from earlier.

The surface spin disorder was first studied in NiFe$_2$O$_4$ nanoparticles and is regarded as an important issue to investigate surface spin behavior in nanomaterials (Figure 10.10a) [65]. Surface spins have

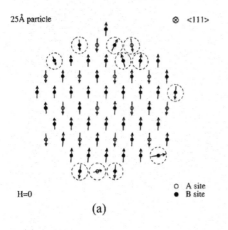 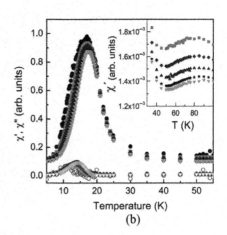

FIGURE 10.10 (a) Computed spin orientation for 2.5 nm diameter of NiFe$_2$O$_4$ NPs at $H = 0$ where dotted circles are used for exceptional silly orientations. (a) Reproduced with permission from [65]. Copyright 1996 American Physical Society. (b) Temperature-dependent magnetic susceptibilities of NiO particles of an average diameter of 6.5 nm in an ac field of $H_0 = 10$ Oe at measuring frequency range of 10 Hz $\leq f \leq$ 10 kHz. The inset shows the increased temperature zone of χ'. Reproduced with permission from [68]. Copyright 2008 IOP Publishing.

achieved importance to control the nanoparticle's magnetism due to surface functionalization that has diverse applications [66]. Even nonmagnetic nanoparticles reveal either ferromagnetism or paramagnetism behavior owing to surface spin magnetization, which has different alignments in contrast to core spins and might increase or diminish the total magnetization of the individual nanoparticle as well. Importantly, surface spins can be disordered and frustrated, resulting in broken bonds on the surface with magnetic anisotropy that differs from that of the core spins. Further, surface disorder and frustration can lead to surface spin-glass behavior and also show interface interactions with the core spins in nanoparticles of ferrites [67]. Figure 10.10b shows temperature-dependent magnetic susceptibilities of NiO particles with an average diameter of 6.5 nm [68].

10.4 APPLICATION-ORIENTED PROPERTIES

10.4.1 Electronic and Optical Properties

The electronic and optical properties of nanomaterials are interconnected with each other broadly. The electronic wavefunctions are restricted by quantum consequences for nanoscale directions, and the crystal structures of NPs are analogous to bulk counterparts with a subclass of the electronic behaviors of that phase. The electronic density of states of one-dimensional carbon nanotubes is regarded as a model scheme to illustrate size-dependent influences on NPs. An absolute relationship between the geometrical structure and electronic properties of carbon nanotubes is investigated. The crystal structure and conductive nature of carbon nanotubes are approximated based on the graphene sheet that imposes the quantization of the wavevector towards the circumferential passage, and it becomes either metallic or semiconducting [69].

The nanoscale particles possess dimensional optical characteristics that show a robust UV-visible extinction band due to the incident photon energy, and the collective conduction of electrons is called localized surface plasma resonance (LSPR), which is different from the spectrum of the bulk one. However, when semiconducting carbon nanotubes are individualized by consistently resolved excitation emission, optical images become very robust, and explicit absorption and emission spectra are observed for the presence of the van Hove singularities in the electronic properties. The explored optical behaviors in nanotubes are explained via excitonic effects [70] to build optical sensors [71]. Metallic nanoparticles that show amusing physical behaviors have wide applications in diverse areas of optics and catalysis. Plasmon resonances, for example, are the almost free conduction electrons of Ag and Au nanoparticles that underpin collected excitation and play a vital role in LSPR. Interestingly, this plasmon frequency vigorously relies on the morphologies of metallic nanoparticles.

10.4.2 Thermal Properties

Thermal characteristics are another fascinating aspect of nanomaterials and metal NPs. When thermal conductivities are compared to fluids and solids, NPs have higher thermal conductivities. For instance, the element Cu and even Al_2O_3 oxide have many times higher thermal conductivity than that of H_2O and engine oil at ambient temperature. Hence, it is said that the solid particles in restoring fluids are desired to have superior thermal conductivities in contrast to those of traditional heat-transforming liquids or fluids. Nanofluids are prepared by dispersing the solid fragments having nanosize into numerous liquids such as water, ethylene glycol, or oil. It is desired and found that nanofluids show much better thermal conducting properties than those of conventional heat passing fluids as well as fluids containing microscopic-sized particles. This auspicious thermal conductivity in the nanoparticle originated from the extensive total surface of nanomaterials. This massive total surface area improves stable suspension and promotes heat conductivity in either H_2O or ethylene fluids, as seen in CuO or Al_2O_3 NPs.

10.4.3 Ferroelectric Properties

The ferroelectricity phenomenon deals with a spontaneous electric polarization moment that occurs due to the structural distortions of atoms or ions in the lattice. The spontaneous alignment of electric dipoles is analogous to ferromagnetism and is seen under a critical temperature, and above it, a structural phase transition occurs. T_c relies on the size of particles in materials that can be calculated using Eq. (7) as,

$$T_c(D) = \frac{T_c^{bulk} - \Phi}{D - D_0} \qquad (7)$$

Herein, D denotes particle size, Φ is used as a fitting parameter, D_0 represents the dimension of particles for T_c at 0 K, and T_c^{bulk} indicates the ferroelectric to paraelectric transformation temperature of the bulk state [72]. According to this model, the polarization that disappears below the critical diameter is called superparaelectricity, which is similar to superparamagnetism. The critical size of several prominent ferroelectric materials such as $PbTiO_3$, $Bi_4Ti_3O_{12}$, and $PbZr_{0.3}Ti_{0.7}O_3$ has been predicted and found on the nm scale with superior properties [73]. However, the polarization-ceasing phenomenon is usually explained based on free energy calculations by incorporating a gradient of polarization and an entity connected to a surface of free energy with preferential cubic (paraelectric phase) symmetry rather than tetragonal (ferroelectric phase) symmetry. The phenomenon of superparaelectricity is key because of its applications and is very crucial in fundamental science as well as applications.

10.4.4 Mechanical Properties

The mechanical properties of nanoparticles are intimately connected to the particle size, and it is investigated that when the particle dimension reduces, the mechanical properties improve. Numerous mechanical entities like elastic modulus, hardness, stress and strain, adhesion, and friction are used to identify the proper mechanical attributes of nanomaterials. Moreover, surface coating, coagulation, and lubrication are also utilized to develop mechanical features as these properties are unlike for NPs in contrast to microparticles and bulk nature. The corresponding stiffness between NPs and the periphery is well maintained when nanomaterials are lubricated or greased, but if the contact pressure is remarkably high, the NPs are subjected to either a plan or imperfect face, which is a significant scenario that manifests the performance of NPs during the contact state. To achieve the effective consequences from this phenomenon, a profound knowledge of the fundamental mechanical characteristics, including elasticity, motion rule, abrasion, and interfacial adherence, as well as dimensional depending features, must be achieved. However, utmost investigations have been performed on the aforementioned mechanical properties using a single structural unit of nanowires or nanorods; it is seen that the mechanical strength of these structural units increases owing to the superior internal excellence of NPs. Dislocation, slides, nano- and microtwins' defects, and so on are regarded as shortcomings of lattice in NPs that are very energetic and must be removed from proper structures to get the best properties. If the cross-section of synthesized nanorods and nanowires is lesser, the number of imperfections will be smaller, as the nanosize dimension vanishes such types of deficiencies from the crystals. The individual mechanical attributes of nanomaterials can be utilized for novel applications in numerous areas (Figure 10.11), such as tribology, surface engineering, nanofabrication, and manufacturing to enhance mechanical characteristics [74] and achieve benefits for mankind.

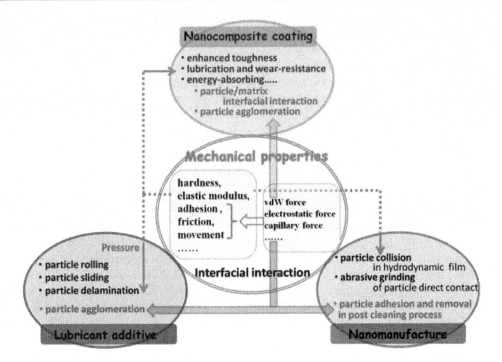

FIGURE 10.11 Diagrammatic presentation of the mechanical properties of nanomaterials for advanced applications. Reproduced from [74]. Copyright 2014 IOP Publishing.

10.4.5 Vibrational Properties

The Vibrational characteristics of nanomaterials are connected to the vibrational or phonon modes, and their characterization is studied by using Raman and FTIR spectroscopic techniques which are considered more efficient and improved methods than many other elemental analytical processes. Importantly, particle size, morphology, and disorders of nanoparticles robustly influence vibrational attributes. Between these especially, Raman spectroscopy is mostly used to measure the vibrational properties of diverse nanoparticles, which are established on the inelastic light scattering principles of phonons in lattices, but for the bulk state of materials, the momentum selection rule, q = 0, where q denotes a phonon wavevector, is followed by the light scattering. When nanoparticles or crystals are formed, this wavevector selection rule is collapsed for the uncertainty principle, and it becomes q ≠ 0 with broad and asymmetric modes in the Raman spectra. The Raman profile is generally illustrated using the phonon confinement model given in the following Eq. (8) [75]:

$$I(\omega) = \int \frac{|C(q)|^2 \, d^3q}{[\omega - \omega(q)]^2 + (\Gamma_0/2)^2} \tag{8}$$

In this equation, Γ_0 means the bar width of the Raman peak, and $\omega(q)$ indicates the dispersion graph of phonon, whereas $C(q)$ is the coefficient related to nanocrystal morphology and size parameters. Dimensions of nanoparticles are measured from this equation via experimental Raman spectra, and the result well agrees with that of TEM analysis as well. However, surface-enhanced Raman spectroscopy is the latest and most advanced technology to know the vibrational attributes and phonon modes of nanostructures as well as zero-dimensional nanomaterials like TiO_2, ZnO, and PbS using the SPR phenomenon.

FTIR is another basic spectroscopic analytical tool that uses fingerprint regions for NPs and delivers significant qualitative and quantitative data on local structures with existing bondings in the materials. The FTIR technique detects the functional groups in organic species of single and double bonds and water at different wavenumbers with vibrational peaks such as carboxylated C–O (2033 cm^{-1}) and O–H peak at 3280 cm^{-1}. It also detects the chemical structure of numerous inorganic compounds and elements, for example, hafnium oxide RE$_2$Hf$_2$O$_7$ (here RE =Y, Pr, La, Gd, Lu, and Er) and nanoparticles of Pt.

10.5 CONCLUSIONS

Magnetic phenomena mainly correlate with the structure and relevant properties based on classical and quantum magnetization. Between these, quantum effects are dominant for the structural alternation and modification from bulk to nanostructure, which consequently influences the origination of novel properties in nanomaterials. In this chapter, numerous prospects of crystal and magnetic structures – properties as well as the importance of micromagnetic simulation on MNPs – are discussed. Several important techniques have been employed for structural and morphological studies, and their outcomes are addressed for a precise understanding of MNPs attributes. However, the basic characteristics intimated to the magnetism of materials, along with adjustable magnetic properties, like superferromagnetism, superparamagnetism, superantiferromagnetism, superspin, surfacespin phase, and super spin glass, are mentioned in detail. Nanomagnetic materials offer a wide range of promising applications, including spintronics, electro-optics, mechatronics, and other fields that exploit the associated properties of MNPs studied in this chapter.

REFERENCES

[1] G. Cao, *Nanostructures and Nanomaterials*, Published by Imperial College Press and Distributed by World Scientific Publishing Co., 2004. https://doi.org/10.1142/p305.

[2] B. Lenk, H. Ulrichs, F. Garbs, M. Münzenberg, The building blocks of magnonics, *Phys. Rep.* 507 (2011) 107–136. https://doi.org/10.1016/j.physrep.2011.06.003.

[3] J. Cantu-Valle, I. Betancourt, J.E. Sanchez, F. Ruiz-Zepeda, M.M. Maqableh, F. Mendoza-Santoyo, B.J.H. Stadler, A. Ponce, Mapping the magnetic and crystal structure in cobalt nanowires, *J. Appl. Phys.* 118 (2015) 024302. https://doi.org/10.1063/1.4923745.

[4] S. Bedanta, A. Barman, W. Kleemann, O. Petracic, T. Seki, Magnetic nanoparticles: A subject for both fundamental research and applications, *J. Nanomater.* 2013 (2013) 1–22. https://doi.org/10.1155/2013/952540.

[5] M.K. Hossain, J. Ferdous, M.M. Haque, A.K.M.A. Hakim, Study and characterization of soft magnetic properties of Fe$_{73.5}$Cu$_1$Nb$_3$Si$_{13.5}$B$_9$ magnetic ribbon prepared by rapid quenching method, *Mater. Sci. Appl.* 6 (2015) 1089–1099. https://doi.org/10.4236/msa.2015.612108.

[6] M.K. Hossain, J. Ferdous, M.M. Haque, A.K.M.A. Hakim, Development of nanostructure formation of Fe$_{73.5}$Cu$_1$Nb$_3$Si$_{13.5}$B$_9$ alloy from amorphous state on heat treatment, *World J. Nano Sci. Eng.* 05 (2015) 107–114. https://doi.org/10.4236/wjnse.2015.54013.

[7] C.P. Bergmann, M.J. de Andrade, eds., *Nanostructured Materials for Engineering Applications*, Springer, Berlin, Heidelberg, 2011. https://doi.org/10.1007/978-3-642-19131-2.

[8] I. Khan, K. Saeed, I. Khan, Nanoparticles: Properties, applications and toxicities, *Arab. J. Chem.* 12 (2019) 908–931. https://doi.org/10.1016/j.arabjc.2017.05.011.

[9] S. Tehrani, E. Chen, M. Durlam, M. DeHerrera, J.M. Slaughter, J. Shi, G. Kerszykowski, High density submicron magnetoresistive random access memory (invited), *J. Appl. Phys.* 85 (1999) 5822–5827. https://doi.org/10.1063/1.369931.

[10] J.R. Childress, R.E. Fontana, Magnetic recording read head sensor technology, *Comptes Rendus Phys.* 6 (2005) 997–1012. https://doi.org/10.1016/j.crhy.2005.11.001.
[11] S. Kaka, M.R. Pufall, W.H. Rippard, T.J. Silva, S.E. Russek, J.A. Katine, Mutual phase-locking of microwave spin torque nano-oscillators, *Nature.* 437 (2005) 389–392. https://doi.org/10.1038/nature04035.
[12] D.A. Allwood, Magnetic domain-wall logic, *Science* 80(309) (2005) 1688–1692. https://doi.org/10.1126/science.1108813.
[13] M.I. Anik, M.K. Hossain, I. Hossain, I. Ahmed, R.M. Doha, Chapter: Biomedical applications of magnetic nanoparticles, in: A. Ehrmann, T.A. Nguyen, M. Ahmadi, A. Farmani, P. Nguyen-Tri (Eds.), *Magn. Nanoparticle-Based Hybrid Mater*, 1st ed., Woodhead Publishing, Elsevier, Sawston, 2021: pp. 463–497. https://doi.org/10.1016/B978-0-12-823688-8.00002-8.
[14] M.I. Anik, M.K. Hossain, I. Hossain, A.M.U.B. Mahfuz, M.T. Rahman, I. Ahmed, Recent progress of magnetic nanoparticles in biomedical applications: A review, *Nano Sel.* 2 (2021) 1146–1186. https://doi.org/10.1002/nano.202000162.
[15] Z.L. Wang, Nanostructures of zinc oxide, *Mater. Today.* 7 (2004) 26–33. https://doi.org/10.1016/S1369-7021(04)00286-X.
[16] G. Casillas, J.J. Velázquez-Salazar, M. Jose-Yacaman, A new mechanism of stabilization of large decahedral nanoparticles, *J. Phys. Chem. C.* 116 (2012) 8844–8848. https://doi.org/10.1021/jp3011475.
[17] J. Zhu, Q. Fu, Y. Xue, Z. Cui, Accurate thermodynamic relations of the melting temperature of nanocrystals with different shapes and pure theoretical calculation, *Mater. Chem. Phys.* 192 (2017) 22–28. https://doi.org/10.1016/j.matchemphys.2017.01.049.
[18] G.K. Williamson, W.H. Hall, X-ray line broadening from filed aluminium and wolfram, *Acta Metall.* 1 (1953) 22–31. https://doi.org/10.1016/0001-6160(53)90006-6.
[19] H.C. Gatos, Structure and properties of solid surfaces, in: *Surfaces Interfaces Glas. Ceram*, Springer, Boston, MA, 1974: pp. 195–240. https://doi.org/10.1007/978-1-4684-3144-5_13.
[20] D. Valim, A.G.S. Filho, P.T.C. Freire, J.M. Filho, C.A. Guarany, R.N. Reis, E.B. Araújo, Evaluating the residual stress in PbTiO 3 thin films prepared by a polymeric chemical method, *J. Phys. D. Appl. Phys.* 37 (2004) 744–747. https://doi.org/10.1088/0022-3727/37/5/015.
[21] X.-D. Zhou, W. Huebner, Size-induced lattice relaxation in CeO$_2$ nanoparticles, *Appl. Phys. Lett.* 79 (2001) 3512–3514. https://doi.org/10.1063/1.1419235.
[22] D. Liu, C. Li, F. Zhou, T. Zhang, H. Zhang, X. Li, G. Duan, W. Cai, Y. Li, Rapid synthesis of monodisperse au nanospheres through a laser irradiation-induced shape conversion, self-assembly and their electromagnetic coupling SERS enhancement, *Sci. Rep.* 5 (2015) 7686. https://doi.org/10.1038/srep07686.
[23] C. Li, K.L. Shuford, M. Chen, E.J. Lee, S.O. Cho, A facile polyol route to uniform gold octahedra with tailorable size and their optical properties, *ACS Nano.* 2 (2008) 1760–1769. https://doi.org/10.1021/nn800264q.
[24] N.G. Khlebtsov, L.A. Dykman, Optical properties and biomedical applications of plasmonic nanoparticles, *J. Quant. Spectrosc. Radiat. Transf.* 111 (2010) 1–35. https://doi.org/10.1016/j.jqsrt.2009.07.012.
[25] J. Ye, P. Van Dorpe, W. Van Roy, G. Borghs, G. Maes, Fabrication, characterization, and optical properties of gold nanobowl submonolayer structures, *Langmuir.* 25 (2009) 1822–1827. https://doi.org/10.1021/la803768y.
[26] J. Chen, J.M. McLellan, A. Siekkinen, Y. Xiong, Z.-Y. Li, Y. Xia, Facile synthesis of gold–silver nanocages with controllable pores on the surface, *J. Am. Chem. Soc.* 128 (2006) 14776–14777. https://doi.org/10.1021/ja066023g.
[27] C.L. Nehl, H. Liao, J.H. Hafner, Optical properties of star-shaped gold nanoparticles, *Nano Lett.* 6 (2006) 683–688. https://doi.org/10.1021/nl052409y.
[28] M. Liu, P. Guyot-sionnest, mechanism of silver(I)-assisted growth of gold nanorods and bipyramids, *J. Phys. Chem. B.* 109 (2005) 22192–22200. https://doi.org/10.1021/jp054808n.
[29] R. Skomski, Nanomagnetics, *J. Phys. Condens. Matter.* 15 (2003) R841–R896. https://doi.org/10.1088/0953-8984/15/20/202.
[30] S.E. Inderhees, J.A. Borchers, K.S. Green, M.S. Kim, K. Sun, G.L. Strycker, M.C. Aronson, Manipulating the magnetic structure of co core/CoO shell nanoparticles: Implications for controlling the exchange bias, *Phys. Rev. Lett.* 101 (2008) 117202. https://doi.org/10.1103/PhysRevLett.101.117202.
[31] J. Fidler, T. Schrefl, Micromagnetic modelling – the current state of the art, *J. Phys. D. Appl. Phys.* 33 (2000) R135–R156. https://doi.org/10.1088/0022-3727/33/15/201.
[32] Ò. Iglesias, A. Labarta, Finite-size and surface effects in maghemite nanoparticles: Monte Carlo simulations, *Phys. Rev. B.* 63 (2001) 184416. https://doi.org/10.1103/PhysRevB.63.184416.
[33] C. Abert, Micromagnetics and spintronics: Models and numerical methods, *Eur. Phys. J. B.* 92 (2019) 120. https://doi.org/10.1140/epjb/e2019-90599-6.

[34] T. Schrefl, G. Hrkac, S. Bance, D. Suess, O. Ertl, J. Fidler, Numerical methods in micromagnetics (finite element method), in: *Handbook of Magnetism and Advanced Magnetic*, John Wiley & Sons, Ltd, Chichester, UK, 2007. https://doi.org/10.1002/9780470022184.hmm203.

[35] W. Granig, C. Kolle, D. Hammerschmidt, B. Schaffer, R. Borgschulze, C. Reidl, J. Zimmer, Integrated gigant magnetic resistance based angle sensor, in: *2006 5th IEEE Confernce Sensors*, IEEE, 2006: pp. 542–545. https://doi.org/10.1109/ICSENS.2007.355525.

[36] S. Blundell, *Magnetism in Condensed Matter*, Oxford University Press, New York, 2001.

[37] D.P. Landau, K. Binder, *A Guide to Monte Carlo Simulations in Statistical Physics*, Cambridge University Press, Cambridge, 2014. https://doi.org/10.1017/CBO9781139696463.

[38] M.A. Hoque, M.R. Ahmed, G.T. Rahman, M.T. Rahman, M.A. Islam, M.A. Khan, M.K. Hossain, Fabrication and comparative study of magnetic Fe and α-Fe2O3 nanoparticles dispersed hybrid polymer (PVA + Chitosan) novel nanocomposite film, *Results Phys.* 10 (2018) 434–443. https://doi.org/10.1016/j.rinp.2018.06.010.

[39] A. Akbarzadeh, M. Samiei, S. Davaran, Magnetic nanoparticles: Preparation, physical properties, and applications in biomedicine, *Nanoscale Res. Lett.* 7 (2012) 144. https://doi.org/10.1186/1556-276X-7-144.

[40] A. Soumyanarayanan, N. Reyren, A. Fert, C. Panagopoulos, Emergent phenomena induced by spin – orbit coupling at surfaces and interfaces, *Nature.* 539 (2016) 509–517. https://doi.org/10.1038/nature19820.

[41] M.W. Wu, J.H. Jiang, M.Q. Weng, Spin dynamics in semiconductors, *Phys. Rep.* 493 (2010) 61–236. https://doi.org/10.1016/j.physrep.2010.04.002.

[42] D.L. Mills, S.M. Rezende, Spin damping in ultrathin magnetic films, in: *Spin Dynamics In Confined Magnetic Structures. II*, Springer, Berlin, Heidelberg, 2003: pp. 27–59. https://doi.org/10.1007/3-540-46097-7_2.

[43] T. McGuire, R. Potter, Anisotropic magnetoresistance in ferromagnetic 3d alloys, *IEEE Trans. Magn.* 11 (1975) 1018–1038. https://doi.org/10.1109/TMAG.1975.1058782.

[44] N. Nagaosa, J. Sinova, S. Onoda, A.H. MacDonald, N.P. Ong, Anomalous hall effect, *Rev. Mod. Phys.* 82 (2010) 1539–1592. https://doi.org/10.1103/RevModPhys.82.1539.

[45] C. Adamo, V. Barone, Toward reliable density functional methods without adjustable parameters: The PBE0 model, *J. Chem. Phys.* 110 (1999) 6158–6170. https://doi.org/10.1063/1.478522.

[46] C.P. Bean, J.D. Livingston, Superparamagnetism, *J. Appl. Phys.* 30 (1959) S120–S129. https://doi.org/10.1063/1.2185850.

[47] D.S. Mathew, R.-S. Juang, An overview of the structure and magnetism of spinel ferrite nanoparticles and their synthesis in microemulsions, *Chem. Eng. J.* 129 (2007) 51–65. https://doi.org/10.1016/j.cej.2006.11.001.

[48] M. Knobel, W.C. Nunes, L.M. Socolovsky, E. De Biasi, J.M. Vargas, J.C. Denardin, superparamagnetism and other magnetic features in granular materials: A review on ideal and real systems, *J. Nanosci. Nanotechnol.* 8 (2008) 2836–2857. https://doi.org/10.1166/jnn.2008.15348.

[49] S. Bedanta, W. Kleemann, supermagnetism, *J. Phys. D. Appl. Phys.* 42 (2009) 013001. https://doi.org/10.1088/0022-3727/42/1/013001.

[50] L. N'eel, Th'eorie du trainagemagn'etique des ferromagn'etiques en grains fins avec applications aux terres cuites, *Ann. Geophys.* 5 (1949) 99–136.

[51] L. Weil, L. Gruner, A. Deschamps, Orientation despr'ecipitations du cobalt dans un alliageCuCo, *Comptes Rendus.* 244 (1957) 2143.

[52] A. Knappwost, Kollektivparamagnetismus und Volumen magnetisierter Aerosole, Zeitschrift Für Elektrochemie, Berichte Der Bunsengesellschaft Für Phys. *Chemie.* 61 (1957) 1328–1334. https://doi.org/10.1002/BBPC.19570611010.

[53] A.E. Berkowitz, P.J. Flanders, precipitation in a beta-brass-fe alloy, *J. Appl. Phys.* 30 (1959) S111–S112. https://doi.org/10.1063/1.2185846.

[54] D.G. Rancourt, magnetism of earth, planetary, and environmental nanomaterials, *Rev. Mineral. Geochemistry.* 44 (2001) 217–292. https://doi.org/10.2138/rmg.2001.44.07.

[55] S. Bedanta, O. Petracic, X. Chen, J. Rhensius, S. Bedanta, E. Kentzinger, U. Rücker, T. Brückel, A. Doran, A. Scholl, S. Cardoso, P.P. Freitas, W. Kleemann, Single-particle blocking and collective magnetic states in discontinuous CoFe/Al$_2$O$_3$ multilayers, *J. Phys. D. Appl. Phys.* 43 (2010) 474002. https://doi.org/10.1088/0022-3727/43/47/474002.

[56] T. Jonsson, J. Mattsson, C. Djurberg, F.A. Khan, P. Nordblad, P. Svedlindh, Aging in a magnetic particle system, *Phys. Rev. Lett.* 75 (1995) 4138–4141. https://doi.org/10.1103/PhysRevLett.75.4138.

[57] P. Jönsson, T. Jonsson, J.L. García-Palacios, P. Svedlindh, Nonlinear dynamic susceptibilities of interacting and noninteracting magnetic nanoparticles, *J. Magn. Magn. Mater.* 222 (2000) 219–226. https://doi.org/10.1016/S0304-8853(00)00557-6.

[58] J.A. De Toro, M.A. López de la Torre, J.M. Riveiro, A. Beesley, J.P. Goff, M.F. Thomas, Critical spin-glass dynamics in a heterogeneous nanogranular system, *Phys. Rev. B.* 69 (2004) 224407. https://doi.org/10.1103/PhysRevB.69.224407.

[59] K. Binder, A.P. Young, Spin glasses: Experimental facts, theoretical concepts, and open questions, *Rev. Mod. Phys.* 58 (1986) 801–976. https://doi.org/10.1103/RevModPhys.58.801.
[60] E. Navarro, M. Alonso, A. Ruiz, C. Magen, U. Urdiroz, F. Cebollada, L. Balcells, B. Martínez, F.J. Palomares, J.M. González, Low temperature superspin glass behavior in a Co/Ag multilayer, *AIP Adv.* 9 (2019) 125327. https://doi.org/10.1063/1.5130158.
[61] G. Parisi, Order parameter for spin-glasses, *Phys. Rev. Lett.* 50 (1983) 1946–1948. https://doi.org/10.1103/PhysRevLett.50.1946.
[62] C. Djurberg, P. Svedlindh, P. Nordblad, M.F. Hansen, F. Bødker, S. Mørup, Dynamics of an interacting particle system: Evidence of critical slowing down, *Phys. Rev. Lett.* 79 (1997) 5154–5157. https://doi.org/10.1103/PhysRevLett.79.5154.
[63] W. Kleemann, O. Petracic, C. Binek, G.N. Kakazei, Y.G. Pogorelov, J.B. Sousa, S. Cardoso, P.P. Freitas, Interacting ferromagnetic nanoparticles in discontinuous $Co_{80}Fe_{20}/AlO_3$ multilayers: From superspin glass to reentrant superferromagnetism, *Phys. Rev. B.* 63 (2001) 134423. https://doi.org/10.1103/PhysRevB.63.134423.
[64] K. Binder, J.D. Reger, Theory of orientational glasses models, concepts, simulations, *Adv. Phys.* 41 (1992) 547–627. https://doi.org/10.1080/00018739200101553.
[65] R.H. Kodama, A.E. Berkowitz, E.J. McNiff, Jr., S. Foner, Surface spin disorder in $NiFe_2O_4$ nanoparticles, *Phys. Rev. Lett.* 77 (1996) 394–397. https://doi.org/10.1103/PhysRevLett.77.394.
[66] F. Zeb, W. Sarwer, K. Nadeem, M. Kamran, M. Mumtaz, H. Krenn, I. Letofsky-Papst, Surface spin-glass in cobalt ferrite nanoparticles dispersed in silica matrix, *J. Magn. Magn. Mater.* 407 (2016) 241–246. https://doi.org/10.1016/j.jmmm.2016.01.084.
[67] Y. Labaye, O. Crisan, L. Berger, J.M. Greneche, J.M.D. Coey, Surface anisotropy in ferromagnetic nanoparticles, *J. Appl. Phys.* 91 (2002) 8715. https://doi.org/10.1063/1.1456419.
[68] E. Winkler, R.D. Zysler, M. Vasquez Mansilla, D. Fiorani, D. Rinaldi, M. Vasilakaki, K.N. Trohidou, Surface spin-glass freezing in interacting core – shell NiO nanoparticles, *Nanotechnology.* 19 (2008) 185702. https://doi.org/10.1088/0957-4484/19/18/185702.
[69] R. Saito, G. Dresselhaus, M.S. Dresselhaus, *Physical Properties of Carbon Nanotubes*, Published by Imperial College Press and Distributed by World Scientific Publishing Co., 1998. https://doi.org/10.1142/p080.
[70] F. Wang, The optical resonances in carbon nanotubes arise from excitons, *Science* 80(308) (2005) 838–841. https://doi.org/10.1126/science.1110265.
[71] D.A. Heller, Optical detection of DNA conformational polymorphism on single-walled carbon nanotubes, *Science* 8(311) (2006) 508–511. https://doi.org/10.1126/science.1120792.
[72] K. Ishikawa, K. Yoshikawa, N. Okada, Size effect on the ferroelectric phase transition in $PbTiO_3$ ultrafine particles, *Phys. Rev. B.* 37 (1988) 5852–5855. https://doi.org/10.1103/PhysRevB.37.5852.
[73] J.F. Meng, R.S. Katiyar, G.T. Zou, X.H. Wang, raman phonon modes and ferroelectric phase transitions in nanocrystalline lead zirconate titanate, *Phys. Status Solidi.* 164 (1997) 851–862. https://doi.org/10.1002/1521-396X(199712)164:2<851::AID-PSSA851>3.0.CO;2-J.
[74] D. Guo, G. Xie, J. Luo, Mechanical properties of nanoparticles: Basics and applications, *J. Phys. D. Appl. Phys.* 47 (2014) 013001. https://doi.org/10.1088/0022-3727/47/1/013001.
[75] R. Singh, J.W. Lillard, Nanoparticle-based targeted drug delivery, *Exp. Mol. Pathol.* 86 (2009) 215–223. https://doi.org/10.1016/j.yexmp.2008.12.004.

Nanomagnetic Materials
Structural and Magnetic Properties

11

P. Maneesha[1,†], Suresh Chandra Baral[1,†],
E. G. Rini[1], and Somaditya Sen[1]

1 Department of Physics, Indian Institute of Technology Indore, Indore, India

† These authors have contributed equally to this work

Contents

11.1	Introduction to Nanomagnetism	208
	11.1.1 History of Nanomagnetism	208
	11.1.2 Correlation to Structure and Morphology	208
11.2	Origin of Nanomagnetism	209
11.3	Features That Dominate the Magnetic Properties in Nanomagnets	209
	11.3.1 Size Effect	209
	11.3.1.1 Sample Dimensions and Characteristic Lengths	210
	11.3.2 Surface Effects	210
	11.3.2.1 Broken Translation Symmetry	210
	11.3.2.2 Nanoscopic Samples and Magnetization Reversal	211
	11.3.3 Morphological Effects	211
11.4	Magnetic Interaction Energies in Nanomagnets	212
	11.4.1 Zeeman Energy	212
	11.4.2 Exchange Energy	212
	11.4.3 Anisotropy Energy	213
	11.4.4 Magnetostatic Energy	213
11.5	Magnetic Structures	214
	11.5.1 Collinear and Non-Collinear Magnetic Structures	214
	11.5.2 Magnetic Space Groups	215
11.6	Structural and Magnetic Characterization	215

DOI: 10.1201/9781003197492-11

11.7	Some of the Important Nanomagnetic Crystal Structures and Their Magnetic Properties		216
	11.7.1 Metallic Compounds		216
	11.7.2 Metal Oxides (MOs)		216
	11.7.3 Metal Chalcogenides		217
	11.7.4 Perovskites		217
	11.7.5 Spinels		218
11.8	Tuning the Crystal Structure and Magnetic Properties		218
	11.8.1 Effects of Method of Preparation		219
	11.8.2 Effects of Doping		219
	11.8.3 Effects of Pressure		220
	11.8.4 Effects of Temperature		220
11.9	Recent Developments in Nanomagnets in Correlation With Structure		221
11.10	Conclusions and Outlook		221
11.11	Acknowledgments		222
References			222

11.1 INTRODUCTION TO NANOMAGNETISM

11.1.1 History of Nanomagnetism

The journey of magnetism started around the sixth century B.C. when the Greek philosopher Thales of Miletus became curious about the attraction between iron and lodestone. The quantum mechanical property of electron spin led to an exciting field of nanomagnetism in solid-state physics. The first initiative to permanent magnetism arose with the discovery of electrons. The understanding of each electron itself as a tiny magnetic dipole in addition to the charge and mass was revolutionary in terms of magnetism. The arrangement of these magnetic dipoles in close proximity can be explained with the help of quantum mechanics. Albert Fert and Peter Grunberg brought this regime of nanomagnetism to another stage of development by winning the 2007 Nobel Prize in Physics for the discovery of giant magnetoresistance. Nowadays, devices developed by using nanomagnetism have more capacity for information storage and transportation. As an alternative to electricity in novel small gadgets like cell phones and computers, research is focusing on magnetricity. Nanotechnology based on nanomagnetism can be applied in diverse fields since these special materials can meet the need for small and powerful devices.

11.1.2 Correlation to Structure and Morphology

Depending on the crystal structure, the magnetism exhibited by the nanomagnets changes. In addition to that, based on the size and morphology of the nanomagnets, the magnetic response also varies. Bond lengths, bond angles, covalency of bonds, coordination number – all these structure parameters will influence the exchange interactions of electrons in the nanometric samples that result in the variations in magnetism of these materials. Morphology of nanoparticles (NPs) is significant; depending on the flatness, sphericity, and aspect ratio, the morphology will be different, and that changes the magnetic properties as well. Controlling the structure, morphology, and size of NPs has a paramount role in exploiting the properties for many novel applications. This chapter mainly deals with the structural and magnetic properties of nanomagnets, the structure correlation of magnetism in detail, and the tuning of the crystal structure and magnetic properties in different known nanomagnets. We also look at recent developments in nanomagnets in their structural aspects.

11.2 ORIGIN OF NANOMAGNETISM

Nanomaterials have different magnetism origins than bulk samples because of lengths that are equivalent to characteristic lengths, like limiting domain sizes and translational symmetry breaks, resulting in reduced coordination numbers, greater exchange bonds, and frustration phenomena. Surface atoms (or interface atoms) are also higher in nanoscopic or mesoscopic objects. Nano-objects also have magnetic properties modified by their close proximity to other physical systems. As the size decreases from micrometer to nanometer, it changes the relation of surface atoms to the internal atoms. Thus, the internal energy is decreased to compensate for the increase in surface energy that leads to the enhanced properties of nanomaterials. The differences of the origins of magnetism in nanomaterials are need to be understood using quantum mechanics.

Moving from a bulk to a nanodomain involves reduction in the dimensionality and sometimes modifies the cationic coordination number due to an increased surface area. Hence, more electrons can exhibit magnetism in the nanodomains. The broken translational symmetry in nanometric samples is another reason. As the size tends to a nano dimension, they are comparable to the fundamental lengths, such as exchange length, magnetic domain wall width, etc. The magnetic properties of these materials change accordingly since the magnetic properties depend on these characteristics' lengths. The magnetic properties of nanomagnets are related to these characteristic lengths. While entering into the world of nanomagnetism, the density of electrons curve also varies, which can also impart the changes in the magnetic properties of the nanomagnets. The naturally occurring magnetic molecules and the artificially structured low dimensional magnetic materials constitute the interesting field of nanomagnetism. Disorder on the surface of nanomaterials becomes significant as the surface-to-volume ratio increases. As a result, exchange anisotropy, spin-glass freezing, and particle interactions are also taken into consideration. Magnetic moment orientation will stiffen as a result of all these phenomena.

11.3 FEATURES THAT DOMINATE THE MAGNETIC PROPERTIES IN NANOMAGNETS

11.3.1 Size Effect

The intrinsic magnetic properties of bulk defect-free materials (e.g., saturation magnetization M_S, coercive force H_C, and Curie temperature T_C) are governed only by chemical and crystallographic structures. It does not matter how big or how small the bulk samples are; if you compare small and big cobalt samples, their M_S, H_C, and T_C values are equal. As compared to their bulk counterparts, magnetic nanoparticles exhibit a wide range of unusual magnetic properties. Surface effects and finite-size effects significantly affect the magnetic characteristics of NPs. Decreasing particle size increases their relevance. As far as the electrons are concerned, finite-size effects are the consequence of quantum confinement. In the simplest terms, surface effects require symmetry breaking of the crystal structure at the particle boundaries but can also be caused by a difference in chemical and magnetic structures of the internal ("core") and surface ("shell") parts of a nanoparticle. Here we discuss the effect of size, shape, and environment on magnetic properties of NPs, including interactions between particles and between particles and matrix particles.

Nanoscale samples of the same constituents behave differently from macroscopic samples. Under normal experimental conditions, thermal fluctuations are a major contributor to this difference. When the thermal energy $k_B T$ matches the particle's anisotropy energy, then magnetic nanoparticles exhibit superparamagnetism, resulting in no magnetic field.

11.3.1.1 Sample Dimensions and Characteristic Lengths

In this case, the magnetic properties of large magnetic particles are affected by the characteristic lengths. Smaller magnetic particles have a bigger effect on magnetic properties. Thus, they have the lowest energy configuration in the single domain. The magnetic domain of a ferromagnetic particle is limited with critical size, D_{cr}, which is the maximum size that it can have before it will generate enough energy to be divided into more than one, which varies from material to material. It can vary in size from about 10 nanometers (or microns) to a few thousand. Defining the critical diameter is as follows: $D_{cr} = \dfrac{72\sqrt{AK}}{\mu_0 M_s^2}$, where A is the exchange stiffness constant, K is the uniaxial anisotropy constant, and M_s is the saturation magnetization; $\mu_0 \sim 4\pi \times 10^{-7}\ Hm^{-1}$ is the vacuum magnetic permeability [1].

11.3.2 Surface Effects

The surface and interface of nanostructures affect the stability of the structure of the NPs. A surface modification can improve the surface functionalization of NPs and stabilize the nanostructure.

11.3.2.1 Broken Translation Symmetry

Crystals have borders where the translational symmetry breaks or disappears. The majority of atoms in nanometric solids are on or close to the surface of these boundaries. Translational symmetry is absent in these systems, which leads to many consequences that affect their physical properties.

As a result of symmetry breaking, the following three aspects are presented:

(1) The relationship between the physical properties of the samples and their dimensionality – samples with quasi-zero dimension (0D), unidimensional (1D), bidimensional (2D), or tridimensional (3D),
(2) How the atoms at the interface are coordinating, and
(3) On the surface (or interface) of nanoscopic samples, the proportion of atoms increases.

11.3.2.1.1 Dimensions and Density of Electronic States
Dimensionality affects the electronic band structure of a solid. A simple example of this is the free electron model of a conducting solid, with electrons treated as if they were gases (called Fermi gases), with boundaries determined by infinite potentials. The density of electronic states in a small space depends on its dimensionality, which is measured by the density of electrons D(E). Dimensionality differences of different D(E) values are shown in Figure 11.1.

11.3.2.1.2 Reduced Coordination Number and Dimensionality
This leads to a lower number of neighbors for atoms on surfaces than atoms in bulk because of the broken translation symmetry. Atomic magnetic moments are also altered on interfaces because the point symmetry is disrupted at their sites, leading to level splitting and modification of their magnitudes. For example, compared to bulk structures, Fe atoms show enhanced magnetic moments when in contact with Cu, Pd, and Ag. In addition, atoms at interfaces have magnetic properties that are affected by defects and impurities, such as adsorbates; strain can also alter these properties, changing the crystal lattice parameters.

11.3.2.1.3 Nanoscopic Samples and Proportion of Surface Atoms
It is widely recognized that atoms on the surface play an important role in catalysis. It is common to prepare catalysts in the form of powder, or porous matrices, due to the fact that they depend on the surface of the particles to interact with the molecules involved in the reaction. The physical properties of nanoscale

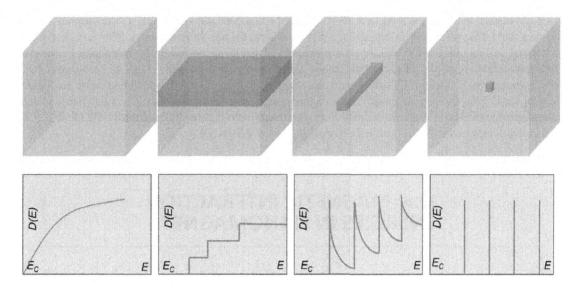

FIGURE 11.1 Density of electronic states D(E) with energy, in bulk, in two dimensions, one dimension, and zero dimensions (from left to right).

samples are enhanced by the surface atoms as the sample size decreases. It is obvious, since the surface area of the samples changes as a function of $\sim r^2$, whereas the volume of the samples changes as a function of $\sim r^3$. As a consequence, the surface-to-volume ratio varies roughly with r^{-1} and so increases with decreasing sample size.

11.3.2.2 Nanoscopic Samples and Magnetization Reversal

Nanomagnets may not behave like macroscopic objects in terms of their dynamic magnetization behavior. In this case, thermal fluctuations are believed to play a more significant role under the usual experimental conditions. As an example, superparamagnetism is observed in nanoscopic magnetic particles: their magnetization inverts spontaneously because their anisotropy energy is comparable to their thermal energy. It is possible for a single-domain magnetic particle to spontaneously invert its magnetization, i.e., the direction of its magnetization can change from +z to −z when T is above a certain blocking temperature T_B. Using the magnetization of these particles for information storage would lead to the loss of information at $T = T_B$ if this effect were to exist. Because of this, there has been an increased focus on thermal stability in magnetic storage as the physical size of the recorded bit decreases.

11.3.3 Morphological Effects

There is only a secondary influence for morphological effects on the magnetic properties of nanomagnets. Saturation magnetization in a homogenous sample is determined only by the structural and chemical order, and it is independent of the shape. For the heterogeneous nanomagnets, especially for core-shell structure, the magnetization of core and shell have different values; hence, the magnetic properties depend on the size and shape of the particle. Based on the local magnetic anisotropy, the coercivity and the hysteresis loop in the nanomagnets are determined. We discuss magnetic anisotropy in the next section. Among the different anisotropies, shape anisotropy is directly related to the shape of the particle. That is, a rod- or needle-shaped NP would have a different magnetic anisotropy than a spherical or cubic one. Magnetocrystalline anisotropy is less in polycrystalline materials due to the preferred orientation of grains. Whereas, for spherical NPs, in every direction the magnetization is possible to the same extent for the same field. Magnetization along the long axis is possible for non-spherical ones. This is called shape anisotropy and can be used to enhance the properties of nanomagnets when the morphology changes. Different morphologies and corresponding shape anisotropy can influence the magnetic properties of the NPs. For example, in a study of Ni nanostructures, the coercivity and saturation magnetization are higher for nanowire (NW) samples in comparison to nanoparticle samples and the mixture of nanoparticle and nanowire samples. The values of coercivity and saturation magnetization change from 119.82 Oe to 139.71 Oe at room temperature for the change in morphology from NP to NW.

11.4 MAGNETIC INTERACTION ENERGIES IN NANOMAGNETS

In this section, we review several sources of energy that exist in magnetic systems. The following discussion is limited to ferromagnetic materials for simplicity of terminology, but all aspects can be extended to other types of materials as well. Micromagnetism is used to describe these energies.

Micromagnetism is the study of the competing energies within a system, which results in magnetic length scales and complex distributions of magnetizations. William Fuller Brown, Jr. outlined its principles in 1940 [2].

Whether analytical or numerical, micromagnetism relies on two basic assumptions.

- There is enough gradual variation in the direction of a magnetic moment from one (atomic) site to another to be able to ignore the discrete nature of matter. In the approximation of continuous media, magnetization M and all other quantities are continuous functions of the variable.
- No matter what material you use, the magnetization vector norm M_s will remain constant and homogeneous. Temperature can be either zero or finite. One way of considering the latter is as the mean-field representation.

These two approximations for magnetization allow us to define magnetization distributions by the unit vector m(r), such that $M_s(r) = M_s\, m(r)$. Ferromagnets have Gibbs free energy defined by $\varphi = \iiint (U - TS - \sigma.\varepsilon - \mu_0 M \cdot H_{ext})\, dV$, where U is the internal energy, T is the temperature, S is the entropy, and σ and ε are the strain and stress tensors. The free energy $U\text{-}TS$ includes the exchange, magnetocrystalline, dipolar, and magnetoelastic energies as well as the Ginzburg–Landau ordering energy. The term $\sigma.\varepsilon$ denotes the elastic energy. The last characteristic describes the interaction between the magnetization and the applied magnetic field H_{ext}.

11.4.1 Zeeman Energy

Energy related to magnetic moments as found in an external magnetic field is the Zeeman energy. Its density is $E_Z = \mu_0 M.H$, where E_Z tends to favor the magnetization to align along the applied field. According to our prior discussion, this term should not be considered as an internal energy contributor but rather as an inducer of magnetic enthalpy in a system.

11.4.2 Exchange Energy

Magnetism is governed by the exchange interaction, which is responsible for setting up magnetic order. Due to the indistinguishability of the electrons, there is no classical analogy for this quantum effect.

A spin-spin exchange interaction between two contiguous spins, S_i and S_j, can be described by the Hamiltonian $H = -2J\, S_i \cdot S_j$, in which J is the exchange constant, which indicates the value of the interaction known as the Heisenberg Hamiltonian. This describes several magnetic aspects of materials, especially insulators. The energy of a pair of spins is $E_{ex}^{pair} = -2J\, S_i \cdot S_j$ and the angle between the two spins i and j as $\theta\,(i, j) = \Delta\theta$, $E_{ex}^{pair} = -2J\, S^2 \cos\Delta\theta \approx J\, S^2 (\Delta\theta)^2$, where the approximation $\cos \Delta\theta \approx 1 - (\Delta\theta)^2/2!$ has been used by neglecting terms independent of θ.

The interaction energy, $E_{ex}^{pair} = -2J\, S_i \cdot S_j = -2J\, S^2\, m_i \cdot m_j$, is the energy resulting from the electrostatic exchange interaction. Electron-electron interactions lead to magnetic ordering. If the exchange interaction in some spins remains constant below a certain critical temperature, then different kinds of the ordering of spins are possible.

In ferromagnetism, the magnetic moments favor parallel alignment as a result of positive exchange interactions. As a result, a spontaneous magnetization M_s develops. M_s is usually in the range of 10^6 amperes/meter, which is a much higher value than the magnetization resulting from diamagnetism or paramagnetism. In the case of antiferromagnetism, the negative exchange energy favors the antiparallel alignment of neighboring moments, resulting in zero net moments magnetic field, and ferrimagnetism results from negative exchange coupling between moments since each is located on a different sublattice, resulting in nonzero net magnetization.

11.4.3 Anisotropy Energy

According to the theory of magnetic ordering, a magnetization can spontaneously occur, regardless of its direction. The internal energy in a real system depends on how M is oriented in relation to the underlying crystal structure. The result is a combination of crystal-field effects (coupling orbital electrons to the lattice) and spin-orbit effects (orbital coupling with spin moments). Its density will be given as E_{mc}, which stands for magnetocrystalline anisotropy energy (MAE). E_{mc} results in the magnetization of a solid aligning itself along certain axes (or planes), called easy directions. In opposition, hard axes (or planes) are those with the highest level of energy.

Several different types of anisotropy are there. Magnetocrystalline anisotropy is caused by a coupling between spin orbits due to crystal symmetry. Other types are anisotropy due to broken symmetry at the surface and anisotropy arising due to stress on the crystal structure as a result of magnetization. This also includes shape anisotropy, which arises due to the shape of individual grains.

It is also worthwhile to consider the magnetoelastic anisotropy energy, written as E_{mel}. It is the energy associated with strain (deformation) of a material, either compressive or extensive or shear. E_{mel} is an extension of E_{mc} in terms of strain. The anisotropy energy in micromagnetism is described in terms of phenomenology, ignoring all the details at the microscopic level. As a result, we may take the sum of E_{mel} and E_{mc} as written E_a or E_K, where A is the anisotropy and K is the anisotropy constant.

11.4.4 Magnetostatic Energy

Magnetic energy, also known as dipolar energy and written E_d, is the mutual Zeeman-type energy that results from the stray magnetic field (itself called a dipolar field and written H_d). In the case of an infinitesimal moment $\delta_\mu = M\delta V$, enthalpy is defined by the Zeeman energy, taking into account that the magnetic body is the origin of all magnetic fields (dipolar field H_d) and magnetic moments, which are responsible for internal energy.

Dipolar energy is a critical contribution to micromagnetism, making it difficult to analyze. Because of its non-local nature, it can only be described analytically in a limited number of simple situations. The numerical evaluation is also very time-consuming since all moments interact with each other; this is also the reason why numerical simulation is a very limited tool. A magnetic dipole creates a non-uniform magnetic field in its direction and magnitude, as a result of which magnetostatic energy plays a key role in the formation of non-uniform magnetizations in bulk and nanostructures, particularly magnetic domains. We discuss this term in detail in the following section due to all these reasons.

11.5 MAGNETIC STRUCTURES

Magnetic structures are the different configurations of magnetic moments in the crystal lattice. They can be simple (collinear or non-collinear) or complex (frustrated, spin density wave, sine wave, canonical, helical, skyrmion, etc.) magnetic structures. The simultaneous existence of crystallographic order and magnetic order are opposite to each other and leads to sequences of transitions. Neutron diffraction is the best experimental method to confirm the magnetic order. Through the analysis of neutron diffraction data with numerical methods involving group theory, one can ensure the order of nanomagnets. Even if these magnetic structures are periodic, both crystallographic periodicity and magnetic periodicity don't need to be the same. These magnetic structures can be explained in two ways. The first one is the simple case in which structure periodicity is an integral multiple of nuclear arrangement. This shows a picture of magnetic orientations within the unit cells. These magnetic unit cells are used to explain the magnetic structures for simple arrangements of magnetic spins. In this way, structures are mainly described as ferromagnetic, antiferromagnetic, or ferrimagnetic. The other methods used to describe the magnetic structures are somewhat complex and are used for complex structures and neutron diffraction analysis in reciprocal space. In this analysis, the structures are described in terms of propagation vectors. Instead of constructing a complete magnetic unit cell with many atoms, this method is based on the nuclear unit cell and propagation vector, k. The relation between the equivalent magnetic atoms in different nuclear unit cells can be done with this propagation vector. This implies general formalism for both simple and complex magnetic structures.

The magnetic moment distribution associated with i_{th} atom is m_i, and for any magnetic structure, this moment can be expanded as a Fourier series.

$$m_i = \sum m_{v,k} e^{-ik \cdot R}$$

where $m_{v,k}$ is the Fourier component of the distribution associated with the propagation vector k.

The k can be chosen from the first Brillouin zone of the Bravais lattice of the nuclear unit cell. Because of the nuclear structure, the atoms are periodic in the lattice and hence the wave vector k is periodic in the reciprocal space.

11.5.1 Collinear and Non-Collinear Magnetic Structures

There are thousands of magnetic structures in compounds showing magnetic behaviors. If the spins are aligned parallel directions, it is ferromagnetic (FM), and if the spins are antiparallel, or non-collinear, it is antiferromagnetic (AFM) or ferrimagnetic structures. Also, if the spins are disordered and not parallel, it forms complex magnetic structures like helical, canonical, etc. Examples of disordered systems such as topological glasses, spin glasses, and substitutional alloys include competing exchange interactions. For example, in a study of core-shell Ni-NiO structure, it exhibits equal probability of ferromagnetic and antiferromagnetic interaction. This competing magnetic interaction leads to spin glasses. Collinear magnetism is the simplest ordered arrangement of magnetic moments in parallel or antiparallel directions. For collinear ferromagnetism, all the magnetic moments are aligned parallel and possess maximum magnetization in the nanomaterial, and for collinear antiferromagnetism, there is an equal amount of parallel and antiparallel magnetic moments, which leads to zero magnetization in the nanomagnetic material. Because of the geometric frustrations of magnetic moments, non-collinear magnetism arises. Another reason for non-collinear magnetism is the preferred direction of magnetization that leads to magnetic anisotropy. It also arises due to the competition between the exchange interaction and the magnetic anisotropy.

11.5.2 Magnetic Space Groups

The magnetic space group explains both space symmetry and the electron spin and is an extension of the crystallographic space group which describes the space symmetry alone. Magnetic symmetry groups consist of the time-reversal symmetry operation in addition to the conventional crystallographic point groups. Just like the conventional crystal lattices, there are magnetic lattices with this symmetry operation.

Magnetic materials are invariant under spatial inversion symmetry but violate time-reversal symmetry. That means magnetic moments are present along with the antisymmetric operator in nanomagnets. The conventional crystallographic point groups (32 point groups) do not contain antisymmetric operators. There are in total 122 point groups (32 + 32 + 58) when an additional antisymmetric operation is added to the crystallographic point groups. This includes *32* conventional point groups without antisymmetric operator and 32 point groups with the antisymmetry operation multiplied with all the point group operations and 58 magnetic point groups with half of the point groups multiplied with the antisymmetry operation. The space group must elaborate to explain the symmetry of magnetic materials. There are in total 1651 magnetic space groups (230 + 230 + 674 + 517) which are called Shubnikov groups. In this 230 are ordinary crystallographic space groups without any additional symmetry (colorless groups), 230 are space groups with an additional antisymmetric operation for all the symmetry operations, 674 are with an additional antisymmetric operation for half of the symmetry operations, and 517 are space groups with translation with time-reversal symmetry [3].

11.6 STRUCTURAL AND MAGNETIC CHARACTERIZATION

Particle size can be determined with a transmission electron microscope (TEM). From the TEM the real particle size is obtained. TEM also provides the particle size distribution and the aggregation and chaining of particles as well. High-resolution TEM (HR-TEM) is useful to investigate the internal structure of NPs. From the X-ray diffraction (XRD) analysis, the crystallite size and microstrain can be determined along with the unit cell parameters. In the nanomagnets, the thickness of the domain is comparable with characteristic dimensions. There are certain particular sizes in which NPs are energetically favorable to become a single domain in it. Magnetic properties of single-domain nanomagnets are mainly determined by blocking temperature, T_B, and superspin or nanoparticle magnetic moment, μ_m. T_B is the temperature at which the thermal fluctuations overcome the energy barrier for the flip between two equilibrium states related to the relaxation time of the nanoparticle superspin. μ_m is the maximum allowed magnetization at a given time. For the 3D characterization of NPs with sub-nanometer resolution, the atomic force microscope (AFM) is a good choice. It can be used for the characterization of nanoparticle size, size distribution, variable geometry NPs, direct visualization of hydrated NPs/liquid medium, and characterization of physical properties of NPs such as response to magnetic fields. To analyze the surface morphology, a scanning electron microscope (SEM) is used. The magnetic structures and the ordering in magnetic structure can be confirmed with the help of neutron diffraction studies. For double perovskite compounds where the A site and B site ordering cannot be determined by the XRD analysis due to the similar scattering factors of the cations, neutron diffraction can provide an exact idea of the ordering pattern.

There are several issues related to the magnetic property measurements of nanomaterials. The magnetic moment is a sensitive function of the number of atoms. Hence, the number of atoms in a magnetic cluster should be known for the measurement of magnetic properties. Coagulation, contamination, and surface oxidation need to be avoided at the time of synthesis and measurement. The temperature needs to be controlled since it plays an important role in the magnetism of these materials. In the case of particles in a substrate or embedded magnetic NPs, the role of the interface and the substrate could be pronounced. Thermo magnetization curves field cooled (FC), zero field cooled (ZFC), and field cooled cooling (FCC) conventions used to study the magnetic properties of magnetic systems. Initially, the sample is cooled in the absence of a magnetic field (ZFC). Then the magnetic field is applied at the lowest possible temperature and the magnetization is measured while increasing the temperature in the presence of the field. After reaching a considerable high temperature, the temperature is gradually reduced in the presence of the field and the magnetization is measured while the temperature reduces; this is known as FCC measurement. After the lowest temperature is attained, the temperature is raised again to the highest possible temperature while the magnetization is measured again; this is termed as FC measurement. The value of T_B can be obtained from the ZFC-FC protocol. Determination of the μ_m can be done with the analysis of the magnetization isotherms in the superparamagnetic state. The magnetic size of a nanomagnet can be determined from values obtained from the magnetization curves. Magnetic size refers to the size of the magnetically ordered part in the nanoparticle. Another magnetic characterization technique is AC magnetic measurements. Here AC moment is measured which arises due to the application of an AC field to a sample. This time-dependent moment shows the magnetization dynamics of the sample. In DC measurements throughout the measurement time, the sample moment is constant and hence the dynamics cannot be obtained.

11.7 SOME OF THE IMPORTANT NANOMAGNETIC CRYSTAL STRUCTURES AND THEIR MAGNETIC PROPERTIES

11.7.1 Metallic Compounds

FePt alloy exhibits two solid-state phases in the phase diagram. One is the face-centered cubic (FCC) disorder phase and the other one is a face-centered tetragonal (FCT) ordered phase. Among this FCC is the high-temperature stable phase. In the final product of the sample preparation, the FCC disordered phase dominates because most of the synthesis methods can be kinetically controlled. Post-deposition annealing is used for the FCT ordered phase. This annealing requires nucleation ordering originating at these sites, and the transformation occurs in the first order. For the phase transformation, the crystal defects and grain boundaries are the nucleation sites in the case of FePt thin films. However, FePt NPs have fewer defects and lack conventional grain boundaries, which make them more difficult to transform [4]

Another example is that of FeAu nanoclusters. The structure of these nanoclusters embedded in a W matrix evolves from an amorphous to a crystalline structure during their formation. At the surface of the nanoclusters, crystallization starts and results in the formation of a polycrystalline structure. A duplex amorphous core-crystalline shell structure is reported for this material [5]. They show low-temperature magnetic properties of ferromagnetism/antiferromagnetism, superparamagnetism, or spin-glass behavior as well.

11.7.2 Metal Oxides (MOs)

The wide applications in the domain of spintronics and optoelectronics make MOs, especially semiconducting metal oxides, suitable for extensive research. The size-related structural changes can change the cell parameters in NPs of CuO, ZnO, SnO_2, Al_2O_3, MgO, ZrO_2, AgO, TiO_2, CeO_2, etc. These MOs nanomagnets can tune their magnetic properties as the size and shape vary. Hydrothermally grown Fe_3O_4

nanoparticles with a diameter of 27 nm have been reported to have ferromagnetic properties [6]. On the other hand, one of the most interesting stories related to magnetism is in the field of polymorphism of iron (III) oxide. The same chemical composition reveals the existence of multiple phases with different crystal structures and physical properties: α-Fe_2O_3 (hematite), β-Fe_2O_3, Γ-Fe_2O_3 (maghemite), and ε-Fe_2O_3 (luogufengite). The phase transformations occur due to the application of a combination of different amounts of heat and pressure. Hexagonal α-Fe_2O_3 is the most thermodynamically stable under ambient conditions. Antiferromagnetic ζ-Fe_2O_3 has a Neel transition temperature of ~69 K and is formed from cubic β-Fe_2O_3 due to application of pressure >30 GPa [7]. Hollow maghemite NPs reveal freezing of disordered spins at the inner and outer surfaces leading to the development of spin-glass-like behavior displaying memory, remanence, and aging effects as well as an exchange bias phenomenon [8]. Another typical example of an antiferromagnetic material is the ilmenite-type $FeGeO_3$, which reveals a Neel temperature of ~79 K [9]. There are more examples to talk about which are not discussed here due to space limitations.

11.7.3 Metal Chalcogenides

The sulfides and selenides of metals fall into a general category of materials commonly termed metal chalcogenides. While MnS and MnSe NPs are paramagnetic [10], CdS and CdSe reveal ferromagnetism. Nickel sulfide exhibits various phases, such as NiS, Ni_3S_2, NiS_2, Ni_3S_4, Ni_7S_{10}, and Ni_9S_8 [11]. So, it is still a challenge to obtain a proper morphology with pure nickel sulfide. Most binary metal sulfide nanostructures have been synthesized based on the Kirkendall effect. This effect includes the vacancy diffusion, which occurs due to the mutual diffusion of two metals through an interface. Depending on the metal-to-chalcogen (M:X) ratio, the structural and magnetic properties of TM chalcogenides with layered crystal structures are defined. Among these Fe-intercalated titanium dichalcogenides, Fe_yTiX_2 and $(Fe)_7X_8$ compounds are of great importance. Most of these 2D transition metal chalcogenides are intrinsically nonmagnetic. Hence, research is going on for introducing intrinsic room-temperature ferromagnetism in these materials. Various strategies can be used to attain this ferromagnetism, such as defect engineering, doping with transition metal elements, and phase transfer. Experimental conditions, such as nucleation temperature, ion irradiation dose, doping amount, and phase ratio, can also induce ferromagnetic ordering.

11.7.4 Perovskites

A novel concept of octahedral tilting is possible in perovskite oxide nanomagnets for tuning the structural and magnetic properties and for designing nanodevices. The octahedral tilt engineering of exchange interactions is a challenging problem. The changing parameters for the modification of octahedra are the strain and the thickness of the thin film samples. In perovskite $SrRuO_3$ (SRO) ultrathin layers tuning of ferromagnetism is done by oxygen coordination of adjacent $SrCuO_2$ (SCO) layers [12]. The infinite-layered CuO_2 exhibits structural transformation from "planar-type" to "chain-type" with reduced film thickness. This will vary the polyhedral connection at the interface, hence resulting in the change of octahedral distortion of SRO. These structural changes lead to the variation in the Ru spin state and strength of hybridization. All these actions result in the change in the magnetoresistance and anomalous Hall resistivity. Perovskite ABO_3 nanoparticles (A = K, Li; B = Ta, Nb or A = Ba, Sr, Pb; B = Ti) at room temperature exhibit unexpected ferromagnetic properties by oxygen vacancies at the surface of the nanocrystalline materials. The ions Ta^{4+} and/or Ta^{3+} (Ti^{3+} and/or Ti^{2+}) appear at the surface with nonzero net spin as a result of these vacancies. This offers a nonzero magnetization which increases with decreasing particle size [13].

The oxide and halide perovskites have the same perovskite structure with the divalent oxygen ions sites in oxide perovskite being replaced with monovalent halide ions in halide perovskites. To satisfy charge neutrality inorganic metal cations, such as Pb^{2+}, Sn^{2+}, and Ge^{2+}, with a valence state of 2+ is possible for halide perovskites. The oxide perovskites can host trivalent and tetravalent cations at the B site, but a limited number of variations in a composition is possible in halide perovskites. Novel materials are formed with the accommodation of organic cations in the halide perovskite structure called inorganic-organic hybrid halide perovskites. Ferromagnetism has been exhibited by the organic-inorganic hybrid lead-based perovskite systems. The low temperature (100 K) ferromagnetic hysteresis loop was stable even at a high temperature of 380 K. This substantiates the fact that the origin of magnetism was embedded in its defective nature. Because of the strong spin-orbit coupling of lead, lead-based oxide and halide perovskites exhibit interesting spin properties. This enables the motion of an electron to its quantum spin. The strength determines to what extent the spin of an electron interacts with the magnetic field. These couplings enhance the magnetic properties of these compounds. Lattice defects in these materials can tune and exhibit exceptional properties. Lead is toxic. However, lead-based perovskites are important materials in applications. Hence, the world is looking forward to replace lead and find lead-free systems that exhibit similar important properties that lead can induce in the crystal structure.

Partial substitution at A and B sites in perovskites leads to doubling of the perovskite unit cell and forming the double perovskite $A'A''B'B''O_6$ structure. This compound shows more complex magnetic behavior at low temperatures as the number of magnetic cations increases in the A', A'', B', and B'' sites. Depending on the ordering of cations, these compounds exhibit in different space groups. A site ordering of these double perovskites has had less interest as these materials are mostly unstable and need high-pressure synthesis. For the B site ordered double perovskite, we have rock salt, columnar, and layered arrangement depending on how the cations are arranged. Different magnetic properties of the bulk and thin-film counterparts for size-dependent double perovskites are reported for La_2BMnO_6 (B = Ni, Co) NPs [14]. The interactions that came to action in the nanoparticle range, like superparamagnetism, super ferromagnetism, and spin-glass behaviors, are related to the changes in the magnetic properties in these NPs. Though similar materials, La_2NiMnO_6 (LNMO) and La_2CoMnO_6 (LCMO) show different results in the magnetic studies. The coercivity, remnant magnetization, and saturation magnetization values are higher for LCMO. There is no observed secondary magnetic transition for LCMO at a lower temperature, which is related to the disordered states in the double perovskite structure. The AC susceptibility values exhibit much more complex magnetic structures than LNMO NPs and may have antisite defects or a second phase with a different transition temperature. Both LNMO and LCMO NPs show the DC moment is lower than the bulk or thin films.

11.7.5 Spinels

Spinel ferrite NPs are attractive nanomagnets, as magnetic properties can be tuned by changing the particle shape and size. In cubic spinel ferrite NPs such as MFe_2O_4 where M = Mg, Mn, Fe, Co, Ni, Cu, and Zn, the magnetic performance can be recognized by the degree of inversion and defects present, a balance between the magnetic ordered and crystallographic ordered fraction of the nanomagnet, size distribution, and the mutual exchange interaction. Iron oxides are widely used in magnetic storage and magnetic resonance imaging applications. In magnetite Fe_3O_4, the coercivity, H_C, attains a maximum value (190 Oe) at critical size (76 nm), and above this critical value, H_C decreases. The cube-like Fe_3O_4 NPs change from a single- to multi-domain structure above this critical size [15]. The Fe_2O_3 nanomagnets show ferromagnetism for 55 nm size and exhibit superparamagnetism without hysteresis at 12 nm particle size, showing the size dependency of these compounds [16].

11.8 TUNING THE CRYSTAL STRUCTURE AND MAGNETIC PROPERTIES

As the size decreases to nano range, the number of surface atoms and interface atoms increases that generate stress or strain in the crystal lattice, which leads to perturbations in the structure. There are various ways of describing the structural variations of nanomagnets. We will discuss the effects of structural and magnetic modifications of the nanomagnets in general in this section. The decrease of particle size will lead to the decrease of magnetic anisotropy that further induces superparamagnetism in the nanomaterials [17]. The effects of doping and temperature, pressure, and method of preparation will change the nanoparticle size and their properties in significant ways. The final history of the crystal structure of nanomagnets is a result of its thermal history of synthesis which involves both kinetic and thermodynamic factors.

11.8.1 Effects of Method of Preparation

The presence of defects, mainly oxygen vacancies, depends on the synthesis condition. These defects can tune the magnetic properties of these nanomagnets. The factors that cause a decrease of the value of saturation magnetization are surface effects, defects, and interparticle interactions [18]. These factors are responsible for the enhancement of effective magnetocrystalline anisotropy, which shows an increase in the hysteresis losses [19]. All these factors are closely linked to the crystalline structure of the nanomagnets and also the synthesis protocol. Polyol, sol-gel, or microemulsion methods are useful and better methods for the doped nanomagnetic samples. These methods confirm the homogeneity of the doping and are also capable of synthesizing NPs with low particle size distribution.

The degree of inversion (δ) of the spinel structure of ferrite nanomagnets depends on the method of preparation of these NPs. A mixed spinel structure is observed with a degree of inversion of δ value ~0.6 in $NiFe_2O_4$ NPs prepared by the sol-gel method [20]. In the perovskite nanomagnets, the sintering temperature and cooling time greatly influence the ordering of ions in the lattice. Also, it induces antisite defects within the lattice. Hence the choice of exact preparation method, the optimization of sintering temperature, and the setting of cooling time are extremely important to attain a highly ordered crystal lattice. Organic ligands can be used to stop crystal growth on the nanometer scale. Similarly, capping ligands can reduce surface defects in the nanomagnetic structure of halide perovskites. In the solution methods of preparation of nanomaterials, many factors must be taken into consideration, such as solubility, solvent compatibility, cost, purity, and toxicity, as well as the choice of anions that are considered inert.

11.8.2 Effects of Doping

Doped, undoped, and capped ZnO NPs exhibit room-temperature ferromagnetism. The crystal structure of a ZnO nanoparticle can be tuned with doping of Co that further changes the observed magnetic properties. The lattice parameters and hence the unit cell volume were found to be decreased with the Co doping. The maximum ferromagnetism observed for the dopant concentration is very low for $x = 0.025$ [21]; however, this concentration of dopant is very low for the justification of room-temperature ferromagnetism by any of the exchange mechanisms. The correlation between structure and magnetism in $Zn_{1-x}Co_xO$ nanomagnets can be identified with the help of EPR. Also, another reason for the change in magnetic properties is the chemical oxidation or reduction of the doped Co species in the ZnO. Magnetic studies on Co-doped ZnS indicated ferromagnetic and diamagnetic behavior. At lower Co concentrations, 0% and 1%, it shows paramagnetic behavior, and mixed ferromagnetic and paramagnetic properties at 5% and 10% Co concentrations in the ZnS matrix [22].

In perovskites and double perovskites, when the A site or B site is doped with other elements, changes are made in the unit cell parameters. As a result of this, octahedral distortions or tilting will happen in the crystal structure. These changes will affect the exchange interactions within the nanomagnetic perovskite structure that will make variations in the Curie temperature and order of spins within the lattice. For example, in the R_2NiMnO_6 double perovskites, as the R site is doped with various rare earth elements from La to Lu, the rare earth ionic radii decrease that will lead to the decrease in the Ni-O-Mn bond angle. This decrease of bond angle will result in a reduction in super-exchange interaction in the material. Hence the magnetic curie temperature will reduce to a lower value. Hence doping will induce changes in the structure as well as magnetic properties of these nanomagnets.

An effective method to increase the magnetic anisotropy of spinel ferrites is to dope it with large cations – for example, La-doped $CoFe_2O_4$ [23] and Ce-doped $NiFe_2O_4$ [24]. Fe doping in SnO_2 increases the crystallite size and leads to the uniformly dispersed spherical-like morphology since Fe ions increase the grain agglomeration and cluster formation. The change in the Fe concentration results in the formation of oxygen vacancies (O_v). The pure SnO_2 is nonmagnetic even in the presence of oxygen vacancies, while Fe-doped SnO_2 NPs exhibit weak ferromagnetism [25]. Bound magnetic polaron mechanism is behind the role of Fe and O_v. Gd doping in magnetite influences the crystal growth and magnetic properties of the magnetite NPs. The 5% of Gd-doped magnetite NPs exhibited ferrimagnetic properties with small coercivity at 260 K compared to the undoped, superparamagnetic magnetite NPs [26].

11.8.3 Effects of Pressure

Since nanomaterials are influenced by environmental factors such as temperature and pressure, the study of nanomagnets under high pressure will expand the properties of these materials. The application of pressure will modify the interatomic interaction and explore the nanoscale physical and chemical interactions which are linked with the magnetic properties. Since high pressure can vary the interatomic distances, one can study relations between the structure and magnetic properties of these materials.

Pressure is an important thermodynamic variable to alter the crystal structure and chemical bonding of nanomagnets. Chemical pressure and hydrostatic pressure will induce modifications in the crystal structure. As the chemical pressure increases, that leads to the change in bond length and bond angle and consequently varies the interparticle exchange interaction, which will result in the change in magnetic properties of double perovskites.

The main effects that occur on nanomagnets due to high pressure are:

(1) Modification of nanostructural elements,
(2) Modifications of the interaction among nano-objects, and
(3) Tuning of interactions among the nano-object and pressure transmitting medium.

In the case of magnetite NPs, pressure influences the size of nanomaterial. This is because of the very high surface-to-volume ratio supersaturation and surface tension of these nanomagnets. At higher pressures, crystallization from a homogeneous supersaturated solution will lead to a change in the Gibbs free energy that leads to an increase of the size of NPs.

11.8.4 Effects of Temperature

The magnetic ordering in nanomagnets is dependent on the annealing temperature and how slowly it is cooled. The mixed-valence manganese oxides $La_{1-x}A_xMnO_3$ (A is a divalent ion like Ca, Sr, Ba) are of special interest because of their magnetocaloric properties and negative magnetoresistance. In this class of compounds, the crystal structure is modified with heat treatment. After annealing at 600°C for 2 h the crystal structure changes from orthorhombic *Pnma* to rhombohedral $R\bar{3}c$. The saturation magnetization and the Curie temperature were observed to increase with annealing. The Curie temperature is more than doubled after annealing at 600°C for 2 h [27]. An FCC study of core-shell Au/Fe_3O_4 NP reveals an exchange bias effect [28]. This exchange bias started to appear at ~40 K. As the temperature decreases below 40 K, the bias increases. The interaction of spins located in the ordered region of the Fe_3O_4 shell and the magnetically disordered regions (in the inner and outer surface of the Fe_3O_4 shell) is responsible for this exchange bias effect.

At low temperatures, interaction of the electron spins can be strong enough to align them parallel to each other (ferromagnetic alignment). There has been little variation in the magnetic susceptibility with increasing temperature when the compounds become metal-like Pauli paramagnetic compounds. During some antiferromagnetic phases, at low temperature, the collinearity disappears, and any misalignment of spins by a minute amount leads to canted orientation.

11.9 RECENT DEVELOPMENTS IN NANOMAGNETS IN CORRELATION WITH STRUCTURE

The technological applications of these nanomagnets are in the growing stage, and various applications of these materials are in the domain of spintronics and magnetic storage devices. In the case of perovskites and double perovskites, most of the interesting and magnetic properties of nanomagnets are exhibited at very low temperatures, which inhibits the application of these materials at room temperature. Recently, researchers are focusing on room-temperature magnetic materials. Chemical co-doping, alloying, and high-pressure synthesis are some of the possible solutions for retaining room-temperature magnetism in these nanomaterials. Understanding the structural correlation to the magnetic properties is important to achieve nanomagnetism-based devices.

Size reduction for smaller devices and understanding the basic mechanisms of magnetic interactions of these nanomagnets are the two important aspects of research today. In-depth study of magnetic configurations and the magnetoresistive response of nanostructures, along with study on novel techniques like magneto-optic microscopy, magnetic force microscopy, polarized photoelectron spectroscopy, or magneto transport, are prime branches of contemporary intense research focus.

Different interesting magnetic structures like nanomagnetic skyrmions etc. give rise to interesting physics and devices. Topological stability and spin-polarized currents provide promising applicability in novel nanodevices. But attaining control over their geometrical positions is highly difficult but needed for implementation of these structures on devices. Recent advancements in the area of opto-magnetic materials, in which optical functionalities are combined with magnetic properties, combine ferromagnetism or spin transitions with luminescence or photo induced phase transitions. Single-molecule magnets (SMMs) are another recent example showing magnetic hysteresis loops due to slow magnetic relaxation.

11.10 CONCLUSIONS AND OUTLOOK

The basics and origin of nanomagnetism and magnetic interactions in NPs have been discussed. This chapter also stresses the structural correlation to the magnetic properties for different crystal structures. Synthesis conditions, temperature, and pressure affect the structure, morphology, size, and surface effects of the NPs. These in turn will influence the magnetic properties exhibited by the nanomagnets in various ways. Such modifications in magnetism due to changes in structure, morphology, size, etc. can be utilized to modify applicability in memory devices, ultra-spintronic devices, magnetic sensing, high-density data storage, high-energy magnetic materials, etc. Recent developments include size reduction, structure correlated magnetic functionalities, magnetricity, etc. A huge possibility of exploration lies in the field of structural engineering and its correlation to magnetism, which leads to novel device fabrication regimes. This chapter is an attempt to guide the reader towards understanding the importance of these areas and how slight variations in the crystal structure affect the magnetism of these novel nanomagnet domains.

11.11 ACKNOWLEDGMENTS

The first author MP acknowledge the Government of India for Prime Minister Fellowship (PMRF - 2101307) and SCB the Department of Science and Technology (DST, Govt. of India) for Inspire fellowship (IF190617). The corresponding author acknowledges the Department of Science and Technology (DST, Govt. of India) for research grant under the AMT project (DST/TDT/AMT/2017/200), and EGR acknowledges the Department of Science and Technology (DST, Govt. of India) for financial support under the Women Scientist Scheme-A (SR/WOS-A/PM-99/2016 (G)).

REFERENCES

[1] Alberto Passos Guimaraes, "The origin of nanomagnetic behavior," in *Principles of Nanomagnetism*, P. Avouris B. Bhushan D. Bimberg K. von Klitzing H. Sakaki R. Wiesendanger, Eds. Springer, 2009.

[2] W. F. Brown, "Theory of the approach to magnetic saturation," *Physical Review*, vol. 58, no. 8, p. 736, Oct. 1940.

[3] C. J. Bradley and A. P. Cracknell, *The Mathematical Theory of Symmetry in Solids. Representation Theory for Point Groups and Space Groups*, Oxford University Press, 1972.

[4] Qiu Jiao Ming, Wang Jian Ping, J. P. Wang, and J. M. Qiu, "Tuning the crystal structure and magnetic properties of FePt nanomagnets," *Advanced Materials*, vol. 19, no. 13, pp. 1703–1706, Jul. 2007.

[5] E. Folcke, J. M. le Breton, W. Lefebvre, J. Bran, R. Larde, and F. Golkar, "Investigation of the magnetic properties of FeAu nanoclusters in a W matrix: Evidence for exchange-bias phenomenon," *Journal of Applied Physics*, vol. 113, no. 18, p. 183903, May 2013.

[6] Laurent Sophie, Forge Delphine, Port Marc, Roch Alain, Robic Caroline, and vander Elst Luce, "Magnetic iron oxide nanoparticles: Synthesis, stabilization, vectorization, physicochemical characterizations and biological applications," *Chemical Reviews*, vol. 108, no. 6, pp. 2064–2110, Jun. 2008.

[7] Jiri Tucek, Libor Machala, Shigeaki Ono, Asuka Namai, Marie Yoshikiyo, and Kenta Imoto, "Zeta-Fe 2 O 3 – A new stable polymorph in iron (III) oxide family," *Scientific Reports*, vol. 5, no. 1, p. 15091, 2015.

[8] Khurshid Hafsa, Lampen Kelley Paula, Iglesias Oscar, Alonso Javier, and Phan Manh-Huong, "Spin-glass-like freezing of inner and outer surface layers in hollow Γ-Fe2O3 nanoparticles," *Scientific Reports 2015 5:1*, vol. 5, no. 1, pp. 1–13, Oct. 2015.

[9] Daisuke Nakatsuka, Takashi Yoshino, Jun Kano, Hideki Hashimoto, Makoto Nakanishi, and Jun Takada, "High-pressure synthesis, crystal structure and magnetic property of ilmenite-type FeGeO3," *Journal of Solidstate Chemistry*, vol. 198, pp. 520–524, 2013.

[10] N Moloto, M J Moloto, M Kalenga, S Govindraju, and M Airo, "Synthesis and characterization of MnS and MnSe nanoparticles: Morphology, optical and magnetic properties," *Optical Materials*, vol. 36, no. 1, pp. 31–35, 2013.

[11] Jayaraman Theerthagiri, K Karuppasamy, Govindarajan Durai, Abu Ul Hassan Sarwar Rana, Prabhakarn Arunachalam, and Kirubanandam Sangeetha, "Recent advances in metal chalcogenides (MX; X= S, Se) nanostructures for electrochemical supercapacitor applications: A brief review," *Nanomaterials*, vol. 8, no. 4, p. 256, 2018.

[12] Lin Shan, Zhang Qinghua, Sang Xiahan, Zhao Jiali, Cheng Sheng, and Huon Amanda, "Dimensional control of octahedral tilt in SrRuO3 via infinite-layered oxides," *Nano Letters*, vol. 21, no. 7, pp. 3146–3154, Apr. 2021.

[13] S. G. Bahoosh and J. M. Wesselinowa, "The origin of magnetism in perovskite ferroelectric ABO 3 nanoparticles (A=K,Li; B=Ta,Nb or A=Ba,Sr,Pb; B=Ti)," *Journal of Applied Physics*, vol. 112, no. 5, p. 53907, Sep. 2012.

[14] Yuanbing Mao, Jason Parsonsa, and John S. McCloy, "Magnetic properties of double perovskite La 2 BMnO 6 (B= Ni or Co) nanoparticles," *Nanoscale*, vol. 5, pp. 4720–4728, 2013.

[15] Qing Li, Christina W Kartikowati, Shinji Horie, Takashi Ogi, Toru Iwaki, and Kikuo Okuyama, "Correlation between particle size/domain structure and magnetic properties of highly crystalline Fe3O4 nanoparticles," *Scientific Reports 2017 7:1*, vol. 7, no. 1, pp. 1–7, Aug. 2017.

[16] Young wook Jun, Jung wook Seo, and Jinwoo Cheon, "Nanoscaling laws of magnetic nanoparticles and their applicabilities in biomedical sciences," *Accounts of Chemical Research*, vol. 41, no. 2, pp. 179–189, Feb. 2008.

[17] Michael T. Klem, Damon A. Resnick, Keith Gilmore, Mark Young, Yves U. Idzerda, and Trevor Douglas, "Synthetic control over magnetic moment and exchange bias in all-oxide materials encapsulated within a spherical protein cage," *Journal of the American Chemical Society*, vol. 129, no. 1, pp. 197–201, Jan. 2007.

[18] A. G. Roca, D. Niznansky, J. Poltierova Vejpravova, B. Bittova, M. A. Gonzalez Fernández, and C. J. Serna, "Magnetite nanoparticles with no surface spin canting," *Journal of Applied Physics*, vol. 105, no. 11, p. 114309, 2009.

[19] M. Vasilakaki, C. Binns, and K. N. Trohidou, "Susceptibility losses in heating of magnetic core/shell nanoparticles for hyperthermia: A Monte Carlo study of shape and size effects," *Nano Scale*, vol. 7, no. 17, pp. 7753–7762, 2015.

[20] Muhammad Atif, Muhammad Kashif Nadeem, and Muhammad Siddique, "Cation distribution and enhanced surface effects on the temperature-dependent magnetization of as-prepared NiFe2O4 nanoparticles," *Applied Physics A: Materials Science and Processing*, vol. 120, no. 2, pp. 571–578, Aug. 2015.

[21] Jordan Chess, Gordon Alanko, Dmitri A. Tenne, Charles B. Hanna, and Alex Punnoose, "Correlation between magnetism and electronic structure of Zn1–xCoxO nanoparticles," *Journal of Applied Physics*, vol. 113, no. 17, pp. 17–302, May 2013.

[22] Sunil Kumar and N. K. Verma, "Room temperature magnetism in cobalt-doped ZnS nanoparticles," *Journal of Super Conductivity and Novel Magnetism*, vol. 28, pp. 137–142, 2015.

[23] S Burianova and Niznansky Daniel, "Surface spin effects in La-doped CoFe2O4 nanoparticles prepared by microemulsion route," *Journal of Applied Physics*, vol. 110, no. 7, p. 073902, Oct. 2011.

[24] Gagan Dixit, J. P. Singh, R. C. Srivastava, and H. M. Agrawal, "Structural, optical and magnetic studies of Ce doped NiFe2O4 nanoparticles," *Journal of Magnetism and Magnetic Materials*, vol. 345, pp. 65–71, Nov. 2013.

[25] Mayuri Sharma, Shalendra Kumar, Rezq Naji Aljawfi, S Dalela, S N Dolia, and Adil Alshoaibi, "Role of fe-doping on structural, optical and magnetic properties of SnO 2 nanoparticles," *Journal of Electronic Materials*, vol. 48, no. 12, pp. 8181–8192, Dec. 2019.

[26] Honghu Zhang, Vikash Malik, Surya Mallapragad, and Mufit Akin, "Synthesis and characterization of Gd-doped magnetite nanoparticles," *Journal of Magnetism and Magnetic Materials*, vol. 423, pp. 386–394, 2017.

[27] Zentkova M, Antonak M, Mihalik M, Mihalik M, Vavra M, and Girman V, "Effect of doping and annealing on crystal structure and magnetic properties of La1-xAgxMnO3 magnetic nanoparticles," *Low Temperature Physics*, vol. 43, no. 8, pp. 990–995, Aug. 2017.

[28] L. Leon Felix, J. A. H. Coaquira, M. A. R. Martínez, G. F. Goya, J. Mantilla, and M. H. Sousa, "Structural and magnetic properties of core-shell Au/Fe 3 O 4 nanoparticles," *Scientific Reports*, vol. 7, p. 41732, 2017.

Magnetism in Monoatomic and Bimetallic Clusters
A Global Geometry Optimization Approach

12

J. L. Morán-López[a,b], A. P. Ponce-Tadeo,[b] and J. L. Ricardo-Chávez[c]

a Advanced Materials Division, Instituto Potosino de Investigación Científica y Tecnológica, San Luis Potosí, S.L.P. México

b Group for Computational Science and Engineering, National Supercomputer Center, Instituto Potosino de Investigación Científica y Tecnológica, San Luis Potosí, S.L.P. México

c Laboratorio Nacional de Supercómputo and Facultad de Ciencias Físico Matemáticas, Benemérita Universidad Autónoma de Puebla, Calle 4 Sur Número 104, Colonia Centro Histórico, Puebla, PUE, México

Contents

12.1 Introduction	226
12.2 Determination of Cluster Structure Using Graph Theory	228
12.2.1 Topological Cluster Structure	229
12.2.2 Generation of a Complete Set of Topological Structures	230
12.2.3 Solving the Distance Geometry Problem	231
12.2.4 Applying the Topological Structure Concept to Larger Clusters	232
12.3 Electronic and Magnetic Structure	232
12.4 Monoatomic Clusters	233
12.5 Bimetallic Clusters	235
12.6 Conclusions	240
12.7 Acknowledgments	241
12.8 References	241

DOI: 10.1201/9781003197492-12

12.1 INTRODUCTION

The understanding of the transition, from atomic to bulk properties, passing through atomic clusters, thin films, layered heterostructures, and surfaces of the physicochemical properties of transition metal systems, continues to be one of the most studied topics, experimentally as well as theoretically. This knowledge is important from fundamental knowledge of nanostructured materials as well as their current and potential technological applications [1–2]. It is well known that systems with dimensions in the nanometer scale show special properties. For example, their use as catalyzers is now a subject of many applications in many industrial processes, and their special magnetic properties are of common use in many applications, such as ultrahigh-density hard disks [3]. Further advances to reach areal recording densities to values above the Tbit/in^2 range are still necessary. For recording at that level, it is required to develop new nanostructured materials with high magnetic anisotropy that guarantee stability at high enough temperatures.

Among the various options, monoatomic and bimetallic transition metal nanoclusters are excellent candidates; one can modify their properties through the number of atoms and the cluster chemical composition. These two features, at a given temperature, define the lowest-energy geometrical structure of the aggregate and the way in which the components spatially distribute. These characteristics, in turn, determine the electronic distribution that depends on their local geometrical and atomic environment [4]. Furthermore, under particular conditions, the systems may show atomic magnetic moments with different magnitudes and orientations. Applications in ultrahigh magnetic storage and medical applications have inspired intense research in ferromagnetic bimetallic nanoparticles. Attractive systems are FePt nanoparticles; they show magnetocrystalline anisotropy energy (MAE) in the range of 7×10^7 ergs/cm^3 [5]. It has been observed that the Pt spin-orbit coupling (SOC) produces such a large MAE [6] that may keep nanoparticles, of less than 10 nm, thermally stable at room temperatures. In a recent publication, the synthesis of monodispersed FePt nanoparticles, produced by a gas phase condensation method, was reviewed [7]. The author accounts for the successful production of self-assembled nanoparticles that order magnetically; this system is known as self-organized-magnetic-array (SOMA) patterned media. The particular bulk properties shown by FePt alloys make bit patterned media of extreme interest [7]. The strong chemical FePt interaction rules the high bulk order-disorder temperature. Thus, highly ordered bimetallic low dimensional systems are also expected. Furthermore, based on the high Curie temperatures observed in the ordered as well as in the disordered bulk phases, one foresees ferromagnetic self-assembled nanoparticle systems to be used in ultrahigh-density recording media. This is further supported by the fact that magnetocrystalline energy, in general, gets enhanced in low-dimensional systems.

The structure stability and magnetic properties of monoatomic transition metal clusters have been extensively studied. Experimentally, the determination of the structure and the measurement of the magnetic properties, as a function of size, are complex task. The report of the magnetic moment of Fe$_n$ nanostructures with sizes between 25 and 700 atoms was reported [8]. The researchers observe small oscillations around $\mu = 3.0$ μ_B in the range of $25 \leq n \leq 120$. For higher values, there is a smooth decrease up to the bulk value. From the theoretical point of view, the calculation of the physicochemical properties of magnetic nanostructures is also a very complicated and computationally demanding problem. Density functional theory (DFT)-based studies are most reliable [9–11]. Within the DFT formalism, one can include static correlations to compute to a good approximation, structural stability, and electronic and magnetic properties. Köhler et al. [10] reported in a tight binding based DFT, the structures and magnetic moments of Fe$_n$, for n $\leq$ 32. They observed that for clusters with n $\leq$ 20, a magnetic moment per atom that oscillates around 3 μB, with a minimum value of 2.6 μ_B in the 13-atom icosahedral cluster. This low value is produced by an antiferromagnetic coupling between the central and surrounding atoms.

Experimental studies of pure Pt clusters, carried out in nanoparticles of several hundreds of atoms embedded in polymers, reported a negligible average magnetic moment of 0.012 μ_B per atom [12]. In smaller clusters, with 13 atoms, an average magnetic moment of 0.65 μ_B per atom was determined [13].

From the theoretical point of view, based on the fact that the bulk crystal structures of most transition metal systems are compact lattice structures, it was expected that transition metal clusters should form polyhedral atomic arrangements [14]. However, it was found that this is not always the case; Pt small clusters form planar structures [15]. Although it is not yet completely clear, there is a consensus that strong spin-orbit interactions favor the existence of planar structures in clusters with a very small number of atoms. The transition between 2D and 3D geometrical structures depends on the theoretical model; by using a density functional formalism with ultrasoft pseudo potentials, the planar structures go up to 4 atoms [16], and it increases to 10 when the augmented projected wave pseudo potentials and plane-wave basis set [17] is implemented. Furthermore, in a recent DFT publication [18], the lowest energy geometrical structure of the Pt heptamer was found to be a planar with an average magnetic moment of 0.571 μB.

More interesting but at the same time much more difficult to handle are bimetallic clusters [19]. By adding a second chemical element, two additional parameters play an important role; the composition and the spatial distribution of the components. They offer the possibility of synthesizing a wider set of novel materials with new physicochemical properties. The scenario gets even richer when bimetallic clusters contain magnetic elements. As mentioned prior, due to their potential applications, those systems are presently the subject of intense research. However, to calculate the properties of bimetallic magnetic nanoparticles, one is confronted with various problems. The most relevant are the following: first, one needs to determine the geometrical structure. The exact location of the atoms rules the physical properties. Due to the smallness of the systems, minor modifications to the geometrical characteristics may change their properties dramatically. In addition, in bimetallic systems, the chemical distribution of the components plays an important role in determining the cluster geometry. Thus, it is necessary to perform self-consistent calculations that take into account simultaneously the structural and chemical degrees of order. With the current computational resources, it is not possible to study clusters with a large number of atoms. The problem arises from the fact that the number of homotops (clusters with the same size, composition, and geometrical structure but with different spatial arrangements of A and B atoms) grows exponentially. To illustrate the magnitude of the complexity, let's take the case of the number of homotops in a bimetallic cluster A_nB_{13-n}. They are, according to the binomial distribution, 13, 78, 286, 715, 1287, and 1716, for n = 1, 2, 3, 4, 5, and 6, respectively. Many of those configurations are equivalent; the numbers of non-equivalent homotops for the same values of n reduce to 2, 4, 8, 13, 16, and 20 for an icosahedral structure and 2, 5, 13, 18, 31, and 34 if the 13-atom bimetallic particle has a cubo-octahedral geometry [20]. It is important to note that the number of non-equivalent homotops depends on the geometrical structure.

In particles with magnetic elements, in addition, one has to calculate the local spin and the orbital magnetic moments. The last one can be ignored if the spin-orbit coupling is weak. Then, furthermore, the energy has to be calculated for the different relative orientations of the magnetic moments: ferromagnetic, antiferromagnetic, or non-collinear. In ferromagnetic nanostructures, the most stable configuration is collinear. However, in clusters in which the lowest energy configuration involves antiferromagnetism and frustration phenomena are present, the non-collinear arrangements are favored [21–23].

Theoretical studies of FePt nanoclusters within the density functional theory, have been published recently [18, 24–25]. Gruner et al. [24] performed the calculation of the largest clusters to date. By means of the Vienna ab initio simulation package (VASP) code, they calculated the total energies of closed shell structures with 13, 55, 147, 309, and 561 sites with icosahedral, cubo-octahedral, and decahedral geometries. In order to perform such a huge calculation, they fixed the geometry and the chemical order and ignored spin-orbit coupling. As a result, they found as ground state icosahedra: for the 13-atom cluster, an ordered arrangement of Pt and Fe atoms, and for the larger clusters, an icosahedral structure with alternating FePt shells. The calculated average atomic magnetic moment per atom of the 13-atom particle was 1.48 μ_B and for the larger clusters approximately 1.72 μ_B per atom, respectively.

In more recent papers, it has been shown that the spin-orbit interaction is important in Pt particles and cannot be ignored. However, to perform such a calculation, one still has to restrict to particles with a small number of atoms. Alvarado-Leyva et al. [18] calculated the Pt-heptamer and FePt$_6$ including SOC. They reported a planar structure for Pt$_7$, but it evolves to a 3D structure when one of the Pt atoms is substituted by Fe. The total average magnetic moment found in those clusters was 0.64 and 0.88 μ_B, respectively.

Furthermore, Rodríguez-Kessler and Ricardo-Chávez [25] used a similar formalism and code to calculate the properties of Fe_nPt_m for N = n + m < 6. They reported the lowest energy structures for the dimer and up to the hexamer for all the compositions. Pure Pt particles adopt planar structures with average magnetic moments that go from 1 μ_B (N = 2) to 0.66 μ_B (N = 6). On the other hand, in the case of pure Fe clusters, the lowest energy structures for N = 4, 5, and 6 are 3D, and the average magnetic moment goes from 3 μ_B to 3.33 μ_B for N = 2 to 6.

Chittari and Kumar [26] published, even more recently, an extensive study of FePt nanoclusters within the VASP formalism but also ignored SOC. Their results for the heptamer $Fe_{7-n}Pt_n$, for all concentrations n are the following. The pure Fe and Pt heptamer geometrical structures are a pentagonal bipyramid and a planar side capped double square, respectively. The reported average magnetic moments of pure Fe and Pt heptamers are 3.2 and 0.57 μ_B, respectively. The geometrical and magnetic moment results for the rest of the $Fe_{7-n}Pt_n$ clusters are: the Fe_6Pt cluster is an octahedral Fe particle capped with the Pt atom on a triangular face. The average Fe magnetic moment is $\bar{\mu}_{Fe}$ = 3.21 μ_B, and the magnetic moment in the Pt atom is μ_{Pt} = 0.7 μ_B. In the case of Fe_5Pt_2, the lowest energy structure is an irregular pentagonal bipyramid where the average magnetic moments are $\bar{\mu}_{Fe}$ = 3.31 and $\bar{\mu}_{Pt}$ = 0.72 μ_B, respectively. For Fe_4Pt_3 the structure is a Fe tetrahedron capped with three Pt atoms; the average magnetic moments are $\bar{\mu}_{Fe}$ = 3.44 and $\bar{\mu}_{Pt}$ = 0.774 μ_B, respectively. For Fe_3Pt_4 a similar structure is found in which one of the Fe atoms is substituted by Pt. In this case, the average magnetic moments are $\bar{\mu}_{Fe}$ = 3.33 μ_B and $\bar{\mu}_{Fe}$ = 0.5 μ_B. In Fe_2Pt_5 a pentagonal bipyramid structure is reported, with $\bar{\mu}_{Fe}$ = 3.39 μ_B and $\bar{\mu}_{Fe}$ = 0.64 μ_B. Finally, for $FePt_6$ a square pyramid capped with two Pt atoms is found. The corresponding average magnetic moments are $\bar{\mu}_{Fe}$ = 3.51 μ_B and $\bar{\mu}_{Fe}$ = 0.53 μ_B.

In this chapter, we discuss the results for the ground state geometries and magnetic moments in monoatomic Pt_n clusters for n ≤ 13; we believe that this is the most complete study up to date. We also present results of particular bimetallic clusters; first, those platinum clusters with one single Fe atom of the type $FePt_n$, n = 2–6, and later the heptamer bimetallic cluster, $Fe_{7-n}Pt_n$ with n = 0, 1, 2, . . . 7. By means of global geometry optimization, we explored all possible configurations and calculated the total energies within the DFT framework, including spin-orbit interactions. Thus, we report the ground state configurations and the electronic and magnetic properties.

12.2 DETERMINATION OF CLUSTER STRUCTURE USING GRAPH THEORY

The stable geometries of atomic clusters of size N are sets of coordinates $R_i = \{r_1^i, r_2^i, \ldots, r_N^i\}$ -with i ranging from 1 to an unspecified number N_g, which locally minimize the potential energy surface (PES) of the collection of atoms [27–28]. One of these geometries, having energy E_0, is the global minimum of the PES, and as energy increases by an amount $\Delta E=k_BT$ with respect to E_0, a certain number of low-lying geometries with energies comparable to $E_0 + \Delta E$ become also relevant for the physical and chemical properties of the system. From a fundamental point of view, there is no known way to compute the whole set of non-equivalent geometries $\{R_i, i=1,2, \ldots, N_g\}$ where N_g is the number N of atoms and their chemical species [29–31]. This is due to the fact that the problem of computing all local minima on the PES can be mapped to the problem of computing all undirected simple graphs of order N, which is known to be non-deterministic polynomial-time hard (NP-hard) for large enough values of N [32–34]. Aside from the mathematical difficulties, the problem of locating the relevant minima on the PES has been shown to be a considerable computational challenge, not only due to a large number of minima to be searched for large N but also due to the lack of clues about the position of these minima on the vast PES. A thorough non-biased and highly non-local sampling of the PES must therefore be carried out in order to discover the global minimum and low-lying minima [35]. There is no known efficient way to perform this task, although many methods for sampling the PES with variable degree of success have been developed in the last decades [29–31]. Nevertheless, in the case of very small clusters with $N = 8$ atoms (and perhaps beyond this number, as we show later), it is possible to take advantage of the mapping between geometrical arrangements and graphs to generate a nearly complete set of meaningful (high-quality, not redundant or impossible) trial geometries for the global optimization of cluster structure on the PES at moderate computational cost.

12.2.1 Topological Cluster Structure

We may disregard the distance dependence of the chemical bond and describe the bonding properties of clusters using the concept of topological structure, which can be defined as follows: a particular geometry having N atoms and M bonds can be represented by a $N \times N$ connectivity or adjacency matrix A whose elements take the values of,

$$A_{ii} = 0, \quad (1)$$

$$A_{ij} = \begin{cases} 1 & \text{if atom } i \text{ is bonded to atom } j \\ 0 & \text{otherwise} \end{cases} \quad (2)$$

Furthermore,

$$\sum_{\substack{i=2 \\ j<i}}^{N} A_{ij} = M \quad (3)$$

Note that all the adjacency matrices are symmetric since $A_{ij} = A_{ji}$ for $i \neq j$, then, only the $N(N-1)/2$ entries of the lower triangular part of the matrix are needed.

An adjacency matrix representing a topological structure of a cluster is equivalent to an undirected simple graph $G(N,M)$ where the atoms are vertices and the bonds are links or edges of the graph [36]. In order to avoid disconnected graphs (clusters with dissociated atoms), M goes from a minimum value $M_{min} = N - 1$ for the least compact clusters, and up to $M_{max} = N(N-1)/2$ for the most compact ones. For each value of M there are up to $(M_{max})! / [(M_{max} - M)! (M)!]$ graphs. However, most of these graphs are either isomorphic, i.e., equivalent under a permutation of the indices of vertices, or disconnected (having entire rows of zeros). Considering the particular case of $N = 4$, Table 12.1 shows the total number and the number of non-equivalent graphs as a function of M, as well as a graphical representation of the corresponding topological structures, using the same distance for all bonds. Note that without making reference to a specific electronic structure method to compute the energy and vibrational frequencies, we cannot determine whether the topological structures are true local minima or transition states or even if the structures are planar or three dimensional. The topological structures do not contain enough information to elucidate these properties. Due to this reason, the graphical representation of the non-equivalent graphs shown in Table 12.1, suggesting that the graphs are highly symmetric, must not be regarded as significant.

For instance, if we consider the tetrahedron ($M = 6$), it is clear that it cannot be in general a local minimum because it depends on the constituent atoms due to the Jahn-Teller effect [36]. Therefore, in order to take into account the possibility of symmetry-breaking deformations, we determine the symmetry of the trial geometries obtained from the topological structures and artificially break them, generating more meaningful lower-symmetry trial geometries in addition to the highly symmetric ones. Through the concept of topological structure, we can easily remove bonds between pairs of atoms i and j by setting the adjacency matrix element A_{ij} to zero. In this way, we generate a parent graph from the original graph. If allowed, i.e., if it's not a disconnected or isomorphic graph, the parent graph constitutes a neighbor of the original graph in topological space. Starting from the most compact structure and subsequently removing bonds, we can discover the paths linking all the topological structures (as an example, the left part of Figure 12.1 shows the links among the topological structures shown in Table 12.1).

TABLE 12.1 Total Number of Graphs and Number of Non-Equivalent Graphs as a Function of M for $N = 4$ Atoms; a Graphical Representation of the Topological Structures Corresponding to the Non-Equivalent Graphs Is Shown in the Last Column

M	TOTAL	NON-EQUIVALENT	GRAPHICAL REPRESENTATION
3	20	2	
4	15	2	
5	6	1	
6	1	1	

We can also employ a simple electronic structure method like the topological Hückel model [36] to infer the relative structural stability, as shown in the right part of Figure 12.1. Nevertheless, as we noted before, working in topological space, it is not possible to determine whether a structure corresponds to a true minimum or a transition state, so it is not possible to compute activation energies through barriers. We will show later that, once we obtain a set of coordinates, the nature of each structure can be found as well as the reaction paths linking them on the PES.

12.2.2 Generation of a Complete Set of Topological Structures

The computational complexity of generating the complete set of non-equivalent graphs of order N [5] can be reduced significantly by using an incremental approach. Starting from the whole set of non-equivalent graphs of order $N - 1$, we take these graphs, one by one, and generate graphs of order N by adding a new vertex and computing the non-equivalent graphs resulting from all possible number of connections the new vertex can have to the existing ones. This process may be accomplished by resizing the adjacency matrix to dimension $N \times N$ while keeping the existing $(N - 1) \times (N - 1)$ core intact and computing all the ways that connections 1, 2, . . ., $N - 1$, connections may be distributed in the $N - 1$ sites of the last row. In the very beginning, we take the graph of order $N = 2$ (the dimer), which has the adjacency matrix and generate from it the following graphs of order $N = 3$:

$$\begin{pmatrix} 0 & 1 \\ 1 & 0 \end{pmatrix}, \tag{4}$$

$$\begin{pmatrix} 0 & 1 & 1 \\ 1 & 0 & 0 \\ 1 & 0 & 0 \end{pmatrix} \begin{pmatrix} 0 & 1 & 0 \\ 1 & 0 & 1 \\ 0 & 1 & 0 \end{pmatrix} \begin{pmatrix} 0 & 1 & 1 \\ 1 & 0 & 1 \\ 1 & 1 & 0 \end{pmatrix} \tag{5}$$

These graphs have to be analyzed for disconnectivity (entire rows of zeros) or isomorphism (equivalence under a permutation of the indices of vertices).

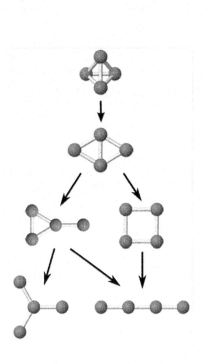

FIGURE 12.1 Left: Links among the non-equivalent topological structures for $N = 4$ atoms through the removal or creation of a single bond. Right: Structural stability according to the topological Hückel model.

One of the most reliable tests for isomorphism performs the comparison of the largest eigenvector of the connectivity matrix of two graphs (Perron-Frobenius theorem). The dominant eigenvector gives the popularity of each vertex and is independent of the labeling of the vertices [38]. Using this test, we conclude that the first two graphs of Eq. (5) are isomorphic, corresponding to the linear trimer, while the third graph is the triangle, as shown by the accompanying figures below each adjacency matrix. Note that the incremental approach to generating the graphs of order N from the non-equivalent graphs of order $N - 1$ reduces the number of graphs to be analyzed and compared since there is no need to generate all permutations of the indices of vertices.

12.2.3 Solving the Distance Geometry Problem

Once the set of N_g non-equivalent topological structures of size N have been computed, the next step consists in finding the coordinates $R_k = \{r_1^k, r_2^k, \ldots, r_N^k\}$ for each $k = 1, 2, \ldots, N_g$, subject to the restriction that the interatomic distances $\{d_{ij}^k = \| r_i^k - r_j^k \|; i = 2, \ldots, N; j = 1, \ldots, i - 1\}$ determined from these coordinates must be consistent with the adjacency matrix of the structure up to a reasonable tolerance. This procedure is called the distance geometry (DG) problem [39], which itself is also NP-hard [40] and may be formulated as a global minimization problem by introducing the objective function

$$f\left(\mathbf{r}_1^k, \mathbf{r}_2^k, \ldots, \mathbf{r}_N^k\right) = \sum_{\langle ij \rangle} w_{ij} \left(d_{ij}^2 - r_{\alpha\beta}^2\right)^2 \tag{6}$$

where $\langle ij \rangle$ denotes the set of i and j values $\{i = 2, \ldots, N; j = 1, \ldots, i - 1\}$ that have the elements of the adjacency matrix $A_{ij} = 1$, and $r_{\alpha\beta}$ are typical interatomic distances for atomic species i_α and j_β, up to a reasonable tolerance. The weights $0 \leq w_{ij} \leq 1$ are introduced for convenience in order to give more flexibility to the minimization process. When problem (6) is formulated using a complete set of interatomic distances, it can be solved exactly by a matrix decomposition method [41], but in general, we must work with a (sometimes very) restricted subset of distances. For instance, in the case of the linear chain of Table 12.1, we know only three distances out of a total of six. To solve the DG problem with a limited set of distances, several approaches have been developed [42]. One of the most successful and flexible approaches is the DGSOL method developed by More and Wu [43–44]. DGSOL is based on the global continuation of the objective function (6) by convolution with a Gaussian function, in order to smooth out the small (local) oscillations of the original function and make it convex. This process may fail if the problem (6) is not consistent, but generally, it ends at a local or global minima after a certain number of steps [43].

We have already mentioned that an advantage of working with topological structures is their loss of information on bond distance and atomic species, i.e., a topological structure may correspond equally to a homogeneous or alloy cluster. The difference is established once we solve the DG problem with appropriate interatomic distances and assign species to the atomic sites. Also, the highly symmetric geometries we may obtain from the topological structures could not be regarded as an inconvenience, since it is always possible to generate a set of lower-symmetry geometries from a highly symmetric one (enforcing the Jahn-Teller effect) by subsequently rescaling one by one every bond of the original geometry and solving the distance geometry problem with the new set of bond lengths. This procedure may be repeated again and again by rescaling with $2, 3, \ldots, M_{max}$ bonds at a time to finally obtain a totally asymmetrical geometry. In exactly the same way, it could be possible to discover the reaction paths connecting a local minimum with its neighbor minima by exploring all possible transition states.

12.2.4 Applying the Topological Structure Concept to Larger Clusters

After a certain limit of, usually $N = 9$ atoms, we cannot work anymore with a complete set of graphs due to the lack of enough man and computer power to enumerate all the graphs for larger values of N. However, we can take the set of stable geometries for $N = 9$ and compute the corresponding topological structures. These structures can then be taken as a starting point to generate the set of non-equivalent graphs for $N = 10$ which, although incomplete, may be good enough to determine the global and low-lying geometries. This process can be repeated again to generate the geometries for $N = 11$ and so on. Using the graph-theory techniques described in this section and with the current computational facilities, it is possible to determine with a high degree of confidence the most stable geometries of clusters up to $N = 20$ atoms.

12.3 ELECTRONIC AND MAGNETIC STRUCTURE

To calculate the energy of any particular cluster, the trial cluster geometries were relaxed self-consistently within the density functional theory without imposing any symmetry constraint and using the conjugated gradient algorithm until the force on each atom was less than 0.005 eV/Å. To solve the Kohn-Sham equations of density functional theory (DFT) using a plane-wave basis set, we used the VASP [45]. To describe the interaction between valence electrons and ionic cores, we performed the calculations by means of the projector-augmented wave (PAW) method [46] and treated the electronic exchange and correlation effects using the Perdew, Burke, and Emzerhof (PBE) gradient-corrected functional. It is well established that this function reproduces the correct structural and magnetic ground state properties of transition metals [47].

The VASP code calculates the electronic structure of infinite periodic systems; thus, in the case of clusters, in order to minimize the interaction between neighbor cluster images, they are placed in a cubic supercell with at least 10 Å of vacuum in all space directions. This isolation, combined with a plane-wave cutoff energy of 450 eV, represents a good compromise between precision and computing time. Furthermore, due to the large size of the supercell, only the Γ point is necessary to sample the Brillouin zone. Although the bulk face-centered cubic (fcc) Pt phase is paramagnetic with the negligible local magnetic moment, the small size and confinement effects increase the electronic interaction and localization; these characteristics of this particular characteristics yield strong magnetic states. In addition, the large Pt nuclear mass produces a strong spin-orbit coupling (SOC), and relativistic effects have to be considered. To take into account these phenomena, we performed the structural relaxation in two steps. First, a relaxation at the scalar relativistic (SR) level was carried out using the collinear spin formulation of DFT as implemented in VASP. We considered a representative set of total S_z values, for both ferromagnetic (FM) and antiferromagnetic (AF) alignments of the local spin moments. Then, at a second stage, a further geometrical relaxation was performed at a fully relativistic level including SOC (RSOC) and employing the non-collinear spin formulation of DFT as implemented by Hobbs et al. [48]. In this case, in order to search for the easy and hard magnetization axes, we tried several non-perpendicular directions of the spin quantization axis.

After the relaxation process for all the relevant magnetic configurations is performed, the relative stability of the local minima is inferred from the analysis of the binding energy per atom for each cluster size and Pt concentration n,

$$E_B = -\frac{E(Fe_n Pt_m) - nE(Fe) - mE(Pt)}{n+m}, \quad (7)$$

Here, E(Fe) and E(Pt) are the energies of Fe and Pt atoms, respectively. The case of pure Pt clusters corresponds to $n = 0$.

The local electronic and magnetic properties, like charge, spin, and orbital moments, at each atomic site depend on the particular environment and were computed in the Wigner-Seitz sphere of the atoms.

12.4 MONOATOMIC CLUSTERS

First, we present the results for the pure platinum clusters Pt_n for $n = 3$ and up to 13 atoms. The geometrical structure results are presented in Figure 12.2. Here, we show only the most stable configurations. For $n = 3$, the most stable geometry is a triangle. In the case of $n = 4$, we obtain a rhombus. For the pentamer, the lowest energy structure is a capped square. For the hexamer, we found as its ground state a planar triangular arrangement. All these clusters are bidimensional; the rest are tridimensional structures.

When we add an atom to the hexamer, two atoms are located at one of the triangle tip sites, displaced up and below the plane. For $n = 8$, the cluster consists of triangles and rectangles. In the case of a Pt cluster with nine atoms, the structure consists mainly of triangular arrangements. For $n = 10$, the ground state corresponds to a perfect triangular pyramid. For the rest of the clusters, the geometries are complex structures.

In Table 12.2, we present the results for the size dependence of the cohesive energy per atom (in electron volts), the range of values adopted by the atomic bonds (in angstroms), and the average total magnetic moment per atom (in Bohr magnetons). The binding energy per atom shows an increasing monotonic

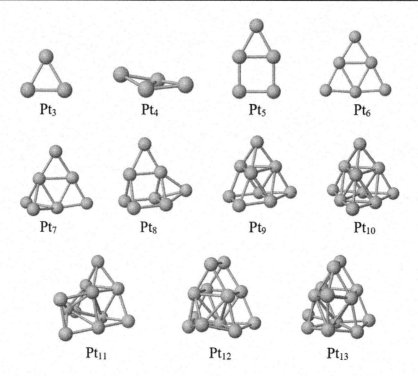

FIGURE 12.2 Lowest energy geometrical structures adopted by the monoatomic Pt clusters Pt$_n$ for n = 3 to 13.

TABLE 12.2 The Binding Energy E$_B$ (in Electron Volts), the Bond Distance Range (in Å), and the Average Total Magnetic Moment (in μ_B) for the Pure Pt$_n$ Clusters n = 3–13; the Calculation Is for the Non-Collinear Magnetic Arrangement Including Spin-Orbit Coupling

CLUSTER	SOC E$_{B(N)}$(eV)	L(Å)	μ_T (μ_B)
Pt$_3$	2.194	2.49–2.5	0.879
Pt$_4$	2.521	2.52–2.53	1.048
Pt$_5$	2.786	2.44–2.50	0.611
Pt$_6$	3.035	2.46–2.70	0.492
Pt$_7$	3.119	2.46–2.8	0.552
Pt$_8$	3.217	2.49–2.59	0.051
Pt$_9$	3.362	2.52–2.79	0.329
Pt$_{10}$	3.505	2.54–2.77	0.606
Pt$_{11}$	3.518	2.51–2.78	0.146
Pt$_{12}$	3.568	2.52–2.84	0.145
Pt$_{13}$	3.635	2.53–2.86	0.116

behavior: from 2.194 in the trimer to 3.635 eV in the 13-atom Pt cluster. The bond length varies from 2.44 Å in the pentamer to 2.86 Å in the 13-atom cluster. On the other hand, the individual magnetic moments of the Pt atoms depend on the cluster size and geometry, but the average total magnetic moment goes from 1.04 μ_B in the tetramer to 0.051 μ_B in the 8-atom aggregate. To our knowledge, this is the most complete and exact report on pure Pt nanoclusters.

12.5 BIMETALLIC CLUSTERS

Now, we treat the case of atomic bimetallic clusters composed of Fe and Pt atoms. A more detailed discussion can be found in [49] and [50]. First, we use the results for the pure Pt clusters reported earlier and substitute one of the Pt atoms with Fe. We present the results for FePt$_3$, FePt$_4$, FePt$_5$, and FePt$_6$. In Figure 12.3, we show the geometrical structures and the orbital and spin magnetic moments. Here we show only the most stable configurations. For the tetramer, the most stable geometry is a distorted rhombus. In the case of the pentamer, the lowest energy structure is not a two-dimensional system, like Pt$_5$, but a tetragonal pyramid. For the hexamer, we found as the ground state an almost planar triangular arrangement, and finally, the ground state for FePt$_6$ is a tetrahedron FePt$_3$ capped with Pt on three of its faces. One notices that the substitution of a Pt atom by Fe, in general, generates new arrangements. In the cases studied, only in FePt$_3$ and FePt$_5$, the Fe atom substitutes the Pt atom, deforming the original structure.

The orbital and spin magnetic moments are particular to each atom; they depend on the geometry and the chemical environment. Thus, in each atom, we show by big and small arrows the orbital and spin magnetic moments, respectively. One observes that these moments are clearly non-collinear, an effect that is produced by the strong spin-orbit coupling of platinum. In each figure, we give in brackets the absolute values of the orbital and spin moments. A more detailed account of the FePt$_{7-n}$ clusters can be found in [49]. We now focus on the heptamer Fe$_{7-n}$Pt$_n$ for $n = 0, 7$, and start with the pure Fe cluster. In Figure 12.4 we show the results for the geometrical structures, with the lowest energy, of the heptamers Fe$_{7-n}$Pt$_n$ with $n = 0, 1, 2, \ldots 7$. It is noticeable that, although distorted, some of the geometrical arrangements on the Fe-rich side are similar to those of the bulk structure. In the case of the Fe$_7$ cluster, we found as the lowest energy structure a distorted capped octahedron. This structure was obtained at 0.31 eV above a distorted pentagonal pyramid-like cluster calculated by using less accurate theories [20, 21].

For Fe$_6$Pt, we found again a distorted capped octahedron, in which the octahedron is formed by Fe atoms. The geometry for Fe$_5$Pt$_2$ is also a distorted capped octahedron, but now one of the Pt atoms occupies one of the octahedron apex. Figure 12.4d contains the results for the Fe$_4$Pt$_3$ cluster, and the results for the Fe$_3$Pt$_4$ are shown in Figure 12.4e. A Pt-pentagonal ring with one Fe atom on each side of the ring is the most stable structure for the Fe$_2$Pt$_5$ cluster, as presented in Figure 12.4f. This is a quasi-planar structure, characteristic of pure small Pt clusters. A triangular structure formed by six Pt atoms and one Fe atom above the central triangle is the lowest energy structure for FePt$_6$. This structure is similar to that reported by Alvarado et al. [18], but they found that the Fe atom takes part of the triangular structure and a Pt atom is above the central triangle. It is important to note that the results given here, for $n = 1-6$, are the most exact reported in the literature [50] up to now. The case of Pt$_7$ was discussed earlier.

In Table 12.3, we show the results for the binding energy $E_B(n)$ in eV, the range of the values adopted by the bonds in the cluster L (in Å), the total spin moment per atom, μ_S (in μ_B), and the total magnetic moments per atom, μ_T (in μ_B), ignoring the spin-orbit coupling (col) and when that interaction is included (SOC), for all the Pt concentrations n. In Figure 12.4, the spin and orbital magnetic moments, in each atom, are shown by red and blue arrows, respectively. The total magnetic moments (spin and orbital) per atom are the vector sums of the individual atoms divided by n, and the total magnetic moment per atom of the cluster is the sum of both contributions divided by the total number of atoms in the particle.

To appreciate the results better, we plot in Figure 12.5a the results for the binding energy as a function of the number of Pt atoms for the two cases, when the spin-orbit interaction is taken into account and when it is ignored. The results for the average total spin magnetic moment are shown in the lower figure. We see that when the SOC is ignored, $E_{B(n)}$ increases as a function of n up to a maximum of 3.58 eV/atom for $n = 5$ and then decreases to 3.28 eV/atom for the pure Pt heptamer. The increase in $E_{B(n)}$ shows clearly that the bonds in Fe-Pt are stronger than Fe-Fe or Pt-Pt. Furthermore, from the pure element heptamers, one observes that the bond Fe-Fe is weaker than the Pt-Pt; this is in agreement with the fact that the melting point of Pt is higher than that of Fe.

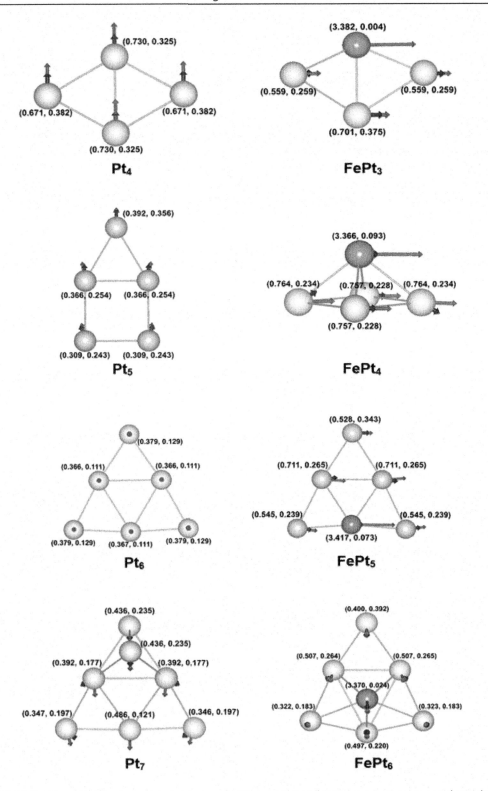

FIGURE 12.3 Ground state geometrical structure for Pt$_n$, n = 4–7 and FePt$_n$, n = 3–6. Here, we show also the spin (large) and orbital (small) local magnetic moments. We give the numerical values in brackets. Reproduced with permission from [49]. Copyright (2019) Elsevier, October 6, 2021.

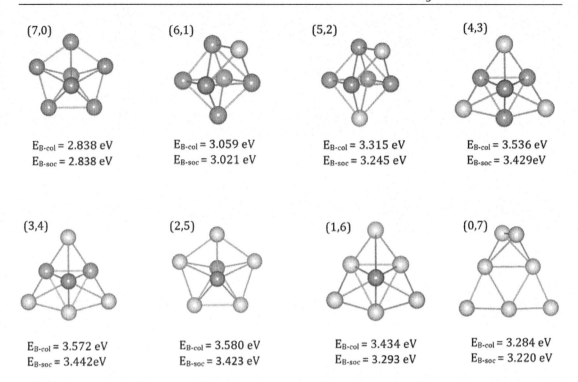

FIGURE 12.4 Lowest energy geometrical structures adopted by the binary heptamers $Fe_{7-n}Pt_n$ with n = 0, 1, 2, ... 7. Reprinted by permission from [50]. Copyright Springer Nature, October 6, 2021.

When the SOC is taken into account, the binding energy is very similar for the heptamers with n = 3, 4, and 5. Furthermore, the binding energies for higher concentrations of Pt differ more as a function of n. This again shows the importance that the SOC plays in Pt clusters. In relation to the dependence of the magnetic moment on n, the total spin (collinear case), and total magnetic moment (spin plus orbital) obtained by including SOC, one notices that it follows a similar behavior. This is shown in the lower panel. The magnetic moments decrease almost linearly with a slight deviation at n = 4. Since the orbital angular moment is very small as compared to the spin contribution, the last one determines the overall cluster behavior.

For small n, the particles are very asymmetrical, and therefore they do not have a well-defined symmetry axis. To get an idea of their magnetic anisotropy, we define as the x-axis the direction in which the total magnetic moment is oriented. Then, we place all the local magnetic moments in a perpendicular direction, let's say y, for example, let them relax, and calculate the cluster energy. We repeat the procedure but placing now the magnetic moments along the perpendicular z-axis and define the magnetic anisotropy energy as the difference between the highest value (y or z) and the one along the easy axis (x). It is important to note that this value gives just an approximation to the magnetic anisotropy. In order to calculate an exact value, it is necessary to perform a sweep over all angles (θ, Φ), which requires a computationally demanding task. From the results one can observe, the expected result is that the binary clusters with the highest anisotropy energy are the most symmetric geometrical structures, $FePt_6$ and Fe_5Pt_2.

We also studied, for each cluster, how much the magnetic moments deviate from collinear arrangements. As mentioned prior, depending on the chemical element and the environment around it, the spin and orbital moments have particular values and orientations, and so the total magnetic moment $\mu_{T,i} = \mu_{s,i} + \mu_{L,i}$. Thus, the angle between the ij pair of the total magnetic moments $\mu_{T,i}$ and $\mu_{T,j}$ is given by $\cos\theta_{ij} = \mu_{T,i} \cdot \mu_{T,j}/|\mu_i||\mu_j|$. However, since the angles differ from one pair to the other, there is a set of different angles in the whole cluster. Thus, to have an idea of the non-collinearity, we compare the absolute value of the total

TABLE 12.3 The Binding Energy per Atom (in eV), $E_{B}(n)$, the Range of the Values Adopted by the Bond Lengths in the Cluster L (in Å), the Spin Magnetic Moment per Atom, μ_S, and the Total Magnetic Moment per Atom, μ_T (in μ_B), as a Function of the Pt Composition n; on the Left Side, the Results with (without) Spin-Orbit Coupling Are Presented

	COL			SOC			
CLUSTER	$E_{B(n)}(EV)$	L(Å)	μ_S (μ_B)	$E_{B(n)}(EV)$	L(Å)	μ_S (μ_B)	μ_T (μ_B)
Fe7	2.838	2.265–2.773	2.889	2.838	2.266–2.773	2.886	2.95
Fe6Pt	3.059	2.286–2.716	2.624	3.021	2.286–2.714	2.615	2.688
Fe5Pt2	3.315	2.299–2.589	2.339	3.245	2.300–2.590	2.317	2.415
Fe4Pt3	3.536	2.361–2.730	2.076	3.429	2.365–2.718	2.041	2.154
Fe3Pt4	3.572	2.379–2.609	1.608	3.442	2.387–2.590	1.609	1.706
Fe2Pt5	3.58	2.317–2.627	1.338	3.423	2.343–2.623	1.313	1.454
FePt6	3.434	2.425–2.777	1.051	3.293	2.369–2.642	0.847	1.066
Pt7	3.284	2.448–2.748	0.525	3.22	2.459–2.796	0.405	0.596

Reprinted by permission from [50]. Copyright Springer Nature, October 6, 2021.

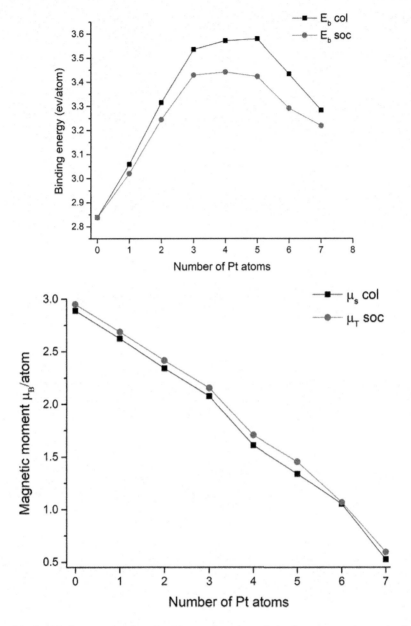

FIGURE 12.5 (a) The binding energy $E_{B(n)}$, in eV, as a function of the Pt concertation n, in the heptamers, $Fe_{7-n}Pt_n$ for $n = 1$–7. The line with circles (squares) represents the results including (excluding) the SOC. (b) The average spin magnetic moment and average total magnetic moment (in μ_B) as a function of n, ignoring (curve with squares) and including SOC (curve with circles). Reprinted by permission from [50]. Copyright Springer Nature, October 6, 2021.

magnetic moment $\left|\mu_T\right| = \left|\sum_{i=0}^{7} \mu_{T,i}\right|$ with the sum of the partial absolute contributions $\Sigma = \sum_{i=0}^{7} \left|\mu_{T,i}\right|$ and define a non-collinearity index ℓ by

$$\ell = 1 - \frac{|\vec{\mu}_T|}{\Sigma}. \tag{8}$$

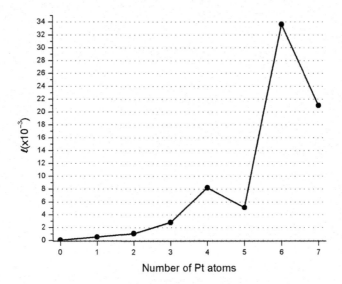

FIGURE 12.6 The non-collinearity index ℓ (see the text for definition) for the bimetallic heptamers as a function of Pt concentration n. Reprinted by permission from [50]. Copyright Springer Nature, October 6, 2021.

Thus, if they are collinear, the two values coincide, $|\mu_T| = \Sigma$ and the index $\ell = 0$; if there is a deviation, ℓ is positive. The largest value that it can take is 1 and corresponds to a case in which the cluster has an even number of atoms and their magnetic moments are coupled antiferromagnetically. The results for ℓ, in the set of clusters FePt, are plotted in Figure 12.6. We see that there are two maxima, for Fe_3Pt_4 and $FePt_7$, due to the particular geometry adopted by the clusters (see Figure 12.2).

12.6 CONCLUSIONS

This chapter contains state-of-the-art calculations on the structural stability and magnetic properties of monoatomic and bimetallic transition metal clusters. We discussed at length the use of graph theory to find the possible geometrical structures of small nanocluster; a complex problem but one that is essential to find the ground state atomic geometrical arrangement. Since the physicochemical properties and the structural geometry are interrelated, it is necessary to carry a long and tedious self-consistent calculation. We calculated the electronic and magnetic properties within the framework of the ab initio VASP density functional theory. Furthermore, to take into account non-collinear magnetic arrangements, we included the spin-orbit interaction

As examples, we presented the results for pure Pt_n clusters for $n = 3$ to 13. This is the most complete set of results for the platinum nanoclusters. It is important to note that the spin-orbit coupling is essential to describe correctly the physicochemical properties of Pt systems. We found planar structures for $n = 3$–6 and tridimensional structures for larger n. The cohesive energy is a monotonic function of the Pt concentration. In contrast, the average total magnetic moment shows oscillations with a maximum value of 1.048 μ_B for the tetramer.

We also discussed the binary nanoalloys $FePt_n$ for $n = 4$–7 and the more complex heptamers $Fe_{7-n}Pt_n$ with $n = 0, 1, 2, \ldots 7$. These results were first discussed in [49] and [50]. In the single Fe atom binary clusters, the Fe atom enters as a substitutional impurity in the cases $FePt_4$ and $FePt_6$. The Fe magnetic moment is much larger than those of Pt atoms and defines the magnetization orientation. In the case of the $Fe_{7-n}Pt_n$ heptamers, we found that for the rich Fe clusters, despite the smallness of the cluster, the geometrical structures resemble units of the bulk fcc ordered alloys. On the other hand, the rich Pt nanoparticles are quasi-bidimensional. The dependence of the binding energy on Pt concentration n shows clearly that to correctly describe the Pt nanostructures, it is necessary to include the spin-orbit coupling. However, since the orbital contribution to the magnetic moment is small, the total magnetic moment does not depend sensitively on the inclusion of spin-orbit interaction. We also analyzed the magnetic anisotropy energy in these small structures and found that the largest value corresponds to the most symmetric structures, like Fe_5Pt_2 and $FePt_6$. The first one is a capped cuboctahedron and the second is a quasi-planar structure. Finally, we addressed the question of how collinear the total atomic magnetic moments in the clusters are. To that aim, we defined an index that depicts this property. We observed that the structure, in the FePt heptamers, that deviates the most from the collinear arrangement is $FePt_6$.

12.7 ACKNOWLEDGMENTS

Computing resources provided by the Centro Nacional de Supercómputo (Instituto Potosino de Investigación Científica y Tecnológica, San Luis Potosí, México) and Laboratorio Nacional de Supercómputo (Benemérita Universidad Autónoma de Puebla, Puebla, México) are gratefully acknowledged.

12.8 REFERENCES

[1] F. Baletto and R. Ferrando. *Structural properties of nanoclusters: Energetic, thermodynamic, and kinetic effects*, Rev. Mod. Phys. Vol. 77, page 371 (2005).
[2] F. W. Payne, W. Jiang, and L. A. Bloomfield, *Magnetism and magnetic isomers in free chromium clusters*, Phys. Rev. Lett. Vol. 97, page 193401 (2006).
[3] S. Sun, C. B. Murray, D. Weller, L. Folks, and A. Mose, *Monodisperse FePt nanoparticles and ferromagnetic FePt nanocrystal superlattices*, Science Vol. 287, page 1989 (2000).
[4] J. M. Montejano-Carrizales, F. Aguilera-Granja, C. Goyheneux, V, Pierron-Bohnes, and J. L. Morán-López, *Structural, electronic, and magnetic properties of Co_nPt_{M-n} for M=13, 19, and 55; from first principles*, J. Mag. Mag. Mat. Vol. 355, pages 215–224 (2014).
[5] D. Weller, A. Moser, L. Folks, M. E. Best, W. Lee, M. F. Toney, M. Schwickert, J. Thiele, and M. F. Doemer, *High Ku materials approach to 100 Gbits/in^2*, IEEE Trans. Magn. Vol. 36, p. 10 (2000).
[6] A. Chen and P. Holt-Hindle, *Platinum-based nanostructured materials: Synthesis, properties, and applications*, Chem. Rev. Vol. 110, page 3767 (2010).
[7] J.-P. Wang, *FePt magnetic nanoparticles and their assembly for future magnetic media*, IEEE, Vol. 96, page 1847 (2008).
[8] I. M. L. Billas, J. A. Becker, A. Chatelain, and W. A. de Heer, *Magnetic moments of iron clusters with 25 to 700 atoms and their dependence on temperature*, Phys. Rev. Lett. Vol. 71, page 4067 (1993).
[9] M. Castro, *The role of the Jahn – Teller distortions on the structural, binding, and magnetic properties of small Fe_n clusters, n≤7*, Int. J. Quantum Chem. Vol. 64, page 223 (1997).
[10] C. Köhler, G. Seifert, and T. Frauenheim, *Density-functional based calculations for Fe(n) (n≤32)*, Chem. Phys. Vol. 309, 23 (2005).
[11] K. Cervantes-Salguero and J. M. Seminario, *Structure and energetics of small iron clusters*, J. MOl. Model Vol. 18, page 4043 (2012).
[12] Y. Yamamoto, T. Miura, Y. Nakae, T. Teranishi, M. Miyake, and H. Hori, *Magnetic properties of the noble metal nanoparticles protected by polymer*, Physica B (Amsterdam) 329–333, 1183 (2003)
[13] X. Liu, M. Bauer, H. Bertagnolli, E. Roduner, J. van Slageren, and F. Phillipp, *Structure and magnetization of small monodisperse platinum clusters*, Phys. Rev. Lett. Vol. 97, page 253401 (2006).

[14] J. A. Alonso, *Electronic and atomic structure, and magnetism of transition-metal clusters*, Chem. Rev. Vol. 100, page 637 (2000).

[15] V. Kumar and Y. Kawazoe, *Evolution of atomic and electronic structure of Pt clusters: Planar, layered, pyramidal, cage, cubic, and octahedral growth*, Phys. Rev. B, Vol. 77, page 205418 (2008), and references therein.

[16] L. Xiao and L. Wang, *Structures of platinum clusters: Planar or spherical?* J. Phys. Chem. A, Vol. 108, page 8605 (2004).

[17] K. Bhattacharyya and C. Majumder, *Growth pattern and bonding trends in Pt_n (n = 2–13) clusters: Theoretical investigation based on first principle calculations*, Chem. Phys. Lett, Vol. 446, page 374 (2007).

[18] P. G. Alvarado-Leyva, F. Aguilera-Granja, A. García-Fuente, and A. Vega, *Spin-orbit effects on the structural, homotop, and magnetic configurations of small pure and Fe-doped Pt clusters*, J. Nanopart. Res. Vol. 16, page 11 (2014).

[19] R. Ferrando, J. Jellinek, and R. L. Johnston, *Nanoalloys: From theory to applications of alloy clusters and nanoparticles*, Chem. Rev. Vol. 108, page 845 (2008).

[20] R. Rodríguez-Alba, S. E. Acosta-Ortiz, and J. L. Morán-López, *Ordered states in binary alloys with one magnetic component: A binomial description*, Solid State Communications Vol. 218, page 6 (2015).

[21] J. Mejía-López, A. H. Romero, M. E. García, and J. L. Morán-López, *Noncollinear magnetism, spin frustration, and magnetic nanodomains in small Mn_n clusters*, Phys. Rev. B Vol. 74, page 140405 (2006).

[22] J. Mejía-López, A. H. Romero, M. E. García, and J. L. Morán-López, *Understanding the elusive magnetic behavior of manganese clusters*, Phys. Rev. B Vol. 78, page 134405 (2008).

[23] F. Muñoz, A. H. Romero, J. Mejía-López, and J. L. Morán-López, *First-principles theoretical investigation of monoatomic and dimer Mn adsorption on noble metal (111) surfaces*, Phys. Rev. B Vol. 85, page 115417 (2012).

[24] M. E. Gruner, G. Rollmann, P. Entel, and M. Farle, *Multiply twinned morphologies of FePt and CoPt nanoparticles*, Phys. Rev. Lett. Vol. 100, page 087203 (2008).

[25] P. L. Rodríguez-Kessler and J. L. Ricardo-Chávez, *Structures of FePt clusters and their interactions with the O_2 molecule*, Chemical Phys. Lett. Vol. 622, page 34 (2015).

[26] B. L. Chittari and V. Kumar, *Atomic structure, alloying behavior, and magnetism in small Fe-Pt clusters*, Phys. Rev. B Vol. 92, page 125442 (2015).

[27] D. G. Truhlar, *Potential Energy Surfaces, Encyclopedia of Physical Science and Technology*, 3rd Ed., Vol. 13, pages 9–17, Academic Press, New York (2001).

[28] J. P. K. Doye, M. A. Miller, and D. J. Wales, *Evolution of the potential energy surface with size for Lennard-Jones clusters*, J. Chem. Phys. Vol. 111, pages 8417–8428 (1999).

[29] J. Maddox, *Crystals from first principles*, Nature Vol. 335, page 201 (1988).

[30] S. M. Woodley and R. Catlow, *Crystal structure prediction from first principles*, Nature Materials Vol. 7, pages 937–946 (2008).

[31] S. Heiles and R. L. Johnston, *Global optimization of clusters using electronic structure methods*, Int. J. Quantum Chem. Vol. 113, pages 2091–2109 (2013).

[32] L. T. Wille and J. Vennik, *Computational complexity of the ground-state determination of atomic clusters*, J. Phys. A Vol. 18, page L419 (1985).

[33] G. W. Greenwood, *Revisiting the complexity of finding globally minimum energy configurations in atomic clusters*, Z. Phys. Chem. Vol. 211, page 105 (1999).

[34] A. B. Adib, *NP-hardness of the cluster minimization problem revisited*, J. Phys. A: Math. Gen. Vol. 38, 8487–8492 (2005).

[35] E. Hückel, *Quantentheoretische Beiträge zum Benzolproblem*, Z. Phys. Vol. 70, pages 204–286 (1931); Vol. 72, pages 310–337 (1931); *Quantentheoretische Beiträge zum Problem der aromatischen und ungesättigten Verbindungen. III*, Vol. 76, pages 628–648 (1932); *Die freien Radikale der organischen Chemic. Quantentheoretische Beiträge zum Problem der aromatischen und ungeslittigten Verbindungen. IV*, Vol. 83, pages 632–668 (1933).

[36] Y. Wang, T. F. George, D. M. Lindsay, and A. C. Beri, *The Hückel model for small metal clusters. I. Geometry, stability, and relationship to graph theory*, J. Chem. Phys. Vol. 86, No. 6, pages 3493–3499 (1987).

[37] H. A. Jahn and E. Teller, *Stability of polyatomic molecules in degenerate electronic states. I. Orbital degeneracy*, Proc. R. Soc. A. Vol. 161 issue A905, 220–235 (1937).

[38] K. Bryan and T. Leise, *The $25,000,000,000 eigenvector: The Linear Algebra behind google*, SIAM Rev. Vol. 48, pages 569–581 (2006).

[39] G. M. Crippen and T. F. Havel, *Distance Geometry and Molecular Con-formation*, John Wiley & Sons, New York (1988).

[40] Nathanael Beeker, Stephane Gaubert, Christian Glusa, and Leo Liberti, *Distance Geometry: Theory, Methods, and Applications*, Chapter 5, Springer Sciences, New York (2013).

[41] L. M. Blumenthal, *Theory and Applications of Distance Geometry*, 2nd ed., Oxford Clarendon Press, London (1970).
[42] A. Mucherino, C. Lavor, L. Liberti, and N. Maculan, Editors, *Distance Geometry – Theory, Methods, and Applications*, Springer Science & Business Media, New York (2013).
[43] J. Moré and Z. Wu, *Distance geometry optimization for protein structures*, J. Glob. Optim. Vol. 15, pages 219–234 (1999).
[44] Leo Liberti, Carlile Lavor, Nelson Maculan and Antonio Mucherino, *Euclidean distance geometry and applications*, SIAM Rev. Vol. 56, pages 3–69 (2014).
[45] G. Kresse and J. Hafner, *Ab initio molecular dynamics for liquid metals*, Phys. Rev. B Vol. 47, page 558 (1993); ibid. Ab initio molecular-dynamics simulation of the liquid-metal – amorphous-semiconductor transition in germanium, Vol. 49, page 14251 (1994); G. Kresse and J. Furthmller, *Efficiency of ab-initio total energy calculations for metals and semiconductors using a plane-wave basis set*, Comput. Mat. Sci. Vol. 6, page 15 (1996); G. Kresse and J. Furthmller, *Efficient iterative schemes for ab initio total-energy calculations using a plane-wave basis set*, Phys. Rev. B Vol. 54, page 11169 (1996).
[46] G. Kresse and D. Joubert, *From ultrasoft pseudopotentials to the projector augmented-wave method*, Phys. Rev. B Vol. 59, page 1758 (1999).
[47] J. P. Perdew, K. Burke, and M. Ernzerhof, *Generalized Gradient Approximation Made Simple*, Phys. Rev. Lett. Vol. 77, pages 3865–3868 (1996).
[48] D. Hobbs, G. Kresse, and J. Hafner, *Fully unconstrained noncollinear magnetism within the projector augmented-wave method*, Phys. Rev. B Vol. 62, page 11556 (2000).
[49] A. P. Ponce-Tadeo, J. L. Morán-López and J. L. Ricardo-Chávez, *Global geometry optimization and magnetic properties of Pt and FePt Clusters*, Materials Today, Proceedings Vol. 14, pages 47–51 (2019).
[50] A.P Ponce-Tadeo, J. L. Morán-López, and J. L. Ricardo-Chavez, *Structural and magnetic properties of $Fe_{7-n}Pt_n$ with n = 0, 1, 2, . . . 7, bimetallic clusters*, J Nanopart. Res. Vol. 18, id.330, pages 1–14 (2016).

Nanoscale Characterization

13

Arvind Kumar[a], Swati[a], Manish Kumar[a], Neelabh Srivastava[b], and Anadi Krishna Atul[b]

a Materials Science Research Lab (Theory and Experimental), Department of Physics, ARSD College, University of Delhi (India)

b Department of Physics, School of Physical Sciences, Mahatma Gandhi Central University, Motihari, India

Contents

13.1	Introduction	246
13.2	Structural Characterization of Nanomaterials	247
	13.2.1 X-Ray Diffraction (XRD)	247
13.3	Morphological Characterization of Nanomaterials	248
	13.3.1 Electron Microscopy	248
	13.3.1.1 Scanning Electron Microscope (SEM)	249
	13.3.1.2 Transmission Electron Microscopy (TEM)	249
13.4	Scanning Probe Microscopy (SPM)	252
	13.4.1 Atomic Force Microscopy (AFM)	252
	13.4.2 Magnetic Force Microscopy (MFM)	252
	13.4.3 Scanning Tunneling Microscopy (STM)	255
13.5	Surface Characterization of Nanomaterials	256
	13.5.1 X-Ray Photoelectron Spectroscopy (XPS)	256
13.6	Spectroscopic Characterization of Nanomaterials	257
	13.6.1 UV-Vis Spectroscopy	257
	13.6.1.1 Bandgap Estimation by Tauc Plot	258
	13.6.1.2 Diffuse Reflectance Spectroscopy (DRS)	259
	13.6.2 Photoluminescence (PL) Spectroscopy	259
	13.6.3 Fourier Transform Infrared (FTIR) Spectroscopy	260
	13.6.4 Raman Spectroscopy	261
13.7	Dielectric Measurements of Nanomaterials	262
	13.7.1 Frequency-Dependent Dielectric Measurements	263
	13.7.2 Temperature-Dependent Dielectric Measurements	264
13.8	Conclusions	265
13.9	Acknowledgments	265
13.10	References	265

DOI: 10.1201/9781003197492-13

13.1 INTRODUCTION

Nanometer-sized (~ 1–100 nm) materials are the focus of the current research in the field of science and technology owing to their exceptional chemical, physical properties, chemical nature, and domain size because of their reduced dimensionality. An increase in surface-to-volume ratio in terms of shape, size, composition, surface properties, molecular weight, solubility, and stability significantly affects its physicochemical properties, which could make it tunable in terms of functional properties, i.e., optical, electronic, and magnetic [1–2]. The physiological behavior of nanomaterials is determined based on their properties, for which the characterization of nanomaterials is indispensable to confirm their quality.

The ongoing research and development at nanoscale dimensions is focused on the search for numerous fabrication methods to achieve the desired properties required for its possible applications in the various fields of material science (e.g., sensors, inks, catalysts, etc.) and life sciences (e.g., drug delivery, cancer therapy, etc.) [3–4].

Characterization of nanomaterials is of vital importance to look for its novel applications. To achieve the unique and novel physicochemical properties of nanomaterials, researchers have developed various characterization techniques which are continuously being upgraded for better analysis [5]. To characterize a material at the nanoscale level, multiple studies need to be done to analyze its various physical and chemical features, i.e., composition, structure size, morphology, surface area, optical properties, surface composition, oxidation state, and electrochemistry. It is impossible to capture all pertinent characteristics of nanomaterials using a single technique [6–7]. This chapter presents an overview of the general characterization techniques that are being used by researchers to understand the physical properties of nanomaterials, such as structural, morphological, optical, and dielectric. To understand the effect of material properties on their composition and structure, investigation of crystal structures with their chemical nature, including their imperfections (e.g., defects), is very crucial.

Different characterization tools are categorized based on the theory behind that technique, the information it can offer, or the materials for which it is used. The structural information up to the three-dimensional arrangement of atoms from sub-angstrom to sub-micron length scales is extracted by X-ray, neutron, and electrons diffraction techniques [8–10]. The investigation of structures at the nanoscale level would become very unique owing to the advent of modern transmission electron microscopes with various techniques such as high-resolution imaging, diffraction, and spectroscopy. Herein, the structural analysis could be investigated via powder X-ray diffraction technique. The microstructural and morphological characterizations could be performed using scanning electron microscopy (SEM), transmission electron microscopy (TEM), atomic force microscopy (AFM), magnetic force microscopy (MFM), and scanning tunneling microscopy (STM) techniques. SEM image displays the information related to the external morphology of the samples whereas the direct visual information such as size, disparity, structure, and morphology of nanoclusters can be obtained using TEM. AFM and MFM are also known to be one of the best techniques to measure quantitatively the surface roughness, nano-textures, and morphology of the non-magnetic and magnetic structures, respectively. To gather the information about electronic states in the materials, the composition of sample and thickness, etc. of thin film or bulk materials, X-ray photoelectron spectroscopy (XPS) could be useful, which is based on the photoelectric effect.

In continuation to these mentioned techniques, optical properties of nanostructured materials could be investigated by characterizing the samples through different spectroscopic techniques, such as UV-Vis spectroscopy, photoluminescence, Fourier transform infrared spectroscopy (FTIR), and Raman spectroscopy. Using FTIR and Raman spectroscopy, detailed information about the nanomaterials, such as chemical structure, phase formations, polymorphism, crystallinity, and molecular interactions, could be deduced. The dielectric properties of the nanomaterials could be measured using dielectric measurement setups such as LCR meter, impedance analyzer, etc.

So, in the present chapter, an attempt has been made to describe the different measurement techniques along with their operational principles for the characterization of nanomaterials in relation to the property studied. This is required to get a better understanding of nanomaterials for their translation into future-generation products and devices.

13.2 STRUCTURAL CHARACTERIZATION OF NANOMATERIALS

13.2.1 X-Ray Diffraction (XRD)

XRD is an extensively used tool to characterize single-crystal or polycrystalline materials. X-rays have a very short wavelength (10^{-8}–10^{-12}m), which is comparable to atomic dimensions, and hence are excellent probes to study nanomaterials (dimensions in the range of 1–100 nm). The atoms in the crystal are periodically arranged with a fixed interatomic spacing on the order of a few angstroms, which acts as a three-dimensional grating. X-rays exhibit the phenomenon of diffraction like visible light when incident on crystals. The powder diffraction method is one of the most commonly used methods in XRD analysis, as it overcomes the difficulty of growing a single crystal, and another advantage is that it results in statistically representative and volume-averaged values. The polycrystalline material is powdered by crushing because the obtained fine powder has many tiny crystallites in random orientations. When X-rays are incident on crystalline substances, maxima or minima in certain directions are obtained following Bragg's law. [11]

$$2d\sin\theta = n\lambda \qquad (1)$$

where d is the interplanar spacing, λ is the wavelength of an X-ray source, and θ is the angle of diffraction.

Figure 13.1a,b shows Bragg's reflection from the atomic planes and basic features of an X-ray diffractometer, respectively. For a particular order, if λ is known, then θ can be obtained from the diffractometer, and hence d, the interplanar spacing, can be calculated for a particular plane.

A crystalline substance produces a diffraction pattern; the same pattern is always obtained for the same substance. In a mixture of substances, each one produces its own pattern independent of the others. The XRD pattern of a crystalline substance is unique to itself; it is like the fingerprints of the material. The diffraction pattern is used to determine the composition of the compound by comparing the position

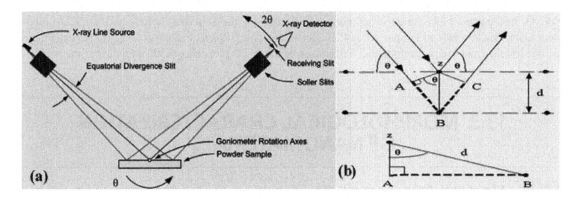

FIGURE 13.1 (a) Schematic of powder X-ray diffractometer and (b) representation of Bragg's law. Adapted with permission from (under open access) [12–13]. Copyright (2015) NIST Research Library, and copyright (2012) Update Publishing House.

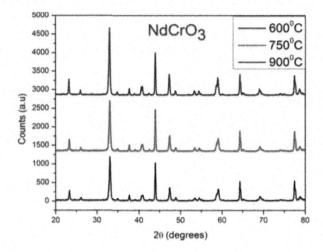

FIGURE 13.2 XRD pattern of NdCrO₃ nanoparticles. Adapted with permission from [13]. Copyright (2016) Elsevier.

and intensity of peaks with the reference patterns available from the International Centre for Diffraction Data (ICDD, previously known as the Joint Committee on Powder Diffraction Standards, JCPDS) database [14]. Figure 13.2 shows the XRD pattern for NdCrO₃ nanoparticles prepared by Jada Shanker et al. [15] at different temperatures.

XRD is also used to determine the crystalline structure, lattice constants, geometry, nature of phase, preferred orientation of polycrystals, identification of the unknown material, defects, etc., and is also useful in determining the interfacial strain [12–16]. Broadening of the XRD peaks is a measure of the average size of nanocrystals by the Scherrer formula [13–17]

$$D = K\lambda / \beta cos\theta \qquad (2)$$

(where D is the average particle size, λ is the wavelength of the X-ray source, and β is known to be the full width at half maximum).

The value of the Scherrer constant (K) in the formula is generally taken to be ~ 0.9 and accounts for the shape of the particle. Smaller nanocrystals have a wider reflection peak, but the size so obtained has the limitation that the Scherrer equation does not consider the existence of a size distribution and defects in the crystalline system and is generally larger than actual values. The average crystallite size for NdCrO₃ nanoparticles and lattice strain were calculated using the Scherrer formula and Rietveld refinement technique at different temperatures [13]. Dorofeev et al. [18] and Prabhu et al. [19] also calculated the particle size and micro-strain using Williamson-Hall (W-H) plot using XRD data of nanoparticles.

13.3 MORPHOLOGICAL CHARACTERIZATION OF NANOMATERIALS

13.3.1 Electron Microscopy

The invention of the electron microscope has opened new avenues in the research area of nanoscale materials. It is possible to image sub-micron-sized objects as single atomic positions (nanoscale). Owing to this, it has become a key technology for material characterization in the world of nanoscale and has enabled significant developments in the nanoscale engineering of micro-sized components such as mobile phones, plasma screen televisions, and materials for cars and airplanes.

13.3.1.1 Scanning Electron Microscope (SEM)

SEM is an instrument that reveals information about the external morphology of a sample. It consists of a high vacuum system with an electron gun, a conducting lens, and an imaging device. Herein, the electron gun emits electrons from the cathode. The emitted electrons are accelerated towards the anode with an energy in the range 1–50 keV. This electron beam is converged by a condenser lens under the effect of the magnetic field. Following are some important parts of this device:

(i) **Electron guns:** Electron guns have the capacity to emit electrons with 1 to 30 kV accelerating voltage. For routine imaging 15–30 keV is used.
(ii) **Vacuum:** Pressure inside the chamber is maintained between 0.1 to 10^{-4} Pascal.
(iii) **Specimen:** According to the specimen stage installed in the chamber, the specimen is prepared, i.e., 3–20 cm in diameter. Usually, an automated or computer-controlled stage is used so that it can perform 3–5 degrees of freedom for the incoming electron beam.
(iv) **Control of electron beam:** The electron beam is adjusted by electromagnetic lenses such that it moves towards the specimen and generates images, although a range of different scan patterns are programmed for the user in modern mechanics.
(v) **Specimen preparation:** The emission of secondary electrons (SEs) occurs from the top surface usually < 20 nm. Thus, it is essential to have a clean environment as hydrocarbons from prolonged exposure to air or grease from fingers can contaminate the sample, thereby affecting the SE imaging [20].

Specimen preparation is performed in clean environments such as sealed boxes, desiccators, and vacuum packs to reduce exposure to water/humidity. Dirty environments such as oil, aqueous solutions, or specimen surfaces should be cleaned by systematic methods such as soaking in solvents and ultrasonic agitation. As-prepared clean samples (non-conducting or poorly conducting) require an ultrathin coating of carbon, gold, or platinum, which can conduct away from the electrostatic charges that accumulate over the surface. Charge accumulation occurs due to the difference in the potential of incoming electrons (negatively charged primary electrons) and outgoing current (number of SEs, backscattered electrons (BSEs), and transmitted primary electrons).

Figure 13.3a–f [21] shows the FE-SEM examination results of four different types of nanostructured samples for their morphology. Figure 13.3a–f represents the formation of nanoparticles, nanorods (with hexagonal top), uniformly grown hexagonal nested structures, an airscape of a maze with hexagonal walls like morphology, and typical inclined examination of the samples 3 and 4, respectively. A high density of columnar growth complex nanostructures with uniform structures in a large area has been found for the nanostructures. In another study, flower-like silver nanostructures with controlled morphology and composition through wet-chemical synthesis have been reported by Zhou et al. (2014) [22].

13.3.1.2 Transmission Electron Microscopy (TEM)

TEM is another versatile tool for characterizing nanoparticles for its direct visual information such as size, disparity, structure, and morphology of nanoclusters [23–24]. In this technique, an electron beam is transmitted through an ultrathin specimen and generated signals are recorded as images that are magnified and focused by an objective lens and shown on a fluorescent screen using sensors as a charge-coupled device (CCD) camera. Herein, the electron beam interacts with the specimen mostly by diffraction, not by absorption. The relative orientation of the electron beam and the atomic plane of the crystal determines the intensity of diffraction. TEM consists of the electron gun, electromagnetic lenses for focusing the

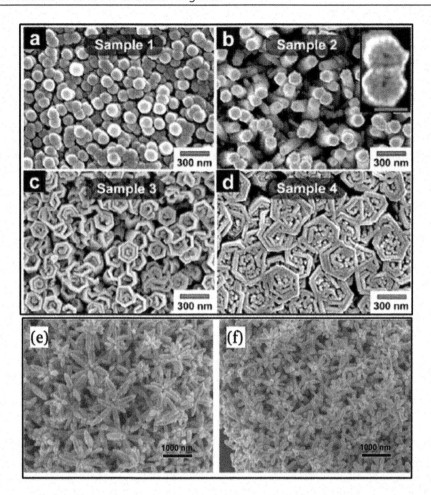

FIGURE 13.3 SEM images of synthesized AlN nanostructures at different positions of the quartz tube: (a) Nanoparticles film of sample 1. (b) Hexagonal nanorod arrays with a small pit on the top of each rod for sample 2, with an inset showing the higher magnification image. (c) Hexagonal nested arrays of sample 3. (d) The hexagonal nanomaze of sample 4. Adapted with permission from [21]. Copyright (2011) Royal Society of Chemistry. Flower-like Ag nanostructures prepared with polyvinylpyrrolidone (PVP) and different amounts of catalyzing agent $NH_3 \cdot 3H_2O$: (e) 200 μL, (f) 400 μL. Adapted with permission (under open access) from [22]. Copyright (2014) Springer Open Journal.

electron beam before and after passing through the specimen, and a transmitted electron detection system. Modern TEM contains automated specimen holders that could be tilted to a range of angles and provide specific diffraction conditions. The electron gun used to emit electrons has sufficient energy to penetrate up to 1 μm within the sample, i.e., 80–300 kV accelerating voltage [25]. For routine imaging, 200–300 keV electrons are used. However, energy less than 100 keV is used to characterize very light elements like carbon to reduce the specimen damage.

TEM specimen preparation is one of the tedious tasks for microscopic characterization. Surface cleaning of the specimen is an important part of the sample preparation. To remove these contaminations, a 0.1 M cacodylic acid buffer (pH 7.3) is used at ambient conditions. Fixation is done by using various fixatives, i.e., aldehydes (glutaraldehyde), osmium tetroxide, tannic acid, or thiocarbohydrazide. The specimen is rinsed carefully to remove the excess fixative. Post fixation is done to achieve a stabilized specimen before TEM investigation by using 1% osmium tetroxide prepared in 0.1 M cacodylic acid buffer (pH 7.3) for 1.5 hrs at room temperature. Thereafter, the dehydration of specimen takes place using ethanol, and this ethanol is immiscible with the plastic embedding medium; thus, infiltration of the specimen is performed using a transitional solvent. The prepared specimen is cut into semithin and ultrathin slices (sections), which are known as microtomy and ultramicrotomy. There are mainly two techniques of experimental crystallography which could be examined inside:

(i) **Selected area electron diffraction (SAED):** SAED is a technique of experimental crystallography which could be examined inside the TEM for the identification of crystal structure and crystal defects. X-ray diffraction is also similar to it but has less sensitivity. It can be examined up to several hundred nanometers, due to which it can be used as a primary tool for characterization in materials science and solid-state physics.
(ii) **High-resolution transmission electron microscopy (HRTEM):** HRTEM is a scientific technique that is based on the imaging mode of a transmitted electron that can image crystallographic structures up to the atomic scale. It is more sensitive than TEM. Herein, the interference of electron waves and the image plane of electron waves provides the contrast.

From this point of view, TEM is a very rich technique. Manchala et al. (2019) have synthesized graphene utilizing a green synthesis approach by using aqueous polyphenol extracts of eucalyptus for its application in high-performance supercapacitors [26]. They performed a TEM examination for its morphological characteristics, whose image is shown in Figure 13.4a,b. It can be seen that the formation of few-layered nanosheets along with scrolled, wrinkled, and transparent features is similar to that of few-layer nanosheets. The HRTEM image (Figure 13.4c) shows the interplanar spacing ($d_{spacing}$) of ~ 0.4 nm,

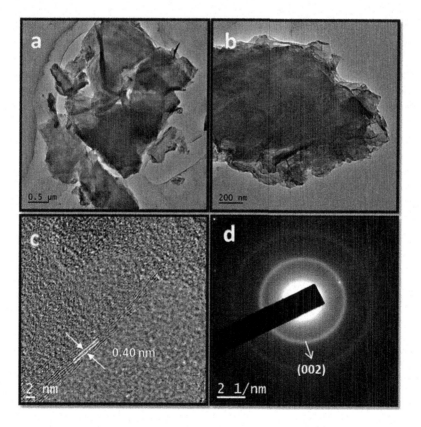

FIGURE 13.4 TEM (a, b), HRTEM (c), and SAED (d) images of synthesized E-graphene. Adapted with permission from [26]. Copyright (2019) American Chemical Society.

which matches closely with the data by powder XRD (PXRD) and the SAED pattern as mentioned in Figure 13.11d, i.e., $d_{spacing}$ (i.e., 0.356 nm). The diffraction dots of the (002) plane preferably match with the PXRD pattern of E-graphene sheets.

13.4 SCANNING PROBE MICROSCOPY (SPM)

In this branch of microscopy, images of the specimen's surface are generated by using physical probe scanning. This technique has enabled researchers to scan images up to nanoscale order. In this imaging technique, a fine probe is used to scan the surfaces rather than a light or electron beam. The use of this technology resolves the limit of a wavelength of light and electrons, and it also enables to provide true 3D maps of surfaces and gives resolution up to the atomic level.

13.4.1 Atomic Force Microscopy (AFM)

AFM is one of the non-destructive techniques used for measuring quantitatively the surface roughness and nano-textures. It consists of a probe at the end of the cantilever with a sharp tip (nominal tip radius of the order of 10 nm) which raster scans the specimen surfaces with the help of piezoelectric sensors. In AFM scanning, a cantilever tip senses the force between surface and tip. Herein, the tip is attached to the cantilever across the free end and is brought towards the surface. Interaction between surface and cantilever shows positive and negative deflections, i.e. indication of attractive and repulsive forces. This laser detection system is bulky and costly. Another easy and convenient method for the detection of deflection of laser and cantilever is performed by using piezoresistive AFM probes, which are made up of piezoresistive elements that act as a strain gauge. This strain gauge is fabricated by Wheatstone Bridge, although it is not like a laser detection system. A constant force is maintained between the probe and sample while raster scanning is done across the surface in AFM. The separation distance between the tip and the sample relies on the interaction forces. If the separation between the tip and the sample is closer, there will be repulsive interaction, whereas a larger separation shows an attractive one. 3D images of the surface can be scanned by monitoring the motion of the probe. This technique is also useful for size measurements or manipulations of nano-objects. AFM imaging could be performed in three different modes: namely, contact mode, tapping mode, and non-contact mode.

Manchala et al. [26] performed the AFM investigation to determine the lateral size and thickness of graphene as shown in Figure 13.5. Based on this result, it was found that the synthesized E-graphene is mostly a combination of monolayer and multiple layers of sheets [26]. In another study, Sun et al. [27] developed a method to convert self-assembled block copolymer nanodomains into high-quality carbon nanostructures with well-defined morphology using direct pyrolysis of poly(styrene-b-2-vinylpyridine) (PS-b-P2VP) block copolymer (BCP) as thin films. AFM investigations are processed at different temperatures and conditions, which are shown in Figure 13.6.

13.4.2 Magnetic Force Microscopy (MFM)

MFM, an offspring of the AFM technique, allows the imaging of magnetic structures on a nanometer scale (50–100 nm) by measuring the response of a sharp magnetic tip mounted on a flexible cantilever. The interaction of the magnetic tip with the stray field of the sample is measured by recording the changes in either the static deflection or the resonant frequency of the cantilever with a sensitive displacement sensor. Raster scanning of the sample with respect to the tip and measured interaction as a

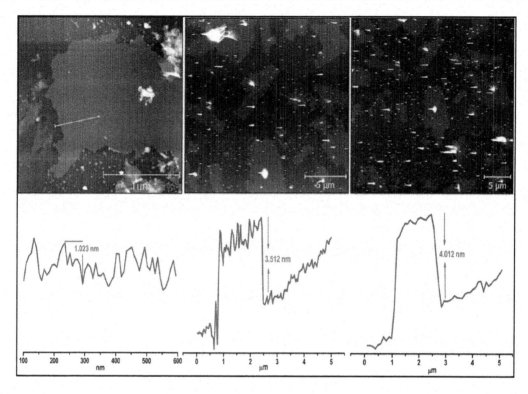

FIGURE 13.5 Typical AFM images and line profiles of E-graphene. Adapted with permission from [26]. Copyright (2019) American Chemical Society.

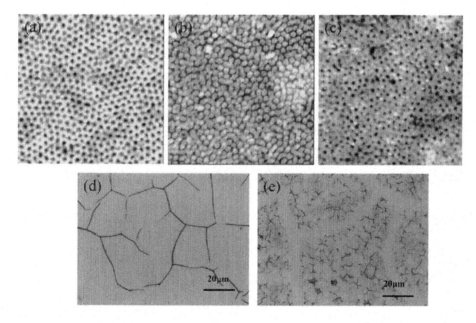

FIGURE 13.6 AFM topographic images of carbon nanostructures with (a) 430°C and (b and c) 900°C pyrolysis of Ct nanodomains within the P(S500002VP165000) thin films with a capping layer of SiO$_2$, thickness (a and b) 40 or (c) 100 nm. (d and e) OM images of sample-b and -c without removal of the capping layer. Adapted with permission from [27]. Copyright (2015) RSC Publishing.

function of position produces the MFM image of the sample. The magnetic force between sample and tip can be described by:

$$F = \mu_0 (m.\nabla) H \tag{3}$$

Where m is the magnetic moment of the tip, H is the magnetic stray field from the sample surface, and μ_0 is the magnetic permeability of the free space. Due to tip-sample interaction, the cantilever behaves as if it had modified spring constant K (K = K_0 + F′), where K_0 is the natural spring constant and $F' = \dfrac{\partial F_z}{\partial Z}$ is the derivative of interaction force gradient along the z-axis [28–29], which leads to a change in the resonant frequency of the mechanical vibration of the cantilever. By measuring the change in resonant frequency via optical interferometry technique, MFM images could be found in alternate bright and dark contrast. In an MFM, two basic detection modes (static and dynamic) can be applied to measure the force derivative.

Figure 13.7 shows the schematic view of the MFM imaging along with the MFM image of the Co_2FeAl thin film [30]. Figure 13.7b,c shows the MFM images for Co_2FeAl/p-Si, with their line scan sectional analysis. Clear bright and dark contrast was observed for the structure; this is indicative of the formation of magnetic domains. Average phase shift, i.e., magnetic signal strength (Figure 13.7c), and magnetic domain size were found to be ~ 0.23° and ~ 10 nm for the CFA film on Si substrate. This represents the magnetic behavior of CFA film.

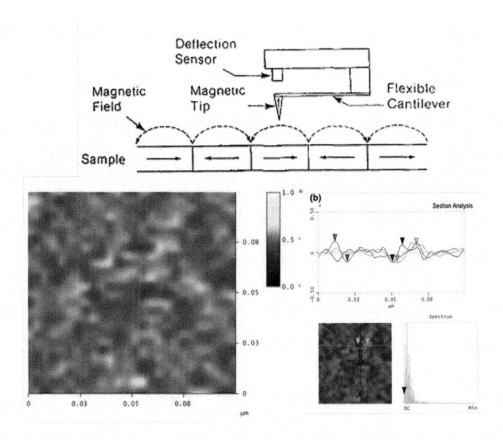

FIGURE 13.7 (a) Schematic view of the MFM imaging. Adapted with permission from [28]. Copyright (1992) Elsevier. (b) MFM image and (c) sectional analysis of Co_2FeAl/p-Si. Adapted with permission from [30]. Copyright (2014) Springer Nature.

13.4.3 Scanning Tunneling Microscopy (STM)

STM is also used for topographical imaging of the samples, in which the surface of the conducting specimen is scanned with the tip of fine probes using piezoelectric crystals at a distance of 0.5–1 nm. It is based on the phenomena of quantum tunneling and is the most suitable technique for imaging a single dot even at very low temperatures, up to 4.2 K [31]. The measurement of a tunneling current between a sharp STM tip and the material's surface is done by STM [31–32]. Each 1 Å reduction in the distance shows the increment in the tunneling current by an order of the current; thus, a piezoelectric scanner is used to control the variation in the distance along the XYZ-direction. The data of tunneling current scanned by the tip of the cantilever precisely provide the morphology of the surface. This examination could be performed in various liquids or gases at a wide range of temperatures even at ultrahigh vacuum (UHV) conditions.

Rieboldt et al. (2014) have studied the nucleation and growth of Pt nanoparticles (NPs) on TiO_2 (1 1 0) surfaces of different oxidation states using STM instruments, as shown in Figure 13.8 [33]. Herein, Pt is observed to be trapped at O on-top atoms and surface O vacancies at room temperature. Evaporation of ~ 2.5% ML Pt at room temperature (RT) is shown in Figure 13.8 by STM examination. Homogeneous NP distributions (15%–20% of Pt NPs at step edges) are estimated to have high densities (0.07 ± 0.01 nm^{-2}) of small Pt NPs at the surfaces of r-TiO_2 and o-TiO_2 although the h-TiO_2 surface has larger Pt NPs with lower density at 0.030 ± 0.005 nm^{-2}. It is concluded that Pt NPs or Pt atoms on the h-TiO_2 surface show higher mobility at the right temperature in comparison to Pt NPs at r- and o-TiO_2 surfaces. It indicates that the larger NPs are formed on h-TiO_2.

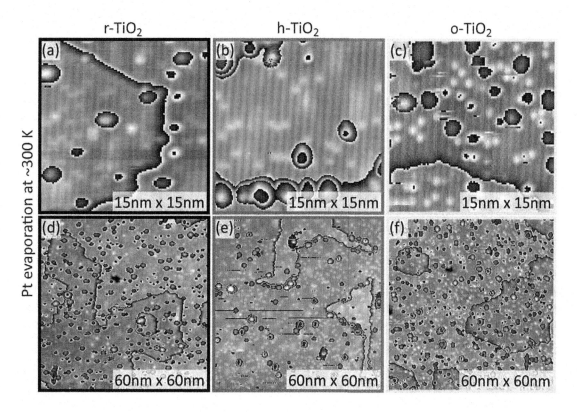

FIGURE 13.8 STM images of the r-TiO_2 (a and d), h-TiO_2 (b and e), and o-TiO_2 (c and f) surfaces after evaporation of Pt at RT. The STM images were acquired at RT. Adapted with permission from [33]. Copyright (2014) AIP Publishing.

13.5 SURFACE CHARACTERIZATION OF NANOMATERIALS

13.5.1 X-Ray Photoelectron Spectroscopy (XPS)

X-ray photoelectron spectroscopy is a powerful characterization technique that is based on the phenomenon of the photoelectric effect and is also known as ESCA (electron spectroscopy for chemical analysis). It is widely used to investigate the electronic state, composition of the sample and thickness, etc. of thin films or bulk materials [34]. The electron beam generated by the X-ray source excites the sample, and photoelectrons are ejected from the sample's surface, which is further detected by an electron energy analyzer. Usually, an XPS concentric hemispherical analyzer (CHA) is used as an analyzer. The electrons coming out of the analyzer are passed through the detector to measure the current generated by photoelectrons. The outer hemisphere is more negative due to the applied potential difference across the two hemispheres. The energy required for electrons to reach the detector from the analyzer is given by,

$$E = e\,\Delta V \tag{4}$$

where E represents the kinetic energy and ΔV is the applied potential difference across the hemisphere of inner and outer radii, R_1 and R_2, respectively. Owing to the constant value of radii, this formula is modified as:

$$E = k\,e\,\Delta V \tag{5}$$

Here k is introduced as a spectrometer constant, and its value is determined by the design of the analyzer.

The electrons with higher energy than the prior expression will follow the path of a larger radius than its mean value; similarly, lower energy electrons follow a smaller radius. The distance traveled by an electron inside the material between successive inelastic collisions is introduced as inelastic mean free path (IMFP). The values of IMFP are unique for each element and substrate or solid material matrix of which the element is a part [35–36].

Tarachand et al. 2018 have fabricated Ag-doped CuS nanocomposites by a simple polyol method using hexagonal nanodisks of CuS. The formation of pure CuS nanodisks is confirmed by XPS investigation due to the absence of any other peak except C, Cu, S, and oxygen [37]. Figure 13.9a shows the high-resolution spectrum of C 1s with the deconvolution of three sub-peaks at 284.6 eV, 286.1 eV, and 287.9 eV is consistent with the literature [38–39]. Sub-peaks at 284.6 eV is indicating the hydrocarbon (C=C/C–C) attached with the surface of CuS nanodisk as an internal reference to nullify the charging effect and the observed peaks at 286.1, 287.9 eV match with the C–OH and C=O bonds, respectively. As shown in Figure 13.9d, peaks of the high-resolution spectrum of O 1s are centered at 532.1 eV representing the reaction by-products due to the C–O and C=O groups present on the surface of the nanodisks. The peaks of Cu 2p are deconvoluted into two pairs of doublets as depicted in Figure 13.9b where the first doublet with major peaks at 931.9 (Cu $2p_{3/2}$) and 951.9 (Cu $2p_{1/2}$) eV appears due to the presence of the Cu^+ state [40]. However, the presence of the Cu^{2+} state in the CuS nanodisks is revealed by the peaks of the second doublet centered at 933.1 eV (Cu $2p_{3/2}$) and 953.0 eV (Cu $2p_{1/2}$) with a separation of 19.9 eV [41]. Two main doublets are also observed in Figure 13.9c. The deconvolution of the S 2p spectrum at 161.7 eV and 162.9 eV is observed due to metal sulfides (M–S) [42] whereas 163.1 eV and 164.5 eV are assigned for the values for metal disulfide (S–S) [43].

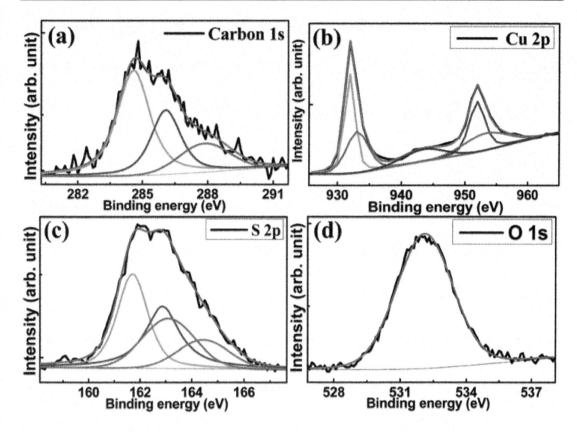

FIGURE 13.9 XPS spectra of CuS nanoparticles. High-resolution spectrum of (a) C 1s, (b) Cu 2p, (c) S 2p, and (d) O 1s. Adapted with permission from [37]. Copyright (2018) Royal Society of Chemistry.

13.6 SPECTROSCOPIC CHARACTERIZATION OF NANOMATERIALS

13.6.1 UV-Vis Spectroscopy

Nanoparticles exhibit optical properties that depend on their size, shape, concentration, agglomeration state, and refractive index near the surface of the nanoparticles [44]. UV-Vis spectroscopy is a valuable characterization tool for studying nanomaterials. It measures the absorption or transmission of UV or visible light through a sample in comparison to a reference or blank sample [45]. The source used is chosen such that it can emit wavelength over a wide range; a single xenon lamp is generally used as a source for both UV and visible ranges. Generally, quartz sample holders are used as they are transparent to a range of wavelengths in the UV region [45]. The detecting system is either a photomultiplier tube (PMT), photodiodes, or charge-coupled devices (CCDs).

The typical absorption spectrum obtained from UV-Vis spectroscopy for AuNPs at various concentrations of MSG is shown in Figure 13.10a [46]. On the y-axis, absorbance, optical density, or transmittance can be plotted as a function of wavelength. Absorbance (A) of the sample is related to the transmittance (T) as $A = -\log_{10}(T)$, where $T = I/I_0$, I is the intensity observed in the detecting system,

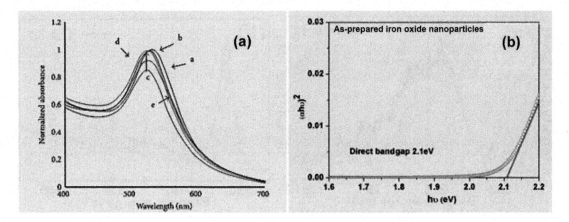

FIGURE 13.10 (a) An example of an absorption spectrum taken from a UV-Vis spectrophotometer for Au nanoparticles at different concentrations of MSG. Adapted with permission (under open access) from [46]. Copyright (2014) Hindawi. (b) Tauc plots for bandgap estimation. Adapted with permission (under open access) from [49]. Copyright (2016) Elsevier.

and I_0 is the incident intensity of light [44]. From the sharpness of absorption peaks, an estimation of size distribution and concentration can be made. The smaller the particles, the higher the percentage of their extinction due to absorption. If the particles in the sample tend to aggregate, then the scattering of light will increase. UV-Vis spectroscopy is also useful in monitoring the stability of nanoparticle solutions; as the particles destabilize, the intensity of the original extinction peak will decrease and broaden in size and intensity. Sometimes, a secondary peak is also observed at longer wavelengths [45]. The bandgap calculation could be performed using UV-Vis spectroscopy via Tauc plots and diffuse reflectance spectroscopy.

13.6.1.1 Bandgap Estimation by Tauc Plot

From the absorption spectra of the samples, the absorption coefficient α can be calculated using the well-known Beer-Lambert relation.

$$\alpha = \frac{1}{d} \ln\left(\frac{1}{T}\right) \tag{6}$$

where d is the path length and T is the transmittance [47–48]. Direct and indirect bandgap can be estimated using the Tauc formula (given by Eq. 7) by plotting the curve between energy ($h\nu$) on x-axis and $(\alpha h\nu)^2$ on y-axis [4].

$$\alpha h\nu = B\left(h\nu - E_g\right)^n \tag{7}$$

where α is absorption coefficient, E_g (eV) is bandgap, hν is the incident photon energy (eV), B is known as the proportionality constant, and n (2 & 1/2) is an exponent depending on the nature of the electronic transition. For calculating direct bandgap, substitute n = 1/2, and for indirect bandgap, n = 2 [48]. By fitting a straight line to the linear portion of the Tauc plots (Figure 13.10b) by least-square fit, its intercept on the x-axis [(αhυ)² = 0] and [(αhυ)^{1/2} = 0] is found, which gives the direct or indirect bandgap value of the material, respectively [49]. Figure 13.10b shows the Tauc plot for the iron oxide nanoparticles that give the direct bandgap value of 2.1 eV [49].

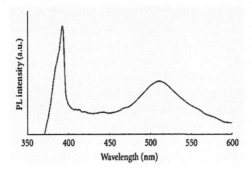

FIGURE 13.11 PL spectrum of ZnO nanoparticles. Adapted with permission (under open access) from [53]. Copyright (2012) Hindawi.

13.6.1.2 Diffuse Reflectance Spectroscopy (DRS)

UV-Vis diffuse reflectance spectroscopy is a widely used characterization method to gather information about the electronic transitions in solids. This technique is useful for the characterization of powder samples that show enhanced scattering. Moreover, there is no need to disperse the sample in a liquid medium, so the material is not contaminated and can be reused. The bandgap, E_g, can be measured by this method by applying the Kubelka-Munk method (K-M or F(R)), as given by:

$$F(R) = \frac{(1-R)^2}{2R} \quad (8)$$

where R is the reflectance; F(R) is proportional to the extinction coefficient.

Using the K-M equation and plotting $(F(R)h\nu)^n$ as a function of energy (where n depends on the type of electronic transition), the value of E_g can be evaluated similarly to the Tauc method discussed prior [50].

13.6.2 Photoluminescence (PL) Spectroscopy

Photoluminescence is the phenomenon exhibited by the sample when it is excited by the absorption of photons and then emits them with a decay time characteristic of the sample environment. In PL spectroscopy, the wavelength at which the sample emits radiation is used to identify the atoms, molecules, and chemical structures [44]. This technique is extremely sensitive to the structural changes of the nanomaterials, which makes it important and widely used to study the dynamic properties of nanomaterials [51–52]. PL spectroscopy is used for determining the bandgap, detecting impurity levels and defects, and studying recombination mechanisms, material quality, molecular structure, and crystallinity.

In the experimental setup, the first monochromator is used to select an excitation wavelength, whereas the second one, usually positioned at 90° to the incident light, is used to observe luminescence. The two monochromators are kept perpendicular to minimize the intensity of scattered light.

When the dimensions of semiconductors are reduced to a nanoscale, their physical properties change significantly. Photoluminescence, unlike XRD, IR, and Raman spectroscopy, is very sensitive to the surface effects or adsorbed species of semiconductor particles. So, PL can be used as a probe of electron-hole surface processes and to study compound semiconductor surfaces. For device fabrication, using semiconductors of smaller dimensions, such studies help check the feasibility of improved technology. PL spectroscopy provides information about the surface state density by intensity variations and width of the spectrum. Surface states are due to the interruption of the periodic arrangement of the atoms or to the deposition of impurities at the surface. This effect is more pronounced in nanoparticles due to their large surface-to-volume ratio. Figure 13.11b shows the PL spectrum of ZnO nanoparticles having an excitation wavelength of 320 nm at RT [53], which depicts two emission peaks at ~ 392 nm (UV region) and around 520 nm. These peaks are identified due to the near bandgap exciton emission and the existence of singly ionized oxygen vacancies, respectively [53]. This emission is caused by the radiative recombination of a photogenerated hole with an electron occupying the oxygen vacancy [53]. The PL spectrum also shows the narrow size distribution of nanoparticles in the ZnO powder as the full width half maximum (FWHM) of the luminescence peak is only in few nanometers.

Time-resolved photoluminescence (TRPL) is also useful for studying fast electronic deactivation processes which result in the emission of photons. The lifetime of a molecule in its lowest excited singlet state depends on the molecular environment (such as solvent, presence of quenchers, or temperature) and also on the interaction with other molecules [54]. Suparna Sadhu et al. [55] studied the photophysical properties of CdS nanoparticles by TRPL spectroscopy. The obtained results suggest that the fast and slow components decay times of capped CdS nanocrystals are due to trapping of carriers in surface states and/or by e–h radiative recombination processes, respectively [55].

13.6.3 Fourier Transform Infrared (FTIR) Spectroscopy

FTIR is a characterization technique used to obtain the infrared spectrum of absorption, emission, and photoconductivity of solids, liquids, and gases [56]. In FTIR spectroscopy, the process of Fourier transform is used to convert the raw data (interferogram) into the actual spectrum. FTIR spectra are important to generally identify and characterize unknown materials, attached functional groups, contaminants in material, etc. FTIR records and collects high spectral resolution data, usually between the wavenumber range 5000 and 400 cm^{-1} for the mid-IR region and between 10,000–4000 cm^{-1} for the near-IR region, having a typical resolution of 4 cm^{-1} [55].

Kusior et al. [57] studied the surface properties of TiO$_2$ nanoparticles using FTIR absorption spectra as shown in Figure 13.12 recorded for the different nanostructures of TiO$_2$. It has been found that bands between 500–800 cm^{-1} correlate well with the vibration modes of the anatase phase. Observed FTIR peaks at 570 cm^{-1}, 690 cm^{-1}, and 800 cm^{-1} correspond to the Ti-O-Ti stretching bond vibrations, Ti-O-O bonds, and due to O-O vibrations, respectively [57]. Other observed broad peaks in between 2600–3700 cm^{-1} and a sharp peak at 1658 cm^{-1} were assigned due to the presence of chemically and physically adsorbed water molecules, respectively [58].

The attenuated total reflectance FTIR (ATR-FTIR) spectroscopy technique is used for in-situ characterization of the liquid-solid or solid/air interface to probe surface adsorption on nanoparticle surfaces while minimizing the drawbacks of sample preparation complexity and lack of spectra reproducibility in the conventional transmission IR mode [58]. ATR-FTIR is based on the principle of total internal reflection of infrared radiation at the boundary of two interfaces. A thin film of the nanoparticles under study is coated on the ATR crystal, which is infrared transparent and has a refractive index higher than the sample, so that the condition of total internal reflection is fulfilled. The radiation penetrates slightly beyond the ATR crystal into the sample at each reflection. So, the incident beam is attenuated as it comes out of the ATR crystal after 10–20 reflections; since the beam has penetrated a small distance through the sample at each reflection, this attenuation gives rise to a reasonable spectrum [59]. The amount of penetration into the sample is called the penetration depth, which is given by the expression:

$$d_p = \frac{\lambda}{2\pi n_1 \sqrt{\left[sin^2\theta - \left(\frac{n_2}{n_1}\right)^2 \right]}} \tag{9}$$

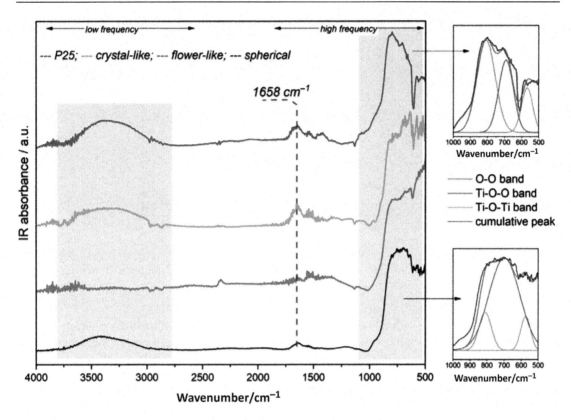

FIGURE 13.12 FTIR spectra recorded for different types of TiO$_2$ nanomaterials. Adapted with permission (under open access) from [57]. Copyright (2018) Elsevier.

where d_p is the penetration depth, n_1 is the refractive index of the optically dense medium (ATR crystal), n_2 is the refractive index of the optically rare medium (sample), and λ is the wavelength. d_p is of the order of 1 μm [60].

This is an excellent tool to study surfaces since it is only the surface of the sample which is penetrated by the radiation. The penetration depth can be varied to some extent by changing the angle of incidence, and hence the composition of the sample surface with varying penetration depth can be studied [59]. This provides a method to study changes in surface properties and chemical composition (e.g., of polymers).

Diffuse reflectance infrared Fourier transform spectroscopy (DRIFTS) is a convenient technique commonly used for the analysis of samples that can be ground into a fine powder (less than 10 microns) and mixed with an IR transparent material like potassium bromide (KBr) [61]. The IR beam incident on the sample cup containing a mixture of IR transparent material and sample is reflected from their surface, causing light to diffuse and scatter as it moves throughout the sample. The detector collects the interferogram signal, which is then used to generate the spectrum. Mei Chen et al. [62] performed an in-situ DRIFTS technique to study the NO$_2$ sensing mechanism of flower-like and tube-like ZnO nanomaterials. A study suggests that the NO$_2$ response process mainly included electron transfer on the donor sites and/or the participation of surface oxygen species to form the nitrate species.

13.6.4 Raman Spectroscopy

Raman spectroscopy is a non-invasive method of characterization materials that provides detailed information about chemical structure, phase and polymorphism, crystallinity, and molecular interactions [63].

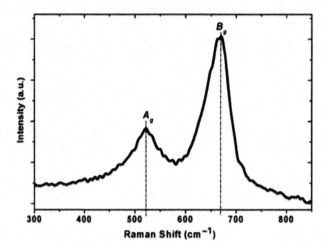

FIGURE 13.13 Raman spectrum of the as-synthesized LNMO nanoparticles. Adapted with permission from [64]. Copyright (2012) Royal Society of Chemistry.

In this method, light from a high-intensity laser source is made to fall on a cell, usually a narrow glass or quartz tube filled with the sample, which scatters the light. Light scattered at the same wavelength as the laser source (Rayleigh scattering) does not provide much information, but a small percentage of light (typically 0.0000001%) is scattered at different wavelengths (Raman scattering) depending on the chemical structure of the analyte. It has several advantages over IR spectroscopy. In a typical Raman spectrometer, light scattered sideways from the sample is collected by a lens and passed through a grating monochromator. The signal is measured by a sensitive photomultiplier, and after amplification, a spectrum is formed and then processed by a computer [59, 63].

Raman spectroscopy is an excellent tool to predict the simultaneously existing phases in a compound. A typical Raman spectrum features many peaks at different wavelength positions, which correspond to a specific molecular bond vibration, lattice vibrations, etc. Generally, the Raman spectrum is like a chemical fingerprint for a particular molecule or material. This is very useful in quickly identifying the material. Figure 13.13 displays the Raman spectrum for synthesized La_2NiMnO_6 nanoparticles [64] having two broad peaks at around 524 and 670 cm^{-1}. These peaks were assigned to the A_g antisymmetric stretching (or Jahn-Teller stretching mode) and B_g symmetric stretching vibrations of the MnO_6 octahedra, respectively. It is noticeable from the Raman spectra that the peak positions shifted towards the higher energy side as compared to the bulk crystal, which is due to surface strain from the nanosized particles. Kumar et al. [65] have studied the spin-phonon coupling in A_2CoMnO_6 (A = La, Pr, and Nd) by the temperature-dependent Raman spectroscopy technique. They have estimated the coupling strength constant for the two modes and found that many Raman modes are consistent with the monoclinic structure of the compound.

13.7 DIELECTRIC MEASUREMENTS OF NANOMATERIALS

A dielectric material is an electrical insulator that can be polarized after applying an electric field across it. Dielectric materials have great importance in our day-to-day life mainly in electronic circuits, which require a dielectric medium to form a circuit. High-frequency electronic circuits are made up of dielectric materials, and all the functioning of that particular circuit depends on the properties of the specific material [66]. The storage capacity and dissipation of electric and magnetic fields in materials are also obtained from the dielectric properties [67]. An electrically insulating dielectric material gets polarized under an applied electric field, and this phenomenon is known as dielectric polarization [68]. The polarizability in terms of permittivity of the materials is a complex number, and the real part of it is known as the dielectric constant. We can say that the materials with zero electrical conductivity are perfectly namely dielectric and all insulators cannot be dielectrics. The knowledge of the properties of dielectric materials namely dielectric constant and loss tangent under the functioning conditions are essential for designing a high-frequency circuit. Apart from it, impedance spectroscopy analysis is also a significant tool for the analysis of complex electrical properties of any dielectric material. It can divide the capacitive and resistive contributions of grains, grain boundaries, and electrode specimen interface in the materials [69]. The grains and grain boundary usually show diverse conducting nature. In the present section, efforts have been made to discuss the dielectric properties, i.e., frequency and temperature-dependent dielectric constant and loss tangent including impedance spectroscopy.

13.7.1 Frequency-Dependent Dielectric Measurements

The frequency-dependent (42 Hz to 5MHz) dielectric constant and loss tangent data are represented in Figure 13.14 for pure BiFeO$_3$ (BFO) and Ni-doped BFO nanoparticles of composition BiFe$_{1-x}$Ni$_x$O$_3$ (0 ≤ x ≤ 0.07) via the sol-gel route [66]. The pure and doped compositions are represented as BFO, BFNO1 (1% Ni

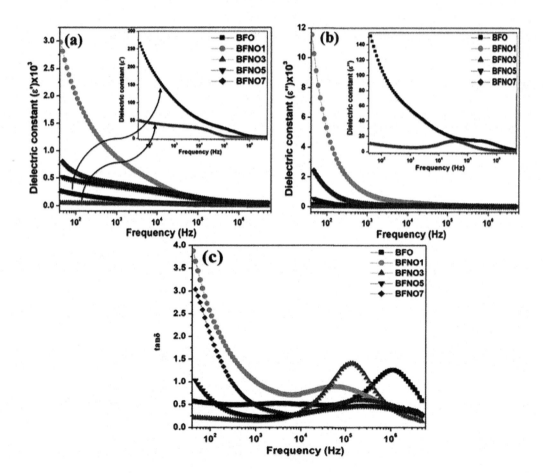

FIGURE 13.14 Frequency-dependent dielectric constant (a) real part, (b) imaginary part, and (c) dielectric loss for pure BFO and Ni-doped BFO nanoparticles, i.e., BFNO1, BFNO3, BFNO5, and BFNO7. Adapted with permission from [66]. Copyright (2018) AIP Publishing.

in BFO), BFNO3 (3% Ni in BFO), BFNO5 (5% Ni in BFO), and BFNO7 (7% Ni in BFO) [65]. Both the dielectric constant and the loss tangent represent a similar kind of decreasing behavior with an increase in frequency. It is basically due to the space charge polarization phenomenon because the strength of the dielectric constant depends on it. However, this type of reduction in the dielectric constant may be understood with the help of the Maxwell-Wagner-type interfacial polarization related to dipolar relaxation [67]. As per this phenomenon, low values of dielectric constant arise in the material due to the space charge carrier alignment at low frequencies, and at high frequencies, dipoles do not follow a rapid deviation of the applied field and result in a drop of dielectric constant. Also, there are many polarization contributions observed in the dielectric constant at a low-frequency regime, such as ionic, electronic, interfacial, and orientational polarization [68]. The electronic polarization in the materials is responsible for the dielectric constant because, with the increase in the frequency, both ionic and orientational polarizability do not contribute and turn out to be irrelevant. In the low-frequency regime, a high dielectric constant is observed because of the existence of the space charge polarization due to the inhomogeneous dielectric structure. On account of this reasoning, it is confirmed that the dielectric properties of $BiFe_{1-x}Ni_xO_3$ ($0 \leq x \leq 0.07$) nanoparticles are enhanced with the Ni doping in the pristine BFO multiferroic material [66].

13.7.2 Temperature-Dependent Dielectric Measurements

Now, apart from the discussion of the frequency-dependent dielectric constant, the temperature-dependent dielectric constant is also very useful for various purposes, such as information about the dielectric constant and loss tangent at various temperatures at fixed frequencies, to study the dielectric T_c, and to investigate the dielectric relaxation in the samples at various temperatures, etc. In this section, Kumar et al. reported about the sol-gel derived nanocomposites of LSMO ($La_{0.67}Sr_{0.33}MnO3$) and BTO ($BaTiO_3$), and Figure 13.15 represents the dielectric constant and loss tangent (inset of Figure 13.15) of one of the nanocomposites LB5 ($La_{0.67}Sr_{0.33}MnO_3\_5\%BaTiO_3$) at various fixed frequencies (10 kHz, 100 kHz, and 1 MHz) [69]. The

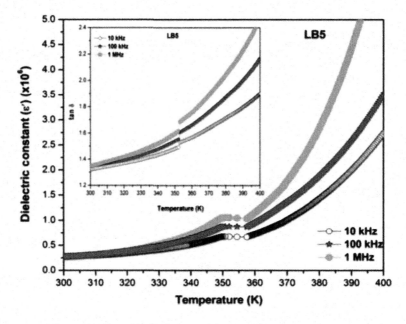

FIGURE 13.15 Temperature-dependent dielectric constant at fixed frequencies (10 kHz, 100 kHz, and 1 MHz) of LB5 nanocomposite. The inset shows dielectric loss variations with a temperature of LB5. Adapted with permission from [69]. Copyright (2015) AIP Publishing.

dielectric constant and loss tangent also show the increasing behavior with the increase of temperature, which may be due to the Maxwell-Wagner-type contribution to the dielectric constant [70]. Interfacial polarization is the region of dielectric constant enhancement in the LSMO-BTO composite, which may be due to the BTO filler particles that are acting as trapping centers at the interface. Apart from this, the metallic nature of the host material (LSMO) is the basic cause of the dissipation factor in the LSMO-BTO composites. Figure 13.15 for the dielectric constant and inset of Figure 13.15 for the loss tangent/dissipation factor clearly show the strong frequency-dependent peak around 350 K, and the suppression of the peak with the increase of frequency confirms the dielectric T_c and relax or type behavior in the samples [69].

13.8 CONCLUSIONS

Different experimental techniques, XRD, Raman, FTIR, PL, and UV-Vis techniques, were explored along with electron microscopy (SEM/TEM) and AFM/MFM techniques in this chapter to characterize the nanomaterials. The dielectric properties measurement technique was also discussed in detail. Overall, this chapter provided an overview of nanomaterials characterization techniques.

13.9 ACKNOWLEDGMENTS

Arvind Kumar wants to acknowledge the financial support received from the UGC in the form of the UGC-BSR Research Start-up Grant (F.30–374/2017(BSR)), New Delhi, India. I am also thankful to Mr. Devendra Pal (McGill University, Montreal, Canada) and Ms. Kavita for their valuable support and suggestions time to time during this piece of work.

13.10 REFERENCES

[1] Hines, M. A., &Guyot-Sionnest, P. (1996). Synthesis and characterization of strongly luminescing ZnS-capped CdSe nanocrystals. *The Journal of Physical Chemistry*, *100*(2), 468–471.
[2] Goesmann, H., & Feldmann, C. (2010). Nanoparticulate functional materials. *AngewandteChemie International Edition*, *49*(8), 1362–1395.
[3] Chen, G., Roy, I., Yang, C., & Prasad, P. N. (2016). Nanochemistry and nanomedicine for nanoparticle-based diagnostics and therapy. *Chemical Reviews*, *116*(5), 2826–2885.
[4] Lohse, S. E., & Murphy, C. J. (2012). Applications of colloidal inorganic nanoparticles: From medicine to energy. *Journal of the American Chemical Society*, *134*(38), 15607–15620.
[5] Mahajan, K. D., Fan, Q., Dorcéna, J., Ruan, G., &Winter, J. O. (2013). Magnetic quantum dots in biotechnology – synthesis and applications. *Biotechnology Journal*, *8*(12), 1424–1434.
[6] Burda, C., Chen, X., Narayanan, R., & El-Sayed, M. A. (2005). Chemistry and properties of nanocrystals of different shapes. *Chemical Reviews*, *105*(4), 1025–1102.
[7] Sau, T. K., Rogach, A. L., Jäckel, F., Klar, T. A., & Feldmann, J. (2010). Properties and applications of colloidal nonspherical noble metal nanoparticles. *Advanced Materials*, *22*(16), 1805–1825.
[8] Billinge, S. J., & Levin, I. (2007). The problem with determining atomic structure at the nanoscale. *Science*, *316*(5824), 561–565.
[9] Gommes, C. J., Prieto, G., Zecevic, J., Vanhalle, M., Goderis, B., de Jong, K. P., & de Jongh, P. E. (2015). Mesoscale Characterization of Nanoparticles Distribution Using X-ray Scattering. *Angewandte Chemie*, *127*(40), 11970–11974.

[10] Gilfrich, J. (1993). *Structural and chemical analysis of materials: X-ray, electron and neutron diffraction, X-ray, electron and ion spectrometry, electron microscopy.* JP Eberhart Published by John Wiley & Sons Ltd, 1991; 545 pages; ISBN 0 471 92977 8.
[11] Cullity, B. D. (1956). *Elements of X-ray diffraction.* Addison-Wesley Publishing.
[12] Mourdikoudis, S., Pallares, R. M., &Thanh, N. T. (2018). Characterization techniques for nanoparticles: Comparison and complementarity upon studying nanoparticle properties. *Nanoscale, 10*(27), 12871–12934.
[13] Shanker, J., Suresh, M. B., &Babu, D. S. (2016). Synthesis, characterization and electrical properties of NdXO$_3$ (X= Cr, Fe) nanoparticles. *Materials Today: Proceedings, 3*(6), 2091–2100.
[14] Schwartz, L. H., & Cohen, J. B. (2013). *Diffraction from materials.* Springer Science & Business Media.
[15] Cline, J. P., Mendenhall, M. H., Black, D., Windover, D., & Henins, A. (2015). The optics and alignment of the divergent beam laboratory X-ray powder diffractometer and its calibration using NIST standard reference materials. *Journal of Research of the National Institute of Standards and Technology, 120,* 173.
[16] Sharma, R., Bisen, D. P., Shukla, U., & Sharma, B. G. (2012). X-ray diffraction: A powerful method of characterizing nanomaterials. *Recent Research in Science and Technology, 4*(8).
[17] Patterson, A. L. (1939). The Scherrer formula for X-ray particle size determination. *Physical Review, 56*(10), 978.
[18] Dorofeev, G. A., Streletskii, A. N., Povstugar, I. V., Protasov, A. V., & Elsukov, E. P. (2012). Determination of nanoparticle sizes by X-ray diffraction. *Colloid Journal, 74*(6), 675–685.
[19] Prabhu, Y. T., Rao, K. V., Kumar, V. S. S., & Kumari, B. S. (2013). Synthesis of ZnO nanoparticles by a novel surfactant assisted amine combustion method. *Advances in Nanoparticles, 2*(01), 45.
[20] Echlin, P. (2011). *Handbook of sample preparation for scanning electron microscopy and X-ray microanalysis.* Springer Science & Business Media.
[21] Ji, X., Li, H., Wu, Z., Cheng, S., Hu, H., Yan, D., & Yan, P. (2011). Growth of AlN hexagonal oriented complex nanostructures induced by nucleus arrangement. *Cryst Eng Comm, 13*(16), 5198–5203.
[22] Zhou, N., Li, D., & Yang, D. (2014). Morphology and composition-controlled synthesis of flower-like silver nanostructures. *Nanoscale Research Letters, 9*(1), 1–6.
[23] Carter, B. A., Williams, D. B., Carter, C. B., & Williams, D. B. (1996). *Transmission Electron Microscopy: A Textbook for Materials Science. Diffraction. II* (Vol. 2). Springer Science & Business Media.
[24] Heydenreich, J. (1995). S. Horiuchi. *Fundamentals of high-resolution transmission electron microscopy.* 1994, 342 pages, 212 figures, 12 tables ISBN 0-444–88744 – X, Preis: Dfl, 375, 00, US $214.25.
[25] Kohl, H., & Reimer, L. (2008). Transmission electron microscopy. *Springer Verlag Springer Series on Optical Sciences, 36.*
[26] Manchala, S., Tandava, V. S. R. K., Jampaiah, D., Bhargava, S. K., & Shanker, V. (2019). Novel and highly efficient strategy for the green synthesis of soluble graphene by aqueous polyphenol extracts of eucalyptus bark and its applications in high-performance supercapacitors. *ACS Sustainable Chemistry & Engineering, 7*(13), 11612–11620.
[27] Sun, Y. S., Huang, W. H., Liou, J. Y., Lu, Y. H., Shih, K. C., Lin, C. F., & Cheng, S. L. (2015). Conversion from self-assembled block copolymer nanodomains to carbon nanostructures with well-defined morphology. *RSC Advances, 5*(128), 105774–105784.
[28] Grütter, P., Rugar, D., & Mamin, H. J. (1992). Magnetic force microscopy of magnetic materials. *Ultramicroscopy, 47*(4), 393–399.
[29] Kumar, A., & Srivastava, P. C. (2015). Magnetic, structural and transport properties across the Heusler alloy (Co$_2$FeAl)/n-Si interfacial structure. *Journal of Materials Science: Materials in Electronics, 26*(8), 5611–5617.
[30] Kumar, A., & Srivastava, P. C. (2014). Electronic and magneto-transport across the Heusler alloy (Co$_2$FeAl)/p-Si interfacial structure. *Journal of Electronic Materials, 43*(2), 381–388.
[31] Kano, S., Tada, T., & Majima, Y. (2015). Nanoparticle characterization based on STM and STS. *Chemical Society Reviews, 44*(4), 970–987.
[32] Binnig, G., Rohrer, H., Gerber, C., & Weibel, E. (1982). Surface studies by scanning tunneling microscopy. *Physical Review Letters, 49*(1), 57.
[33] Rieboldt, F., Vilhelmsen, L. B., Koust, S., Lauritsen, J. V., Helveg, S., Lammich, L., & Wendt, S. (2014). Nucleation and growth of Pt nanoparticles on reduced and oxidized rutile TiO$_2$ (110). *The Journal of Chemical Physics, 141*(21), 214702.
[34] Moulder, J. F. (1995). Handbook of X-ray photoelectron spectroscopy. *Physical Electronics,* 230–232.
[35] Briggs, D., & Grant, J. T. (Eds.). (2012). *Surface analysis by Auger and X-ray photoelectron spectroscopy.* SurfaceSpectra.
[36] Ray, S., & Shard, A. G. (2011). Quantitative analysis of adsorbed proteins by X-ray photoelectron spectroscopy. *Analytical Chemistry, 83*(22), 8659–8666.

[37] Tarachand, Hussain, S., Lalla, N. P., Kuo, Y. K., Lakhani, A., Sathe, V. G., Deshpande, U., & Okram, G. S. (2018). Thermoelectric properties of Ag-doped CuS nanocomposites synthesized by a facile polyol method. *Physical Chemistry Chemical Physics, 20*(8), 5926–5935.

[38] Laidani, N., Calliari, L., Speranza, G., Micheli, V., & Galvanetto, E. (1998). Mechanical and structural properties of Ni-C films obtained by RF sputtering. *Surface and Coatings Technology, 100*, 116–124.

[39] Mattevi, C., Eda, G., Agnoli, S., Miller, S., Mkhoyan, K. A., Celik, O., & Chhowalla, M. (2009). Evolution of electrical, chemical, and structural properties of transparent and conducting chemically derived graphene thin films. *Advanced Functional Materials, 19*(16), 2577–2583.

[40] Riha, S. C., Johnson, D. C., & Prieto, A. L. (2011). Cu_2Se nanoparticles with tunable electronic properties due to a controlled solid-state phase transition driven by copper oxidation and cationic conduction. *Journal of the American Chemical Society, 133*(5), 1383–1390.

[41] Zhang, J., Yu, J., Zhang, Y., Li, Q., & Gong, J. R. (2011). Visible light photocatalytic H2-production activity of CuS/ZnS porous nanosheets based on photoinduced interfacial charge transfer. *Nano letters, 11*(11), 4774–4779.

[42] Folmer, J. C. W., & Jellinek, F. (1980). The valence of copper in sulphides and selenides: An X-ray photoelectron spectroscopy study. *Journal of the Less Common Metals, 76*(1–2), 153–162.

[43] Kundu, M., Hasegawa, T., Terabe, K., & Aono, M. (2008). Effect of sulfurization conditions on structural and electrical properties of copper sulfide films. *Journal of Applied Physics, 103*(7), 073523.

[44] Rihn, B. H. (Ed.). (2017). *Biomedical application of nanoparticles*. CRC Press.

[45] Tom, J. Analysis & separations. www.technologynetworks.com/analysis/articles/uv-vis-spectroscopy-principle-strengths-and-limitations-and-applications-349865

[46] Mohd Sultan, N., & Johan, M. R. (2014). Synthesis and ultraviolet visible spectroscopy studies of Chitosan capped gold nanoparticles and their reactions with analytes. *The Scientific World Journal, 2014*.

[47] Patel, P. C., Srivastava, N., & Srivastava, P. C. (2013). Synthesis of wurtzite ZnS nanocrystals at low temperature. *Journal of Materials Science: Materials in Electronics, 24*(10), 4098–4104.

[48] Tauc, J. (Ed.). (2012). *Amorphous and liquid semiconductors*. Springer Science & Business Media.

[49] Deotale, A. J., &Nandedkar, R. V. (2016). Correlation between particle size, strain and band gap of iron oxide nanoparticles. *Materials Today: Proceedings, 3*(6), 2069–2076.

[50] López, R., & Gómez, R. (2012). Band-gap energy estimation from diffuse reflectance measurements on sol–gel and commercial TiO_2: A comparative study. *Journal of Sol-Gel Science and Technology, 61*(1), 1–7.

[51] Mohapatra, S., Nguyen, T. A., & Nguyen-Tri, P. (Eds.). (2018). *Noble metal-metal oxide hybrid nanoparticles: Fundamentals and applications*. Elsevier.

[52] Harris, D. C. (2010). *Quantitative chemical analysis*. Macmillan.

[53] Talam, S., Karumuri, S. R., &Gunnam, N. (2012). Synthesis, characterization, and spectroscopic properties of ZnO nanoparticles. *International Scholarly Research Notices, 2012*.

[54] Nevin, A., Cesaratto, A., Bellei, S., D'Andrea, C., Toniolo, L., Valentini, G., & Comelli, D. (2014). Time-resolved photoluminescence spectroscopy and imaging: New approaches to the analysis of cultural heritage and its degradation. *Sensors, 14*(4), 6338–6355.

[55] Sadhu, S., Chowdhury, P. S., & Patra, A. (2008). Synthesis and time-resolved photoluminescence spectroscopy of capped CdS nanocrystals. *Journal of Luminescence, 128*(7), 1235–1240.

[56] Ghosh, S., & Basu, R. N. (2019). Nanoscale characterization. In *Noble metal-metal oxide hybrid nanoparticles* (pp. 65–93). Woodhead Publishing.

[57] Kusior, A., Banas, J., Trenczek-Zajac, A., Zubrzycka, P., Micek-Ilnicka, A., & Radecka, M. (2018). Structural properties of TiO_2 nanomaterials. *Journal of Molecular Structure, 1157*, 327–336.

[58] Song, K. (2017). Interphase characterization in rubber nanocomposites. In *Progress in rubber nanocomposites* (pp. 115–152). Woodhead Publishing.

[59] Banwell, C. N. (1972). *Fundamentals of molecular spectroscopy*. McGraw Hill Publishing Company.

[60] Beasley, M. M., Bartelink, E. J., Taylor, L., & Miller, R. M. (2014). Comparison of transmission FTIR, ATR, and DRIFT spectra: Implications for assessment of bone bioapatite diagenesis. *Journal of Archaeological Science, 46*, 16–22.

[61] Pandey, K. K., &Theagarajan, K. S. (1997). Analysis of wood surfaces and ground wood by diffuse reflectance (DRIFT) and photoacoustic (PAS) Fourier transform infrared spectroscopic techniques. *HolzalsRoh-und Werkstoff, 55*(6), 383–390.

[62] Chen, M., Wang, Z., Han, D., Gu, F., &Guo, G. (2011). High-sensitivity NO_2 gas sensors based on flower-like and tube-like ZnO nanomaterials. *Sensors and Actuators B: Chemical, 157*(2), 565–574.

[63] Das, R. S., & Agrawal, Y. K. (2011). Raman spectroscopy: Recent advancements, techniques and applications. *Vibrational Spectroscopy, 57*(2), 163–176.

[64] Mao, Y. (2012). Facile molten-salt synthesis of double perovskite La$_2$BMnO$_6$ nanoparticles. *RSC Advances, 2*(33), 12675–12678.

[65] Kumar, D., Kumar, S., & Sathe, V. G. (2014). Spin – phonon coupling in ordered double perovskites A$_2$CoMnO$_6$ (A= La, Pr, Nd) probed by micro-Raman spectroscopy. *Solid State Communications, 194*, 59–64.

[66] Nadeem, M., Khan, W., Khan, S., Husain, S., & Ansari, A. (2018). Tailoring dielectric properties and multiferroic behavior of nanocrystalline BiFeO$_3$ via Ni doping. *Journal of Applied Physics, 124*(16), 164105.

[67] Kaur, M., & Uniyal, P. (2017). Study on structural, multiferroic, optical and photocatalytic properties of ferroelectromagnetic nanoparticles: Bi$_{0.9}$Ba0.1Fe$_{0.8}$Ti$_{0.2}$O$_3$. *Journal of Superconductivity and Novel Magnetism, 30*(2), 431–439.

[68] Chakrabarti, K., Das, K., Sarkar, B., Ghosh, S., De, S. K., Sinha, G., & Lahtinen, J. (2012). Enhanced magnetic and dielectric properties of Eu and Co co-doped BiFeO$_3$ nanoparticles. *Applied Physics Letters, 101*(4), 042401.

[69] Kumar, M., Shankar, S., Dwivedi, G. D., Anshul, A., Thakur, O. P., & Ghosh, A. K. (2015). Magneto-dielectric coupling and transport properties of the ferromagnetic-BaTiO$_3$ composites. *Applied Physics Letters, 106*(7), 072903.

[70] Kamba, S., Nuzhnyy, D., Savinov, M., Šebek, J., Petzelt, J., Prokleška, J., & Kreisel, J. (2007). Infrared and terahertz studies of polar phonons and magnetodielectric effect in multiferroic BiFeO$_3$ ceramics. *Physical Review B, 75*(2), 024403

Mathematical Modeling and Simulation of Exchange Coupling Constant (*J*) and Zero-Field Splitting Parameters (*D*)

14

Satadal Paul
Department of Chemistry, Bangabasi Morning College, Kolkata, WB, India

Contents

14.1	Exchange Coupling of Spins	269
	14.1.1 Mathematical Modeling	270
	14.1.1.1 Broken Symmetry Approach	273
	14.1.2 Computational Methods	276
14.2	Magnetic Anisotropy	278
	14.2.1 Mathematical Modeling	280
	14.2.2 Computational Methods	282
14.3	Summary and Outlook	284
14.4	References	285

14.1 EXCHANGE COUPLING OF SPINS

Magnetism in a system stems from the ordering of magnetic moments arising from the spin angular momenta of unpaired electrons, which is an intrinsic quantum chemical property and does not have any classical analogue. The coupling between spins centered at different magnetic sites is crucial for understanding the properties of magnetic materials. There exists a miscellany of exchange mechanisms, such as direct exchange, indirect exchange, double exchange, superexchange, and so on, through which the spins can interact.[1] The direct exchange operates through space between two localized spin moments, which are close enough to have sufficient overlap of their wave functions. It gives a strong but short-range coupling, which decreases rapidly as the spin sites are separated. Superexchange describes the interaction between spins on sites far enough apart to be connected by direct exchange but coupled over a relatively long distance through a non-magnetic bridge. This mechanism acquired the name "superexchange" because of the relatively large distance between the magnetic sites. Indirect exchange is the phenomenon of the coupling of conduction electrons with localized spins in a metal. This is the dominant exchange interaction in metals, where there is little or no direct overlap between neighboring electrons. It therefore acts through an intermediary, in which metals are the conduction electrons (itinerant electrons). On the other hand, the coupling of two localized spins through an itinerant electron is defined as a double exchange. In the case of mixed-valence systems, where metal centers are present with different oxidation states, Anderson and Hasegawa introduced the concept of double exchange.[2] Irrespective of the exchange mechanism, the exchange coupling among the spins is usually quantified through the phenomenological Heisenberg-Dirac-van Vleck (HDVV) spin Hamiltonian,

$$\hat{H} = -J_{ij} \hat{S}_i \cdot \hat{S}_j \qquad (1)$$

where, $\hat{S}_i$ and $\hat{S}_j$ are the spin angular momentum operators on magnetic sites i and j and J_{ij} is the exchange coupling constant between them. The first spin Hamiltonian in the literature was presented by van Vleck in his famous book in 1932.[1] This Hamiltonian became popular as the "spin Hamiltonian" since it involves only the spin degrees of freedom.

14.1.1 Mathematical Modeling

There is a one-to-one correspondence between the eigenvalues of the HDVV Hamiltonian (Eq. 1) and those of the exact Hamiltonian because both the Hamiltonian (HDVV and exact) commute with the spin square operator. For example, in a diradical, the possible lowest energy electronic states are singlet ($S = 0$) and triplet ($S = 1$), which are the eigen states of the Heisenberg Hamiltonian. Let, in the dimer A–B, the spins are localized in the orbitals ϕ_a and ϕ_b, and the molecular orbitals (MOs) formed through a linear combination of these orbitals are

$$\phi_1 = \frac{1}{\sqrt{2}}(\phi_a + \phi_b) \text{ and } \phi_2 = \frac{1}{\sqrt{2}}(\phi_a - \phi_b). \qquad (2)$$

There are several possibilities of electron distribution in these two Mos, giving rise to multiple configurations (Figure 14.1), where the overbar denotes the occupation with spin-down electrons. The lowest singlet state of the system (ψ_s) is an approximately equal mixture of S_1 and S_2, while the lowest triplet state is ψ_T. Since S_3 is in a different symmetric form compared to S_1 and S_2, this state corresponds to an excited state much higher in energy. Hence, singlet-triplet splitting becomes

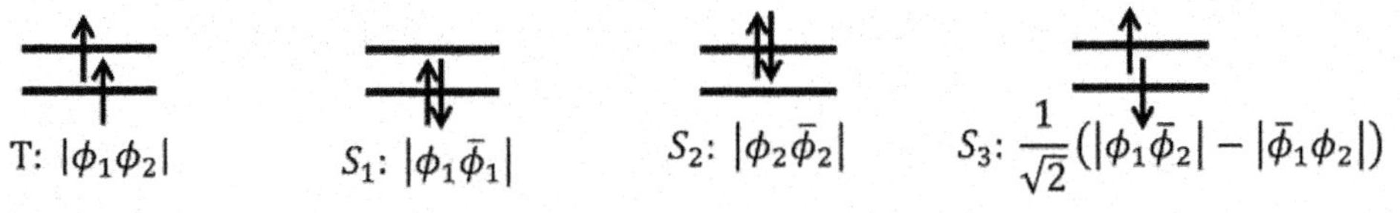

T: $|\phi_1\phi_2|$ S_1: $|\phi_1\bar{\phi}_1|$ S_2: $|\phi_2\bar{\phi}_2|$ S_3: $\frac{1}{\sqrt{2}}(|\phi_1\bar{\phi}_2| - |\bar{\phi}_1\phi_2|)$

FIGURE 14.1 Multi-electronic configuration for two electrons in two MOs.

$$E_T - E_S = J_{12} - \frac{1}{2}(J_{11} + J_{12}) + \frac{(h_1 - h_2)^2}{2K_{12}} \tag{3}$$

where,

$$h_i = \int \phi_i^*(1)\hat{h}\phi_i(1)dr_1$$

$$J_{ij} = \int \phi_i^*(1)\phi_j^*(2)\frac{1}{r_{12}}\phi_i(1)\phi_j(2)dr_1dr_2$$

$$K_{ij} = \int \phi_i^*(1)\phi_j^*(2)\frac{1}{r_{12}}\phi_j(1)\phi_i(2)dr_1dr_2$$

Now, in terms of the orthogonal localized MOs, defined as $\phi_a = \frac{1}{\sqrt{2}}(\phi_1 + \phi_2)$ and $\phi_b = \frac{1}{\sqrt{2}}(\phi_1 - \phi_2)$,

$$J_{11} = \tfrac{1}{2}(J_{aa} + J_{ab}) + K_{ab} - 2\langle aa|bb\rangle$$

$$J_{22} = \tfrac{1}{2}(J_{bb} + J_{ab}) + K_{ab} - 2\langle aa|bb\rangle$$

$$J_{12} = \tfrac{1}{2}(J_{aa} + J_{ab}) - K_{ab}$$

$$K_{12} = \tfrac{1}{2}(J_{aa} - J_{ab})$$

and the singlet-triplet energy splitting can be expressed in terms of the localized orbitals as,[3]

$$E_T - E_S = -2K_{12} + \frac{(h_1 - h_2)^2}{2(J_{aa} - J_{ab})} = J \tag{4}$$

For the degenerate case $h_1 = h_2$, the triplet state is the ground state, whereas a large energy gap between ϕ_1 and ϕ_2 will lead to a singlet ground state. The direct contribution of the first part to the total exchange coupling constant J is always positive and referred to as "potential" exchange by Anderson.[4] He expressed the second part in Eq. 4 in the form of t^2/U, where t is the charge transfer integral and U is the onsite repulsion energy (difference between covalent and ionic configurations), and referred this part as "kinetic" exchange.[4] However, due to the large energy separation between ionic and neutral singlet configurations, Eq. 4 often underestimates the kinetic part of the exchange.[5] Through an extensive dynamic correlation treatment, Malrieu and coworkers allowed the energetic ionic singlets to relax.[6–7] This causes a better mixing of ionic states with low-lying neutral singlet states, ultimately leading to an enhanced antiferromagnetic coupling. In a different approach, Loth et al.[8] considered the energy difference between ferromagnetic and antiferromagnetic terms, i.e., the eigenvalues of ψ_T and ψ_{S_3}, to be equal with the potential exchange, J

$$\Delta E_0^{ST} = -2K_{ab} = -2J \tag{5}$$

In case ψ_{S_3} is considered to be the only singlet state representing antiferromagnetic nature, the ground state will always be ferromagnetic due to the positive value of K_{ab} within Hartree-Fock theory. Hence, other determinants such as S_1 and S_2, corresponding to ionic configuration, also need to be considered along with the S_3 state to construct a configuration state function (CSF) representing the singlet state. Mixing of these configurations with S_3 stabilizes the antiferromagnetic state (Figure 14.2) and refines the value of the zeroth-order singlet-triplet splitting in Eq. 4. However, the ionic states which are eigenfunctions of the spin square operator will occur mainly due to the kinetic energy of electrons and give rise to

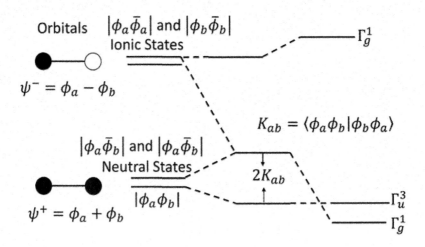

FIGURE 14.2 Stabilization of antiferromagnetic state through configuration interaction of neutral and ionic singlet states.

the kinetic part of the exchange, according to Anderson.[4] Configuration interaction of these two CSFs (S_1 and S_2) with ψ_{S_3} leads to an energy lowering of the singlet state, which can be estimated from second-order perturbation theory as,

$$\Delta E_{SE}^2 = \frac{4F_{ab}^2}{E_0 - E_{|a\bar{a}|}}. \tag{6}$$

Here, F_{ab} is the Fock operator of the system. This part in Eq. 6 necessarily has a negative contribution towards the exchange coupling constant J and adds to "kinetic exchange". Therefore, similar to the expression given by Hay et al. (Eq. 4), the total coupling constant can be represented as follows,

$$E_{S=0} - E_{S=1} = 2K_{ab} - \frac{4F_{ab}^2}{E_0 - E_{|a\bar{a}|}} = 2J_{HDVV}. \tag{7}$$

Therefore, in essence, the overall sign of J and the concomitant nature of magnetism can be considered to be an ensemble of two competitive terms.[9]

Yamaguchi and coworkers proposed an entirely different formalism, which uses an ab initio symmetry projected unrestricted Hartree-Fock (PUHF) approximation to estimate J.[10] In the UHF approximation, the singlet states ($M_S = 0$) of a magnetic molecule is generally given by,

$$\psi_n^{\pm} = cos\theta_n \phi_n \pm sin\theta_n \phi_n^* \tag{8}$$

where, φ_n and φ_n^* denote the n-th bonding and antibonding approximate natural orbitals. Therefore, taking the superposition of $(n + 1)$ PUHF solutions, the singlet UHF solution is

$$\psi_{UHF}^1 = \left| \psi_1^+ \bar{\psi}_1^- \psi_2^+ \bar{\psi}_2^- ... \psi_n^+ \bar{\psi}_n^- \right| = \sum_{S=0}^{n} C(2S+1) \phi^{2S+1}(PUHF) \tag{9}$$

where S is the total spin angular momentum of the molecule and $C(2S + 1)$ and $\phi^{2S+1}(PUHF)$ are the expansion coefficient and wave function of the spin eigen state $(2S + 1)$. The total energy and the spin square expectation values corresponding to the total wave function in Eq. 9 can thus be expressed as

$$E_{UHF}^1 = \sum_{S=0}^{n} C^2(2S+1)E^{2S+1}(PUHF) = \sum_{S=0}^{n} E^{2S+1}(PUHF)$$

and,

$$\langle \hat{S}^2 \rangle_{UHF}^1 = \sum_{S=0}^{n} C^2(2S+1)S(S+1) = \sum_{S=0}^{n} S(S+1) \quad (10)$$

Now, for a dimeric system with spin angular momentums S_a and S_b at each radical site and where $S = S_a + S_b$, the spin Hamiltonian $\hat{H} = -J_{ij}\hat{S}_i\hat{S}_j$ can be rewritten as,

$$\hat{H} = -J_{ab}(\hat{S}^2 - \hat{S}_a^2 - \hat{S}_b^2). \quad (11)$$

Therefore, using Eq. 10 and 11, the energy of singlet state can be written as,

$$E_{UHF}^1 = -J_{ab}\left(\langle \hat{S}^2 \rangle_{UHF}^1 - \hat{S}_a^2 - \hat{S}_b^2\right). \quad (12)$$

and similarly, the energy for the high spin state with total spin angular momentum S,

$$E_{UHF}^{2S+1} = -J_{ab}\left(\langle \hat{S}^2 \rangle_{UHF}^{2S+1} - \hat{S}_a^2 - \hat{S}_b^2\right). \quad (13)$$

Now, since spin contamination is negligible in the high spin state, from Eq 10,

$$\langle \hat{S}^2 \rangle_{UHF}^{2S+1} \cong \langle \hat{S}^2 \rangle_{PUHF}^{2S+1} = S(S+1). \quad (14)$$

If the energy and spin expectation values of singlet state and $(2S + 1)$ eigen state are defined as those of low spin (LS) and high spin (HS) states, then, from Eq. 12, 13, and 14, the effective exchange integral J_{ab} can be described as,[10–11]

$$J_{ab} = \frac{E_{LS} - E_{HS}}{\langle \hat{S}^2 \rangle_{HS} - \langle \hat{S}^2 \rangle_{LS}}. \quad (15)$$

However, in the case of many magnetic centers, such as a periodic system, it is not always possible to find the eigenfunctions of the HDVV Hamiltonian. In these cases, the common approach is to use the simplified version of the HDVV Hamiltonian, known as the Ising model Hamiltonian, where the total spin operators are replaced by their z-components:[12]

$$\hat{H}_{Ising} = -J_{ij}\hat{S}_i^z\hat{S}_j^z \quad (16)$$

14.1.1.1 Broken Symmetry Approach

The preceding discussion clarifies the need for proper accounting of interaction among different configurations with $M_S = 0$ to correctly describe the open-shell singlet state, i.e. the antiferromagnetic state. The configuration state function formed through several such determinants is an eigenfunction of a spin square operator.[13] To obtain such multideterminant wave functions, the best policy is to resort to correlated multireference ab initio approaches. However, these approaches have to encounter serious challenges

FIGURE 14.3 Representation of broken symmetry state.

in the simultaneous treatment of static and dynamic correlation in such systems.[13] Moreover, with the increase in the dimension of the system, the consideration of extended configuration interaction to account for correlation and a huge basis set renders the post–Hartree Fock method almost unfeasible and painstakingly resource intensive.[6–7] To overcome these problems of ab initio approaches, the best alternative has been the density functional theory (DFT) method, which, in conjunction with "broken symmetry" (BS) formalism, produces a fair estimate of the spin-exchange coupling constant.[14] This BS approach, coined by Noodleman, makes use of an unrestricted or spin-polarized formalism, where a weak antiferromagnetically coupled system is taken as the reference to represent the singlet system. The wave function for a BS state is artificially constructed through two magnetic orbitals (a and b), belonging to two different irreducible representations and having up-spin α and down-spin β at separate magnetic sites.[15] Hence, for a diradical system where there is up-and down-spin density around the magnetic centers A (having orbital a) and B (having orbital b) (Figure 14.3), the guess function for the BS state can be represented as,

$$\phi_{BS}^{Guess} = |(core)a\bar{b}| \tag{17}$$

It is important to note that the regio-specific α and β spin density distribution is qualitatively wrong since a proper singlet wave function has zero spin density at each point. The orbitals of this guess thus need to be re-optimized through the variational principle.[16] At the stationary state, the true BS wave function has the form.[13]

$$\psi_{BS} = |(core)' a'\bar{b}'| \tag{18}$$

where a' and b' are relaxed to their final form. The condition of orthogonality, as maintained by a and b, is no longer applicable to these new sets of optimized functions. Since these spin orbitals are always orthogonal in their spin part, there is no further orthogonality restriction on their space part. This flexibility is used to lower the energy of the system, leading to an antiferromagnetic ground state. Noodleman and coworkers showed that the BS state can be expressed as a weighted average of the different possible pure spin states $|\psi_S\rangle$ of the system,[17–18]

$$\psi_{BS} = \sum_S A(S)|\psi_S\rangle. \tag{19}$$

where $A(S)$ is the Clebsh-Gordon coefficient. Now, from Eq. 1 and 19, the energy splitting between the highest spin state and BS state appears as, [17–18]

$$E(S_{max}) - E(BS) = \left[S_{max}(S_{max}+1) - \sum_S A(S)S(S+1) \right] \frac{J}{2}. \tag{20}$$

Using $\sum_S A(S)S(S+1) = S_{max}$, the following expression was obtained for orthogonal magnetic orbitals by Ginseberg, Noodleman and Davidson:[17–18]

$$J_{GND} = \frac{E_{BS} - E_{HS}}{S_{max}^2}. \tag{21}$$

For $S_A \neq S_B$, this expression changes into

$$J_{GND} = \frac{E_{BS} - E_{HS}}{2S_A S_B}. \tag{22}$$

However, Eq 21 and 22 are constructed assuming magnetic orbitals to be orthogonal. Taking the overlap of magnetic orbitals into account, this expression is corrected as follows:[17–18]

$$J = \frac{E(S_{max}) - E(BS)}{2S_A S_B (1 + S_{AB}^2)}. \tag{23}$$

Noodleman and Davidson also showed that in the context of spin-polarized configuration interaction (CI) treatment, the coupling constant can be separated into the sum of terms, $J = J_F + J_{AF} + J_{LSP} + J_R$, where J_F is the ferromagnetic contribution, J_{AF} is the contribution from superexchange, J_{LSP} is due to ligand spin polarization, and J_R takes the other contributions into account. They further claimed that the coupling constant estimated through Eq. 23 includes all of the J_F, J_{AF}, and J_{LSP} terms.[17–18] Using the localized orbitals a and b (Figure 14.3), Noodleman defined the BS state as,[19–21]

$$\phi_{BS}^1 = |a\bar{b}| \text{ or } \phi_{BS}^2 = |\bar{a}b|. \tag{24}$$

Though BS wave functions are not the eigen function of spin square operator, an adequate singlet function $\phi_{|S=0, M_S=0\rangle}$ can be formed through their combination as follows,[19–21]

$$\phi_{|S=0, M_S=0\rangle} = \frac{\phi_{BS}^1 - \phi_{BS}^2}{\sqrt{2 - 2\langle \phi_{BS}^1 | \phi_{BS}^2 \rangle}} \tag{25}$$

Another possible combination of these two BS wave functions gives the $M_S = 0$ component of the triplet state (T') as,

$$\phi_{|S=1, M_S=0\rangle} = \frac{\phi_{BS}^1 + \phi_{BS}^2}{\sqrt{2 + 2\langle \phi_{BS}^1 | \phi_{BS}^2 \rangle}} \tag{26}$$

By subtracting the energies (E_{BS} and $E_{T'}$) of these prior wave functions (Eq. 25 and 26), the following expression for J can be obtained, [19–21]

$$J = \frac{2(E_{BS} - E_{T'})}{1 + S_{ab}^2}. \tag{27}$$

where S_{ab} is the overlap integral between the spatial parts of spin-polarized α and β orbitals in the BS solution. If spin polarization is neglected for core electrons, the overlap integral can be considered to involve only two singly occupied molecular orbitals (SOMOs), a and b.[19] At this point, it can be recalled that the spin square expectation value for a singlet state is[22]

$$\langle \hat{S}^2 \rangle = M_S(M_S + 1) + N_\beta - \sum_{a,b}^{N_\alpha, N_\beta} \left(S_{ab}^{\alpha\beta} \right)^2 \tag{28}$$

where $S_{ab}^{\alpha\beta}$ is the overlap between the magnetic orbitals referring to opposite spins.

However, the use of BS solution to get a proper estimate of the energy of a singlet state is a crucial problem. To escape this problem, Baerends showed that the single determinant wave function includes all the electron correlation contributions to the energy,[23] while Perdew, Savin, and coworkers suggested that the energy of a singlet state can be approximated through the BS single determinant state in DFT calculation.[24] They further showed that such a BS state can produce an accurate electron and electron pair density even if the spin density distribution is unrealistic. With this lead, Ruiz et al. developed an alternative method in the platform of DFT where they used the broken symmetry approach, coined by Noodleman.[20, 25–26] In this method, the exchange coupling constant for a system with local spins S_1 and S_2 (where $S_2 < S_1$) is expressed as,

$$J = \frac{2(E_{BS} - E_{T'})}{2S_1 S_2 + S_2}. \tag{29}$$

where non-spin projected energies are used. With the assumption $S_1 = S_2$ and $S_{max} = S_1 + S_2$, this expression changes into

$$J = \frac{2(E_{BS} - E_T)}{S_{max}(S_{max} + 1)}. \tag{30}$$

In spite of various approaches to estimate the exchange coupling constant, it becomes obvious from the preceding section that the applicability of these equations depends on the appropriate choice of overlapping limit.[27] The problem of taking the appropriate weight of overlap between magnetic orbitals has been tackled by Yamaguchi et al. through the approximate spin projection technique. Use of this technique and Perdew's prescription that the singlet state can be represented by a single determinantal BS state within the DFT framework allowed Yamaguchi et al. to modify Eq. 15 as,[28–29]

$$J_Y = \frac{E_{BS}^{DFT} - E_{HS}^{DFT}}{\langle \hat{S}^2 \rangle_{HS} - \langle \hat{S}^2 \rangle_{BS}}. \tag{31}$$

The expression of J_Y given by Yamaguchi (Eq. 31) has an appealing aspect which can be understood through the dependence of $\langle \hat{S}^2 \rangle$ on the overlap of magnetic orbitals (Eq. 28).[22, 30] In case of the overlap among all pairs of α and β orbitals, the sum in Eq. 28 is reduced to a summation over $N\beta$ with individual terms all equal to 1. Therefore, the sum equals to $N\beta$ and the total spin expectation value indicates a pure spin state with $<S^2>_{BS} = 0$, and the denominator in Eq. 31 transforms to $S_{HS}(S_{HS} + 1)$, which resembles Eq. 30. On the other hand, if magnetic orbitals do not overlap (BS determinant), the sum in Eq. 28 becomes $N\beta - 2S_A$, where S_A is the sum of α and β magnetic orbitals. In this weakly coupled limit, $<S^2>_{BS} = 2S_A = S_{HS}$ and resembles Noodleman's original expression (Eq. 21). Thus Eq. 31 is superior to others in fair description of magnetic interaction. Yamaguchi et al. further claim that Eq. 21 and Eq. 30 do not work well for large magnetic systems, whereas the approximate spin projected technique (Eq. 31) reliably deals with the magnetic properties of systems with trinuclear or even larger clusters of linear or annular shape.[31–33] In fact, based on Eq. 31, they gave another expression for exchange coupling constant, which is specifically applicable for polynuclear magnetic systems,[32]

$$J_{ab} = \frac{E_{BS} - E_{HS}}{4(N-1)S_a S_b}. \tag{32}$$

14.1.2 Computational Methods

The preceding section elaborates that a reliable value of the exchange coupling constant (J) can only be produced if the spin-state energetics are estimated accurately. Yamaguchi suggested the need for configuration interaction methods such as unitary coupled cluster-based methods such as UCCSD (T) or UQCISD (T) or complete active space (CAS) self-consistent field (CASSCF) techniques to reproduce experimental values of J.[34] However, the CASSCF or CASSCF-based configuration interaction methods (CASCI) produce only 30% of the experimental value of J.[35] The underestimation of J in CAS methods is generally attributed to the neglect of effects like ligand-spin polarization, dynamic spin polarization, double spin polarization, etc., which are important physical mechanisms contributing to the exchange phenomenon.[36-37] One can find these effects in the second-order terms, which entails the application of second-order perturbation theory based upon the UHF or CASSCF wave function.[36-37] The UMP2, being this kind of method, is found to significantly improve the value of J, though the magnitude is still one-half of the experimental value. Another similar method, complete active space second-order perturbation theory (CASPT2) or N-electron valence state theory at second order (NEVPT2), which imposes a second-order correction to the CAS wave functions, is also found useful in producing J close to experimental values.[38] The second-order correction applied to the contracted CASSCF wave function is not enough to include all the important dynamic correlation effects. For example, the effect of charge transfer configurations on isotropic coupling cannot be fully accounted for in second-order correction. Though this problem can be solved by enlarging the CAS, such an enlargement causes a significant loss of interpretive power.[39] In this respect, consideration of the "external correlation" through multireference configuration interaction (MRCI) tools has emerged as an efficient method.[40] Among different MRCI techniques, the difference dedicated CI (DDCI) approach by Miralles et al. has been particularly successful in producing the desired degree of accuracy.[41] Different variants of DDCI wave function have also been formulated which can account for second-order mechanisms such as double spin polarization, kinetic exchange, etc.[42] However, the main limitation of DDCI is its demand for a high basis set, without which the value of J is found to be underestimated.[27] The computational rigor for such ab initio computation urged the requirement of a new technique, which can estimate J with less rigor and comparable accuracy as that of first-principle calculations.

The BS ansatz is indeed a very clever idea to simulate the effects of configuration interaction without actually resorting to such expensive calculations. However, the Achilles' heel of DFT has been the proper choice of exchange correlational (XC) functional during the estimation of any electronic property. Apart from the selection of proper XC functionals, the unrestricted formalism used in DFT brings about an additional problem of spin contamination, particularly in the BS state.[43] Local spin density approximation (LSDA) such as VWN is found to underestimate the spin localization as evidenced by a very low value of spin population in the magnetic site. This causes a huge high spin–low spin energy gap and hence overestimates J.[20, 25] Though the generalized gradient approximation (GGA) is a little improvement over the LDA functional, the problem of overdelocalization cannot be avoided, which again leads to an overestimated exchange coupling constant.[44] Hybrid DFT functionals are believed to impose marked improvement over LSDA or GGA functionals. In general, spin-unrestricted Kohn-Sham Slater function with correlation corrections (X = P86, PL, VWN, LYP) followed by an approximate spin projection technique of Yamaguchi give a reasonable account of singlet-triplet splitting.[45] However, the value of the coupling constant is found to be sensitive to the percentage of Fock exchange in the hybrid XC functional.[43] In this regard, UB2LYP is found reasonable, probably because this functional can properly take the spin-polarization and spin-delocalization effects into account.[46] Martin and Illas noted that a 50% admixture of Fock exchange (UBHandHLYP) appears necessary to get a reasonable agreement in J value with the experiment.[35] Zhao and Truhlar have developed a suite of M06 functionals with the facility to change the fraction of HF exchange from 0% to 100%.[47] Among four different variants of this functional, M06, the one containing 27% HF exchange, produces a J value closer to the experimental value.[48] Not only the exchange effect but also the electron correlation effects play a crucial role in describing magnetic coupling. To confront the short- and long-range interelectronic interaction, another new suite of range-separated functionals has been introduced by Scuseria and coworkers, which appears to be a better performer than usual hybrid functionals in the estimation of J.[49-50] The coupling constant also shows sensitivity to the range-separation parameter in the weight function of the range-separated hybrids.[50]

14.2 MAGNETIC ANISOTROPY

The preferential direction of the permanent magnetic moment with respect to the crystalline axes and/or to the external shape of the magnetic system is defined as magnetic anisotropy. Generally speaking, three major types of magnetic anisotropy can prevail in any magnetic system: the exchange anisotropy, where the interaction between neighboring spins is anisotropic; the magnetic field dependent g–anisotropy; and the single-ion anisotropy which strictly depends upon the interplay between spin-orbit coupling and the ligand field. The latter type plays a key role in the development of single-molecular magnets (SMMs), which are widely applied in memory storage devices, quantum computing, etc. The efficacy of magnetic molecules as quantum bits mostly depends on the existence of large magnetic anisotropy, since a high spin cluster without magnetic anisotropy is merely an ordinary paramagnet that cannot retain magnetization due to the rapid relaxation process. The preferential direction in the molecular frame, along which the net magnetic moment of a system is oriented in the absence of an external magnetic field, is called the easy axis. Consequently, a small (large) magnetic field is required to reach the saturation magnetization in a system with the magnetic moment vector aligned along the easy (hard) axis. The energy required to reverse the orientation of the magnetization at zero applied magnetic fields is defined as the magnetic anisotropic energy (MAE) barrier and is generally represented by U.[51] This MAE in SMMs is experimentally determined by Arrhenius plots of the magnetic relaxation data. The magnetic anisotropy in systems with more than one unpaired spin physically originates from the spin-orbit coupling (SOC) and the spin-spin coupling (SSC). In contrast with the electron spin (S), the orbital angular momentum can interact with the axial crystal field directly, and hence the magnetic anisotropy appears even in the absence of SOC. To illustrate, an orbital triplet state with an orbital magnetic moment ($l = 1$) can split into the singlet and doublet associated with the projections $m_l = 0$ and $m_l = \pm 1$, respectively, in an anisotropic crystal field (Figure 14.4, upper panel). On the contrary, the electron spin (S) of an open-shell metal ion can sense the shape of the molecule only through coupling with the orbital angular momentum. In general, the expectation value of the orbital angular momentum (L) of the electron becomes zero or "quenched" due to a sufficiently strong rotational hindrance posed by the coordinating ligand environment, which is a common feature of the $3d/4d$ transition metal complexes. Thus, in such cases, the single-ion magnetic moment is dominated by the spin-only angular momentum. However, these complexes with an orbital-singlet electronic state can also exhibit magnetic anisotropy, which originates from the mixing of electronically excited states with the $L = 0$ ground state, induced by spin-orbit coupling. On the other hand, orbital degeneracy yields an unquenched or partially quenched orbital angular momentum leading to strong magnetic anisotropy through LS coupling. Generally speaking, an odd number of electrons in the degenerate pairs (d_{xy} and $d_{x^2-y^2}$; $m_l = \pm 2$) or (d_{xz} and d_{yz}; $m_l = \pm 1$) would contribute with orbital angular momentum $L = 2$ and $L = 1$ respectively. For example, a high spin Co(II) ion with d^7 electron configuration carries a total first-order orbital angular momentum $L = 3$ due to $\left(d_{xy}, d_{x^2-y^2}\right)^3 \left(d_{xz}, d_{yz}\right)^3 \left(d_{z^2}\right)^1$ ground state besides the total spin angular momentum $S = 3/2$. Nevertheless, the appearance of the nonzero orbital angular momentum and resulting spin-orbit coupling leads to a large spontaneous splitting of magnetic sublevels M_s, even in the absence of external magnetic field, and thus is aptly defined as zero-field splitting (ZFS). The ZFS removes the degeneracy of the $2S + 1$ spin microstates associated with a single, isolated paramagnetic ion, for which the ground state can be represented as $|SM_s\rangle$, where the S is the total spin of the multiplet and $M_s = S, S-1, S-2, \ldots, -S$ ($2S + 1$ levels) are the magnetic quantum numbers, generated as eigenvalues of the vector operator $\hat{S}_z$ (Figure 14.4, lower panel). The gap between the split magnetic sublevels M_s quantifies the magnetic anisotropy of any system and acts as a barrier for demagnetization with the barrier height $\approx |D|S^2$, where the D is the axial ZFS parameter. The axial ZFS parameter D is a tensor with six different parameters in an arbitrary frame and depends on the symmetry of the system. For example, in the O_h point group, the diagonal values of the tensor in three major directions are equal $D_{xx} = D_{yy} = D_{zz}$. However, in C_1 symmetry, all six parameters of the ZFS tensor may have

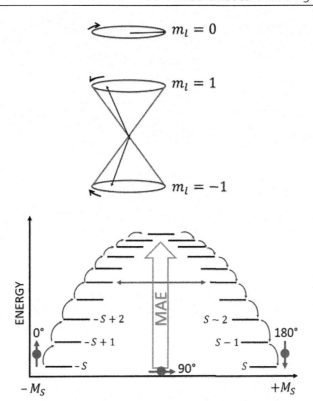

FIGURE 14.4 (Upper panel) Splitting of orbital triplet into a singlet ($m_l = 0$) and a doublet ($m_l = \pm 1$) by an axial crystal field. Adapted with permission from[52]. Copyright (2010) Taylor & Francis Online. (Lower panel) Qualitative energy level diagram of a $|SM_S\rangle$ state showing the thermal relaxation mechanism between the $-M_s$ and $+M_s$ levels through the top of the energy barrier (curved arrows) and the quantum tunneling relaxation mechanism through the thermally activated states (double arrow). The arrow with circle represents the direction of magnetic moment, and the associated digits are the angle between magnetic moment and the easy axis. Adapted with permission from[53]. Copyright (2015) Elsevier.

different values. The magnetization may be relaxed thermally (Orbach or Raman process), by quantum tunneling process or thermally assisted tunneling, and the demagnetization rate depends on temperature following Arrhenius-type law. Tunneling between magnetic sublevels $\langle \hat{S}_z \rangle$ in the ground spin state is very slow for large M_s but significantly faster between small M_s values. The tunneling rate is enhanced due to coupling between different $|SM_S\rangle$ levels through rhombic ZFS term (E, vide infra) leading to decreased effective barrier U_{eff}, which is much smaller than the MAE $|D|S^2$.

The sign of D is a crucial factor in determining the magnetic stability of any system. A positive and negative sign of D implies oblate and prolate spin distribution respectively. A negative value of D is a prerequisite for the magnetic bistability associated with the magnetization reversal barrier in single molecular magnets.[54] For example, a negative value of D can be obtained for a system undergoing an electronic transition between the d-orbitals with the same $\pm m_l$ (for e.g. d_{xy} and $d_{x^2-y^2}$ with $\pm m_l = \pm 2$ or d_{xz} and d_{yz} with $\pm m_l = \pm 1$).[55] In systems with negative D, the M_s magnetic sublevels with the largest magnetic moments ($M_s = \pm S$) are lowest in energy, and hence the system shows a large net magnetization. For example, a system with $S = \pm 5/2$ would split into doubly degenerate $M_S = \pm\frac{5}{2}, \pm\frac{3}{2},$ and $\pm\frac{1}{2}$ levels, which are known as Kramers doublets for such a half-integer spin system. Among these, the $M_S = \pm\frac{5}{2}$ represents the ground state, separated from other $\pm M_s$ levels by energy gaps proportionate to D. While the net magnetization of a system with negative D is aligned along an easy axis, that with positive D is aligned within an easy plane. Systems with positive D values are most stable in their lowest $\pm M_s$ states and don't have any magnetic bistability. However, the effective energy barriers for spin reversal (U_{eff}) have been shown to be approximately independent of the total spin S as $D \propto S^{-2}$, which invokes D to be a more influential parameter than S in enhancing U_{eff} and magnetic blocking temperature in low-dimensional magnets.

14.2.1 Mathematical Modeling

The effective Hamiltonian for a single-ion anisotropy, which is often called the zero-field splitting or the fine structure Hamiltonian, can be expressed as,

$$\hat{H}_{aniso} = \hat{S}_i \bullet D \bullet \hat{S}_i \tag{33}$$

where the $\hat{S}_i$ refers to the vector operator for the effective spin at magnetic center i and D is the second-order symmetric ZFS tensor of rank 3×3. The Hamiltonian in Eq. 33 opeartes on the $2S + 1$ dimensional manifold of spin functions $|SM_S\rangle$ associated with a single, isolated paramagnetic ion. It is always possible to find an orthogonal coordinate system in which the diagonal elements of the D tensor are only the non-vanishing elements. Such a molecular coordinate system that would diagonalize the D is called the magnetic axes frame, where the anisotropic Hamiltonian can be re-written as,

$$\hat{S}_i \bullet D \bullet \hat{S}_i = D_{xx}\hat{S}_x^2 + D_{yy}\hat{S}_y^2 + D_{zz}\hat{S}_z^2 \tag{34}$$

Subtracting the constant $\frac{1}{2}(D_{xx}+D_{yy})(\hat{S}_x^2+\hat{S}_y^2+\hat{S}_z^2)$, one can obtain the following form of the anisotropic Hamiltonian neglecting the higher order terms, as was developed by Abragam and Pryce,[56]

$$\hat{H}_{aniso} = D\left[\hat{S}_z^2 - \frac{1}{3}S(S+1)\right] + E\left(\hat{S}_x^2 - \hat{S}_y^2\right) \tag{35}$$

$$\text{where } D = D_{ZZ} - \frac{1}{2}(D_{XX}+D_{YY}) \text{ and } E = \frac{1}{2}(D_{XX}-D_{YY}) \tag{36}$$

In the model Hamiltonian in Eq. 35, z is considered to be the highest-symmetry axis, leading to the largest absolute value for the $|D_{zz}|$ component of the D tensor, while $|D_{xx}|$ is the smallest one.[53] The two independent components D and E in Eq. 36 are defined as *axial* and *rhombic* (or *transverse*) parameters of the ZFS, respectively, and they have the units of energy. In the absence of any rhombic anisotropy and at zero applied magnetic fields, the energies of magnetic sublevels become equal, i.e. $E(M_s) = E(-M_s)$. The spin Hamiltonian in Eq. 35 can also be represented in the notation of Orbach, in terms of the Stevens even-ordered operators $\hat{O}_2^0, \hat{O}_2^2$, where

$$\hat{O}_2^0 = 3\hat{S}_z^2 - S(S+1) \text{ and } \hat{O}_2^2 = \frac{1}{2}(\hat{S}_+^2 + \hat{S}_-^2). \tag{37}$$

With these Stevens operators, Eq. 35 takes the following form:

$$\hat{H}_{aniso} = \frac{1}{3}D\hat{O}_2^0 + E\hat{O}_2^2. \tag{38}$$

The ratio of E to |D| is called the *rhombicity parameter* and ranges from *0* (the "axial limit") to *1/3* (the "rhombic limit"). In cubic symmetry, i.e., $D = E = 0$ makes the system isotropic. In uniaxial symmetry, $E = 0$ for $D_{xx} = D_{yy}$, and the Hamiltonian reduces to,

$$\hat{H}_{aniso} = D\hat{S}_z^2. \tag{39}$$

and therefore, the energies of the spin multiplets can be expressed as

$$E = DM_S^2. \tag{40}$$

Therefore, a non-zero value for D can be realized for a system with a symmetry lower than cubic or octahedral. Lowering the symmetry from axial leads to a non-zero value of E, which describes anisotropy in the xy plane. The anisotropic Hamiltonian discussed so far applies well to systems with axial or rhombic symmetry. However, real systems may have even lower symmetry, and in such cases, the ZFS operator can be represented by a more general Hamiltonian ($\hat{H}_{CF}$), as was developed by Abragam and Bleaney.[55]

$$\hat{H}_{CF} = \sum_{n,k} B_k^n \cdot \hat{O}_k^n. \tag{41}$$

Here, $\hat{O}_k^n$ is the even-ordered Stevens operators as mentioned earlier, integer k is the order of spherical tensor which satisfies the relation $-n \leq k \leq +n$ and the value of k cannot be larger than $2S$, n is the type of anisotropy and can have values $0, 2, 4, 6, \ldots, 2S$ where $n = 0, 2,$ and 4 represents axial, rhombic, and tetragonal anisotropy, respectively. B_k^n is the ZFS parameter associated with Stevens operators. It follows from Eqs. 39 and 40 that the existence of an energy barrier between the $|SM_S\rangle$ and $|S-M_S\rangle$ states require a negative value of D parameter, whereas for positive D no such energy barrier can exist.

Apart from the single-ion anisotropy discussed prior, anisotropic interaction between neighboring spins in oligonuclear systems leads to the exchange anisotropy. The exchange coupling among neighboring spins, as expressed through the HDVV Hamiltonian (Eq. 1), uses the scalar products of spin operators and thus remains invariant under any rotations. Therefore, the exchange interaction in the framework of the HDVV model does not implicate magnetic anisotropy. However, the HDVV terms of the exchange Hamiltonian can be annexed with anisotropic contributions, which can be written in terms of the $\sum_{ij} \hat{S}_i T_{ij} \hat{S}_j$, where the second-order tensor T_{ij} is composed of one symmetric (D_{ij}) part, as discussed before, and one antisymmetric part (d_{ij}). The corresponding terms are not rotationally invariant and thus depend on the point symmetry of the system, giving rise to magnetic anisotropy. To address the spin anisotropy in polynuclear systems, two model Hamiltonians are commonly used, which are the multispin and the giant spin Hamiltonians. The former Hamiltonian takes the intersite interactions and the local anisotropies into account, whereas the latter considers the ZFS only in the ground state. Introduction of the antisymmetric intersite interactions by Dzyaloshinskii followed by explicit derivation by Moriya using the inclusion of SOC in Anderson's theory of superexchange gives rise to the following multispin model Hamiltonian for binuclear complexes with one unpaired spin in each magnetic site:[57]

$$H = -\sum_{i,j} 2J_{ij} \hat{S}_i \hat{S}_j + \hat{S}_i \bullet D_{ij} \bullet \hat{S}_i + d_{ij} \cdot \hat{S}_i \times \hat{S}_j \tag{42}$$

where $\hat{S}_i$ is the local spin on site i, the J_{ij} is the isotropic exchange coupling constant, the symmetric anisotropic tensor is the D_{ij}, and $d_{ij} = -d_{ji}$ is the Dzyaloshinskii-Moriya pseudovector, corresponding to the antisymmetric part of the total second-order anisotropic tensor. The second term addresses the interaction of local spins with the environment whereas the third term describes the anisotropic exchange interactions between the spins. This antisymmetric exchange appears as a second-order perturbation through the combined effect of the SOC (λ) and isotropic exchange coupling (J_{ij}), as follows,

$$d_{ij} \propto J_{ij} \left(\frac{\lambda}{\Delta E} \right). \tag{43}$$

Here, ΔE refers to the energy of an appropriate excited state in the given ligand field. On the other hand, the symmetric part of the anisotropic exchange, D_{ij} relates to the isotropic exchange coupling constant as,[52]

$$D_{ij} \propto J_{ij}\left(\frac{\lambda}{\Delta E}\right)^2 \qquad (44)$$

However, the crystal field splitting (ΔE) is usually greater than the SOC (λ) for the transition metal ions, and thus the isotropic exchange becomes more significant than the anisotropic contribution in determining the magnetic nature of a polynuclear transition metal complex.

The second model Hamiltonian, known as the giant spin Hamiltonian, is generally expressed as,

$$\hat{H}_{aniso} = D\hat{S}_Z^2 + E\left(S_x^2 - S_y^2\right) + \sum_{n,k\geq 4} B_k^n \cdot \hat{O}_k^n \qquad (45)$$

where $4 \leq k \leq 2S$, S being the spin of the ground state.

14.2.2 Computational Methods

Unlike the exchange coupling constant, which has been evaluated almost accurately and mapped with experimental data, estimation of the value and sign of the ZFS parameter is not straightforward, since it requires accurate inclusion of two different contributions: (i) direct dipolar interaction between unpaired spins (SSC, to first order in perturbation theory) and (ii) the spin-orbit coupling of electronically excited states and the ground state (SOC, to second order in perturbation theory). The SSC contribution of the D tensor can be calculated on the basis of a single ground-state determinant through,

$$D_{kl}^{SSC} = \frac{g_e^2}{4}\frac{\alpha^2}{S(2S-1)}\sum_{\mu\nu}\sum_{\kappa\tau}\left(P_{\mu\nu}^{\alpha-\beta}P_{\kappa\tau}^{\alpha-\beta} - P_{\mu\kappa}^{\alpha-\beta}P_{\nu\tau}^{\alpha-\beta}\right) \times \left\langle \mu\nu \left| \frac{\left(3r_{12,k}r_{12,l} - \delta_{kl}r_{12}^2\right)}{r_{12}^5} \right| \kappa\tau \right\rangle. \qquad (46)$$

Here, g_e is the free electron g value, α (~1/137) is the fine structure constant, $P_{\alpha-\beta}$ is the spin density matrix in the atomic orbital basis, and μ, ν, κ, and τ are the basic functions. The operator $\frac{\left(3r_{12,k}r_{12,l} - \delta_{kl}r_{12}^2\right)}{r_{12}^5}$ represents the spin-spin coupling between a pair of spins. The SSC, being of the first order, presents no particular difficulties. On the contrary, the calculation of SOC is more difficult since it couples states of different total spins. The SOC contribution through second-order perturbation theory can be obtained through an approximated version of the Breit-Pauli operator. However, the Breit-Pauli approximation is composed of one- and two-electron contributions and is difficult to handle. Hence, an effective one-electron operator of the following form is applied to get the SOC contribution,

$$\hat{H}_{SOC} = \sum_i \sum_A \lambda(r_{iA})\hat{l}_A(i) \cdot \hat{s}(i). \qquad (47)$$

where A sums over all nuclei, $\hat{l}_A(i)$ is the orbital angular momentum operator of electron i relative to atom A, and $\hat{s}(i)$ refer to the spin angular momentum operator for the ith electron respectively. The $\lambda(r_{iA})$ is a radial operator of the following form,[58]

$$\lambda\left(\left|\vec{r}_i - \vec{R}_A\right|\right) = \frac{\alpha^2}{2}\frac{Z_{eff}^4}{\left|\vec{r}_i - \vec{R}_A\right|^3}. \qquad (48)$$

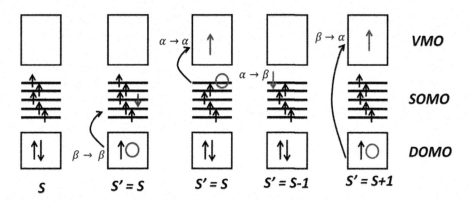

FIGURE 14.5 Schematic representation of four types of excitations in a d^5 system considered in the computation of D_{SOC}. The ground spin state S remains the same in the first two excitations and changes to $S - 1$ and $S + 1$ in two other excitations. VMO, SOMO, and DOMO refer to virtual, singly occupied, and doubly occupied molecular orbital respectively, and the α and β refer to spin-up and spin-down electrons. Adapted with permission from [62]. Copyright (2016) Royal Society of Chemistry.

where $\vec{r}$ and $\vec{R}_A$ refer to the position of the ith electron and A^{th} nucleus and α is the fine structure constant (≈1/137). Since $\lambda(r_{iA})$ is proportional to the inverse third power of the distance between the ith electron and A^{th} nucleus, the two- and three-center integrals become negligible and only the determination of one-center integrals needs to be evaluated. The expectation value of the operator $\lambda(r_{iA})$ may either be approximated as the SOC constants of the appropriate free atoms/ions or can be evaluated theoretically. The effective SOC constant λ is slightly reduced from the free ion value due to the effect of covalency. Several methods ranging from DFT to wave-function theory (WFT) have been developed to calculate the ZFS parameters. However, the requirement of intensive resources limits the application of WFT-based multireference methods, for the determination of D in large polynuclear complexes, and for such systems, methods based on DFT, including the effect of spin-orbit coupling, seem to be the only practical alternative. Pederson and Khanna developed the first DFT-based method to compute ZFS through perturbative treatment of the spin-unrestricted ground state determinant.[59] Later on, Neese developed an alternative linear response approach through solving the couple-perturbed equations, which avoids the truncation problem inherent to the previously applied perturbation approaches.[60] In this approach, the SOC contribution to the ZFS (D_{SOC}) is evaluated, considering four major types of excitations. Among these, excitations of one down-spin (β) from the doubly occupied molecular orbital (DOMO) to a singly occupied molecular orbital (SOMO) and one up-spin (α) from SOMO to virtual molecular orbital (VMO) remain spin-conserved, whereas spin-flip excitations ($\alpha \rightarrow \beta$) within SOMO lead to a $S - 1$ state and DOMO to VMO ($\beta \rightarrow \alpha$) excitation results in a $S + 1$ state (Figure 14.5). Taking all these excitations into consideration, the $\hat{D}$ tensor can be approximated from second-order perturbation theory as,

$$D_{ab} = -\lambda^2 \sum_{K \neq 0} \frac{\langle 0|\hat{L}_a|K\rangle\langle K|\hat{L}_b|0\rangle}{E_K - E_0}. \tag{49}$$

where λ is the atomic SOC constant, $|0\rangle$ and $|K\rangle$ are the spin-orbit free ground state and K^{th} excited state respectively, and $\hat{L}_{a,b}$ (a,b = x,y,z) is the orbital angular momentum operator.[61]

The couple-perturbed DFT (CP-DFT) method by Neese has been found to produce a more reliable estimate of the ZFS parameter, mainly due to revised prefactors for spin-flip excitations that have been derived more rigorously in the CP theory.[62] However, a reliable estimation of the rhombicity parameter (*E/D*) in DFT has been found to be problematic, particularly for compounds with small ZFS.

In spite of the success of the CP-DFT method, one must admit that the single reference character of the DFT is unable to describe the multireference character of the spin eigenfunctions involved in the determination of ZFS, which necessitates employment of the multireference methods in the evaluation of *D*. The complete active space self-consistent field has emerged as one of the central methods for dealing with the spin properties of magnetic molecules. Generally, WFT-based methods compute ZFS in two steps. At first, the electronic states are determined from state-averaged multi-configurational SCF calculations using the CASSCF scheme for selecting relevant electronic configurations. The resulting configuration provides a good representation of the static electron correlation, though this is not accurate enough for evaluating spin Hamiltonian parameters. Hence, dynamic correlation is added using perturbative methods, which consider the effects of the configurations external to the complete active space, and the ground state energy is refined through perturbation of second order. In this regard, the most popular methods have been the complete active space perturbation theory and the N-electron valence perturbation theory. Both methods include all single and double excitations involving at least one inactive or one virtual orbital. Another reliable method to increase the accuracy of the computed spin Hamiltonian parameters is the multireference configuration interaction (MRCI) method. However, numerous numbers of single and double excitations, as should be inevitably considered in the MRCI method (MRCI-SD), are not computationally feasible, and thus it becomes necessary to truncate the MRCI-SD space. In this regard, this is to recall that CASSCF computation is based on the partition of the MO into three subspaces: the inactive orbitals (which remain doubly occupied in all the configurations), the active orbitals (singly or doubly occupied), and the virtual orbitals (which remain unoccupied in all the configurations). An annihilation of an electron (one hole, 1h) in the inactive set and the creation of an electron (one particle, 1p) in a virtual set of orbitals correspond to a degree of freedom. For example, single-electron excitation from the inactive to active set (1h) or from active to virtual set (1p) corresponds to one degree of freedom, whereas double electron excitation (2h/2p) within the same set defines two degrees of freedom. Similarly, double electronic excitation from inactive to virtual (2h-2p) is associated with four degrees of freedom and so on. Truncation of the MRCI-SD space leads to different methods such as DDCI3, DDCI2, etc. where the suffixed number indicates the maximum number of degrees of freedom in the method concerned. It has been shown that the 2h-1p and 1h-2p excitations play a crucial role in the calculation of the energy difference between different spin states.[63]

14.3 SUMMARY AND OUTLOOK

Low-dimensional magnets have emerged as promising materials for different technological applications, ranging from storage devices to quantum computers. Understanding the physical origin and quantification of magnetism in such materials not only helps to explain the magneto-structural correlation but also guides a systematic synthesis of such materials. This chapter has attempted to discuss different mathematical tools which can address interaction between spin and orbital angular momentum, which is responsible for the overall magnetic behavior of open-shell systems. In particular, the exchange coupling constant (*J*), which dictates the ground spin state of any system, and the zero-field splitting parameter (*D*), which is correlated to the magnetic bistability, have been chosen for the present discussion. The isotropic exchange coupling constant (*J*) splits the states into different spin multiplets, whereas the zero-field splitting (*D*) causes lifting of the $(2S + 1)$-fold degeneracy of a particular spin state even in the absence of an external magnetic field. The phenomenological spin Hamiltonian, which represents the isotropic coupling between spins and also the zero-field splitting has the form,

$$\hat{H} = -J_{ij}\hat{S}_i\hat{S}_j + D\left[S_z^2 - \frac{1}{3}S(S+1)\right] + E\left(S_x^2 - S_y^2\right). \tag{50}$$

The energy difference between the highest and lowest $\pm M_s$ states is equated with the multiples of axial anisotropy D and provides a quantitative description of the magnetization reversal barrier, whereas the rhombic ZFS term (E) couples different $|SM_s\rangle$ states leading to a dramatic lowering of MAE through quantum mechanical tunneling.

Starting from the spin Hamiltonian, different effective Hamiltonians have been discussed to derive the numerical values of J and D. Accurate estimation of the tiny energy difference between spin states is crucial in the reliable estimation of the exchange coupling constant, which undoubtedly requires the use of rigorous wave-function-based methods. However, the application of density functional theory coupled with the broken symmetry approach has appeared as an alternative yet efficient tool in reproducing the experimental value of J. However, the value of the exchange coupling constant largely depends on the choice of the DFT exchange-correlational functional. On the other hand, magnetic anisotropy in terms of the ZFS parameters can be assessed through consideration of the spin-orbit coupling and the spin-spin coupling. The success of the couple-perturbed DFT method in the reliable estimation of D does not rule out the requirement of consideration of the multireference spin eigenfunction. Thus, a more accurate account of the ZFS can be obtained from a balanced consideration of the SOC-free ground states, through performing state-averaged CASSCF calculation for all the states, followed by post-CASSCF perturbation or variational treatment. Information about the SOC-free wave function and energy is used to calculate the spin-orbit coupling. The SOC Hamiltonian, an approximate version of the Breit-Pauli Hamiltonian, is considered as a perturbation of the spin-orbit free Hamiltonian, and hence the coefficients of configuration state functions corresponding to the ground state wave function obtained in the first step are not varied under the action of the SOC Hamiltonian.

14.4 REFERENCES

(1) Kahn, O. *Molecular Magnetism*, Illustrate.; **1993**, Wiley-VCH, New York.
(2) Anderson, P. W.; Hasegawa, H. Considerations on Double Exchange. *Phys. Rev.*, 100, **1955, 675**. https://doi.org/10.1103/PhysRev.100.675.
(3) Hay, P. J.; Thibeault, J. C.; Hoffmann, R. Orbital Interactions in Metal Dimer Complexes. *J. Am. Chem. Soc.*, 97, **1975, 4884**. https://doi.org/10.1021/ja00850a018.
(4) Anderson, P. W. New Approach to the Theory of Superexchange Interactions. *Phys. Rev.*, 115, **1959, 2**. https://doi.org/10.1103/PhysRev.115.2.
(5) Fink, K.; Wang, C.; Staemmler, V. Superexchange and Spin-Orbit Coupling in Chlorine-Bridged Binuclear Cobalt(II) Complexes. *Inorg. Chem.*, 38, **1999, 3847**. https://doi.org/10.1021/ic990280n.
(6) Calzado, C. J.; Cabrero, J.; Malrieu, J. P.; Caballol, R. Analysis of the Magnetic Coupling in Binuclear Complexes. I. Physics of the Coupling. *J. Chem. Phys.*, 116, **2002, 2728**. https://doi.org/10.1063/1.1430740.
(7) Calzado, C. J.; Cabrero, J.; Malrieu, J. P.; Caballol, R. Analysis of the Magnetic Coupling in Binuclear Complexes. II. Derivation of Valence Effective Hamiltonians from Ab Initio Cl and DFT Calculations. *J. Chem. Phys.*, 116, **2002, 3985**. https://doi.org/10.1063/1.1446024.
(8) de Loth, P.; Cassoux, P.; Daudey, J. P.; Malrieu, J. P. Ab Initio Direct Calculation of the Singlet-Triplet Separation in Cupric Acetate Hydrate Dimer. *J. Am. Chem. Soc.*, 103, **1981, 4007**. https://doi.org/10.1021/ja00404a007.
(9) Kollmar, C.; Couty, M.; Kahn, O. A Mechanism for the Ferromagnetic Coupling in Decamethylferrocenium Tetracyanoethenide. *J. Am. Chem. Soc.*, 113, **1991, 7994**. https://doi.org/10.1021/ja00021a028.
(10) Nagao, H.; Nishino, M.; Shigeta, Y.; Soda, T.; Kitagawa, Y.; Onishi, T.; Yoshioka, Y.; Yamaguchi, K. Theoretical Studies on Effective Spin Interactions, Spin Alignments and Macroscopic Spin Tunneling in Polynuclear Manganese and Related Complexes and Their Mesoscopic Clusters. *Coord. Chem. Rev.*, 198, **2000, 265**. https://doi.org/10.1016/S0010-8545(00)00231-9.
(11) Yamaguchi, K.; Fukui, H.; Fueno, T. Molecular Orbital (MO) Theory For Magnetically Interacting Organic Compounds. Ab-Initio Mo Calculations Of The Effective Exchange Integrals For Cyclophane-Type Carbene Dimers. *Chem. Lett.*, 15, **1986, 625**. https://doi.org/10.1246/cl.1986.625.
(12) Illas, F.; De P R Moreira, I.; De Graaf, C.; Barone, V. Magnetic Coupling in Biradicals, Binuclear Complexes and Wide-Gap Insulators: A Survey of Ab Initio Wave Function and Density Functional Theory Approaches. *Theor. Chem. Acc.*, 104, **2000, 265**. https://doi.org/10.1007/s002140000133.

(13) Neese, F. Prediction of Molecular Properties and Molecular Spectroscopy with Density Functional Theory: From Fundamental Theory to Exchange-Coupling. *Coord. Chem. Rev.*, 253, **2009, 526**. https://doi.org/10.1016/j.ccr.2008.05.014.

(14) Noodleman, L. Valence Bond Description of Antiferromagnetic Coupling in Transition Metal Dimers. *J. Chem. Phys.*, 74, **1981, 5737**. https://doi.org/10.1063/1.440939.

(15) Ciofini, I.; Daul, C. A. DFT Calculations of Molecular Magnetic Properties of Coordination Compounds. *Coord. Chem. Rev.*, 238, **2003, 187**. https://doi.org/10.1016/S0010-8545(02)00330-2.

(16) Neese, F. Definition of Corresponding Orbitals and the Diradical Character in Broken Symmetry DFT Calculations on Spin Coupled Systems. *J. Phys. Chem. Solids.*, 65, **2004, 781**. https://doi.org/10.1016/j.jpcs.2003.11.015.

(17) Noodleman, L.; Norman, J. G. The Xα Valence Bond Theory of Weak Electronic Coupling. Application to the Low-Lying States of Mo2Cl84-. *J. Chem. Phys.*, 70, **1979, 4903**. https://doi.org/10.1063/1.437369.

(18) Noodleman, L.; Davidson, E. R. Ligand Spin Polarization and Antiferromagnetic Coupling in Transition Metal Dimers. *Chem. Phys.*, 109, **1986**, 131. https://doi.org/10.1016/0301-0104(86)80192-6.

(19) Caballol, R.; Castell, O.; Illas, F.; Moreira, I. D. P. R.; Malrieu, J. P. Remarks on the Proper Use of the Broken Symmetry Approach to Magnetic Coupling. *J. Phys. Chem. A*, 101, **1997**. 7860, https://doi.org/10.1021/jp9711757.

(20) Ruiz, E.; Cano, J.; Alvarez, S.; Alemany, P. Broken Symmetry Approach to Calculation of Exchange Coupling Constants for Homobinuclear and Heterobinuclear Transition Metal Complexes. *J. Comput. Chem.*, 20, **1999, 1391**. https://doi.org/10.1002/(SICI)1096-987X(199910)20:13<1391::AID-JCC6>3.0.CO;2-J.

(21) Noodleman, L.; Peng, C. Y.; Case, D. A.; Mouesca, J. M. Orbital Interactions, Electron Delocalization and Spin Coupling in Iron-Sulfur Clusters. *Coord. Chem. Rev.*, 144, **1995, 199**. https://doi.org/10.1016/0010-8545(95)07011-L.

(22) Szabo, A.; Ostlund, N. S. *Modern Quantum Chemistry: Introduction to Advanced Electronic Structure Theory*; **1996**, Dover Books on Chemistry; Dover Publications, New York.

(23) Gritsenko, O. V.; Schipper, P. R. T.; Baerends, E. J. Exchange and Correlation Energy in Density Functional Theory: Comparison of Accurate Density Functional Theory Quantities with Traditional Hartree-Fock Based Ones and Generalized Gradient Approximations for the Molecules Li2, N2, F2. *J. Chem. Phys.*, 107, **1997, 5007**. https://doi.org/10.1063/1.474864.

(24) Perdew, J. P.; Ernzerhof, M.; Burke, K.; Savin, A. On-Top Pair-Density Interpretation of Spin Density Functional Theory, with Applications to Magnetism. *Int. J. Quantum Chem.*, 61, **1997, 197**. https://doi.org/10.1002/(SICI)1097-461X(1997)61:2<197::AID-QUA2>3.0.CO;2-R.

(25) Ruiz, E.; Rodríguez-Fortea, A.; Cano, J.; Alvarez, S.; Alemany, P. About the Calculation of Exchange Coupling Constants in Polynuclear Transition Metal Complexes. *J. Comput. Chem.*, 24, **2003, 982**. https://doi.org/10.1002/jcc.10257.

(26) Ruiz, E. *Principles and applications of Density Functional Theory in Inorganic Chemistry II, Theoretical Study of the Exchange Coupling in Large Polynuclear Transition Metal Complexes Using DFT Methods*; **2012**, Springer, https://doi.org/10.1007/b97942.

(27) Moreira, I. D. P. R.; Illas, F. A Unified View of the Theoretical Description of Magnetic Coupling in Molecular Chemistry and Solid State Physics. *Phys. Chem. Chem. Phys.*, 8, **2006, 1645**. https://doi.org/10.1039/b515732c.

(28) Yamaguchi, K.; Takahara, Y.; Fueno, T. Ab-Initio Molecular Orbital Studies of Structure and Reactivity of Transition Metal-OXO Compounds. In: Smith V.H.; Schafer, H.F.; Morokuma, K. (eds) *Applied Quantum Chemistry*; **1986**, Springer, Dodrecht. https://doi.org/10.1007/978-94-009-4746-7_11.

(29) Soda, T.; Kitagawa, Y.; Onishi, T.; Takano, Y.; Shigeta, Y.; Nagao, H.; Yoshioka, Y.; Yamaguchi, K. Ab Initio Computations of Effective Exchange Integrals for H-H, H-He-H and Mn 2 O 2 Complex: Comparison of Broken-Symmetry Approaches. *Chem. Phys. Lett.*, 319, **2000**, 223. https://doi.org/10.1016/S0009-2614(00)00166-4.

(30) Herrmann, C.; Yu, L.; Reiher, M. Spin States in Polynuclear Clusters: The [Fe2O2] Core of the Methane Monooxygenase Active Site. *J. Comput. Chem.*, 27, **2006, 1223**. https://doi.org/10.1002/jcc.20409.

(31) Onishi, T.; Yamaki, D.; Yamaguchi, K.; Takano, Y. Theoretical Calculations of Effective Exchange Integrals by Spin Projected and Unprojected Broken-Symmetry Methods. I. Cluster Models of K2NiF4-Type Solids. *J. Chem. Phys.*, 118, **2003, 9747**. https://doi.org/10.1063/1.1567251.

(32) Onishi, T.; Yamaguchi, K. Theoretical Calculations of Effective Exchange Integrals by Spin Projected and Unprojected Broken-Symmetry Methods. III. Cluster Models of Three-Dimensional KNiF3solid. *J. Chem. Phys.*, 121, **2004, 2119**. https://doi.org/10.1063/1.1766294.

(33) Onishi, T.; Yamaguchi, K. Theoretical Calculations of Effective Exchange Integrals by Spin Projected and Unprojected Broken-Symmetry Methods II: Cluster Models of Jahn-Teller Distorted K2CuF4 Solid. *Polyhedron*, 28, **2009, 1972**. https://doi.org/10.1016/j.poly.2008.11.037.

(34) Okumura, M.; Takada, K.; Maki, J.; Noro, T.; Mori, W.; Yamaguch, K. Theoretical Approaches to Molecular Magnetisms: Through-Bond Couplings between Triplet Carbenes and Related Species. *Mol. Cryst. Liq. Cryst. Sci. Technol. Sect. A. Mol. Cryst. Liq. Cryst.*, 233, **1993, 41**. https://doi.org/10.1080/10587259308054946.
(35) Illas, F.; Martin, R. L. Magnetic Coupling in Ionic Solids Studied by Density Functional Theory. *J. Chem. Phys.*, 108, **1998, 2519**. https://doi.org/10.1063/1.475636.
(36) Mouesca, J. -M; Noodleman, L.; Case, D. A. Density-functional Calculations of Spin Coupling in [Fe4S4]3+ Clusters. *Int. J. Quantum Chem.*, 56, **1995, 95**. https://doi.org/10.1002/qua.560560710.
(37) Willett, R. D.; Gatteschi, D.; Kahn, O. *Magneto-Structural Correlations in Exchange Coupled Systems*; **1985**, D Reidel Publishing Co., Hingham, MA .
(38) Malrieu, J. P.; Caballol, R.; Calzado, C. J.; De Graaf, C.; Guihéry, N. Magnetic Interactions in Molecules and Highly Correlated Materials: Physical Content, Analytical Derivation, and Rigorous Extraction of Magnetic Hamiltonians. *Chem. Rev.*, 114, **2014, 429**. https://doi.org/10.1021/cr300500z.
(39) De Graaf, C.; Sousa, C.; De P. R. Moreira, I.; Illas, F. Multiconfigurational Perturbation Theory: An Efficient Tool to Predict Magnetic Coupling Parameters in Biradicals, Molecular Complexes, and Ionic Insulators. *J. Phys. Chem. A*, 105, **2001, 11371**. https://doi.org/10.1021/jp013554c.
(40) Calzado, C. J.; Angeli, C.; Caballol, R.; Malrieu, J. P. Extending the Active Space in Multireference Configuration Interaction Calculations of Magnetic Coupling Constants. *Theor. Chem. Acc.*, 126, **2010, 185**. https://doi.org/10.1007/s00214-009-0642-9.
(41) Miralles, J.; Castell, O.; Caballol, R.; Malrieu, J. P. Specific CI Calculation of Energy Differences: Transition Energies and Bond Energies. *Chem. Phys.*, 172, **1993, 33**. https://doi.org/10.1016/0301-0104(93)80104-H.
(42) Muñoz, D.; Illas, F.; de Moreira, I. P. R. Accurate Prediction of Large Antiferromagnetic Interactions in High-Tc HgBa2Can-1CunO2n+2+δ (n = 2, 3) Superconductor Parent Compounds. *Phys. Rev. Lett.*, 84, **2000, 1579**. https://doi.org/10.1103/PhysRevLett.84.1579.
(43) de P. R. Moreira, I.; Illas, F.; Martin, R. L. Effect of Fock Exchange on the Electronic Structure and Magnetic Coupling in NiO. *Phys. Rev. B – Condens. Matter Mater. Phys.*, 65, **2002, 155102**. https://doi.org/10.1103/PhysRevB.65.155102.
(44) Martin, R. L.; Illas, F. Antiferromagnetic Exchange Interactions from Hybrid Density Functional Theory. *Phys. Rev. Lett.*, 79, **1997, 1539**. https://doi.org/10.1103/PhysRevLett.79.1539.
(45) Yamaguchi, K.; Yoshioka, Y.; Takatsuka, T.; Fueno, T. Extended Hartree-Fock (EHF) Theory in Chemical Reactions. *Theor. Chim. Acta*, 48, **1978, 185**. https://doi.org/10.1007/bf00549018.
(46) Bartlett, R. J.; Musiał, M. Coupled-Cluster Theory in Quantum Chemistry. *Rev. Mod. Phys.*, 79, **2007, 291**. https://doi.org/10.1103/RevModPhys.79.291.
(47) Zhao, Y.; Truhlar, D. G. A New Local Density Functional for Main-Group Thermochemistry, Transition Metal Bonding, Thermochemical Kinetics, and Noncovalent Interactions. *J. Chem. Phys.*, 125, **2006, 194101**. https://doi.org/10.1063/1.2370993.
(48) Valero, R.; Costa, R.; De P. R. Moreira, I.; Truhlar, D. G.; Illas, F. Performance of the M06 Family of Exchange-Correlation Functionals for Predicting Magnetic Coupling in Organic and Inorganic Molecules. *J. Chem. Phys.*, 128, **2008, 114103**. https://doi.org/10.1063/1.2838987.
(49) Peralta, J. E.; Melo, J. I. Magnetic Exchange Couplings with Range-Separated Hybrid Density Functionals. *J. Chem. Theory Comput.*, 6, **2010, 1894**. https://doi.org/10.1021/ct100104v.
(50) Phillips, J. J.; Peralta, J. E.; Janesko, B. G. Magnetic Exchange Couplings Evaluated with Rung 3.5 Density Functionals. *J. Chem. Phys.*, 134, **2011, 214101**. https://doi.org/10.1063/1.3596070.
(51) Waldmann, O. A Criterion for the Anisotropy Barrier in Single-Molecule Magnets. *Inorg. Chem.*, 46, **2007, 10035**. https://doi.org/10.1021/ic701365t.
(52) Palii, A.; Tsukerblat, B.; Clemente-Juan, J. M.; Coronado, E. Magnetic Exchange between Metal Ions with Unquenched Orbital Angular Momenta: Basic Concepts and Relevance to Molecular Magnetism. *Int. Rev. Phys. Chem.*, 29, **2010, 135**. https://doi.org/10.1080/01442350903435256.
(53) Atanasov, M.; Aravena, D.; Suturina, E.; Bill, E.; Maganas, D.; Neese, F. First Principles Approach to the Electronic Structure, Magnetic Anisotropy and Spin Relaxation in Mononuclear 3d-Transition Metal Single Molecule Magnets. *Coord. Chem. Rev.*, 289–290, **2015, 177**. https://doi.org/10.1016/j.ccr.2014.10.015.
(54) Zein, S.; Duboc, C.; Lubitz, W.; Neese, F. A Systematic Density Functional Study of the Zero-Field Splitting in Mn(II) Coordination Compounds. *Inorg. Chem.*, 114, **2008, 10750**. https://doi.org/10.1021/ic701293n.
(55) Abragam, A.; Bleaney, B. *Electron Paramagnetic Resonance of Transition Ions*; **1970**, Oxford University Press, Oxford.
(56) Maurice, R.; Broer, R.; Guihéry, N.; de Graaf, C. Zero-Field Splitting in Transition Metal Complexes: Ab Initio Calculations, Effective Hamiltonians, Model Hamiltonians, and Crystal-Field Models. In *Handbook of Relativistic Quantum Chemistry*; **2016**, Springer. https://doi.org/10.1007/978-3-642-40766-6_37.

(57) Boča, R. Zero-Field Splitting in Metal Complexes. *Coord. Chem. Rev.*, 248, **2004, 757**. https://doi.org/10.1016/j.ccr.2004.03.001.
(58) Neese, F.; Solomon, E. I. Calculation of Zero-Field Splittings, g-Values, and the Relativistic Nephelauxetic Effect in Transition Metal Complexes. Application to High-Spin Ferric Complexes. *Inorg. Chem.*, 37, **1998, 6568**. https://doi.org/10.1021/ic980948i.
(59) Pederson, M. R.; Porezag, D. V.; Kortus, J.; Khanna, S. N. Theoretical Calculations of Magnetic Order and Anisotropy Energies in Molecular Magnets. *J. Appl. Phys.*, 87, **2000, 5487**. https://doi.org/10.1063/1.373380.
(60) Neese, F. Calculation of the Zero-Field Splitting Tensor on the Basis of Hybrid Density Functional and Hartree-Fock Theory. *J. Chem. Phys.*, 127, **2007, 164112**. https://doi.org/10.1063/1.2772857.
(61) Damgaard-Møller, E.; Krause, L.; Tolborg, K.; Macetti, G.; Genoni, A.; Overgaard, J. Quantification of the Magnetic Anisotropy of a Single-Molecule Magnet from the Experimental Electron Density. *Angew. Chemie Int. Ed.*, 59, **2020, 21203**. https://doi.org/10.1002/ange.202007856.
(62) Duboc, C. Determination and Prediction of the Magnetic Anisotropy of Mn Ions. *Chem. Soc. Rev.*, 45, **2016, 5834**. https://doi.org/10.1039/c5cs00898k.
(63) Calzado, C. J.; Angeli, C.; Taratiel, D.; Caballol, R.; Malrieu, J. P. Analysis of the Magnetic Coupling in Binuclear Systems. III. the Role of the Ligand to Metal Charge Transfer Excitations Revisited. *J. Chem. Phys.*, 131, **2009, 044327**. https://doi.org/10.1063/1.3185506.

Novel Magnetism in Ultrathin Films With Polarized Neutron Reflectometry

15

Saibal Basu and Surendra Singh
Solid State Physics Division Bhabha Atomic Research Centre Mumbai, India

Contents

15.1	Introduction	289
15.2	Optical Theory of Reflection	290
	15.2.1 Fresnel Reflectivity for an Ideal Surface	294
	15.2.2 Reflectivity From a Film of Finite Thickness	296
	15.2.3 Reflectivity From a Periodic Multilayer	297
	15.2.4 Data Analysis Technique for Specular Reflectometry	297
15.3	Investigation of Interfacial Magnetization in Thin Films	299
	15.3.1 $3d$ Transition Metal: A Co Film	299
	15.3.1.1 Co Films Grown by Different Techniques: Sputtering and Electrodeposition	299
	15.3.1.2 High-Density and Non-Magnetic Co Film at Interfaces	301
	15.3.2 Helical Magnetic Structure in Co/Gd Multilayers	302
	15.3.3 Proximity Effect in Superconductor and Ferromagnet Oxide Heterostructures	303
15.4	Conclusion	305
15.5	References	306

15.1 INTRODUCTION

A rapidly increasing number of research and applications in science and engineering are in thin films and multilayers down to the sub-nanometer scale. Interfaces between two materials can give rise to novel physical phenomena and functionalities, not exhibited by either of the constituent materials alone. To understand the physical and magnetic properties of interfaces in nanostructures, detailed and microscopic structural and magnetic characterization is required. Polarized neutron reflectometry (PNR) is currently one of the most popular non-destructive tools for characterizing magnetism in thin films of interest. Historically, Fermi and Zinn, in 1946, were the first to present neutron reflectivity measurements for finding out the coherent nuclear scattering cross-section of various materials [1]. Approximately a decade later, the first report on x-ray reflectivity (XRR) for thin-film characterization by L.G. Parratt appeared [2]. Presently, both techniques are routinely used for characterization of thin films at a mesoscopic length scale. The reincarnation of neutron reflectometry (NR) as a useful tool for studying thin films and their magnetic structures is due to G. P. Felcher of Argonne National Laboratory in the 1980s [3]. Later, his comments in the popular articles are worth mentioning [4]: "Ten years ago, at a modest spallation-neutron source at Argonne National Laboratory, a 'gizmo' was installed which was later christened as a reflectometer. The popularity of the gizmo spread like wildfire: now virtually all neutron sources possess at least one of them. Some are graced by splendid names: CRISP and SURF at Rutherford, TOREMA at Julich, DESIR in Saclay, and EVA in Grenoble."

Atomic force microscopy (AFM), transmission electron microscopy (TEM), and cross-sectional TEM (XTEM) are some of the complementary and competing techniques, which are widely used for studying surfaces and interfaces in thin-film heterostructures (single layers, multilayers, etc.). These techniques usually provide the image over a small area of the sample (typically a few microns), and some of them are destructive. On the other hand, neutron (and x-ray) reflectometry are non-destructive tools that can provide information with sub-nanometer depth resolution [5–13]. In addition, the depth-dependent information with such resolution is typically averaged over a few centimeters of a sample area, thus bridging two wide-apart length scales. Also, a neutron is highly penetrating and can provide information about deeply buried interfaces, unlike any other technique. For neutron reflectometry, one uses thermal neutrons (or cold neutrons) extracted from a research reactor or a spallation neutron source. Thermal neutrons are in equilibrium with moderators at typically in the temperature range of 30°C to 50°C in a reactor or a spallation neutron source, having energy in the range of tens of milli electron volt (meV). By cold neutron, we refer to neutrons with energy typically less than 5 meV, and they are available from special devices, known as cold neutron sources, that re-thermalize the neutrons from a reactor (or spallation source) to lower energy. PNR provides both physical and magnetic structures of thin films with resolution at a sub-nanometer length scale, which has made PNR such an important technique. The popularity of neutron reflectometry is well augmented by that of XRR.

Reflectometry (XRR and PNR) uses the phenomena of reflection of waves from thin-film surfaces and interfaces. Both techniques depend on the optical properties of the medium for neutron or x-ray waves. Since the x-rays interact with the atomic electron cloud and the neutrons with the nuclei, the refractive index for x-rays depends on the electron density of the materials and the refractive index for neutrons depends on the coherent scattering length density for neutron-nuclear interaction in the material. This basic difference makes them complementary techniques for the study of a layered structure in thin films. In addition, the neutron possesses a magnetic moment of -1.91 μ_N (nuclear magneton), and PNR is ideally suited for understanding the magnetic structure of thin films also. PNR has gained importance because of the rapidly growing interest in thin-film multilayers of magnetic and non-magnetic materials, which give rise to various important phenomena like giant magnetoresistance (GMR) and spin-valve effects. In the present chapter, we attempt to highlight the importance of this extremely important technique with the help of several recent examples, emphasizing the investigation and understanding of structure and magnetism in thin-film heterostructures. First, we briefly discuss the theory of reflectometry and related data analysis techniques. Results from $3d$ transition metal films of interest and a ferromagnet/superconductor trilayer are discussed next as examples.

15.2 OPTICAL THEORY OF REFLECTION

We present the theory of neutron reflectometry briefly in this section before we attempt to discuss applications in various magnetic thin films of interest. Several publications and books have dealt extensively with this topic [2–3, 6–8, 14–20]. The theory of neutron/x-ray reflectivity closely follows the treatment of reflection and transmission of optical waves in a medium [21]. A neutron wave at grazing incident on a medium experiences an average potential, arising from neutron-nucleus interaction:

$$V(r) = \frac{2\pi \hbar^2}{m} \rho(r) b_{coh} \quad (1.1)$$

where $\rho(r)$ is the number density of the scattering centers (atoms/molecules) in the medium, and b_{coh} is the coherent scattering length for each scattering center. The product $\rho(r) b_{coh}$ is termed the nuclear scattering length density (NSLD). The potential seen by the neutron in the medium translates into a refractive index n for a medium as given in the following:

$$n = 1 - \frac{\lambda^2}{2\pi} \rho(r) b_{coh} \quad (1.2)$$

where λ is the wavelength of the neutron. This simple expression also tells us an interesting fact that depending on the sign of b_{coh}, the refractive index can be less than or more than 1. For most materials, b_{coh} is positive and the refractive index for thermal neutrons is marginally less than 1, except for a few elements, e.g., Ti. The expression of the refractive index indicates its dependence on the scattering length density of the medium. One may write a generalized expression for the refractive index that dictates the propagation of the neutron (also x-ray) wave in the medium:

$$n = 1 - \delta + i\beta \quad (1.3)$$

The imaginary part takes care of the absorption of the neutron/x-ray propagation in the medium. Most of the materials have nearly zero absorption for neutrons. The δ and β for neutron and x-ray are defined here:

$$\delta_n = \frac{\rho \lambda^2}{2\pi} b_{coh}; \quad \beta_n = \frac{\rho \lambda^2}{2\pi} b_{abs} \quad (1.4)$$

For x-rays:

$$\delta_x = \frac{\rho_e r_0 \lambda^2}{2\pi} \left[f_0 + f^d \right]; \quad \beta_x = \frac{\rho_e r_0 \lambda^2}{2\pi} f^{abs} \quad (1.4a)$$

where r_0 is the classical electron radius 2.818 fm. For x-rays, the reflection is caused by the interaction of x-ray with electron and the value of $f_0 = Z$, the atomic number. The product $\rho_e r_0 f_0$ is termed as electron scattering length density (ESLD). f^d and f^{abs} are dispersion and absorption terms.

The values of both δ and β depend on the density of the medium for both x-rays and neutrons. δ is typically about 10^{-5}–10^{-6}. Usually, the refractive index is less than 1 and neutrons undergo total external reflection up to a critical incident angle θ_c:

$$\theta_c = \sqrt{\frac{\rho b_{coh}}{\pi}} \lambda \quad (1.5)$$

Since the critical angle of reflection is quite small, compared to its optical counterpart, neutron (or x-ray) reflectivity takes place at near-grazing incidence to the reflecting surface. This demands the preparation of a well-collimated neutron beam for reflectometry experiments and an elaborate arrangement for placing the sample in a reflecting position. As an example, for Ni, which has a relatively larger value of b_{coh} among the metallic elements, the value of wavelength-dependent critical angle θ_c is 6 arc minutes/Å of neutron wavelength. As an example, this gives a critical angle of 24 arc minutes for 4 Å neutrons. Ni has been a preferred choice of neutron mirror coatings for a long time because of its large critical angle of reflection. So far, the treatment assumed that the reflecting medium is without any magnetism. For a medium in which the scattering centers have magnetic moments, neutrons, because of their inherent magnetic moment, will experience additional potential energy in a magnetic field $B(r)$, other than the nuclear potential, given by:

$$V_{mag}(r) = -\vec{\mu}.\vec{B}(r) \quad (1.6)$$

where $\vec{\mu}$ is the magnetic moment of the neutron and $\vec{B}(r)$ is the magnetic field, a function of position. It is clear that depending on the relative orientation of the neutron magnetic moment and the local field (whether parallel or antiparallel), the magnetic potential $V_{mag}(r)$ can have positive or negative value with respect to the nuclear potential. One may now write the total potential for the magnetic film as a summation of the nuclear and the magnetic parts as given in the following:

$$V_{tot}(r) = \frac{2\pi\hbar^2}{m}\rho(r)\left(b_{coh} \pm b_{mag}\right) \quad (1.7)$$

In the case of an unpolarized neutron beam as well as non-magnetic (including a magnetic film with net-zero magnetization, e.g. paramagnetic, antiferromagnetic) thin films, the potential reduces to the nuclear part alone since the magnetic part averages to zero. For a long-range magnetic (ferromagnetic and ferrimagnet) thin film and incident polarized neutron beam, the magnetic part either adds up to or is subtracted from the nuclear potential, depending on whether the sample is magnetized parallel or antiparallel to the neutron beam. In the magnetic part of the potential, the $\rho(r)b_{mag}$ is termed as magnetic scattering length density (MSLD). The magnetization in emu/cc can be obtained from MSLD (in Å^{-2}) using a relation: M (emu/cc) = MSLD (Å^{-2}) / 2.9109 × 10^{-9} [7]. Accordingly, there will be different refractive index and critical angle for spin up, R$^+$ (parallel) and spin down, R$^-$ (antiparallel) neutrons, which produces two different reflectivity patterns for PNR. The difference between the spin-dependent reflectivity (R$^+$ − R$^-$) can be used to extract a detailed depth profile of magnetization density. Figure 15.1(a) shows a schematic of the nuclear and magnetic potential for neutrons in a magnetic medium (film).

Primarily, whenever there is a refractive index mismatch at an interface (air/film or between layers of a multilayer), a reflection of the incident wave takes place. The reflected *amplitudes* come out at the film-air interface and constitute the reflected *beam*. These waves interfere to produce the reflected intensity that is recorded. The reflected intensity is an interference pattern of the amplitudes from various interfaces. It contains information regarding all the layers and interfaces in a thin film. The motion of a neutron of mass m and energy E in a medium with potential V described prior is given by the Schrödinger for neutron wavefunction ψ:

$$-\frac{\hbar^2}{2m}\nabla^2\psi + V\psi = E\psi \quad (1.8)$$

The schematic of the reflection of a wave from a medium is shown in Figure 15.1(b). The incident plane wave, i, gets reflected from the film-air (surface) interface, with r depicting the reflected amplitude at the film-air interface. Part of the beam, t gets transmitted into the medium. The diagram shows the plane of reflection as the x-z plane, and the thin film's depth is along the z-direction. The potential V depends on the variable z, and the wavefunction ψ is separable into z and x variables as:

$$\psi(x,z) = e^{iKx}\psi(z) \quad (1.9)$$

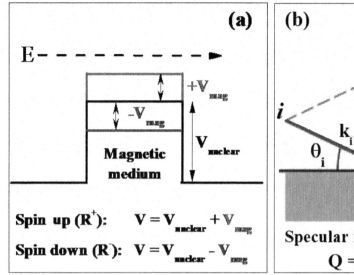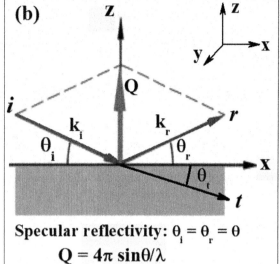

FIGURE 15.1 (a) The potential seen by a neutron of energy E. The nuclear potential $V_{nuclear}$ is as given by Eq. 1.1. The magnetic potential V_{mag} gets added or subtracted from the nuclear potential depending on neutron beam polarization. (b) Geometry of specular neutron reflectivity. The incident angle θ_i = angle of reflection $\theta_r = \theta$. The momentum transfer is given by $Q = \frac{4\pi}{\lambda}\sin\theta$. The incident beam ($i$) gets reflected ($r$) from the surface, and a part of the beam gets transmitted (t) in the medium. The z-x and x-y planes represent the plane of reflection and film surface, respectively. The z is the depth in the reflecting thin-film medium.

Thus Eq. 1.8 reduces to a one-dimensional Schrödinger equation for ψ:

$$-\frac{\hbar^2}{2m}\frac{d^2\psi}{dz^2} + q^2\psi = 0 \tag{1.10}$$

where $q^2(z) = \frac{2m}{\hbar^2}\left[E - V(z)\right] - K^2$ is the z component of the incident wavevector normal to the interface, as shown in Figure 15.1(b). K is the x component of the wavevector, and it is an invariant of motion. In case of reflection of electromagnetic waves, the same equation one obtains for continuity of "Z" component of electric field at an interface. The corresponding equations for an electromagnetic wave are shown in the box (Eq. 1.10a).

The electric field E for the reflection geometry [Figure 15.1 (a)] can be written as $E(x,z,t) = e^{i(Kx-\omega t)}E(z)$ where $E(z)$ follows a one-dimensional Schrödinger similar to equation (1.8):

$$\frac{d^2E}{dz^2} + q^2 E = 0 \tag{1.10a}$$

where ω is the angular frequency of the electromagnetic wave and $e^{i\omega t}$ gives the time variation of the wave. This expression clearly underlines the similarity between neutron and x-ray reflectometry and their theoretical treatment.

The reflection amplitude r and the transmission amplitude t are defined in terms of the solutions at the interface with the normal components of the wavevector q_1 and q_2 in a vacuum and in the medium, respectively, at the interface. The amplitudes across the interface are

$$1 + re^{iq_1 z} \to \psi(z) \leftarrow te^{iq_2 z} \qquad (1.11)$$

With proper boundary conditions at the interface, one obtains the reflectance (reflection amplitude, r) and the reflectivity, $R = r\,r^*$, where r^* is the complex conjugate of r. A similar formalism can be applied to XRR using Maxwell equations. The need to indicate the similarity between NR and XRR is two-fold. One, the theoretical treatment for both process are very similar and can be translated from one to another by a suitable choice of parameters. The other issue is related to planning the experiments. Usually, neutrons are available only at major facilities like research reactors or spallation neutron sources. The XRR technique can be established as a laboratory facility. Often, as soon as a thin-film sample is deposited, one may carry out the characterization of its parameters using XRR, locally, and then decide on further NR experiments based on the quality and structure of the sample. Experimentally, specular (angle of incidence = angle of reflection, Snell's law of reflection) reflectivity (Figure 15.1(b)) is measured as a function of momentum transfer Q_z normal to the sample surface, given by:

$$Q_z = Q = \frac{4\pi}{\lambda} \sin\theta \qquad (1.12)$$

15.2.1 Fresnel Reflectivity for an Ideal Surface

A flat surface with a sharp boundary at the air film interface (Figure 15.1(b)), and the incident beam, reflected beam, and the transmitted beam, are shown as i, r, and t, respectively. The interface between film and air is the x-y plane with $z = 0$. If we consider the incident amplitude as unity, the reflected (r) and transmitted (t) amplitudes at the interface [$z = 0$] can be estimated from Eq. 1.11 using continuity of the wavefunction ψ and its derivative $d\psi/dz$, and one gets

$$1 + r = t;\ q_1(1 - r) = q_2 t \qquad (1.13)$$

Solving these two equations, we obtain the reflection and transmission amplitude at the air-medium interface:

$$r = \frac{q_1 - q_2}{q_1 + q_2};\ t = \frac{2q_1}{q_1 + q_2} \qquad (1.14)$$

where q_1 and q_2 are the normal component of the incident and transmitted momentum vector and given by: $q_1 = \frac{2\pi}{\lambda}\sin\theta_i$ and $q_2^2 = q_1^2 - 4\pi\rho b_{coh} = q_1^2 - q_c^2$. The Fresnel reflected intensity for an ideally flat surface can be defined as:

$$R_F = |r|^2 = \left|\frac{\sin\theta_i - \sqrt{n^2 - \cos^2\theta_i}}{\sin\theta_i + \sqrt{n^2 - \cos^2\theta_i}}\right|^2 \qquad (1.15)$$

when $\cos\theta_i > \cos\theta_c$, or $\theta_i < \theta_c$ the reflected amplitude is complex and the reflected intensity becomes unity. In the angular range $0 < \theta_i < \theta_c$, total reflection takes place. Away from the critical angle, or when the wavevector transfers $Q \gg Q_c$, the prior expression indicates that the reflected Fresnel intensity, $R \propto Q^{-4}$. The expression for reflectivity has a one-to-one correspondence between x-rays and neutrons, except for the fact that the critical angles are dictated by the corresponding electron density (for x-rays) or coherent scattering length density (for neutrons) in the sample. In addition, for neutrons, one finds different reflectivities for spin up (R^+) and spin down (R^-) polarizations from a magnetized heterostructure. Figure 15.2(a) shows the experimental NR (unpolarized) pattern of a glass substrate with the flat reflectivity (= 1) below the critical angle, and the intensity falls with Q^{-4} at large Q. The simulated reflectivity profile for a magnetized Ni substrate, using polarized neutrons, is shown in Figure 15.2(c). The critical angles for the spin up and spin down neutrons are distinctly different, as well as the reflectivity profiles ($R^\pm$) for the magnetized sample. One obtains the magnetization density profile from the difference in spin-dependent reflectivity ($R^+ - R^-$).

Usually, a surface or an interface is never ideally flat and has roughness associated with it. The role of roughness is to make the interface *fuzzy* or smear it out. Drawing an analogy from the effect of thermal vibration on the intensity of diffraction peaks in x-ray diffraction, one can take care of the roughness by multiplying the Fresnel reflectivity with a "Debye-Waller like" factor [22]. The reflected intensity for a rough surface can be represented by:

$$R(Q) = R_F(Q) e^{-Q^2 \sigma^2} \tag{1.16}$$

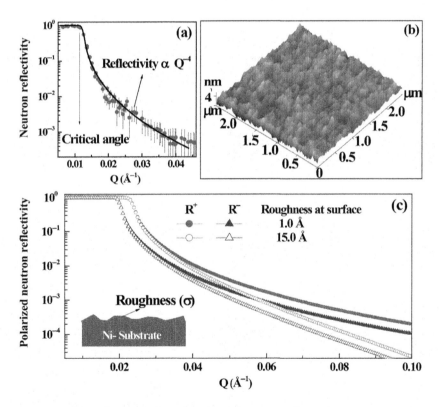

FIGURE 15.2 (a) Experimental PNR pattern of a glass substrate. Reflectivity is unity up to the critical angle. Away from the critical angle at larger Q, the Fresnel reflectivity falls as $\sim Q^{-4}$. (b) AFM image of a Ni surface, indicating the roughness. (c) PNR (spin up R^+ and spin down R^-) profile simulated for Ni substrate with two values of roughnesses (closed and open symbols for Ni substrate with surface roughnesses of 1.0 Å and 15 Å, respectively). Reflectivity falls rapidly at larger Q for higher surface roughness.

where σ is the roughness parameter of the surface/interface. Figure 15.2(b) shows the AFM image of a Ni film with roughness at the film-air interface. It is clear from the prior expression that the role of the "Debye-Waller like" factor is to pull down the reflected intensity at higher momentum transfer Q. Figure 15.2(c) shows the simulated PNR profiles from a Ni substrate for two values of surface roughness, and it is evident that the reflectivity falls faster at higher Q for larger surface roughness.

15.2.2 Reflectivity From a Film of Finite Thickness

We have discussed the reflectivity from an interface of a medium with an infinite thickness (substrate-air interface). In the case of the finite thickness of a film, this gets modified. The neutron beam gets reflected from the surface (film-air interface), as well as from the bottom of the film (substrate-film interface), as shown in Figure 15.3(a). Interference between these two rays causes interference fringes in the reflected beam known as "Kiessig oscillations". In a generalized approach, the reflected intensity can be represented by the Fourier transform of the derivative of a scattering length density profile $\rho(z)$ [20], multiplied by Fresnel reflectivity:

$$R = R_F \left| \int \frac{d\rho(z)}{dz} e^{iQz} dz \right|^2 \tag{1.17}$$

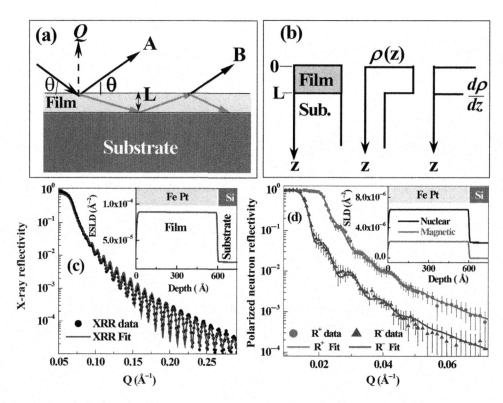

FIGURE 15.3 (a) The neutron beam gets reflected from the surface (beam A) and bottom (beam B) of the film. These two beams interfere and create oscillations in the reflectivity pattern, known as "Kiessig oscillations." (b) A film of a finite thickness (L) on a substrate. The derivative is made up of two δ functions at the surface ($z = 0$) and the substrate-film interface ($z = L$). (c) XRR data (symbol) and corresponding fit (solid line, blue) for a FePt film on Si substrate. Inset (c) shows the ESLD depth profile obtained from XRR data. (d) PNR data (red circles, spin up and blue triangles, spin down) and corresponding fits (solid lines) from the same FePt thin film. The inset of (d) shows the nuclear and magnetic SLD profiles of FePt film obtained from PNR data.

If the density is an ideal step function for a film, as shown in Figure 15.3(b), then its derivative in the z-direction comprises two δ functions: one at $z = 0$, the film-air interface, and the other at $z = L$, the thickness of the film (Figure 15.3(b)). This converts the above expression to:

$$R = R_F \left(1 + e^{iQL}\right) \tag{1.18}$$

The expression indicates that whenever Q is equal to $2\pi/L$ or its even multiple, there will be peaks in the interference pattern. The peaks will lie beyond the falling edge of the Fresnel reflectivity since there is no transmitted beam below the critical angle.

Figure 15.3(c) shows experimental XRR data from a FePt film grown on a Si substrate. The Kiessig oscillations for the thickness of the film are seen in the data. The fit to the data provides thickness (~600 Å), density profile, and interface roughness in the film. The inset of Figure 15.3(c) shows the ESLD depth profile of the film obtained from the best fit (solid line in Figure 15.3(c)) to XRR data. The spin-dependent PNR data for the same film is shown in Figure 15.3(d). The Kiessig oscillations and the splitting between spin up (R+) and spin down (R−) data are evident in Figure 15.3(d). The difference in the PNR data for the two (+ and −) polarization is due to the magnetism in the film. The nuclear and magnetic SLD (NSLD and MSLD) depth profiles of the film obtained from fits to PNR data are shown in the inset of Figure 15.3(d). While the difference in Q between two consecutive Kiessig oscillations gives information about the total thickness of the FePt film, the difference in the reflectivity profiles (R+ − R−) provides its magnetic information. A uniform magnetization with an MSLD of $\sim 2.10 \times 10^{-6}$ Å^{-2} was obtained for the entire FePt layer.

15.2.3 Reflectivity From a Periodic Multilayer

One may also deposit a multilayer sample that consists of periodic bilayers of two materials with the thickness of the layers repeating. PNR contributes uniquely to the investigation of interface magnetization at the buried interfaces of such multilayers [17]. Inset (i) of Figure 15.4(a) shows the schematic of such a periodic bilayer with two components (Ni and Ti) with the individual layer thicknesses d_1 and d_2. The multilayer behaves like a one-dimensional "crystal" in a reflectivity experiment with a lattice parameter $d_1 + d_2$ (sum of the repeat thickness of the bilayer components). One observes a "Bragg peak" from the periodic multilayer in reflectometry, albeit at a much lower Q compared to peaks from atomic layers in a conventional x-ray diffraction experiment, because of the repeating period $(d_1 + d_2)$ is much larger than atomic periodicity. Figure 15.4 shows the spin-dependent experimental PNR data from a periodic bilayer (multilayer) of Ni/Ti with 10 bilayers. The intensity of the Bragg peak depends on the quality of the bilayer (interface roughness and SLD contrast). Inset (ii) of Figure 15.4 shows the NSLD and MSLD depth profile of the Ni/Ti multilayer obtained from the PNR data. Thus, using PNR, one can simultaneously obtain the depth-dependent structural and magnetic properties of the multilayer. One can also study mixing or alloying at the interfaces of such periodic multilayer samples from the intensity of the Bragg peaks.

15.2.4 Data Analysis Technique for Specular Reflectometry

Data analysis for the reflectometry technique is an extremely important task since the process of getting the required physical parameters from reflectometry data through a model-based analysis is a non-trivial issue. One has to move from reflectometry data in "momentum space", obtained in an experiment, to information like SLD, thickness, magnetic moment density, and roughness in "real space", which are Fourier inverses of each other. Parratt's formalism [2] is used to generate the reflectivity profile for a layered (or multilayer) structure. Using the model, we simulate the neutron reflectivity and calculate the difference between experimental and simulated data. The model can accordingly be adjusted by some optimization method to get a closer agreement with the experimental data. The original formalism was

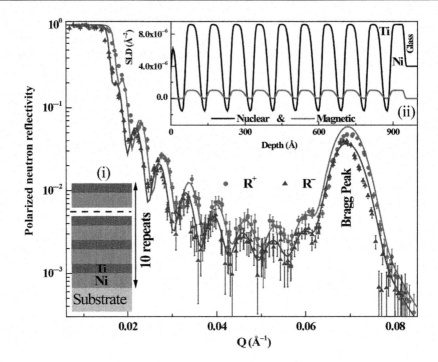

FIGURE 15.4 PNR data (symbols) and fits (solid lines) from a Ni/Ti periodic multilayer with 10 bilayers. The Bragg peak is from the periodicity of the bilayer sequence. Inset (i) shows the schematic of a periodic bilayer with two components, Ni and Ti, with different thicknesses on a substrate. Inset (ii) shows the NSLD and MSLD depth profile of the multilayer extracted from the fit to PNR data.

developed for fitting x-ray reflectivity data [2], which can be easily adapted for PNR data with suitable substitution of x-ray scattering length density with neutron scattering length density and, in addition, magnetic scattering length density for magnetic samples and polarized neutrons.

One starts with a model histogram that approximates the real SLD profile. The thickness of each slice of the layer depends on the resolution of the experimental setup. Starting with a model SLD profile histogram, the next step is to generate the reflectivity profile for the histogram, using the recursive relation suggested by Parratt [2]. Once we have a simulated reflectivity pattern, it is matched with the experimentally measured pattern through an error function (χ^2) minimization process by varying the histogram density profile to improve the fit of the generated reflectivity profile with the measured profile. It is very similar to the fitting of x-ray diffraction data for crystallographic structures. One needs to use the right potential given in Eqs. 1.1–1.6 for neutrons (and also for XRR) to generate the reflectivity pattern. The result of the fit to experimental data provides the nuclear or coherent NSLD profile for unpolarized neutrons and also the magnetic moment density profile for PNR (Figure 15.4). A variety of data fitting and parameter optimization strategies have been developed, and there is freely available software on the internet, to fit reflectivity data [23–25]. We have developed a genetic algorithm-based optimization program for the analysis of reflectivity (XRR and PNR) data [17, 26]. However, the reflected intensity detected in an experiment is a convolution with an instrumental resolution function. Therefore, for optimization (comparing the experimental data with a theoretical profile for a model), we also have to either convolute the theoretical profile with instrument resolution or de-convolute the experimental data for instrument resolution. For comparing the experimental specular reflectivity data with the calculated intensity for a model, we usually convolute the theoretical intensity with an instrumental resolution function [8]. We demonstrate the strength of the non-destructive PNR technique with the help of a few examples in the following section.

15.3 INVESTIGATION OF INTERFACIAL MAGNETIZATION IN THIN FILMS

Investigation of interfacial magnetization and exchange interaction in different magnetic/non-magnetic heterostructures is crucial, and the use of PNR for such studies constitutes a large fraction of experiments in thin films. The obvious reason has been that the interaction between layers of magnetic and non-magnetic metals, ferromagnets and semiconductors, and superconductors and ferromagnets is at the forefront of research both from the point of view of basic understanding of such interaction and also on the application front for many possible devices. We highlight a few of such studies where PNR was used as a primary technique.

15.3.1 3*d* Transition Metal: A Co Film

Co is a well-studied 3*d* transition metal ferromagnet that shows interesting properties in thin-film form. We present two examples of ultrathin Co films, which show properties quite different from their bulk counterparts and that can only be seen using PNR with an excellent depth resolution of a few nanometers! The examples highlight the unique and inherent quality of PNR to reveal the magnetic nature of interfaces. We include two instances of different magnetic moments at the interfaces of Co thin films deposited by different techniques.

15.3.1.1 Co Films Grown by Different Techniques: Sputtering and Electrodeposition

Two Co thin films of thickness of about 350 Å each were grown on Si substrate by two techniques: electrodeposition (ED) and magnetron sputtering. Using PNR results, different magnetic properties of Co films grown by these two techniques were observed [27]. The Co film, grown by ED, henceforth named as sample S1, was electrodeposited on a 250 Å thick Cu seed layer on top of a buffer layer of Ti of thickness 150 Å. Both buffer and seed layers were a requirement for getting good adhesion and quality of the Co layer by ED. The nominal structure of sample S1 can be represented as Si(substrate)/Ti(150 Å)/Cu(250 Å)/Co(350 Å). First, both seed and buffer layers were grown by magnetron sputtering at a base vacuum of 5×10^{-7} Torr. The Co layer in one sample (S1) was electrodeposited on the Si/Ti/Cu buffer using an electrolyte containing 1.5 M $CoSO_4$, $7H_2O$ and keeping temperature and pH constant at 298 K and 2.7, respectively, during the deposition [27]. Another Co film (S2) of the same nominal structure was deposited by sputtering on the buffer layer as mentioned prior. The ED film (S1) and the sputter-deposited film (S2) both were studied using PNR for their magnetization depth profile. X-ray diffraction (XRD) data from both films suggested a crystalline structure [27]. The layered structure of the film was first determined by XRR, which indicated that S1 and S2 have Co layers of thickness 350 Å and 360 Å, respectively. Unlike S2, which has a Co film of uniform density, a part of the Co layer in S1 (grown by ED) at the film-air interface showed lower density compared to the rest of the film. It was demonstrated that this structural difference of the surface Co layer leads to its magnetic property, different from the other sputter-deposited film (S2). Macroscopic magnetization measurements indicated that both the films (S1 and S2) were ferromagnetic with a well-defined magnetization hysteresis loop. PNR data for both the samples were measured at the PNR instrument at Dhruva, BARC [8, 27]. The spin-up (+) and spin-down (−) reflectivity data (symbols) with the corresponding fit (solid lines) are shown in Figures 15.5(a) and (b) for the two samples S1 and S2, respectively. The extracted NSLD from the fits is shown in Figures 15.5(c) and (d) for S1 and S2, respectively. The MSLD depth profiles extracted from PNR data for S1 and S2 are shown in Figures 15.5(e) and (f), respectively. The NSLD obtained from PNR data and the ESLD obtained from XRR (not shown) both relate to the physical density profile of the layers in the thin film heterostructures. The parameters obtained from PNR and XRR match closely. This is a confirmation regarding reliability of the fits from two independent measurements. Both in PNR and XRR, one could see the presence of a low-density layer ~100 Å, near the surface in S1, unlike S2. In addition, from the extracted density profile from XRR and PNR, it was confirmed that the low-density layer is not CoO. Also, in both S1 and S2, the overall density of the Co layers was less than that of bulk Co.

The most interesting part of the result is concerning the MSLD profile for the two samples. In the case of S1, it was observed that while the entire film was ferromagnetic, the surface low-density layer has an MSLD that is negative (Figure 15.5(e)). This conclusion was drawn after attempting several models for the MSLD profile in this sample and observing that a negative magnetization on the surface gives the best fit. This was unlike sample S2, which shows a uniform magnetization throughout the Co layer. Also, the interface roughness parameters for the electrodeposited sample are systematically higher than those for the sputter-deposited sample. The study outlines the microscopic difference that comes between two films deposited by two different techniques. The most interesting difference in the structure and magnetism at the microscopic level was the presence of a low-density surface layer of Co that was antiferromagnetically coupled to the rest of the Co layer in the case of the ED grown sample (S1).

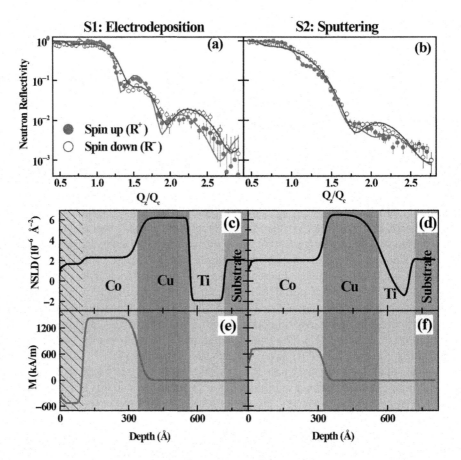

FIGURE 15.5 PNR profile for (a) S1 (grown by electrodeposition) and (b) S2 (grown by sputtering). Spin up and spin down intensities are shown. NSLD depth profiles were obtained from PNR for S1 (c) and S2 (d) and extracted MSLD for the samples S1 (e) and S2 (f). For electrodeposited Co (S1), we observed reversed magnetization at the film-air interface (the hatched region in (e)) due to different growth and morphology of the film. Adapted with permission from [27]. Copyright (2016) Royal Society of Chemistry.

15.3.1.2 High-Density and Non-Magnetic Co Film at Interfaces

3d transition metals, e.g. Fe, Co, and Ni, show interesting magnetic properties both in their bulk and thin-film forms. Theory predicted that a dense face-centered-cubic (*fcc*) phase of cobalt can be non-magnetic [28–29]. However, there was no experimental study in the literature for Co showing highly dense non-magnetic properties. We studied a Co thin film, as thin-film polycrystalline materials have shown the formation of a film with compressive stress, which can increase the density of the film. A Co film was deposited on Piranha cleaned and hydrofluoric acid (HF) etched Si [111] substrate using an electron-beam deposition technique [29]. The layer structure at the mesoscopic length scale was obtained using XRR and PNR. The thickness of the film as obtained from XRR was 270 Å, which corroborates well with the thickness obtained from PNR. The data from both XRR and PNR experiments confirmed that there was a high-density Co layer present on both interfaces (film-air and substrate-film) of the film. The PNR experiments were first carried out at Dhruva BARC and later repeated at SNS Oakridge to confirm the unique result. Figure 15.6(a) shows PNR data (symbols) along with fits (solid lines) from the Co film at room temperature (RT). Figure 15.6(b) shows the NSLD and MSLD depth profile of the film, which best fitted the PNR data at RT. The ESLD depth profile of the film obtained from XRR data (not shown) indicated the presence of high-density Co layers at the film-air and film-substrate interface. This was also confirmed from NSLD data obtained from PNR (Figure 15.6(b)). These high-density layers were about 25 Å thick at the interfaces. Most interestingly, the high-density layers at the interfaces were non-magnetic, as seen in the MSLD (Figure 15.6(b)) obtained from PNR. It was also confirmed from

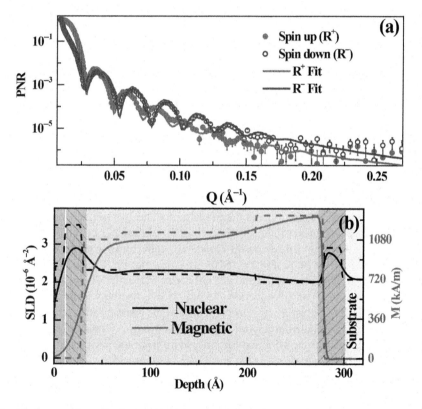

FIGURE 15.6 (a) The PNR data for the Co film grown on the Si substrate. Spin up (red circles) and spin down (blue circles) are shown together with the fits (solid lines). (b) fitted NSLD and MSLD profiles for the same data. The presence of high-density Co layers at the film-air and substrate-film interfaces is evident (shaded regions). This high-density Co layer is non-magnetic (zero magnetization), as can be seen in the MSLD profile. Adapted with permission from [29]. Copyright (2017) Springer Nature.

high-resolution cross-sectional transmission electron microscopy that these high-density layers were *fcc* structure, unlike bulk Co, which has a hexagonal close-packed (*hcp*) structure. Such a high-density Co phase had only been observed under high pressure. It was conjectured that the migration of Co atoms along grain boundaries during deposition might cause such high pressure and give rise to such a phase. This unique result vindicated the strength of PNR in detecting such ultrathin layers and their magnetism.

These results on Co films grown by different techniques show that the deposition technique has a very important role in dictating the properties of a thin film and also demonstrates the fact that PNR plays an important role in obtaining such properties with excellent spatial resolution in sub-nanometer length scales. Presently researchers are looking for novel properties in thin-film heterostructures of various components for obtaining (or tailoring) properties that are not seen in bulk materials. PNR plays a key role in deciphering structure and magnetism in such heterostructures.

15.3.2 Helical Magnetic Structure in Co/Gd Multilayers

Rare earth (RE) transition metal (TM) heterostructures, viz. Gd-Fe, Gd-Co, etc., are known to exhibit several magnetic structures at different temperatures and magnetic fields due to strong antiferromagnetic (AF) interaction at the interfaces. Co and Gd show long-range magnetic order with a Curie temperature of ~1400 K and 293 K, respectively. In the present study, the detailed magnetic structure of the Gd/Co multilayer at different temperatures and fields was investigated using PNR with spin polarization analysis. PNR results suggested the evolution of a unique helical structure in the Gd/Co multilayer [30–32]. This study is important because recently, theoretical and experimental studies [33–34] suggested that helical magnetic structures may be exploited for magnetic storage. The study also indicates the role of PNR in such studies with planar or two-dimensional magnetism. A Gd/Co multilayer of nominal structure Si(100)/[Gd(140 Å/Co(70 Å)] × 8 was grown by DC magnetron sputtering. X-ray diffraction indicated that the Gd and Co layers were polycrystalline with *fcc* and *hcp* structures respectively. In this study, we investigated the magnetic structure, which requires measuring PNR with spin polarization analysis, i.e., measuring reflected neutron beam as a function of neutron polarization. One measures the pattern of four reflectivities (R^{++}, R^{--}, R^{+-}, and R^{-+}) in this case, where the first and second symbols in the superscripts indicate polarization of the incident and reflected neutrons, respectively. The R^{++} and R^{--} measurements are known as non-spin-flip (NSF) measurements, and the R^{+-} and R^{-+} patterns are called spin-flip (SF) measurements. This is one step ahead of the study we presented in the previous example, where we measured the magnetization moment density in the sample from reflectivity R^{+} and R^{-} without polarization analysis of the reflected beam. It must be mentioned here that the NSF data provides the in-plane magnetization that is parallel to the neutron polarization and the SF data provides in-plane magnetization normal to the previous (Figure 15.7(f)). A vectorial sum of the two gives the value and direction of the in-plane moment.

PNR measurements were carried out on the OFFSPEC reflectometer at the ISIS neutron source in Rutherford Appleton Laboratory, UK. Figure 15.7(a–d) shows the PNR data at different temperatures from the sample. At 300 K, Co is ferromagnetic and Gd is paramagnetic. This is clear from the PNR data at 300 K as there is no indication of SF reflection and only NSF reflectivity is present due to ferromagnetism in the Co layers. As the temperature was lowered, around the compensation temperature (T_{comp}, where the Gd and Co moments are aligned antiparallel due to the strong AF interaction and the net moment tends to zero) ~125 K there is strong NSF and SF reflectivity from the sample. The NSF, R^{++}, and R^{--} in Figure 15.7(a–d) and the averaged SF data (R^{+-} + R^{-+}) / 2 in Figure 15.7(a–d) were fitted (solid lines) to get the in-plane magnetic structure in the samples. The data at 300 K (~RT) and 200 K have been fitted to obtain the NSLD and MSLD (Figure 15.7 (e)) in the sample. It was seen that at 300 K, Co has a magnetic moment of 1.52 μ_B per atom. Gd showed a zero magnetic moment since it was in a paramagnetic state at 300 K. PNR data at 200 K could be fitted with antiferromagnetic coupling between Co and Gd layers with Co having +1.65 μ_B per atom and Gd having −1.40 μ_B. The + and − signs indicate antiparallel coupling between the moments. The PNR data at 125 K showed a large SF component, and it was fitted with a helical in-plane magnetic structure in the Co and Gd layers. The moments had in-plane 2π rotation in a layer

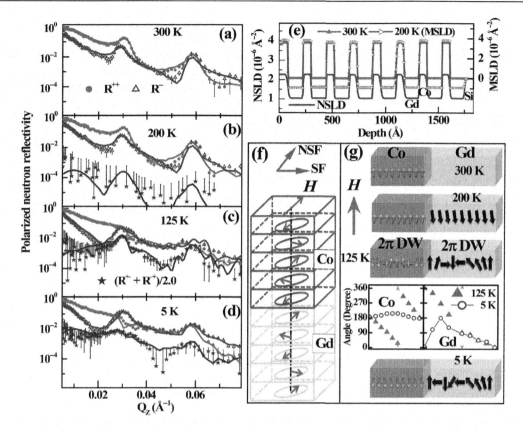

FIGURE 15.7 (a–d) PNR data with polarization analysis (R++, R−−, R+−, and R−+) for Gd/Co multilayer at different temperatures (300 K to 5 K). (e) NSLD and MSLD (at 300 and 200 K) depth profile of the multilayer, where Gd moment is zero at 300 K and small but negative (aligned antiferromagnetically with Co) at 200 K. (f) The helical magnetic structure model for a bilayer of Gd/Co, used for fitting PNR data from the multilayer at low temperatures. (g) The magnetic structure along with the thickness of Gd/Co bilayer in the multilayer at different temperatures. At low temperatures, a helical magnetic structure for both Gd and Co layers was obtained, which are aligned antiferromagnetically at the interfaces. Adapted with permission from [30]. Copyright (2019) American Physical Society.

and were coupled antiferromagnetically at Co-Gd interfaces. The helical structure is shown schematically in Figure 15.7(f). To fit the data, in-plane moments were broken into thin slices of layers with each layer following an in-plane rotation as indicated. The fit indicates that each layer behaved like a 2π domain wall and was antiferromagnetically coupled to the next layer. At 5 K, Co was aligned opposite to the applied field with a small fluctuation: 180° ± 10° and Gd continued to be a helix, but not with a 0–2π rotation, rather a 0–π–0 rotation in a plane. This study indicates the strength of PNR in obtaining such detailed magnetic structure in ultrathin layers of magnetic material and its need for deciphering the structure in two-dimensional magnets.

15.3.3 Proximity Effect in Superconductor and Ferromagnet Oxide Heterostructures

It is well known that superconductors are perfect diamagnets that do not coexist with ferromagnets. This fact brings an interesting question to the fore: what happens if a superconductor (SC) and a ferromagnet

(FM) are brought in proximity? In recent days this problem has attracted a large number of theoretical and experimental studies on a phenomenon known as the "proximity effect" [35–39], which showed suppression of magnetism on the FM side with the formation of a magnetic depleted (MD) layer. The proximity effect is explained in terms of leakage of the spin-triplet Cooper pairs into the FM, leading to a modulation in the magnetism of the interfacial FM layer. This indicates that if the FM-SC layers are interleaved with an insulator (I), the leakage of Cooper pairs should stop and there will be an abrupt break in the long-range order of the triplet spin-paring. However, the presence of the MD layer at FM interfaces for the system where the FM and SC layers were separated by an insulator clearly suggests tunneling of Cooper pairs in the system. Here we describe the observation of suppression of ferromagnetism in a $La_{2/3}Ca_{1/3}MnO_3$ (LCMO) layer, separated from a $YBa_2Cu_3O_{7-\delta}$ (YBCO) layer by a thin oxide insulator $SrTiO_3$ (STO) [37]. A trilayer with the structure STO (substrate)/YBCO (200 Å)/STO (50 Å)/LCMO (200 Å) was grown by pulsed laser deposition technique [37]. The trilayer was grown on a single-crystal STO substrate and the growth was highly textured in the [001] direction as seen in XRD. From macroscopic magnetization (SQUID) measurements, it was found that LCMO became ferromagnetic ~140 K and the YBCO layers went superconducting at a superconducting transition temperature (T_{sc}) ~60 K. The T_{sc} for the YBCO layers in the heterostructure was much lower than that of the bulk phase (~90 K). For the present sample, there were three interesting temperature regions for understanding the coupling between layers: (a) RT (~300 K) when YBCO is not superconducting and LCMO is paramagnetic. (b) 60 K < T < 140 K when the LCMO is ferromagnetic but YBCO is not superconducting, and (c) T < 60 K when LCMO is ferromagnetic and YBCO is superconducting.

The depth-dependent structure of the heterostructure was obtained from XRR data shown in Figure 15.8(a). The inset of Figure 15.8(a) shows the ESLD depth profile. PNR measurements at different temperatures (300 K, 100 K, 50 K, and 10 K) were carried out at the MARIA reflectometer in the MLZ research reactor in Garching Munich. PNR data at 300 K provided an NSLD depth profile, which matches the ESLD depth profile (obtained from XRR) of the heterostructure. The PNR data at 10 K from the heterostructure is shown in Figure 15.8(b). The inset of Figure 15.8(b) shows the magnetization depth profile of the heterostructure which best fitted (solid lines in Figure 15.8(b)) the PNR data at 10 K. It is evident from the magnetization profile at 10 K that an MD layer of thickness ~30 Å was present at the STO/LCMO interface. It is noted that on reducing the thickness of the insulator (STO) layer from 50 Å to 25 Å, the MD layer thickness increased to ~70 Å. The coupling between the superconductor and the ferromagnet was demonstrated from the PNR data at 50 K (below T_{sc}) and 100 K (above T_{sc}), shown in Figure 15.8(c) and (d), respectively. The PNR data in Figure 15.8(c) and (d) were plotted as normalized spin asymmetry (NSA), defined as NSA = $Q^4(R^+ - R^-)/2\pi$ to highlight the difference between different magnetization depth profiles from best fits to PNR data at these temperatures. Two magnetization depth profiles – (a) uniform magnetization in LCMO layer and (b) uniform magnetization with MD layer at the interface – are considered and shown in Figure 15.8(e) and (f) for 50 and 100 K, respectively. The PNR data suggested the formation of the MD layer below T_{sc} in SC/I/FM heterostructure, which can be understood if there is tunneling of Cooper pairs. The justification for tunneling of Cooper pairs was further vindicated by another PNR experiment [40] that correlates the absence/presence of superconductivity with the disappearance/appearance of the dead layer in the ferromagnetic layer [40]. Instead of temperature, a magnetic field of 1 Tesla was used to destroy the superconductivity in the YBCO layer. Using PNR measurements at 5 K from FM/I/SC heterostructure at 1 Tesla showed the absence of the MD layer, suggesting that the Cooper pairs are destroyed at this field and the tunneling of them from the superconductor to the ferromagnet vanishes. The examples discussed in this part demonstrate the need for PNR to understand the interplay of superconductivity and magnetism in such hybrid systems.

Here we presented several examples of magnetism, especially at the interfaces of ultrathin heterostructures, highlighting the unique capability of PNR for resolving outstanding issues in these pseudo-two-dimensional systems. PNR, with its inherent interaction with nuclear potential and atomic spin magnetic moment of the system and its very good depth sensitivity, has become an ideal tool to investigate the structure and magnetic properties of the ultrathin films and buried interfaces in a multilayer system. The presence of external perturbations, e.g. temperature, magnetic field, electric field, applied stress,

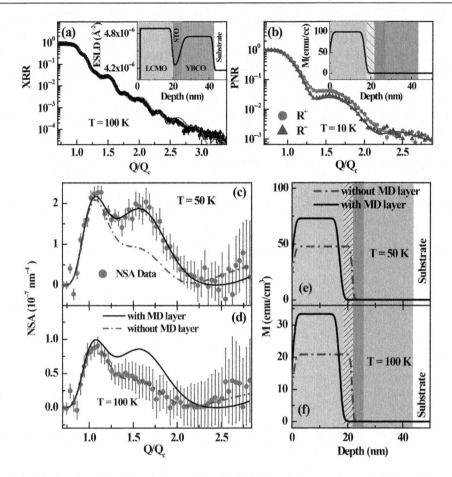

FIGURE 15.8 (a) XRR data from STO (substrate)/YBCO/STO/LCMO heterostructure. The x-axis has been normalized to the critical angle Q_c. The inset of (a) shows the ESLD depth profile of the heterostructure obtained from XRR data. (b) PNR data for heterostructure at 10 K. The inset of (b) shows the magnetization depth profile of the heterostructure obtained from PNR data. An MD layer (marked as a black line with a pin background) is observed at the STO/LCMO interface below superconducting transition temperature (T_{sc}). (c) and (d) The NSA (please see text) data (PNR data) for the heterostructure at 50 K (below T_{sc}) and 100 K (above T_{sc}) with fitted profiles assuming different magnetization depth profiles shown in (e) and (f) at 50K and 100 K, respectively. Adapted with permission from [37]. Copyright (2016) Royal Society of Chemistry.

light, etc., also influences the magnetism in complex oxide heterostructures. The external perturbations that influence the interface magnetism in ultrathin films can be employed easily in the case of the PNR experiment, as it is sometimes difficult to adapt the required sample environment for the macroscopic magnetization technique.

15.4 CONCLUSION

In summary, we briefly discussed the PNR technique and its successful use for the investigation of magnetism occurring in films of ultrathin $3d$ transition metal (Co film) grown by different techniques, antiferromagnetically coupled ultrathin Gd/Co multilayers, and complex oxide interfaces, which are important systems and could find interesting functionalities in devices with operation controlled by the interfaces. It is fair to stress that PNR, due to its inherent sensitivity to reveal small magnetism with sub-nanometer depth resolution, has been and will continue to play a unique and major role in this exciting research area of two-dimensional magnets.

15.5 REFERENCES

1. Fermi, E. and Zinn, W. (1946) Reflection of neutrons on mirrors. *Physical Review* 70: 103A.
2. Parratt, L.G. (1954) Surface studies of solids by total reflection of x-rays. *Physical Review* 95 (2): 359–369.
3. Felcher, G.P. (1988) Principles of neutron reflection. SPIE's 32nd Annual International Technical Symposium on Optical & Optoelectronic Applied Science & Engineering, San Diego, CA.
4. Felcher, G.P. (1994) Of butterflies and terraces. *Neutron News* 5 (4): 18–22.
5. Richardson, R.M., Webster, J.R.P. and Zarbakhsh, A. (1997) Study of off-specular neutron reflectivity using a model system. *Journal of Applied Crystallography* 30 (6): 943–947.
6. Majkrzak, C.F. (1991) Polarized neutron reflectometry. *Physica B: Condensed Matter* 173 (1): 75–88.
7. Fitzsimmons, M.R. and Majkrzak, C.F. (2005) *Application of Polarized Neutron Reflectometry to Studies of Artificially Structured Magnetic Materials, Modern Techniques for Characterizing Magnetic Materials* (Springer, New York).
8. Basu, S. and Singh, S. (2006) A new polarized neutron reflectometer at Dhruva for specular and off-specular neutron reflectivity studies. *Journal of Neutron Research* 14: 109–120.
9. Singh, S., Basu, S., Bhatt, P. and Poswal, A.K. (2009) Kinetics of alloy formation at the interfaces in a Ni-Ti multilayer: X-ray and neutron reflectometry study. *Physical Review B* 79 (19): 195435.
10. Singh, S., Basu, S., Gupta, M., Majkrzak, C.F. and Kienzle, P.A. (2010) Growth kinetics of intermetallic alloy phase at the interfaces of a Ni/Al multilayer using polarized neutron and x-ray reflectometry. *Physical Review B* 81 (23): 235413.
11. Singh, S., Basu, S., Bhattacharya, D. and Poswal, A.K. (2010) Physical and magnetic roughness at metal-semiconductor interface using x-ray and neutron reflectometry. *Journal of Applied Physics* 107 (12): 123903.
12. Singh, S., Fitzsimmons, M.R., Lookman, T., Thompson, J.D., Jeen, H., Biswas, A., Roldan, M.A. and Varela, M. (2012) Magnetic nonuniformity and thermal hysteresis of magnetism in a manganite thin film. *Physical Review Letters* 108 (7): 077207.
13. Singh, S., Haraldsen, J.T., Xiong, J., Choi, E.M., Lu, P., Yi, D., Wen, X.D., Liu, J., Wang, H., Bi, Z., Yu, P., Fitzsimmons, M.R., MacManus-Driscoll, J.L., Ramesh, R., Balatsky, A.V., Zhu, J.-X. and Jia, Q.X. (2014) Induced magnetization in $La_{0.7}Sr_{0.3}MnO_3$/$BiFeO_3$ superlattices. *Physical Review Letters* 113 (4): 047204.
14. Zabel, H. and Theis-Br hl, K. (2003) Polarized neutron reflectivity and scattering studies of magnetic heterostructures. *Journal of Physics: Condensed Matter* 15 (5): S505-S517.
15. Blundell, S.J. and Bland, J.A.C. (1992) Polarized neutron reflection as a probe of magnetic films and multilayers. *Physical Review B* 46 (6): 3391–3400.
16. Sinha, S.K., Sirota, E.B., Garoff, S. and Stanley, H.B. (1988) X-ray and neutron scattering from rough surfaces. *Physical Review B* 38 (4): 2297–2311.
17. Singh, S., Swain, M. and Basu, S. (2018) Kinetics of interface alloy phase formation at nanometer length scale in ultra-thin films: X-ray and polarized neutron reflectometry. *Progress in Materials Science* 96: 1–50.
18. Singh, S. and Basu, S. (2017) Investigation of interface magnetism of complex oxide heterostructures using polarized neutron reflectivity. *Current Applied Physics* 17 (5): 615–625.
19. Zabel, H. (1994) Spin polarized neutron reflectivity of magnetic films and superlattices. *Physica B: Condensed Matter* 198 (1): 156–162.
20. Zabel, H. (1990) X-ray and neutron scattering at thin films. *FestkSrperprobleme* 30: 197.
21. Born, M. and Wolf, E. (1959) *Principles of Optics: Electromagnetic Theory of Propagation* (Pergamon Press, London).
22. Névot, L. and Croce, P. (1980) Caractérisation des surfaces par réflexion rasante de rayons X. Application à l'étude du polissage de quelques verres silicates. *Physical Review Applied (Paris)* 15 (3): 761–779.
23. Bjorck, M. and Andersson, G. (2007) GenX: An extensible X-ray reflectivity refinement program utilizing differential evolution. *Journal of Applied Crystallography* 40 (6): 1174–1178.
24. Kienzle, P.A., Maranville, B.B., O'Donovan, K.V., Ankner, J.F., Berk, N.F. and Majkrzak, C.F. (2017) www.nist.gov/ncnr/reflectometry-software

25. Vignaud, G. and Gibaud, A. (2019) REFLEX: A program for the analysis of specular X-ray and neutron reflectivity data. *Journal of Applied Crystallography* 52 (1): 201–213.
26. Singh, S. and Basu, S. (2016) Simultaneous parameter optimization of x-ray and neutron reflectivity data using genetic algorithms. *AIP Conference Proceedings* 1731 (1): 080007.
27. Singh, S., Prajapat, C.L., Bhattacharya, D., Ghosh, S.K., Gonal, M.R. and Basu, S. (2016) Antiferromagnetic coupling between surface and bulk magnetization and anomalous magnetic transport in electro-deposited cobalt film. *RSC Advances* 6 (41): 34641–34649.
28. Banu, N., Singh, S., Basu, S., Roy, A., Movva, H.C.P., Lauter, V., Satpati, B. and Dev, B.N. (2018) High density nonmagnetic cobalt in thin films. *Nanotechnology* 29 (19): 195703.
29. Banu, N., Singh, S., Satpati, B., Roy, A., Basu, S., Chakraborty, P., Movva, H.C.P., Lauter, V. and Dev, B.N. (2017) Evidence of formation of superdense nonmagnetic cobalt. *Scientific Reports* 7 (1): 41856.
30. Singh, S., Basha, M.A., Prajapat, C.L., Bhatt, H., Kumar, Y., Gupta, M., Kinane, C.J., Cooper, J., Gonal, M.R., Langridge, S. and Basu, S. (2019) Antisymmetric magnetoresistance and helical magnetic structure in a compensated Gd/Co multilayer. *Physical Review B* 100 (14): 140405.
31. Basha, M.A., Prajapat, C.L., Bhatt, H., Kumar, Y., Gupta, M., Kinane, C.J., Cooper, J.F.K., Caruana, A., Gonal, M.R., Langridge, S., Basu, S. and Singh, S. (2020) Helical magnetic structure and exchange bias across the compensation temperature of Gd/Co multilayers. *Journal of Applied Physics* 128 (10): 103901.
32. Basha, M.A., Prajapat, C.L., Bhatt, H., Kumar, Y., Gupta, M., Kinane, C.J., Cooper, J.F.K., Langridge, S., Basu, S. and Singh, S. (2020) Field dependent helical magnetic structure in a compensated Gd/Co multilayer. *Journal of Magnetism and Magnetic Materials* 516: 167331.
33. Dzemiantsova, L.V., Meier, G. and Röhlsberger, R. (2015) Stabilization of magnetic helix in exchange-coupled thin films. *Scientific Reports* 5 (1): 16153.
34. Fust, S., Mukherjee, S., Paul, N., Stahn, J., Kreuzpaintner, W., Böni, P. and Paul, A. (2016) Realizing topological stability of magnetic helices in exchange-coupled multilayers for all-spin-based system. *Scientific Reports* 6 (1): 33986.
35. Hoppler, J., Stahn, J., Niedermayer, C., Malik, V.K., Bouyanfif, H., Drew, A.J., Rössle, M., Buzdin, A., Cristiani, G., Habermeier, H.U., Keimer, B. and Bernhard, C. (2009) Giant superconductivity-induced modulation of the ferromagnetic magnetization in a cuprate – manganite superlattice. *Nature Materials* 8 (4): 315–319.
36. Satapathy, D.K., Uribe-Laverde, M.A., Marozau, I., Malik, V.K., Das, S., Wagner, T., Marcelot, C., Stahn, J., Brück, S., Rühm, A., Macke, S., Tietze, T., Goering, E., Frañó, A., Kim, J.H., Wu, M., Benckiser, E., Keimer, B., Devishvili, A., Toperverg, B.P., Merz, M., Nagel, P., Schuppler, S. and Bernhard, C. (2012) Magnetic Proximity Effect in $YBa_2Cu_3O_7/La_{2/3}Ca_{1/3}MnO_3$ and $YBa_2Cu_3O_7/LaMnO_{3+\delta}$ Superlattices. *Physical Review Letters* 108 (19): 197201.
37. Prajapat, C.L., Singh, S., Paul, A., Bhattacharya, D., Singh, M.R., Mattauch, S., Ravikumar, G. and Basu, S. (2016) Superconductivity-induced magnetization depletion in a ferromagnet through an insulator in a ferromagnet – insulator – superconductor hybrid oxide heterostructure. *Nanoscale* 8 (19): 10188–10197.
38. Singh, S., Bhatt, H., Kumar, Y., Prajapat, C.L., Satpati, B., Kinane, C.J., Langridge, S., Ravikumar, G. and Basu, S. (2020) Superconductivity-driven negative interfacial magnetization in $YBa_2Cu_3O_{7-\delta}/SrTiO_3/La_{0.67}Sr_{0.33}MnO_3$ heterostructures. *Applied Physics Letters* 116 (2): 022406.
39. Kumar, Y., Bhatt, H., Prajapat, C.L., Singh, A.P., Singh, F., Kinane, C.J., Langridge, S., Basu, S. and Singh, S. (2021) Suppression of the superconducting proximity effect in ferromagnetic-superconducting oxide heterostructures with ion-irradiation. *Journal of Applied Physics* 129 (16): 163902.
40. Paull, O.H.C., Pan, A.V., Causer, G.L., Fedoseev, S.A., Jones, A., Liu, X., Rosenfeld, A. and Klose, F. (2018) Field dependence of the ferromagnetic/superconducting proximity effect in a YBCO/STO/LCMO multilayer. *Nanoscale* 10 (40): 18995–19003.

Magnetosomes
Biological Synthesis of Magnetic Nanostructures

16

Marta Masó-Martínez, Paul D Topham, and Alfred Fernández-Castané
Energy and Bioproducts Research Institute & Aston Institute of Materials Research, Aston University, Birmingham, B4 7ET, United Kingdom

Contents

16.1	Magnetotactic Bacteria	310
16.2	Magnetosomes	311
	16.2.1 Structure and Composition	311
	16.2.2 Genetics of Magnetosome Biogenesis	312
	16.2.3 Magnetosome Biogenesis	312
	16.2.3.1 Invagination of the Magnetosome Membrane	313
	16.2.3.2 Protein Sorting	314
	16.2.3.3 Iron Uptake	315
	16.2.3.4 Magnetite Biomineralization	315
	16.2.3.5 Assembly of Magnetosome Chains	315
16.3	Cultivation of Magnetotactic Bacteria	316
	16.3.1 Factors Affecting Magnetosome Formation	316
	16.3.1.1 Nutrients and media composition	316
	16.3.1.2 Iron Source	317
	16.3.1.3 Oxygen Concentration	317
	16.3.1.4 pH and Temperature	317
	16.3.1.5 Redox Conditions	317
	16.3.2 Cultivation Strategies for Mass Production of Magnetosomes	318
16.4	Magnetosome and Magnetotactic Bacteria Applications	319
	16.4.1 Gene and Drug Delivery	319
	16.4.2 Diagnostic and Imaging Tools	320
	16.4.3 Biosensors	320

	16.4.4 Magnetic Hyperthermia	320
	16.4.5 Bioremediation	321
16.5	Conclusions	321
16.6	References	321

16.1 MAGNETOTACTIC BACTERIA

Magnetotactic bacteria (MTB) are a group of gram-negative bacteria that present the distinct ability to orient themselves in a magnetic field. This unique behavior is possible due to the presence of intracellular magnetic nanoparticles called magnetosomes. Magnetosomes are nanosized magnetic crystals of greigite (Fe_3S_4) or magnetite (Fe_3O_4) arranged in a needle-like chain that is naturally synthesized by MTB through a biomineralization process. MTB species exhibit a wide range of diversity in morphology, physiology, and phylogeny. Cell morphologies include rods, vibrios, spirilla, cocci, and ovoid bacteria, as well as giant and magnetotactic multicellular bacteria (Figure 16.1). In addition, MTB

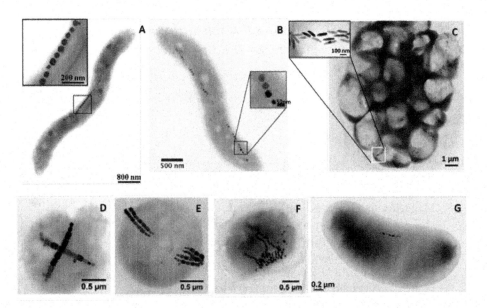

FIGURE 16.1 Electron micrographs of various magnetotactic bacteria that show different cell and magnetosome morphologies as well as different magnetosome chain arrangements. (A) *Magnetospirillum gryphiswaldense* MSR-1 presents spirillum cell morphology and octahedral-shaped magnetosomes. Adapted with permission from [1]. Copyright (2011) Elsevier. The article was printed under a CC-BY license. (B) *Magnetospirillum magneticum* AMB-1 presents spirillum cell morphology and octahedral-shaped magnetosomes. Adapted with permission from [2]. Copyright (2020) American Society for Microbiology. The article was printed under a CC-BY license. (C) Spherical mulberry-like magnetotactic multicellular bacteria containing bullet-shaped magnetosomes. Adapted with permission from [3]. Copyright (2018) Frontiers. The article was printed under a CC-BY license. (D) Magnetotactic cocci LLTC-1 containing elongated prismatic magnetosomes, (E) magnetotactic cocci DMHC-8 containing elongated prismatic magnetosomes, and (F) magnetotactic cocci DMHC-6 with non-chain structure. (D–F) Adapted with permission from [4]. Copyright (2021) John Wiley and Sons. (G) *Desulfovibrio magneticus* RS-1 containing elongated bullet shaped magnetosomes. Adapted with permission from [5]. Copyright (2015) PLOS. The article was printed under a CC-BY license. All CC-BY licensed articles were distributed under a Creative Commons Attribution License 4.0 https://creativecommons.org/licenses/by/4.0/.

motility is made possible by the presence of flagella. The different MTB species present tufts and polar or bipolar flagella. Usually, bacteria swim in run-and-tumbling movements in all directions. However, MTB can passively orient in the earth's magnetic field, and consequently, their movements can be described by a run-and-reverse mechanism. It is hypothesized that the orientation in a magnetic field can present an advantage over non-magnetic bacteria as it helps to locate the optimal position within a vertical concentration gradient.

MTBs are ubiquitous and globally distributed in almost any aquatic habitat. There is the presence of these prokaryotes in the Northern and Southern Hemispheres, in both fresh and salty waters. Geographic location does not seem to be a conditional factor for MTB, but the chemical composition of the environment they usually live in has some common features. All known MTBs are microaerophiles, anaerobes, or facultative anaerobes and are mostly found in sediments or chemically stratified water columns, specifically around the oxic-anoxic interface [6]. MTB have the capacity to swim in either direction along the magnetic field and maintain their position at their preferred oxygen concentration owing to a unique tactic response called magneto-aerotaxis. MTBs are a phylogenetically diverse group of microorganisms affiliated to various groups within the *Proteobacteria* phylum (including *Alphaproteobacteria*, *Deltaproteobacteria*, and *Gammaproteobacteria*), *Nitrospirae* phylum, and candidate division OP3 [6]. Only a few MTB strains, mainly found in the *Alphaproteobacteria* class, such as *Magnetospirillum gryphiswaldense* MSR-1 (Figure 16.1a) and *Magnetospirillum magneticum* AMB-1 (Figure 16.1b), have been successfully isolated and cultivated. The *Magnetospirillum* genus is one of the most studied MTB groups, and both MSR-1 and AMB-1 are considered model strains to study magnetosome biomineralization due to the existence of genetic engineering tools and laboratory cultivation protocols.

16.2 MAGNETOSOMES

Magnetosomes can be found in all MTB species and are essential for their magnetotactic lifestyle. While the use of magnetosomes as a navigational system seems to be the main function, some studies suggest other roles. For instance, the protective role of magnetosomes against metal stress [7], reactive oxygen species (ROS) [8], and UV-B radiation has been studied [9].

16.2.1 Structure and Composition

Magnetosomes consist of an inorganic core (either magnetite or greigite) enveloped in an organic membrane (Figure 16.2). The chemical composition of the mineral phase is species specific. In fact, the chemical composition of the crystal core remains the same even when greigite-producing conditions are provided to a magnetite-producing MTB strain and vice versa. Only a small number of MTB species have been known to produce both iron oxide and iron sulfide nanoparticles within the same cell [3].

Magnetite-producing MTBs, such as the cubo-octahedral shaped magnetosomes from the *Magnetospirillum* species, are the most common and well-studied crystals among MTB. In contrast, greigite-producing MTBs have not yet been grown in a laboratory setting. Magnetite is a mixed-valence iron oxide mineral that contains both ferrous and ferric iron in a 1:2 ratio, respectively [$Fe(II)Fe(III)_2O_4$]. Magnetosomes exhibit high crystallographic perfection. Examination of the magnetosome morphology using electron microscopy techniques revealed the wide diversity of magnetosome crystal shapes, such as square-like, hexagonal, elongated prismatic, tooth-shaped, arrowhead-shaped, rectangular, or bullet-shaped morphologies. Besides the magnetosome morphology, crystal size is also a species-specific feature, ranging from 35 to 120 nm.

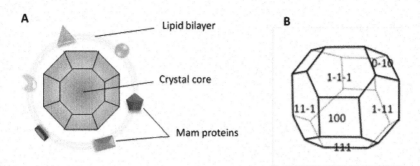

FIGURE 16.2 (A) Schematic structure of a single magnetosome. It is composed of a magnetic mineral crystal core encapsulated by a lipid bilayer. Magnetosome membrane contains Mam proteins. Variation of shape and color denotes the variability of the Mam proteins. (B) Diagram of octahedral magnetite crystal. Adapted with permission from [1]. Copyright (2011) Elsevier. The article was distributed under a Creative Commons Attribution License 4.0 (CC-BY) https://creativecommons.org/licenses/by/4.0/.

Magnetosome crystals are wrapped in an organic membrane called the magnetosome membrane (MM). MM is a lipid bilayer composed of various types of phospholipids and proteins, having a diameter in the range of 10 to 70 nm and a thickness of 5 to 6 nm [10]. MM lipid distribution is very similar to that observed in the cytoplasmic membrane (CM). However, the protein content differs from those that are usually found in the CM. There are more than 25 magnetosome-associated membrane (Mam) proteins.

16.2.2 Genetics of Magnetosome Biogenesis

The genes involved in magnetosome formation are localized in a specific region of the bacterial chromosome called the genomic magnetosome island (MAI). It is estimated that around 30 *mam* and magnetosome membrane-specific (*mms*) genes are involved in magnetosome biogenesis. The MAI comprises approximately 2% of the *Magnetospirillum* sp. genome. This region presents many features associated with genomic islands, such as putative insertion sites near transfer ribonucleic acid (tRNA) genes, distinct levels of guanine-cytosine (GC) enrichment compared to the rest of the genome, and a large number of mobile genetic elements. These features are caused by genetic instability, and the MAI tends to suffer frequent genetic rearrangements that lead to different mutations and new phenotypes (including non-magnetic) when the bacteria are under stress [11].

Although gene size may vary between MTB species, the MAI is generally conserved and grouped in four different operons: *mamGFDC*, *mms6*, *mamXY*, and *mamAB*. The *mamAB* operon is sufficient for some rudimental biomineralization, but the remaining operons encode non-essential proteins involved in controlling magnetosome size, morphology, and arrangement [11]. Table 16.1 lists the proteins encoded by the aforementioned operons and their function.

16.2.3 Magnetosome Biogenesis

The mechanism of magnetosome formation is not yet fully understood. However, it is widely accepted that four steps in MTB magnetosome formation occur: (i) invagination of the cytoplasmic membrane, (ii) sorting of magnetosome proteins into the magnetosome membrane, (iii) iron transport into the magnetosome vesicle and mineralization into magnetite crystals, and (iv) magnetosome chain formation. Figure 16.3 illustrates the routes of iron uptake and biomineralization in MTB.

TABLE 16.1 Function of the Magnetosome-Associated Membrane (Mam) and Magnetosome Membrane-Specific (Mms) Proteins in the Process of Magnetite Biomineralization

FUNCTION	PROTEINS	ENCODING OPERON	REFERENCE
Membrane invagination	MamB	*mamAB*	[11]
	MamI	*mamAB*	[11–13]
	MamL	*mamAB*	[11–13]
	MamQ	*mamAB*	[11–13]
	MamY	*mamXY*	[14]
Protein sorting	MamA	*mamAB*	[15]
	MamE	*mamAB*	[11, 16]
Iron transport	MamB	*mamAB*	[12, 17]
	MamH	*mamAB*	[12, 18]
	MamM	*mamAB*	[17, 19]
	MamZ	*mamXY*	[18]
Redox control	MamE	*mamAB*	[20]
	MamP	*mamAB*	[11]
	MamT	*mamAB*	[11]
	MamX	*mamXY*	[18]
	MamZ	*mamXY*	[18]
Crystal nucleation	MamO	*mamAB*	[21]
	MamE	*mamAB*	[21]
	MamM	*mamAB*	[17]
	MamI	*mamAB*	[22]
pH control	MamN	*mamAB*	[11, 12]
Crystal size and shape control	MamD	*mamGFDC*	[23]
	MamG	*mamGFDC*	[24]
	MamR	*mamAB*	[11]
	MamS	*mamAB*	[11]
	FtsZm	*mamXY*	[25]
	Mms6	*mms6*	[1]
	MmsF	*mms6*	[25]
Chain assembly	MamK	*mamAB*	[26]
	MamJ	*mamAB*	[27]

16.2.3.1 Invagination of the Magnetosome Membrane

Compartmentalization of cellular processes prevents the mixing of contents between two disparate strictly regulated biochemical environments. The formation of the MM is a prerequisite for magnetite biogenesis [11]. Magnetosome membrane invagination involves five Mam proteins: MamB, MamI, MamL, MamQ and MamY. Although the precise roles of these proteins remain unclear, it is believed that MamI, MamL and MamQ participate in the membrane-bending process as they exhibit structural similarities with proteins involved in membrane-remodeling mechanisms in other organisms. MamY can bind and deform liposome membranes, which suggests that it might play a similar role in controlling MM deformation. MamB seems to be the most important protein involved in this process as its suppression heavily affects MM formation. It is thought that MamB may be a landmark protein that

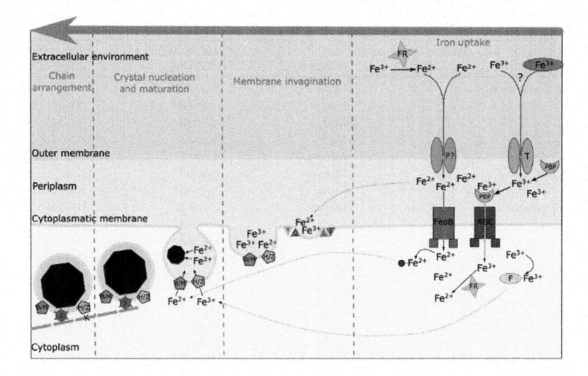

FIGURE 16.3 Schematic representation of cellular iron uptake and magnetosome biogenesis. Firstly, iron is up taken from the extracellular environment with the help of specific outer membrane transporters (T) or porin (P) transporters into the periplasm. In there, periplasmic binding proteins (PBPs) sequester ferric iron and transport it to ABC transporters (for Fe^{3+}). In the case of Fe^{2+}, FeoB transporters are in charge of delivering the iron into the cytoplasm. Secondly, the cytoplasmic membrane is invaginated and forms a vesicle with the help of Mam proteins. Thirdly, in the vesicle lumen iron is accumulated and crystal nucleation starts. Finally, once the crystal is formed, magnetosomes align themselves into a needle-like chain arrangement. If not specified, single capital letters denote the respective Mam proteins. (F), ferritin; (FR), ferric reductase; (O), unknown organic compound; (?), unknown mechanism.

facilitates protein-protein interactions on the MM. Despite having an important role, MM is incapable of invagination alone, suggesting that additional proteins are needed to form larger protein complexes to produce lateral pressure and thus generate the membrane bending required for membrane invagination.

16.2.3.2 Protein Sorting

Most Mam proteins are localized on the MM or the cytoplasmic side of the MM, but little is known about how these proteins are transported into the MM. No specific sequences or localization signals have been identified to date. However, MamA and MamE are considered to be involved in the sorting process.

The MamA structure is highly conserved among MTB and facilitates the binding of other Mam proteins to the surface to form protein complexes on the MM that participate in later steps of biomineralization. MamA-defective mutant showed the presence of empty magnetosome vesicles and the mislocalization of some proteins, such as MamC. MamE contains one transmembrane region and, when removed, some Mam proteins also become mislocalized. The findings suggest that interaction between MamE or MamA with other Mam proteins can help in directing these peptides toward the MM.

16.2.3.3 Iron Uptake

Accumulation of intracellular iron is fundamental for the synthesis of magnetite crystals. However, the mechanisms that MTB employs to absorb iron from the media and its transportation into the magnetosome vesicles are still poorly understood.

Currently, there are three proposed models (Figure 16.3). The first one suggests that iron is directly transported by diffusion into magnetosome vesicles from the periplasm when the vesicle lumen is still in contact with the CM. The second model proposes the transportation and release of iron into magnetosome vesicles by ligation of cytosolic Fe^{2+} to an unknown organic substrate and Fe^{3+} to ferritin proteins [28]. The third model suggests that iron is transported through the CM via general iron uptake systems and transported into magnetosome vesicles with the help of magnetosome-specific transporters. The suggested Mam proteins involved in the transport mechanisms are MamB and MamM for Fe^{2+} and MamH and MamZ for Fe^{3+}.

General iron uptake and homeostasis mechanisms vary between species. For instance, in some MTB strains, siderophore synthesis has been observed, whereas such activity has not been observed in other strains like MSR-1. Even though there is evidence of general iron uptake systems being used to transport iron into the intracellular space, the existence of an additional iron uptake pathway specific for biomineralization has been suggested. This hypothesis is based on the fact that suppression of general iron uptake routes does not impair magnetite formation [29].

It is believed that there are distinct iron pools inside the cell to (i) carry general biochemical reactions that require iron and (ii) be used for magnetite formation. Until relatively recently, it was assumed that magnetite crystals represented 99.5% of the intracellular iron pool. However, recent studies suggest that only 25%–45% of intracellular iron is incorporated into magnetosomes [2, 30].

16.2.3.4 Magnetite Biomineralization

The nucleation of magnetite crystals is more complex than the formation of other iron oxides which exclusively contain either ferrous or ferric iron because magnetite is composed of both forms. An alkaline (> pH 7) and low redox potential environment, as well as a 30 mM minimal iron concentration inside the magnetosome vesicle, are needed to initialize nucleation.

There are two models for magnetite nucleation. The first one suggests the direct co-precipitation of Fe^{2+} and Fe^{3+} to create magnetite, whereas the second model involves the formation of intermediate precursor mineral phases before obtaining magnetite [31]. Regardless of the pathway, nucleation of magnetite occurs close to the MM, and this is a highly regulated process whereby a single crystal is assembled in each vesicle. Notably, deletion of candidate Mam proteins involved in crystal nucleation produced poorly crystalline structures or empty vesicles.

Following crystal nucleation, the crystal grows until it reaches a certain size and shape. The control of its maturation is also strict and is species-specific where several proteins encoded in the *mamGFDC* operon are involved. Some of these proteins only regulate crystal growth or size (MamG, MamR, MamD), whereas others are involved in regulating both crystal size and morphology (MmsF, MamS, Mms6). MamR is also involved in the control of particle numbers.

16.2.3.5 Assembly of Magnetosome Chains

The permanent magnetic dipole moment of each magnetite crystal is not enough to sense magnetic fields, but the sum of each crystal results in a larger single magnetic dipole moment that enables cell alignment in a magnetic field [32]. To this end, MTB can align their magnetosomes into a linear chain with the help of at least two known Mam proteins: MamK and MamJ.

Magnetospirillum spp. present a cytoskeletal filament that goes along the cell from one pole to the other, passing nearby the magnetosome chain. This actin-like filament is formed by MamK. When MamK is deleted in AMB-1, crystal formation is not affected but no cytoskeleton filaments are synthesized and magnetosomes are scattered throughout the cell. When MamK is deleted in MSR-1, magnetosome chains are shorter and fragmented and the cytoskeleton is no longer visible.

MamJ is localized on the MM and helps to attach MamK filaments onto the magnetosomes. MamJ-deficient mutants present aggregated and disorganized magnetosomes throughout the cell. However, MamJ has only been found in *Magnetospirillum* spp., indicating that other MTB species may use different systems to align magnetosome chains.

16.3 CULTIVATION OF MAGNETOTACTIC BACTERIA

Despite the wide diversity of MTB species and their ubiquity in natural environments, the cultivation of pure cultures in a laboratory setting has been difficult due to the particular growth conditions that these bacteria require [6]. Although obligate anaerobes exist, most MTB species can tolerate limited exposure to oxygen, thus easing cell manipulation. MTB species grow at slower rates than other bacteria due to their long cell-dividing cycles, which add to the tediousness of cultivating these bacteria.

MTB species occupy specific positions within the oxic-anoxic interface which represent different specific chemical conditions at that depth. For instance, some magnetite-producing bacteria can live in microaerobic to anaerobic conditions, whereas greigite producers require the presence of sulfide under anaerobic conditions [6]. Such oxygen and redox gradients are difficult to recreate in the laboratory – hence the difficulties in isolating and cultivating many MTB species.

MTB axenic cultures are limited. The most relevant strains employed to study magnetite biomineralization, molecular mechanisms, and magnetosome bioproduction are AMB-1 and MSR-1. Other relevant strains include MS-1 and *Magnetovibrio* MV-1, which have been used for cultivation studies, although MSR-1 remains the preferred choice by the majority of researchers because this strain exhibits the highest oxygen tolerance and magnetosome yield production compared to other cultured MTB [33].

16.3.1 Factors Affecting Magnetosome Formation

The growth of MTB and magnetosome formation are highly influenced by environmental factors. These factors include growth media, redox conditions, temperature, and pH (Figure 16.4).

16.3.1.1 Nutrients and media composition

Several growth media have been developed for MTB growth at both small and large scales (Figure 16.4). Nutrient composition in the media is important in microbial cultures. Carbon, nitrogen, oxygen, and hydrogen are the most essential elements needed for growth. While nitrogen is implicated in essential biosynthetic processes such as protein synthesis, carbon plays a crucial role as an energy source. Cell metabolism is affected when the carbon source is limited. Cells decrease their growth rate and stop non-essential energy-consuming processes, such as magnetite synthesis. Each MTB species has a different preferred carbon source, and notably, none of the species use carbohydrates as carbon sources. For example, MSR-1 can use organic acids (as electron donors) such as succinate, acetate, pyruvate, or lactate, where the latter is the most commonly employed [34–37].

The nitrogen source is also an important factor in MTB growth, not only because it is required for the formation of essential biomolecules such as amino acids, but also because it can be used as a terminal electron acceptor when oxygen is not available [36].

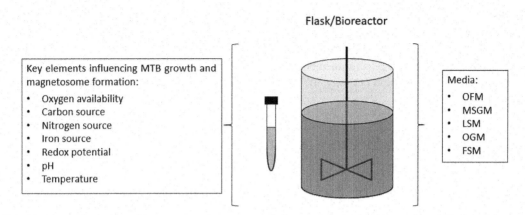

FIGURE 16.4 Key elements influencing cellular and magnetosome yields and commonly used media used to culture MTB. OFM, optimized flask medium; MSGM, magnetic spirillum growth medium; LSM, large scale medium; OGM, optimized growth medium; FSM, flask standard medium.

16.3.1.2 Iron Source

MTB can tolerate extracellular iron concentrations in the micromolar to the low-millimolar range and have the capacity to accumulate large amounts of iron (up to 4% of their dry weight). In nature, the highest occurrence of magnetite-producing MTB corresponds to extracellular iron concentrations of 20–50 µM [6]. Commonly used iron sources in the laboratory include ferric citrate, ferric quinate, ferric malate, and ferrous sulfate. Iron concentrations greater than 20–50 µM do not significantly increase the magnetite yield in cells. In fact, iron concentrations greater than 200 µM can inhibit cell growth.

16.3.1.3 Oxygen Concentration

One of the most important factors that affect magnetite biomineralization is oxygen concentration. For example, MSR-1 can grow in aerobic, microaerobic, and anaerobic conditions. Under aerobic growth, magnetite formation is inhibited, and higher cell densities are obtained, whereas microaerobic conditions induce magnetosome synthesis and lower cell densities are achieved [33]. Therefore, a balance between magnetosome formation and cell growth must be met when designing cultivation strategies.

Interestingly, MSR-1 can either use oxygen or nitrate ions as terminal electron acceptors. For instance, when MSR-1 uses nitrate instead of oxygen, larger magnetite crystals are formed, and there are more of them [36]. However, cell growth can be limited by the accumulation of toxic intermediates originating from the denitrification process. To mitigate this growth limitation, microaerobic growth conditions with a supplementary nitrogen source are usually combined.

16.3.1.4 pH and Temperature

pH and temperature are two key parameters that affect bacterial growth and magnetosome formation. MSR-1's optimal growth temperature is 28°C–30°C with an optimum pH value of 6.8–7. When the pH of the media changes significantly, aberrant magnetosome crystals can be observed. On the other hand, bacterial growth at temperatures between 10°C and 20°C does not affect magnetosome formation, but lower temperatures inhibit their formation [38].

16.3.1.5 Redox Conditions

The oxidation-reduction potential is an important parameter during bacterial cell growth, as small changes can alter growth. In biomineralization, redox potential is also important. Reducing conditions were found to be optimal for magnetosome formation (−250 mV to −500 mV), and perturbation of these conditions prevented magnetosome biomineralization [39]. Contrary to other biological processes, the genetic regulation of MAI is not altered by changes in redox conditions. However, changes in environmental conditions can alter other general metabolic processes involved in magnetosome biogenesis, such as the dissimilatory denitrification pathway, the respiratory chain, and iron uptake mechanisms, impacting magnetosome production.

16.3.2 Cultivation Strategies for Mass Production of Magnetosomes

The use of bioreactors enables the growth of MTB at relatively high cell densities. MSR-1 flask cultures yield a maximum cell growth of 1–1.5 OD_{565}, whereas bioreactor cultures can yield approximately tenfold higher biomass concentrations. Bioprocessing parameters such as oxygen levels, nutrient supply, and pH can only be controlled using bioreactors. Table 16.2 shows a summary of biomass and magnetosome yields obtained using different culture strategies carried out at the bioreactor scale. Most laboratory-scale fermenter studies are performed in volumes between 1–7.5 L. However, there are a small number of reports in which the process is scaled up to 42 L to demonstrate that cultivation can be applied to the pilot plant production scale [30, 34–35].

Heyen and Schüler carried out what is considered to be the first bioreactor-type study that optimized cultivation conditions for various MTB strains by applying different dissolved oxygen (dO_2) concentrations and optimizing FSM media [33]. Oxygen was regulated by establishing an automatic cascade control system by gassing nitrogen and air separately. The authors determined that magnetite biomineralization is optimal under a 20 mbar O_2 threshold. Sun et al. opted for a microoxic fed-batch oxystat strategy in a 42 L fermenter [34]. In this aeration approach, a fixed airflow rate was used, and regulation of dO_2 was manually controlled by changing the agitation speed when a decrease of cell growth was detected. Liu et al. improved the feeding strategy by proposing a pH-stat feeding system to maintain optimal carbon, nitrogen, and iron concentrations [40]. In this strategy, a feeding solution containing carbon, nitrogen, and iron sources was employed to regulate and maintain the pH. Different sodium lactate concentrations and nitrogen sources were tested, and the work concluded that MSR-1 growth yield was higher when low concentrations of sodium lactate were used. Zhang et al. proposed a semicontinuous nutrient-balance feeding culture strategy [35]. OFM media was modified to try to reduce

TABLE 16.2 Summary of Magnetosome and Biomass Yields of Various MTB Strains Cultured Using Different Strategies and Growth Media

STRAIN	CULTIVATION STRATEGY	MEDIUM	SCALE (L)	CELL YIELD (G DCW L^{-1})	OD_{565}	MAGNETOSOME YIELD (MG L^{-1})	REFERENCE
MSR-1	Batch	FSM	4	0.4	1.4	7.9	[33]
AMB-1	Batch	FSM	4	0.48	2.5	4.7	[33]
MSR-1	Fed-batch	OFM	42	2.2	7.2	41.7	[34]
MSR-1	pH-stat	OFM	42	n/a	12.2	83.2	[40]
MV-1	Batch	OGM	5	n/a	n/a	64.3	[41]
MSR-1	Semicontinuous	OFM	42	9.6	42	356.5	[35]
MSR-1	Semicontinuous	OFM	7.5	7.6	30.4	225.5	[35]
MSR-1	Fed-batch	MGM	40	2.4	8	10	[30]
MSR-1	pH-stat	FSM	5	4.2	16.6	54.3	[37]

OFM, optimized flask medium; OGM, optimized growth medium; MGM, minimal growth medium; FSM, flask standard medium; n/a, data not available.

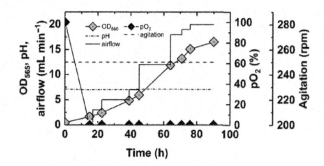

FIGURE 16.5 A representative pH-stat fermentation. MSR-1 was grown in FSM medium with a feed comprising 100 g·L^{-1} lactic acid and 25 g·L^{-1} sodium nitrate. Adapted with permission from [37]. Copyright (2018) Elsevier. The article was distributed under a Creative Commons Attribution License 4.0 (CC-BY) https://creativecommons.org/licenses/by/4.0/.

the accumulation of Na$^+$ and Cl$^-$ ions, which reduced the osmotic potential and consequently inhibited magnetosome formation and cellular growth. Following this strategy, the authors achieved the highest magnetosome and cellular growth yields shown to date. Our group reported the second highest production yields after Zhang et al. by developing a simple pH-stat cultivation strategy without the need for sophisticated control tools, such as extremely sensitive dO$_2$ probes [37]. dO$_2$ was maintained below 1% air saturation by manually adjusting the airflow and agitation between 100 and 500 rpm. Figure 16.5 provides an example to show time profiles of cell growth (OD$_{565}$), stirrer speed, pH, pO$_2$, and airflow when following this cultivation strategy. The effect of lactic acid and nitrate was also studied: cell growth was enhanced at lower lactic acid concentrations, whereas magnetosome production increased at high lactic acid concentrations. Therefore, a balance must be reached between biomass and magnetosome production when designing new fermentation strategies.

16.4 MAGNETOSOME AND MAGNETOTACTIC BACTERIA APPLICATIONS

Magnetosomes exhibit unique properties that exceed those offered by synthetic magnetic nanoparticles (MNPs). Most notably, advantageous features include narrow size distribution, uniform morphology, high crystal purity, high heating capacity, low aggregation tendency, ready dispersion in aqueous media, facile functionalization, high biocompatibility, low toxicity, and high specific absorption rates. Owing to their characteristics, the use of magnetosomes and MTB cells have become of interest in a wide range of fields.

16.4.1 Gene and Drug Delivery

Nanomedicine, and particularly the use of nanoparticles to deliver drugs or other compounds, is an emerging field. The use of external magnetic forces allows the controlled and targeted release of drugs, potentially reducing side effects and increasing efficacy. Magnetosomes are considered promising drug carriers because of their biocompatibility and ability to be magnetically guided to target tissues.

Due to the chemical composition of the MM where carboxyl and amino groups are exposed, drugs and other therapeutical agents can be conjugated to the magnetosome surface. On the other hand, the MM is negatively charged at neutral pH, enabling electrostatic interactions with desired compounds.

For example, Sun et al. conjugated the commonly used antitumor drug doxorubicin to the MM. The magnetosome-drug complex exhibited high stability and potent antitumor activity by inhibiting cancerogenic cellular growth [42]. Deng et al. used magnetosomes as carriers for an antitumor drug to treat leukemia [43]. Magnetosome-gene delivery systems have been developed as an alternative strategy to fight cancer and other medical conditions where foreign genetic material, such as plasmids or small interfering ribonucleic acids, can be carried by magnetosomes using varied binding strategies [44].

Finally, MTB cells have been explored as drug carriers. Owing to their magneto-aerotactic behavior, these bacteria can reach tumoral hypoxic regions where other nanocarriers have difficulty penetrating [45].

16.4.2 Diagnostic and Imaging Tools

Magnetic resonance imaging (MRI) is a powerful technique that uses strong magnetic fields and radiofrequencies to obtain 3D images. Despite being a high spatial resolution technique, conventional MRI has poor sensitivity. The use of MNPs as contrast agents enhances MRI imaging and sensitivity. However, additional processing steps are required to render synthetic MNPs biocompatible. In contrast, magnetosomes can be used directly as biocompatible MRI contrast agents as they present a higher uniform size and regularity of the magnetic mono-crystals [46]. In addition, magnetosomes have the advantage that specific organs or tissues can be targeted by binding specific proteins on the magnetosome surface. The potential use of magnetosomes as MRI contrast agents has been demonstrated in cell cultures as well as in animal models [46].

16.4.3 Biosensors

Magnetosomes can be used in immunologic diagnosis by detecting proteins or DNA. For example, Ceyhan et al. used the biotin-streptavidin strategy to link several functional biomolecules, including biotinylated DNA, oligonucleotides, and biotinylated antibodies, by functionalizing the MM with biotin [47]. Another report proposed an innovative modification of the classical immuno-polymerase chain reaction (PCR) by using magnetosomes functionalized with antibodies for the detection of hepatitis B in human serum. The authors called this a "Magneto Immuno-PCR," and, compared to its homologue Magneto-ELISA, it showed higher sensitivity and a lower detection limit [48].

Magnetosomes can also be used as detection tools in food industry applications. PCR-based techniques have been developed to detect food-borne pathogens in food products. However, the PCR reaction can be inhibited by other components present in the food samples. To avoid this complication, magnetosomes have been proposed as a tool to magnetically separate and concentrate the food-borne population present in samples before PCR is performed. Some pathogens that have been detected using these immune-magnetic separation techniques are *Salmonella*, *Staphylococcal enterotoxins*, and *Vibrio parahaemolyticus* [49].

16.4.4 Magnetic Hyperthermia

Magnetosomes can be employed using hyperthermia treatment. This type of therapy consists of applying heat in tumor cells to induce apoptosis and reduce tumor size. Usually, the working temperature range is around 40°C–45°C. When this technique was first developed, not only carcinogenic cells were affected, but also healthy tissues were heated up and destroyed around the treated area. To avoid these side effects, the use of MNPs that can transform electromagnetic energy into heat was proposed. The efficacy of magnetic hyperthermia using magnetosomes has been demonstrated using various cell lines as well as in mice to treat several types of cancers such as breast cancer [50] and intracranial glioblastomas [51]. It was also revealed that magnetosome chains are more efficient than single magnetosomes or synthetic MNPs in hyperthermia therapies [50].

16.4.5 Bioremediation

As a result of industrial and agricultural activities, wastewaters are generated. Wastewater treatment is a global environmental concern that needs to be addressed. The use of microorganisms to remove contaminants from wastewater represents a more environmentally friendly approach compared to conventional techniques. The major challenge with bio-recovery is that large-scale processing of the collected biomass relies on unit operations such as filtration or centrifugation, which represents an engineering and economic burden. The use of MTB offers an advantage over other microorganisms, as they can be removed by magnetic separation and can adsorb heavy metals on their cell surface due to electrostatic interactions with functional groups present on the cell membrane (i.e. carboxylate, phosphoryl, and hydroxyl groups) [52]. In addition, magnetosomes can be doped with other transition metals, such as Cu, Mn, or Co, when grown in their presence. Therefore, there is a promising future for MTB in bioremediation. However, some of the heavy metals incorporated into MTB cells can impair magnetosome formation and, consequently, affect the efficacy of magnetic separation.

Besides wastewater treatment, *Magnetospirillum* strains have been used to recover heavy metals from disposed circuit boards [53], suggesting a potential role of MTB in e-waste processing. Despite the wide range of potential magnetosome applications, their availability for industrial realization remains a challenge because of the relatively low biomass and magnetosome yields that limit large-scale processing.

16.5 CONCLUSIONS

Magnetosomes are functional magnetic nanoparticles generated by magnetotactic bacteria and are arranged as single-domain magnetic crystals individually wrapped in a phospholipid membrane. They have advantageous properties when compared to synthetic (chemical) MNPs: they are ferrimagnetic; have a narrow size distribution; are coated in organic material, which prevents aggregation; and can be functionalized in vivo using genetic engineering tools, allowing one-step manufacture of functionalized particles. Their biosynthesis is a clean process that is carried out at mild temperatures and generates safe waste. Magnetosomes have highly attractive prospects as "smart materials" for biotechnology and nanomedicine applications, such as cancer therapies, drug delivery, magnetic separation, and metal recovery. To ensure the future deployment of magnetosome-based technologies in industrial settings, fundamental research to unlock the mechanisms of growth and magnetosome formation in more varied MTB species is still needed. In addition, the combination of bioengineering and bioprocessing disciplines will be critical to the development of more robust and intensified bioprocesses that can be translated into real-world applications.

16.6 REFERENCES

[1] Tanaka, M.; Mazuyama, E.; Arakaki, A.; Matsunaga, T. MMS6 Protein Regulates Crystal Morphology during Nano-Sized Magnetite Biomineralization in Vivo. *J. Biol. Chem.*, **2011**, *286* (8), 6386–6392. https://doi.org/10.1074/JBC.M110.183434.

[2] Amor, M.; Ceballos, A.; Wan, J.; Simon, C. P.; Aron, A. T.; Chang, C. J.; Hellman, F.; Komeili, A. Magnetotactic Bacteria Accumulate a Large Pool of Iron Distinct from Their Magnetite Crystals. *Appl. Environ. Microbiol.*, **2020**, *86* (22), 1–51. https://doi.org/10.1128/aem.01278-20.

[3] Teng, Z.; Zhang, Y.; Zhang, W.; Pan, H.; Xu, J.; Huang, H.; Xiao, T.; Wu, L.-F. Diversity and Characterization of Multicellular Magnetotactic Prokaryotes From Coral Reef Habitats of the Paracel Islands, South China Sea. *Front. Microbiol.*, **2018**, *9*, 2135. https://doi.org/10.3389/FMICB.2018.02135.

[4] Liu, P.; Liu, Y.; Zhao, X.; Roberts, A. P.; Zhang, H.; Zheng, Y.; Wang, F.; Wang, L.; Menguy, N.; Pan, Y.; et al. Diverse Phylogeny and Morphology of Magnetite Biomineralized by Magnetotactic Cocci. *Environ. Microbiol.*, **2021**, *23* (2), 1115–1129. https://doi.org/10.1111/1462-2920.15254.

[5] Rahn-lee, L.; Byrne, M. E.; Zhang, M.; Sage, D. Le; Glenn, D. R.; Milbourne, T.; Walsworth, R. L.; Vali, H.; Komeili, A. A Genetic Strategy for Probing the Functional Diversity of Magnetosome Formation. *PLoS Genet. Genet.*, **2015**, *11* (1). https://doi.org/10.1371/journal.pgen.1004811.

[6] Lefevre, C. T.; Bazylinski, D. A. Ecology, Diversity, and Evolution of Magnetotactic Bacteria. *Microbiol. Mol. Biol. Rev.*, **2013**, *77* (3), 497–526. https://doi.org/10.1128/mmbr.00021-13.

[7] Muñoz, D.; Marcano, L.; Martín-Rodríguez, R.; Simonelli, L.; Serrano, A.; García-Prieto, A.; Fdez-Gubieda, M. L.; Muela, A. Magnetosomes Could Be Protective Shields against Metal Stress in Magnetotactic Bacteria. *Sci. Rep.*, **2020**, *10* (1), 1–12. https://doi.org/10.1038/s41598-020-68183-z.

[8] Guo, F. F.; Yang, W.; Jiang, W.; Geng, S.; Peng, T.; Li, J. L. Magnetosomes Eliminate Intracellular Reactive Oxygen Species in *Magnetospirillum gryphiswaldense* MSR-1. *Environ. Microbiol.*, **2012**, *14* (7), 1722–1729. https://doi.org/10.1111/j.1462-2920.2012.02707.x.

[9] Wang, Y.; Lin, W.; Li, J.; Pan, Y. Changes of Cell Growth and Magnetosome Biomineralization in *Magnetospirillum magneticum* AMB-1 after Ultraviolet-B Irradiation. *Front. Microbiol.*, **2013**, *4* (397). https://doi.org/10.3389/fmicb.2013.00397.

[10] Gorby, Y. A.; Beveridge, T. J.; Blakemore, R. P. Characterization of the Bacterial Magnetosome Membrane. *J. Bacteriol.*, **1988**, *170* (2), 834–841. https://doi.org/10.1128/jb.170.2.834-841.1988.

[11] Murat, D.; Quinlan, A.; Vali, H.; Komeili, A. Comprehensive Genetic Dissection of the Magnetosome Gene Island Reveals the Step-Wise Assembly of a Prokaryotic Organelle. *Proc. Natl. Acad. Sci. U. S. A.*, **2010**, *107* (12), 5593–5598. https://doi.org/10.1073/pnas.0914439107.

[12] Nudelman, H.; Zarivach, R. Structure Prediction of Magnetosome-Associated Proteins. *Front. Microbiol.*, **2014**, *5* (9), 1–17. https://doi.org/10.3389/fmicb.2014.00009.

[13] Lohße, A.; Borg, S.; Raschdorf, O.; Kolinko, I.; Tompa, É.; Pósfai, M.; Faivre, D.; Baumgartner, J.; Schülera, D. Genetic Dissection of the *mamAB* and *mms6* Operons Reveals a Gene Set Essential for Magnetosome Biogenesis in *Magnetospirillum gryphiswaldense*. *J. Bacteriol.*, **2014**, *196* (14), 2658–2669. https://doi.org/10.1128/JB.01716-14.

[14] Tanaka, M.; Arakaki, A.; Matsunaga, T. Identification and Functional Characterization of Liposome Tubulation Protein from Magnetotactic Bacteria. *Mol. Microbiol.*, **2010**, *76* (2), 480–488. https://doi.org/10.1111/j.1365-2958.2010.07117.x.

[15] Zeytuni, N.; Ozyamak, E.; Ben-Harush, K.; Davidov, G.; Levin, M.; Gat, Y.; Moyal, T.; Brik, A.; Komeili, A.; Zarivach, R. Self-Recognition Mechanism of MamA, a Magnetosome-Associated TPR-Containing Protein, Promotes Complex Assembly. *Proc. Natl. Acad. Sci. U. S. A.*, **2011**, *108* (33). https://doi.org/10.1073/pnas.1103367108.

[16] Quinlan, A.; Murat, D.; Vali, H.; Komeili, A. The HtrA/DegP Family Protease MamE Is a Bifunctional Protein with Roles in Magnetosome Protein Localization and Magnetite Biomineralization. *Mol. Microbiol.*, **2011**, *80* (4), 1075–1087. https://doi.org/10.1111/j.1365-2958.2011.07631.x.

[17] Uebe, R.; Junge, K.; Henn, V.; Poxleitner, G.; Katzmann, E.; Plitzko, J. M.; Zarivach, R.; Kasama, T.; Wanner, G.; Pósfai, M.; et al. The Cation Diffusion Facilitator Proteins MamB and MamM of *Magnetospirillum gryphiswaldense* Have Distinct and Complex Functions, and Are Involved in Magnetite Biomineralization and Magnetosome Membrane Assembly. *Mol. Microbiol.*, **2011**, *82* (4), 818–835. https://doi.org/10.1111/j.1365-2958.2011.07863.x.

[18] Raschdorf, O.; Müller, F. D.; Pósfai, M.; Plitzko, J. M.; Schüler, D. The Magnetosome Proteins MamX, MamZ and MamH Are Involved in Redox Control of Magnetite Biomineralization in *Magnetospirillum gryphiswaldense*. *Mol. Microbiol.*, **2013**, *89* (5), 872–886. https://doi.org/10.1111/mmi.12317.

[19] Zeytuni, N.; Uebe, R.; Maes, M.; Davidov, G.; Baram, M.; Raschdorf, O.; Nadav-Tsubery, M.; Kolusheva, S.; Bitton, R.; Goobes, G.; et al. Cation Diffusion Facilitators Transport Initiation and Regulation Is Mediated by Cation Induced Conformational Changes of the Cytoplasmic Domain. *PLoS One*, **2014**, *9* (3). https://doi.org/10.1371/journal.pone.0092141.

[20] Siponen, M. I.; Adryanczyk, G.; Ginet, N.; Arnoux, P.; Pignol; David. Magnetochrome: A c-Type Cytochrome Domain Specific to Magnetotatic Bacteria. *Biochem. Soc. Trans.*, **2012**, *40* (6), 1319–1323. https://doi.org/10.1042/BST20120104.

[21] Yang, W.; Li, R.; Peng, T.; Zhang, Y.; Jiang, W.; Li, Y.; Li, J. *mamO* and *mamE* Genes Are Essential for Magnetosome Crystal Biomineralization in *Magnetospirillum gryphiswaldense* MSR-1. *Res. Microbiol.*, **2010**, *161* (8), 701–705. https://doi.org/10.1016/j.resmic.2010.07.002.

[22] Uebe, R.; Schüler, D. Magnetosome Biogenesis in Magnetotactic Bacteria. *Nat. Rev. Microbiol.*, **2016**, *14* (10), 621–637. https://doi.org/10.1038/nrmicro.2016.99.

[23] Scheffel, A.; Gärdes, A.; Grünberg, K.; Wanner, G.; Schüler, D. The Major Magnetosome Proteins MamGFDC Are Not Essential for Magnetite Biomineralization in *Magnetospirillum gryphiswaldense* but Regulate the Size of Magnetosome Crystals. *J. Bacteriol.*, **2008**, *190* (1), 377–386. https://doi.org/10.1128/JB.01371-07.

[24] Arakaki, A.; Yamagishi, A.; Fukuyo, A.; Tanaka, M.; Matsunaga, T. Co-Ordinated Functions of Mms Proteins Define the Surface Structure of Cubo-Octahedral Magnetite Crystals in Magnetotactic Bacteria. *Mol. Microbiol.*, **2014**, *93* (3), 554–567. https://doi.org/10.1111/mmi.12683.

[25] Murat, D.; Falahati, V.; Bertinetti, L.; Csencsits, R.; Körnig, A.; Downing, K.; Faivre, D.; Komeili, A. The Magnetosome Membrane Protein, MmsF, Is a Major Regulator of Magnetite Biomineralization in *Magnetospirillum magneticum* AMB-1. *Mol. Microbiol.*, **2012**, *85* (4), 684. https://doi.org/10.1111/J.1365-2958.2012.08132.X.

[26] Katzmann, E.; Scheffel, A.; Gruska, M.; Plitzko, J. M.; Schüler, D. Loss of the Actin-like Protein MamK Has Pleiotropic Effects on Magnetosome Formation and Chain Assembly in *Magnetospirillum gryphiswaldense*. **2010**, *77* (1), 208–224. https://doi.org/10.1111/j.1365-2958.2010.07202.x.

[27] Scheffel, A.; Gruska, M.; Faivre, D.; Linaroudis, A.; Plitzko, J. M.; Schüler, D. An Acidic Protein Aligns Magnetosomes along a Filamentous Structure in Magnetotactic Bacteria. *Nature*, **2006**, *440* (7080), 110–114. https://doi.org/10.1038/nature04382.

[28] Faivre, D.; Böttger, L. H.; Matzanke, B. F.; Schüler, D. Intracellular Magnetite Biomineralization in Bacteria Proceeds by a Distinct Pathway Involving Membrane-Bound Ferritin and an Iron(II) Species. *Angew. Chemie – Int. Ed.*, **2007**, *46* (44), 8495–8499. https://doi.org/10.1002/anie.200700927.

[29] Rong, C.; Huang, Y.; Zhang, W.; Jiang, W.; Li, Y.; Li, J. Ferrous Iron Transport Protein B Gene (*feoB1*) Plays an Accessory Role in Magnetosome Formation in *Magnetospirillum gryphiswaldense* Strain MSR-1. *Res. Microbiol.*, **2008**, *159* (7–8), 530–536. https://doi.org/10.1016/j.resmic.2008.06.005.

[30] Berny, C.; Le Fèvre, R.; Guyot, F.; Blondeau, K.; Guizonne, C.; Rousseau, E.; Bayan, N.; Alphandéry, E. A Method for Producing Highly Pure Magnetosomes in Large Quantity for Medical Applications Using *Magnetospirillum gryphiswaldense* MSR-1 Magnetotactic Bacteria Amplified in Minimal Growth Media. *Front. Bioeng. Biotechnol.*, **2020**, *8* (16). https://doi.org/10.3389/fbioe.2020.00016.

[31] Faivre, D.; Godec, T. U. From Bacteria to Mollusks: The Principles Underlying the Biomineralization of Iron Oxide Materials. *Angew. Chemie – Int. Ed.*, **2015**, *54* (16), 4728–4747. https://doi.org/10.1002/anie.201408900.

[32] Frankel, R. B.; Bazylinski, D. A. How Magnetotactic Bacteria Make Magnetosomes Queue Up. *Trends in Microbiology.*, 2006, *14* (8), 329–331. https://doi.org/10.1016/j.tim.2006.06.004.

[33] Heyen, U.; Schüler, D. Growth and Magnetosome Formation by Microaerophilic *Magnetospirillum* Strains in an Oxygen-Controlled Fermentor. *Appl. Microbiol. Biotechnol.*, **2003**, *61* (5–6), 536–544. https://doi.org/10.1007/s00253-002-1219-x.

[34] Sun, J. B.; Zhao, F.; Tang, T.; Jiang, W.; Tian, J. S.; Li, Y.; Li, J. L. High-Yield Growth and Magnetosome Formation by *Magnetospirillum gryphiswaldense* MSR-1 in an Oxygen-Controlled Fermentor Supplied Solely with Air. *Appl. Microbiol. Biotechnol.*, **2008**, *79* (3), 389–397. https://doi.org/10.1007/s00253-008-1453-y.

[35] Zhang, Y.; Zhang, X.; Jiang, W.; Li, Y.; Li, J. Semicontinuous Culture of *Magnetospirillum gryphiswaldense* MSR-1 Cells in an Autofermenter by Nutrient-Balanced and Isosmotic Feeding Strategies. *Appl. Environ. Microbiol.*, **2011**, *77* (17), 5851–5856. https://doi.org/10.1128/AEM.05962-11.

[36] Li, Y.; Katzmann, E.; Borg, S.; Schüler, D. The Periplasmic Nitrate Reductase Nap Is Required for Anaerobic Growth and Involved in Redox Control of Magnetite Biomineralization in Magnetospirillum Gryphiswaldense. *J. Bacteriol.*, **2012**, *194* (18), 4847–4856. https://doi.org/10.1128/JB.00903-12.

[37] Fernández-Castané, A.; Li, H.; Thomas, O. R. T.; Overton, T. W. Development of a Simple Intensified Fermentation Strategy for Growth of *Magnetospirillum gryphiswaldense* MSR-1: Physiological Responses to Changing Environmental Conditions. *N. Biotechnol.*, **2018**, *46* (11), 22–30. https://doi.org/10.1016/j.nbt.2018.05.1201.

[38] Moisescu, C.; Ardelean, I. I.; Benning, L. G. The Effect and Role of Environmental Conditions on Magnetosome Synthesis. *Front. Microbiol.*, **2014**, *5* (49), 1–12. https://doi.org/10.3389/fmicb.2014.00049.

[39] Olszewska-Widdrat, A.; Schiro, G.; Reichel, V. E.; Faivre, D. Reducing Conditions Favor Magnetosome Production in *Magnetospirillum magneticum* AMB-1. *Front. Microbiol.*, **2019**, *10* (582), 1–10. https://doi.org/10.3389/fmicb.2019.00582.

[40] Liu, Y.; Li, G. R.; Guo, F. F.; Jiang, W.; Li, Y.; Li, L. J. Large-Scale Production of Magnetosomes by Chemostat Culture of *Magnetospirillum gryphiswaldense* at High Cell Density. *Microb. Cell Fact.*, **2010**, *9*, 1–8. https://doi.org/10.1186/1475-2859-9-99.

[41] Silva, K. T.; Leão, P. E.; Abreu, F.; López, J. A.; Gutarra, M. L.; Farina, M.; Bazylinski, D. A.; Freire, D. M. G.; Lins, U. Optimization of Magnetosome Production and Growth by the Magnetotactic Vibrio *Magnetovibrio blakemorei* Strain MV-1 through a Statistics-Based Experimental Design. *Appl. Environ. Microbiol.*, **2013**, *79* (8), 2823–2827. https://doi.org/10.1128/AEM.03740-12.

[42] Sun, J. B.; Duan, J. H.; Dai, S. L.; Ren, J.; Guo, L.; Jiang, W.; Li, Y. Preparation and Anti-Tumor Efficiency Evaluation of Doxorubicin-Loaded Bacterial Magnetosomes: Magnetic Nanoparticles as Drug Carriers Isolated from *Magnetospirillum gryphiswaldense*. *Biotechnol. Bioeng.*, **2008**, *101* (6), 1313–1320. https://doi.org/10.1002/bit.22011.

[43] Deng, Q.; Liu, Y.; Wang, S.; Xie, M.; Wu, S.; Chen, A.; Wu, W. Construction of a Novel Magnetic Targeting Anti-Tumor Drug Delivery System: Cytosine Arabinoside-Loaded Bacterial Magnetosome. *Mater. (Basel, Switzerland)*, **2013**, *6* (9), 3755–3763. https://doi.org/10.3390/MA6093755.

[44] Wang, X.; Wang, J. gui; Geng, Y. yuan; Wang, J. jiao; Zhang, X. mei; Yang, S. shuang; Jiang, W.; Liu, W. quan. An Enhanced Anti-Tumor Effect of Apoptin-Cecropin B on Human Hepatoma Cells by Using Bacterial Magnetic Particle Gene Delivery System. *Biochem. Biophys. Res. Commun.*, **2018**, *496* (2), 719–725. https://doi.org/10.1016/j.bbrc.2018.01.108.

[45] Felfoul, O.; Mohammadi, M.; Taherkhani, S.; Lanauze, D. de; Xu, Y. Z.; Loghin, D.; Essa, S.; Jancik, S.; Houle, D.; Lafleur, M.; et al. Magneto-Aerotactic Bacteria Deliver Drug-Containing Nanoliposomes to Tumour Hypoxic Regions. *Nat. Nanotechnol.*, **2016**, *11* (11), 941–947. https://doi.org/10.1038/nnano.2016.137.

[46] Zhang, Y.; Ni, Q.; Xu, C.; Wan, B.; Geng, Y.; Zheng, G.; Yang, Z.; Tao, J.; Zhao, Y.; Wen, J.; et al. Smart Bacterial Magnetic Nanoparticles for Tumor-Targeting Magnetic Resonance Imaging of HER2-Positive Breast Cancers. *ACS Appl. Mater. Interfaces*, **2019**, *11* (4), 3654–3665. https://doi.org/10.1021/acsami.8b15838.

[47] Ceyhan, B.; Alhorn, P.; Lang, C.; Schüler, D.; Niemeyer, C. M. Semisynthetic Biogenic Magnetosome Nanoparticles for the Detection of Proteins and Nucleic Acids. *Small*, **2006**, *2* (11), 1251–1255. https://doi.org/10.1002/SMLL.200600282.

[48] Wacker, R.; Ceyhan, B.; Alhorn, P.; Schueler, D.; Lang, C.; Niemeyer, C. M. Magneto Immuno-PCR: A Novel Immunoassay Based on Biogenic Magnetosome Nanoparticles. *Biochem. Biophys. Res. Commun.*, **2007**, *357* (2), 391–396. https://doi.org/10.1016/J.BBRC.2007.03.156.

[49] Xu, J.; Hu, J.; Liu, L.; Li, L.; Wang, X.; Zhang, H.; Jiang, W.; Tian, J.; Li, Y.; Li, J. Surface Expression of Protein A on Magnetosomes and Capture of Pathogenic Bacteria by Magnetosome/Antibody Complexes. *Front. Microbiol.*, **2014**, *5* (136). https://doi.org/10.3389/FMICB.2014.00136.

[50] Alphandéry, E.; Faure, S.; Seksek, O.; Guyot, F.; Chebbi, I. Chains of Magnetosomes Extracted from AMB-1 Magnetotactic Bacteria for Application in Alternative Magnetic Field Cancer Therapy. *ACS Nano*, **2011**, *5* (8), 6279–6296. https://doi.org/10.1021/nn201290k.

[51] Mannucci, S.; Tambalo, S.; Conti, G.; Ghin, L.; Milanese, A.; Carboncino, A.; Nicolato, E.; Marinozzi, M. R.; Benati, D.; Bassi, R.; et al. Magnetosomes Extracted from *Magnetospirillum gryphiswaldense* as Theranostic Agents in an Experimental Model of Glioblastoma. *Contrast Media Mol. Imaging*, **2018**, *2018*. https://doi.org/10.1155/2018/2198703.

[52] Ali, I.; Peng, C.; Khan, Z. M.; Naz, I.; Sultan, M. An Overview of Heavy Metal Removal from Wastewater Using Magnetotactic Bacteria. *J. Chem. Technol. Biotechnol.*, **2018**, *93* (10), 2817–2832. https://doi.org/10.1002/jctb.5648.

[53] Sannigrahi, S.; Suthindhiran, K. Metal Recovery from Printed Circuit Boards by Magnetotactic Bacteria. *Hydrometallurgy*, **2019**, *187* (8), 113–124. https://doi.org/10.1016/j.hydromet.2019.05.007.

Theory and Modeling of Spintronics of Nanomagnets

17

Mehmet C. Onbaşli,[1] Ahmet Avşar,[2]
Saeedeh Mokarian Zanjani,[1] Arash Mousavi Cheghabouri,[1]
and Ferhat Katmis[3]

1 Department of Electrical and Electronics Engineering,
  Koç University, Sarıyer, Istanbul, Turkey

2 Newcastle University, School of Mathematics, Statistics and
  Physics, NE1 7RU Newcastle, United Kingdom

3 Department of Physics, Massachusetts Institute
  of Technology, Cambridge, MA, USA.

Contents

17.1	Introduction	326
17.2	Theories	327
	17.2.1 Micromagnetism in 2D Nanomagnets and Emerging Heterostructures	331
17.3	Magnetotransport Studies in Low-Dimensional Novel Materials	331
17.4	Optical Control of Magnetization Dynamics	333
	17.4.1 Ultrafast Dynamics of Elemental Metallic Magnetic Films Under a fs Pulse	333
	17.4.1.1 Magnetization Dynamics for Fe (Type I) and Gd (Type II)	335
	17.4.2 Short Laser Pulse on Quantum-Confined Metallic Magnetic Thin Films	335
	17.4.2.1 Confinement Effects	336
	17.4.2.2 M3TM and Magnetization Dynamics in Quantum-Confined Ni Film	337
	17.4.2.3 Effect of Film Thickness on DOS_F and M3TM Coefficients	337
	17.4.2.4 Magnetization Dynamics in Quantum-Confined Film	337
	17.4.2.5 Film Thickness Dependence of Magnetization Dynamics	337
17.5	Conclusion	339
17.6	Open Questions in the Theory and Modeling of Nanomagnets	339
17.7	References	340

17.1 INTRODUCTION

Foundations of electronics started with the invention of transistors in 1948 [1], and many developments in silicon and germanium materials growth and processing as well transistor device design followed afterward. In 1958, the first integrated circuit made up of multiple transistors was invented [2]. Since the invention of integrated circuits, the number of elements on a chip has doubled every 18 months [3] over the following six decades. This progress of elements, feature size reduction, and operation frequency scaling is known as Moore's Law [3]. The size of the circuits has been reduced from several cm^3-size light bulbs to very small-size transistors – today such integrated circuits are constructed with $\approx 10^6$–10^8 transistors per cm^2. For comparison, in 1949 a single chip contained about 50 components [3], while today a typical modern desktop computer's central processing unit (CPU) has more than several billion transistors per chip, which makes the modern-day computer much more powerful. Although significantly more powerful than their predecessors, today's CPU chips have more room for growth. There has been a growing disparity between CPU speed and memory capacities. From 1986 to 2000, CPU speed has increased by 55% while memory has increased by only 10% [4]. On the memory side, the data storage density scaled rapidly thanks to shrinking transistors without a parallel scaling in read/write performance. Due to the explosive growth of data-driven machine and deep learning applications, high-definition audio and video streaming, internet of things, and 5G as well as 6G and beyond, memory bandwidth and performance became a central bottleneck. Therefore, it is crucial for the focus to be on the improvement of the CPU's memory rather than its speed. One of the difficulties in reaching adequate CPU memory with current technology is the explosive growth in data storage systems. Data are being created at an increasing speed, and new technologies also require data to be stored with long retention times and to be accessible with high energy efficiency and bandwidth.

The technological progress in the last few decades has enabled many more bits to be stored on each square inch of storage media. Starting with a few kbit per square inch in the 1980s, today's storage systems hold up to several Gbit per square inch. The most important requirements of present information technology are higher data storage capacity, faster broadband memory access, and lower power consumption. However, the state-of-the-art memory systems do not fulfill all the requirements. Therefore, universal memory is still needed for future information technology. Recent research in material science and spintronics achieved notable accomplishments in finding convenient materials or material systems for universal memory.

The understanding of new types of materials has grown at an unprecedented rate during the last several decades. Future technology will be able to satisfy the aforementioned criteria not only by discovering new dimensions of materials but also by comprehending low-dimensional interaction mechanics, such as their charges and spins. The most important requirement is to develop a long-lasting, resilient, wideband, dense, and inexpensive memory that can withstand any external disturbance. As a result, device size and dissipation due to Joule heating can only be reduced by employing magnetic moments, or spin degrees of freedom, rather than charges [5]. Controlling and manipulating such spin populations allows for spintronic devices to be used in daily life. Despite the promise of numerous exotic quantum characteristics [6], newly discovered material systems offer significant potential for replacing or complementing state-of-the-art technological applications, such as high-speed electronics, efficient high-density storage, and quantum information technology. The most exotic and resilient approach for generating a bit of information comes along with using a chiral magnetic structure, where the energy minima are protected by the topology, instead of the spatial domain. Spintronic effects have already impacted our lives and become a part of the microelectronic revolution through highly useful applications such as giant magnetoresistance, hard disk drives, current and Hall sensors, and inductively coupled sensors such as RFID (radio-frequency identification) cards. A thorough understanding of the spintronic phenomena might aid in the development of new applications, and the identification of gaps is necessary for novel logic, sensor, and memory systems.

Magnetic data storage, ultra-low noise sensing, and other advanced applications require the dynamic control or modification of magnetic order in nanostructured materials. However, a unifying magnetism framework is not fully established, and a variety of material-dependent underlying processes need to be modeled separately for different applications. To facilitate data recording or spin manipulation in magnetic materials, optical and voltage control are some of the promising high speed and energy-efficient approaches. Control of magnetic moment with all optical techniques in Angstrom-thin elemental magnetic metals holds a lot of promise for fast and high-throughput recording, in which magnetization is reversed in a picosecond timeframe with femtosecond (fs) laser pulses [7]. These inherently multiscale magnetic properties which span timescales of many orders of magnitude rely on the competing energy terms in the total Hamiltonian and the effective magnetic field across nanomagnets.

In this chapter, we present a generalized theoretical formalism of modeling the dynamics and equilibrium of spins in nanomagnets. We mainly focus on numerical micromagnetics based on the microscopic Landau-Lifshitz-Gilbert (LLG) equation and the quantum confinement of phononic and electronic density of states in nanomagnets. After covering the fundamental concepts in magnetism, we investigate the characteristics of thin-film magnets as well as 2D and topological materials. Next, we present the theory and experimental validation for current, voltage, magnetic field, and femtosecond laser control of magnetism in quantum-confined magnetic elemental metals. We exclude pure quantum-based spin modeling approaches and density functional theory–based numerical models due to their prohibitively high computational cost and limited scalability, although these approaches may provide crucial insights into emerging quantum phenomena not captured otherwise. Finally, we discuss some of the gaps in theory and invite both experimental and theoretical spintronics researchers to answer these riddles, which might eventually lead to near-dissipation-free spin transport, information storage, and sensing near thermodynamic limitations.

17.2 THEORIES

The LLG equation accurately predicts the reversal of a magnetic bit of information from picosecond to millisecond timescales. LLG is the fundamental equation for describing the precession of a magnetic moment subject to a magnetic field. Magnetism in matter arises from unpaired valence electrons' spin and/or orbital angular momentum terms. Magnetic field **H** induces magnetic moment precession around **H**:

$$\dot{\mathbf{M}}_p = \gamma \mathbf{H} \times \mathbf{M} \tag{1}$$

where gyromagnetic factor γ is the ratio of the magnetic moment of a particle to its angular momentum. For an orbiting electron, this value is $|\gamma_e| = 1.78 \times 10^{11}$ rad·(Ts)$^{-1}$. In physical systems, the precession term of equation 1 is accompanied by a damping term that aligns the direction of magnetic moment **M** with the field **H**. To account for this alignment, Landau and Lifshitz (LL) introduced a phenomenological term with a non-adiabatic damping term λ:

$$\dot{\mathbf{M}}_d = -\lambda \mathbf{M} \times \mathbf{M} \times \mathbf{H} \tag{2}$$

where $\dot{\mathbf{M}}_d$ is the damping-induced change in the magnetization vector. The LL equation with damping parameter is rewritten as:

$$\dot{\mathbf{M}} = \gamma \mathbf{M} \times \mathbf{H} - \lambda \times (\mathbf{M} \times \mathbf{H}) \tag{3}$$

While this equation captures the essential characteristics of damping, there is a more accurate representation of damping that directly depends on the change of magnetization, as Gilbert added [8]:

$$\dot{\mathbf{M}} = -\gamma \mathbf{M} \times \mathbf{H} + \eta \mathbf{M} \times \dot{\mathbf{M}} \qquad (4)$$

Equation 4 simplifies the LL equation, captures the physical damping phenomenon more accurately, and makes the computational micromagnetism much easier. Equation 4 contains the magnitudes of both terms of the right-hand side of equation 4, and these vector terms are both perpendicular to $\mathbf{M}$. For computational simplicity, dividing both sides of equation 4 by the magnitude of the magnetic moment and writing the equation in terms of the reduced magnetization unit vector $\mathbf{m} = \dfrac{\mathbf{M}}{|\mathbf{M}|}$ is common:

$$\dot{\mathbf{m}} = -\gamma \mathbf{m} \times \mathbf{H} + \alpha \mathbf{m} \times \dot{\mathbf{m}} \qquad (5)$$

where $\alpha = |\mathbf{M}|\eta$ is the Gilbert damping term.

In describing the time evolution of the spatial spin profiles in magnetic materials, using a quantum mechanical description of magnetism and exchange effects is prohibitively expensive due to the computational memory and calculation requirements for any model that might describe a micrometer or a nanometer magnetic material. On the other hand, classical electromagnetic theory fails to account for the exchange and other microscopic quantum interactions that occur in magnetic materials. To fill this gap, micromagnetic theory was established as a classical continuum theory of magnetization, which has been accurate and computationally accessible. In solving the spin profiles, relevant electromagnetic boundary conditions can be applied. Tangential magnetic field H in one region equals the tangential magnetic field H in the adjacent region added with a surface current on a metallic interface $H_{2t} - H_{1t} = J_s$. The perpendicular component of magnetic flux density B is continuous across an interface ($B_{1n} = B_{2n}$), and, in addition, one may impose open or periodic boundary conditions. Because the microscopic LLG equation is derived classically, it has the potential to bridge the gap between the classical Maxwell equations and the quantum mechanical description of magnetism. Many magnetic phenomena such as hysteresis, magnetic anisotropy, formation, and motion of domain walls, skyrmions, and magnetic bubbles may all be explained by micromagnetism with high quantitative accuracy verified by experiments [9–10].

In computational micromagnetism, the geometry is broken down into a grid with microscopic sites each at a position $\mathbf{r}_i$ with the reduced magnetization vector $\mathbf{m}_i$. Magnetization at each site is calculated using the LLG equation. The effective magnetic field $\mathbf{H}_{eff}$ at each site must be calculated and used in the LLG equation to compute the magnetization direction at the next time step dt. After a time step dt, because of the change in the local magnetic directions, the $\mathbf{H}_{eff}$ must be calculated again for the next time step. The effective field is calculated by the following equation:

$$\mathbf{H}_{eff} = -\dfrac{1}{\mu_0 M_s} \nabla_m \phi \qquad (6)$$

where the operator ∇_m is calculated with respect to the component of magnetization vectors, and φ is the total magnetic energy term calculated by integration of the magnetic energy density ξ over the volume of the magnetic material. For a ferromagnetic material with out-of-plane uniaxial anisotropy, the energy density reads:

$$\xi = \left[A_{ex}\left(\nabla \mathbf{m}(\mathbf{r})\right)^2 - \mu_0 M_s \mathbf{m}(\mathbf{r}).\mathbf{H}_{external} - \mu_0 M_s \mathbf{m}(\mathbf{r}).\mathbf{H}_{demag} - K_u \left(\left(\mathbf{m}(\mathbf{r}).\mathbf{e}_z\right)^2\right) \right] \qquad (7)$$

where saturation magnetization M_s is the magnitude of the magnetization vector. At room temperature, common values for M_s are 1738 kA·m^{-1} for α-Fe, 493 kA·m^{-1} for Ni, and 1424 kA·m^{-1} for Co. A_{ex}

is the exchange interaction energy coefficient, which is 20.7–22.8 pJ·m^{-1} for α-Fe, 7.2–8.5 pJ·m^{-1} for Ni, and 30.2–31.4 pJ·m^{-1} for Co. K_u is the first-order uniaxial energy coefficient, which is 48 kJ·m^{-3} for α-Fe, −4.5 kJ·m^{-3} for Ni, and 453 kJ·m^{-3} for Co [11]. **H$_{demag}$** is the demagnetizing vector field arising from dipolar interactions and constitutes the demagnetization energy of the system. **H$_{external}$** is the applied magnetic field and leads to Zeeman energy. Other energy terms such as bulk and interfacial Dzyaloshinskii-Moriya interaction (DMI), exchange bias, exchange, and cubic anisotropy can also exist in different systems, as shown in Figure 17.1. In such cases, their related energy densities must be calculated and added to equation 7 as separate terms. The main state-of-the-art numerical micromagnetic solvers are the Object Oriented MicroMagnetic Framework (OOMMF) developed by the National Institute of Standards and Technology (NIST) in the US and MuMax3 developed by Ghent University in Belgium [9–10]. Using graphics processing units with high computational power, solving

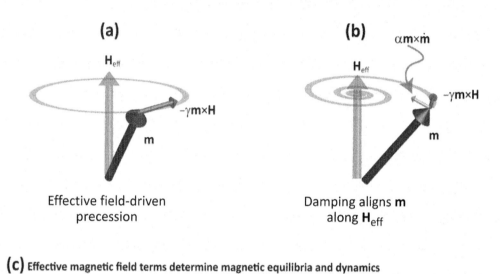

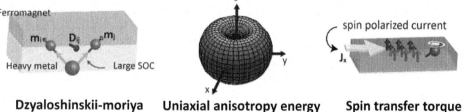

FIGURE 17.1 (a) Micromagnetism shows that effective field drives precession, (b) while damping aligns the magnetic moment towards the effective field. (b) The effective field consists of Heisenberg exchange, demagnetizing field [17], spin-orbit torque, DMI, uniaxial magnetic anisotropy [18], and spin-transfer or spin-orbit torque terms. Inset adapted with permission from [15]. Copyright (2017) Springer Nature.

for the spin profiles with demanding effective fields over large geometries with fine mesh sizes (about 1–3 nm mesh on micron-scale 3D features) and time steps (10^{-15}–10^{-12} steps over a few µs) became common in spintronics research.

In addition to the effect of the internal energy on the development of the magnetization, the introduction of spin-polarized electric current or pure spin current will directly add a torque term to the LLG equation. Two main terms are describing these transferred torques: Slonczewski and Zhang-Li torque. Slonczewski's term describes the spin transfer torque of a spin-polarized current flowing perpendicular to a magnetic slab, and this term is calculated by drawing an analogy between the magnetic spin transfer theory and circuit theory [12]. In the Slonczewski configuration, the spin-polarized electrons enter a magnetic slab from another magnetic layer underneath with the fixed magnetization. The micromagnetic representation of Slonczewski term is written as:

$$\tau_{sl} = \frac{\beta(\epsilon + \alpha \epsilon')}{(1+\alpha^2)}(\mathbf{m} \times (\mathbf{m}_p \times \mathbf{m})) - \frac{\beta(\epsilon - \alpha \epsilon')}{(1+\alpha^2)}(\mathbf{m} \times \mathbf{m}_p) \quad (8)$$

where $\mathbf{m}_p$ is the current polarization direction and α is the Gilbert damping parameter, while

$$\beta = \frac{j_z \hbar}{M_s e d} \quad (9)$$

where j_z is the magnitude of the spin-polarized current in the perpendicular direction to the slab, e is the elementary charge, d is slab thickness,

$$\epsilon = \frac{P\Lambda^2}{(\Lambda^2 + 1) + (\Lambda^2 - 1)(\mathbf{m} \cdot \mathbf{m}_p)} \quad (10)$$

where P is the polarization of electrons entering from the adjacent magnetic layer with fixed magnetization, $\mathbf{m}_p$ is the direction of the magnetic moment unit vector of the fixed layer, and $\Lambda^2 = GR$ where G is the conductance of the stack of the spacer and fixed layers and R is the average of the resistance experienced by spin up and spin down electrons when they traverse the fixed and spacer layers.

For a spin-polarized current flowing along a magnetic slab's surface plane, the interaction between the spin current and magnetization is best described by the Zhang-Li torque term [13]. The micromagnetic representation of the Zhang-Li term [10] is:

$$\tau_{ZL} = \frac{1}{1+\alpha^2}\left[(1+\xi\pm)\mathbf{m}\times(\mathbf{m}\times(\mathbf{u}.\nabla)\mathbf{m}) + (\xi-\alpha)\mathbf{m}\times(\mathbf{u}.\nabla)\mathbf{m}\right] \quad (11)$$

where α is the Gilbert damping parameter, ξ is the non-adiabatic spin transfer coefficient, and

$$\mathbf{u} = \frac{\mu_B}{2e\gamma_0 M_s (1+\xi^2)}\mathbf{j} \quad (12)$$

where μ_B is Bohr's magneton, γ_0 is the electron gyromagnetic ratio, and $\mathbf{j}$ is the spin polarized current density.

Emerging memory elements might rely on lateral spin valves using Heusler alloys [14], current-driven spin-orbit torques along with nonmagnetic heavy metals, magnetic insulators with in-plane or perpendicular magnetic anisotropy [15], and other spin current generation mechanisms. These spin injection mechanisms might be modeled using micromagnetism [16].

17.2.1 Micromagnetism in 2D Nanomagnets and Emerging Heterostructures

Compared with conventional electronics, spintronics is expected to result in significant energy savings and higher information speed and storage density in devices. Proposed spin-based in-memory logic architectures [19] have certain components in common, such as a magnetic contact (e.g., to generate and detect spin) and a spin conducting channel (e.g., to transport spin current). Achieving such functional spintronic devices requires the development of new materials. Furthermore, novel materials for long-distance spin communication are also required to convey spin information between input and output magnets in functional spintronic devices. Time-reversal symmetry is broken when 2D materials are used in proximity with a ferromagnet. Thus, a gap opens on the Dirac cone of the surface's electronic energy band diagram by introducing a robust interfacial magnetic anisotropy with a unique spin texture [20]. Introducing a certain spin texture into such a system may produce a topologically protected phase, like topological skyrmions, which might provide near dissipation-free information storage with long retention. The spin chirality-induced non-coplanar magnetic order phase is due to DMI [21–24], which originates from the broken inversion symmetry in thin films or non-centrosymmetric bulk crystals [25].

Different magnetic phenomena such as spin Hall effect, spin diffusion under in-plane or out-plane magnetic anisotropy, magnetoelectric effect, spin-transfer torque, and domain wall motion have various speeds and energy consumptions [19, 26–30]. Current-driven processes based on spin diffusion under perpendicular magnetic anisotropy are becoming more important due to their higher energy efficiencies and rates [31]. The spin-transfer torque memory elements are some of the most promising contenders for commercial use in terms of performance, while the others are also emerging [27, 31].

From a micromagnetic modeling point of view, these emerging new materials and their new physics highlight the aspect that might not be captured by the LLG equation due to the quantum nature of the electronic band structures of the thin films, their interfaces, and the intrinsic ultrashort time and length scales. The LLG equation does not contain any terms related to (1) the ultrashort timescales that involve scattering of electrons, phonons, and spins[32]; (2) the quantum confinement of electronic, phononic, or spin density of states[32] and the modulation of the topological band structure [20]; and (3) relativistic domain wall (DW) velocities achieved under ultrahigh drive current densities [33]. As a result, a unifying theoretical approach that captures each of these quantum-based phenomena in spintronics that can accommodate scalable and accurate numerical modeling is needed. In the next two sections, we present two example cases, i.e., magnetotransport measurement results and all-optical switching modeling results, to discuss that the new physics of emerging 2D or quantum-confined spintronics needs to be unified in a theoretical framework.

17.3 MAGNETOTRANSPORT STUDIES IN LOW-DIMENSIONAL NOVEL MATERIALS

2D magnetic materials such as CrI_3 have recently been attracting tremendous attention from the spintronics device community thanks to their unique layer-dependent magnetic properties [34]. However, implementation of these intrinsic magnets into spintronic devices such as magnetic memories faces serious challenges due to their high air sensitivity, low Curie temperature, and scalability issues [35]. Intrinsic magnets are rare in nature as well. An appealing alternative to this intrinsic magnet is the use of air-stable, layered materials in which magnetism is extrinsically induced. For this purpose, one promising approach is to investigate the impact of crystal imperfections (e.g., vacancies, substitutions, and crystal edges) on the magnetic properties of monolayer materials such as WS_2, $MoTe_2$, and $PtSe_2$, which can be grown also at a large scale [36]. The formation energy of metallic Pt vacancies in PtSe$_2$ is much lower than those of other layered materials [37], making it particularly attractive for this purpose. The recent experimental investigations suggest that defective PtSe$_2$ exhibits long-range defect-induced magnetism in both metallic (if PtSe$_2$ thickness > 3 layers) [38] and semiconducting (if thickness ≤ 3 layers) [39] films (Figure 17.2(a)). Equally interesting, the pioneering study shows that adding one extra layer of PtSe$_2$ is enough to change the alignment of moments from antiferromagnetic to ferromagnetic coupling (AFM-to-FM crossover) [38]. Similarly, defect-induced magnetism in monolayer WSe$_2$ [40] and monolayer VSe$_2$ [41] were also reported. These findings pave the way for extending the 2D magnet library to include air-stable materials.

There have been many advances in the field of metal and semiconductor spintronics with traditional materials (e.g., Cu and GaAs), such as evidence of spin injection and transport in the non-local geometry by utilizing an electron or a hole as a carrier [42]. Highly promising semiconductors such as GaAs and Si show remarkable spin transport properties, especially below liquid nitrogen temperatures. Spin transport in these materials at room temperature (RT) is observed only in state-of-the-art devices with improved channel/contact interfaces [43]. There are several metallic channel materials such as Cu and Ag that support spin transport at RT [44]. However, probed spin lifetimes and signals are typically too small to be used in complex device architectures. Compared to such bulk materials, 2D spin conducting channel materials such as graphene and black phosphorus (BP) (Figure 17.2(B)) have several advantages for long-distance spin communication, including exhibiting superior spin lifetimes (τ_S) at RT [45]. Experimental results in graphene achieve τ_S on the 0.1–10 ns scale, two orders of magnitude smaller than the theoretically estimated values based on spin-orbit coupling and realistic charge transport values [46]. The reason is that spin relaxation in graphene is highly *extrinsic*. On the other hand, the first generation of spin transport measurements in layered, ultrathin black phosphorus–based spin

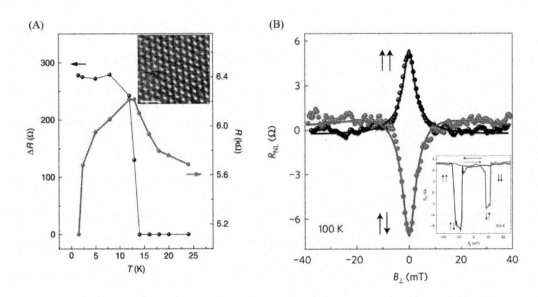

FIGURE 17.2 All-electrical transport measurements in PtSe$_2$ and black phosphorous (BP). (A) Temperature dependence of the PtSe$_2$-based device resistance (R) and its change under a magnetic field (ΔR). The arrows indicate the relevant axes. Inset: Raw high-resolution transmission electron micrograph of bilayer PtSe$_2$ (scale bar: 1 nm). (B) Hanle spin precession curves taken in a ~5 nm thick black phosphorus-based spin valve. The inset shows the spin-valve effect probed in the same device by sweeping an in-plane external magnetic field parallel to the easy axis of its ferromagnetic contacts. The arrows represent the relative magnetization directions of the injector and detector magnets. Adapted with permission from [38–39, 47–48]. Copyright (2017), (2019), (2020) Springer Nature.

valves demonstrate several nanosecond spin lifetimes with spin relaxation lengths over 6 μm [47]. The observation of a direct correlation between spin and charge transport indicates that the spin relaxation mechanism behind this observed spin relaxation time is Elliott-Yafet (EY) and supports the following relation:

$$\frac{1}{\tau_S} \approx \alpha \frac{b^2}{\tau_P} \tag{13}$$

where α is a prefactor, τ_p is the momentum relaxation time, and b^2 is the spin-mixing probability [48]. The excellent agreement between experimental results and systematic first-principles calculations within the framework of EY theory involving only the spin-orbit coupling of the host lattice further confirms that the obtained results are truly *intrinsic* transport properties of BP. The observation of gate tunable enhanced spin accumulation and spin transport parameters in this material makes it appealing for realizing many novel phenomena such as pure spin current–based spin-transfer torque effect.

17.4 OPTICAL CONTROL OF MAGNETIZATION DYNAMICS

In this section, we present two more cases where LLG-based micromagnetic modeling has limitations: (1) ultrafast dynamics of electron-phonon and spin-phonon scattering in the extremely short timescales from low femtoseconds (fs) to a few ns, and (2) quantum-confined magnetic nanometals with reduced density of phononic and electronic states. These cases hint that micromagnetism might need to be extended to cover quantum confinement, electron-phonon, and other scattering types over extremely short timescales.

The use of an ultrashort laser pulse to control magnetization in metallic magnetic thin films with minimal spin-orbit coupling has been the topic of extensive research in recent decades [49–51]. After being illuminated with an ultrashort laser pulse of a 60 fs width, the Ni film magnetization was diminished to ~50% of its maximum in less than picoseconds [50]. This fast process is an order of magnitude higher than a traditional electromagnetic model's predictions. Thus, traditional electromagnetism is not sufficient to describe the ultrafast magnetization dynamics. The interaction of fs laser pulses with the associated phonon, spin, and electron baths allow for laser-driven magnetization control and all-optical switching (AOS) of magnetization [52]. Such effects might pave the way for potential future spintronic applications based on THz spin waves created by ultrafast magnetization control [53], such as high-precision sensing and spectroscopy as well as on-chip communications [54]. Figure 17.3(A) shows the timescales of ultrafast magnetization dynamics.

17.4.1 Ultrafast Dynamics of Elemental Metallic Magnetic Films Under a fs Pulse

Ultrafast magnetization reversal in elemental magnetic metals with minimal spin-orbit coupling is described and modeled by femtomagnetism [32, 56–57]. When a fs laser pulse strikes a magnetic elemental metal, the electron excitation process occurs in a few fs, causing the equilibrium state to be disturbed (cooling process based on the Curie-Weiss law). The electron bath's non-thermal distribution originates from the non-equilibrium conditions. Next, hot electrons thermalize following the Fermi-Dirac distribution and attain equilibrium with each other in a few femtoseconds. At this stage, the electron temperature can reach 1000–2000 K, which is far above the ambient temperature. The non-equilibrium state among

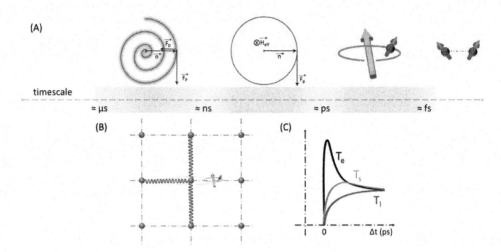

FIGURE 17.3 (A) The characteristic timescales for magnetization dynamics (damping: nanoseconds, precession: ps-ns, exchange: fs-ps). (B) Schematic representation of the three-temperature model (3TM) describing the players of magnetization dynamics in fs timescale. (C) Modeling curves for transient changes in the temperatures of an electron, phonon, and spins. Inspired from [55].

the electron, phonon, and spin baths approaches equilibrium via energy exchange among the baths and via Elliott-Yafet scattering (electron scattering by phonon which flips its spin) in the following few ps (Figure 17.3(B, C)).

The extended energy balance model in the following rate equations (equations 14–17) explains the change in magnetization and temperatures of electrons, phonons, and spins (T_e, T_p, T_s). Here, fs laser pulse energy is injected into the electron bath [32, 57]:

$$C_e \dot{T}_e = -G_{ep}(T_e - T_p) - G_{es}(T_e - T_s) + P(t) \tag{14}$$

$$C_p \dot{T}_p = -G_{ep}(T_p - T_e) - G_{ps}(T_p - T_s) \tag{15}$$

$$C_s \dot{T}_s = -G_{es}(T_s - T_e) - G_{ps}(T_s - T_p) \tag{16}$$

$$\dot{m} = Rm \frac{T_p}{T_c}\left(1 - m \coth\left(m \frac{T_p}{T_e}\right)\right) \tag{17}$$

where, T_e, T_s, T_p, and T_c are electron, spin, phonon, and Curie temperatures, respectively. R (spin-flip ratio) affects the kinetics of transient changes magnetization, m [32]. The kinetics between the baths is determined by the coefficients of energy transfer among electrons and phonons (G_{ep}), electrons and spins (G_{es}), and spins and phonons (G_{sp}) [32, 57]. C_p and C_s are phonon and spin heat capacities, respectively. The electron heat capacity depends on temperature as $C_e = \gamma T_e$, where γ (Sommerfeld coefficient) depends on the free electron density and Fermi energy level [32, 58]. Here, the input Gaussian laser pulse energy density reads,

$$P(t) = \frac{P_0}{\sqrt{2\pi}} \exp\left(-\frac{1}{2}\left(\frac{t}{t_0}\right)^2\right) \tag{18}$$

Here, $P_0 = I_0 / t_0 d$. I_0 is the laser pulse fluence (J·m^{-2}). t_0 is the pulse width (fs). Normalizing the laser fluence to the film thickness (d) captures the thickness dependence of fs pulse-driven magnetization dynamics. Table 17.1 lists the constants used in the solutions to equations 14–17 for Fe and Gd.

TABLE 17.1 Material Parameters Used in the Extended Energy Balance Model

FILM	$C_P \times 10^6$	$C_S \times 10^6$	$G_{EP} \times 10^{18}$	$G_{ES} \times 10^{18}$	$G_{PS} \times 10^{18}$	T_C	$R \times 10^{12}$	$\gamma = C_P/(5T_C)$
Ni	2.33	0.2	4.05	0.6	0.03	627	17.2	743.22
Fe	3.46	0.17	0.7	0.06	0.03	1043	1.86	663.47
Gd	1.78	0.2	0.21	0.6	0.03	297	0.092	1200

C_p, C_s are in (J·m^{-3}·K^{-1}); G_{ep}, G_{es}, G_{ps} are in (W·m^{-3}·K^{-1}), T_c: the Curie temperature (K), R: spin-flip ratio (s^{-1}), γ in (J·m^{-3}·K^{-2}) [32, 50, 57–58].

Due to electron's smaller heat capacity, the electron temperature rises quickly in 200 fs. The time axis in Figure 17.3(C) starts with the laser pulse incidence on the film surface.

17.4.1.1 Magnetization Dynamics for Fe (Type I) and Gd (Type II)

Fe and Ni are classified as type I ($T_c \gg$ room temperature, Fe: 1043 K, Ni: 627 K), while Gd is a type II (T_c of Gd: 297 K). As a result, after a fs laser pulse hits film with a fluence of 70 J·m^{-2}, type I films never undergo thermal demagnetization, while type II films reach zero magnetization in a few ps.

Ongoing research has not yet reached a consensus on the origin of rapid demagnetization [32, 59–60]. Understanding the underlying mechanisms necessitates insights into the interactions among photons, charges, spins, and lattice as well as the angular momentum transfer among them. Terahertz spin waves can be generated on surfaces because of this interaction. Due to exchange coupling, hot electrons lead to transient magnetization drop and the excitation of spin waves. Spin waves in the lattice are inelastically scattered by charges and phonons. The lattice absorbs the damped spin waves which decay evanescently. Because of different coupling strengths among lattice, phonon, and electron temperatures, magnetization recovers longer than electron-lattice relaxation time.

Fe (type I) thin film magnetism: The magnetization dynamics of 20 nm thick Fe films irradiated with a 100-fs pulse with 70 J·m^{-2} fluence are shown in Figure 17.4(A). Iron's magnetization starts switching at approximately 300 fs. After 12 ps, magnetization starts recovering and reaches 78% of its saturation moment in a 20 nm Fe film, according to type I dynamics. The electron, spin, and lattice temperatures all rise due to the fs laser pulse. Although the electron temperature rises abruptly to above Fe's Curie temperature (T_c), it rapidly drops due to electron heat capacity. Thus, the equilibrium lattice temperature remains below T_c. As a result, the magnetization of 20 nm Fe films is preserved.

Gd (type II) thin film magnetism: Type II dynamics emerge, and magnetization is lost altogether in a few tens of ps when films with low C_p and T_c are exposed to a fs laser pulse. The magnetization dynamics of type II film, Gd, are depicted in Figures 17.4(C). Gd has a lower spin-flip ratio (R = 0.092 × 10^{12} s^{-1}) than that for type I metallic films. Thus, magnetization fades slowly after 2.5 ps in Gd films. The equilibrium temperatures T_e, T_s, and T_p all exceed T_c (297 K), and the film undergoes complete thermal demagnetization.

17.4.2 Short Laser Pulse on Quantum-Confined Metallic Magnetic Thin Films

In this subsection, fs laser pulse control of magnetization dynamics in metallic thin films with sub-10 nm thicknesses is studied. These films have strong quantum confinement and low electronic and phononic density of states with respect to those for the bulk. The density of states near the Fermi level (DOS$_F$) and electron-phonon coupling coefficients (G_{ep}) in ultrathin metals have high sensitivity to film thickness within a few Å. When DOS$_F$ and G_{ep} are dependent on thickness, totally different magnetization dynamics features arise when compared to bulk metals. With a lower density of phononic and electronic states,

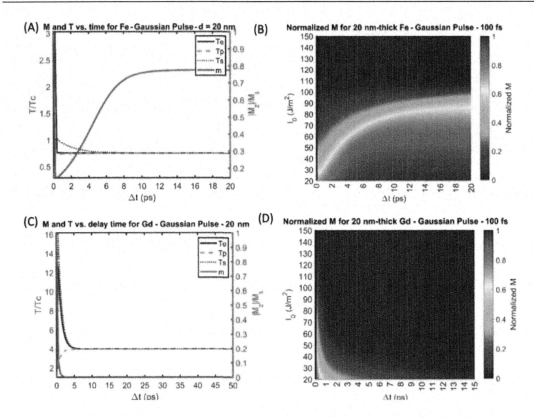

FIGURE 17.4 Magnetization dynamics and lattice temperature of Fe as functions of film thickness, fs laser fluence I_0, transient temperatures of electrons (T_e), phonons (T_p), spins (T_s), and magnetization (m = $|M_z|/M_s$) of (A) 20 nm thick Fe, under a 100 fs pulse and 70 J·m^{-2} fluence. (B) Magnetization in a 20 nm thick Fe film under a 100 fs pulse as a function of fluence, (C) magnetization and bath temperatures of 20 nm thick Gd film as functions of time, and (D) magnetization of a 20 nm thick Gd film irradiated with 100 fs pulse as a function of fluence. Adapted with permission from [61]. Copyright (2020) Arxiv Preprint (2005.03493).

energy losses due to spin scattering with electrons and phonons might be significantly reduced. Thus, quantum confinement might lead to a very efficient energy transfer from fs laser photons to spin waves.

17.4.2.1 Confinement Effects

Because of the change in the DOS$_F$ caused by quantum confinement, metallic magnetic ultrathin films (t < 10 nm) behave differently under fs pulses [58, 62–63]. The quantum behavior may be understood using the free electron theory of metals, and the thin-film quantum well model can be used to describe quantum size effects. Quantum confinement alters the electron heat capacity constant and electron-phonon coupling. Hence, the term "all-optical quantum manipulation of magnetization" is used to describe this quantum behavior.

The magnetization dynamics of sub-10 nm metallic films exposed to fs pulse are modeled in the following. The interaction of laser light with magnetic material has been studied previously [50, 55], but quantum effects related to film thickness on magnetization dynamics may suggest significantly different characteristics. The electron bath receives laser pulse power directly, and T_e rises rapidly in sub-picoseconds. The electron DOS$_F$, electron-phonon coupling coefficient (G_{ep} = 4.05 × 10^{18} W·m^{-3}·K^{-1} × DOS$_F$), and Sommerfeld coefficient (γ = 743.22 J·m^{-3}·K^{-2} × DOS$_F$) are computed, and using a modified three-temperature model (M3TM) [64], the magnetization dynamics in Å-thick magnetic metals are modeled to study the effect of quantum confinement. The effect of quantum confinement on magnetization dynamics is analyzed in the following.

17.4.2.2 M3TM and Magnetization Dynamics in Quantum-Confined Ni Film

Based on Koopmans's model [32], the M3TM is solved with magnetization dynamics. The energy transfer from a fs laser pulse to electron, phonon, and magnetization is described by the following differential equations:

$$C_e \dot{T}_e = -G_{ep}(T_e - T_p) + P(t) \quad (19)$$

$$C_p \dot{T}_p = -G_{ep}(T_p - T_e) \quad (20)$$

$$\dot{m} = Rm \frac{T_p}{T_c}\left(1 - m \coth\left(m\frac{T_c}{T_e}\right)\right) \quad (21)$$

When electrons are excited, two competing processes occur immediately after the illumination by the fs laser pulse. Excited electrons thermalize into Fermi-Dirac distribution [65]. Electrons heat up much faster than the pulse width. Most of the electron's energy is subsequently transmitted to the lattice, while the rest diffuses to the electrons in the deeper layers. The laser pulse ends before thermal equilibrium between electrons and lattice as electron-lattice dynamics take longer than the pulse duration. Due to the extremely low film thicknesses, longitudinal temperature gradients and transverse heat conduction are neglected on the ps timescale [50]. Due to the quantum confinement in the DOS_F, G_{ep} and γ cannot be considered constant.

17.4.2.3 Effect of Film Thickness on DOS_F and M3TM Coefficients

The film thickness dependence of chemical potential and DOS at the Fermi level are shown in Figures 17.5(A) and (B), respectively. Figure 17.5(C) shows how the electron-phonon coupling and Sommerfeld coefficient vary dramatically with increasing film thickness owing to changes in the DOS_F for sub-5 nm (50 Å) thicknesses. This maximum length is a characteristic value given by $d = 10 \lambda_F$, where λ_F is the Fermi wavenumber for the bulk. The graphs are plotted for the electronic temperature at the Bohr energy of $E_B = 13.6$ eV.

17.4.2.4 Magnetization Dynamics in Quantum-Confined Film

The effect of quantum confinement on Ni magnetization dynamics is discussed here. Figure 17.5(D) depicts the change in m, T_e, and T_p for 20 Å Ni film over time until equilibrium. Ni magnetization drops within 100 fs when irradiated with a Gaussian laser pulse of $I_0 = 28$ mJ·m^{-2}, as shown in Figure 17.5(D). In the non-equilibrium condition, the T_e rises rapidly in a few fs due to the lower C_e. On the other hand, T_e cools to an equilibrium temperature (0.5 T_c) with phonon in 200 fs, and magnetization stabilizes near its initial value in 14 ps.

17.4.2.5 Film Thickness Dependence of Magnetization Dynamics

The magnetization dynamics based on the extended M3TM are shown in Figure 17.5(F) for different thicknesses of Ni film. The results demonstrate that layer thickness has little impact on demagnetization and recovery timeframes. It does, however, alter the demagnetization ratio. Figure 17.5(G) depicts the influence of thickness on the magnetization dip for various L_z values of 10–50 Å. The dips in Figure 17.5(F) are at the same position, but their amplitudes are different (Figure 17.5(G)).

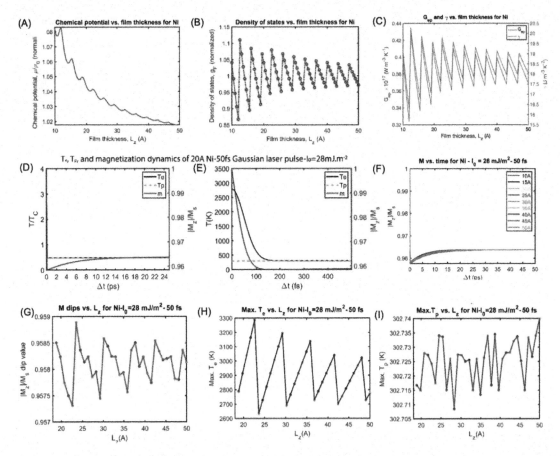

FIGURE 17.5 Quantum-confinement effects on all-optical magnetization control. Thickness (L_z) dependence of (A) chemical potential μ/μ_0 (finite temperature Fermi energy), (B) density of states at Fermi energy g_F, (C) electron-phonon coupling coefficient G_{ep} and Sommerfeld coefficient γ (μ/μ_0 and g_F are dimensionless variables that have been normalized to their bulk values). (D) Transient T_e, T_p, and normalized magnetization (m = $|M_z|/M_s$) for a 20 Å thick Ni film illuminated with a single Gaussian laser pulse of 50 fs width at I_0 = 28 mJ·m^{-2}. (E) 500 fs after illumination, zoomed-in version of (D). (F) The effect of Ni film thickness L_z on magnetization dynamics, (G) demagnetization dip, (H) maximum electron temperature T_e, and (I) maximum phonon temperature T_p, for L_z = 10–50 Å. Adapted with permission from [56]. Copyright (2021) Springer Nature.

This means that thickness-dependent change in the rate (G_{ep}) causes a greater loss/recovery of magnetization.

The dependence of T_e and T_p maxima on film thickness is shown in Figures 17.5(H) and (I), respectively. The characteristics of G_{ep} (Figure 17.5(c)) and the demagnetization dip (Figure 17.5 (G)) are comparable to the change of electron temperature peaks with increasing layer thickness. For specific thicknesses, such as 22 to 22.5 Å (or 25.5 to 26Å, 29 to 29.5 Å, 32.5 to 33 Å, 36 to 36.6 Å, 39.5 to 40 Å, 43 to 43.5 Å, 46.5 to 46 Å), the T_e and m dips are changed dramatically. Depending on L_z, the electron temperature can reach 3300 K. The high sensitivity of electron temperature (i.e., 3300 to 2620 K) to L_z (22 to 22.5 Å) demonstrates that changing thickness might be useful for controlling "hot electrons." The fact that the electron temperature may temporarily exceed T_c does not prevent nanomagnets from regaining magnetization.

17.5 CONCLUSION

We reviewed many theoretical approaches to nanomagnetism. Classical micromagnetic models based on the Landau-Lifshitz-Gilbert equation predict a wide range of experimentally reproducible spin dynamics and equilibria. The modified LLG equation with spin-orbit torques and spin-torque transfer allows for accurate estimation of spin current flow and current-driven domain wall or spin-wave dynamics. As a result, these models are increasingly being utilized to build spintronic nanodevices. Hence, micromagnetic models could assist in the development of new spintronic devices in the future.

A major shortcoming of micromagnetic models is their lack of consideration of quantum confinement, and hence Elliott-Yafet scattering mechanisms or other magnetotransport phenomena unique to low-dimensional magnetic materials cannot be captured by the LLG equation. An important manifestation of micromagnetism's inadequacies in the quantum regime emerged in the ultrafast optical control of magnetization dynamics. As a consequence, a three-temperature model and related approaches were developed. These approaches explain the ultrafast electron-phonon or spin scattering events with high accuracy, but a consensus among researchers on a completely robust quantum description of ultrafast control of magnetization dynamics has not yet been reached.

The energy transfer characteristics from a fs laser pulse to electron, phonon, spin, and magnetization in Fe, Ni (type I films), and Gd (type II film) have been studied using a modified three-temperature model. With $T_c \gg RT$, type I metallic thin films were initially exposed to a single Gaussian fs laser pulse. The magnetization vector then drops for 130–200 fs before returning to its original orientation. Magnetization reaches its steady-state orientation in the next 1–2 ps. The film reaction time and recovered fraction of magnetization rise with increasing the fluence. After interacting with a Gaussian laser pulse, type II films undergo progressive thermal demagnetization in a few ps. Greater laser fluences result in faster reaction times in nm-thick metallic magnetic thin films. THz spin waves might be generated with ultrafast fs laser pulses applied on metallic magnetic thin films.

The calculations presented based on the free-electron model reveal that the electron-phonon coupling coefficient (G_{ep}) in the M3TM is very sensitive to film thickness and cannot be regarded as constant due to quantum-confinement Angström-thick films. This impact alters the magnetization loss upon receiving a fs laser pulse but not the magnetization timescales. The ultrafast exchange-driven magnetization oscillations may be generated by laser-induced magnetization dynamics. Besides, because of the decreased density of phonons, the quantum confinement effect reduces the lattice temperature change, requiring lower laser fluences for magnetization control in Ni ultrathin films. As a result, the energy efficiency of stimulating spin waves using lasers might be improved.

New magnetotransport and ultrafast magnetization phenomena in quantum-confined, 2D, topological, or ultrahigh current-driven nanomagnets prompt the following open questions for capturing the physics of magnetism across different time and length scales.

17.6 OPEN QUESTIONS IN THE THEORY AND MODELING OF NANOMAGNETS

1. How can the quantum confinement in thin films or nanostructures be included in micromagnetic models?
2. How could one capture the new magnetotransport characteristics of emerging 2D magnetic thin films and topological layers?
3. Relativistic domain wall propagation reaches saturation velocities at ultrahigh currents [33]. A different regime is needed to account for the characteristics of relativistic spin waves.

4. Developing a generalized quantum nano/micromagnetic formalism could help capture the physics of spintronics over multiple time and length scales. Hence, a scalable quantum micromagnetic solver might allow for breakthrough discoveries in materials design as well as next-generation spintronic devices, sensors, and in-memory computational elements.

17.7 REFERENCES

[1] J. Bardeen and W. H. Brattain, "The transistor, a semi-conductor triode," *Physical Review*, vol. 74, no. 2, pp. 230–231, Jul. 1948, doi: 10.1103/PhysRev.74.230.

[2] J. S. Kilby, "Miniaturized electronic circuits [US Patent No. 3,138, 743]," *IEEE Solid-State Circuits Society Newsletter*, vol. 12, no. 2, pp. 44–54, 2007, doi: 10.1109/N-SSC.2007.4785580.

[3] G. E. Moore, "Cramming more components onto integrated circuits, Reprinted from Electronics, volume 38, number 8, April 19, 1965, pp. 114 ff.," *IEEE Solid-State Circuits Society Newsletter*, vol. 11, no. 3, pp. 33–35, Feb. 2009, doi: 10.1109/n-ssc.2006.4785860.

[4] W. A. Wulf and S. A. Mckee, *Hitting the Memory Wall: Implications of the Obvious*. Association for Computing Machinery, 1994.

[5] X. Z. Yu *et al.*, "Skyrmion flow near room temperature in an ultralow current density," *Nature Communications*, vol. 3, no. 1, p. 988, 2012, doi: 10.1038/ncomms1990.

[6] X.-L. Qi and S.-C. Zhang, "Topological insulators and superconductors," *Reviews of Modern Physics*, vol. 83, no. 4, pp. 1057–1110, Oct. 2011, doi: 10.1103/RevModPhys.83.1057.

[7] A. v Kimel and M. Li, "Writing magnetic memory with ultrashort light pulses," *Nature Reviews Materials*, vol. 4, no. 3, pp. 189–200, 2019, doi: 10.1038/s41578-019-0086-3.

[8] T. L. Gilbert, "A phenomenological theory of damping in ferromagnetic materials," *IEEE Transactions on Magnetics*, vol. 40, no. 6, pp. 3443–3449, 2004, doi: 10.1109/TMAG.2004.836740.

[9] D. Porter and M. Donahue, "OOMMF Documentation Overview," *OOMMF User's Guide*, Sep. 29, 2021, https://math.nist.gov/oommf/doc/userguide20a3/userguide.pdf.

[10] A. Vansteenkiste, J. Leliaert, M. Dvornik, M. Helsen, F. Garcia-Sanchez, and B. van Waeyenberge, "The design and verification of MuMax3," *AIP Advances*, vol. 4, no. 10, p. 107133, Oct. 2014, doi: 10.1063/1.4899186.

[11] H. Kronmüller and M. Fähnle, *Micromagnetism and the Microstructure of Ferromagnetic Solids*. Cambridge University Press, 2003.

[12] J. C. Slonczewski, "Current-driven excitation of magnetic multilayers," *Journal of Magnetism and Magnetic Materials*, vol. 159, no. 1, pp. L1–L7, 1996, https://doi.org/10.1016/0304-8853(96)00062-5.

[13] S. Zhang and Z. Li, "Roles of nonequilibrium conduction electrons on the magnetization dynamics of ferromagnets," *Physical Review Letters*, vol. 93, no. 12, p. 127204, Sep. 2004, doi: 10.1103/PhysRevLett.93.127204.

[14] T. Kimura, N. Hashimoto, S. Yamada, M. Miyao, and K. Hamaya, "Room-temperature generation of giant pure spin currents using epitaxial Co_2FeSi spin injectors," *NPG Asia Materials*, vol. 4, no. 3, pp. e9, 2012, doi: 10.1038/am.2012.16.

[15] C. O. Avci *et al.*, "Current-induced switching in a magnetic insulator," *Nature Materials*, vol. 16, no. 3, pp. 309–314, 2017, doi: 10.1038/nmat4812.

[16] M. Baumgartner *et al.*, "Spatially and time-resolved magnetization dynamics driven by spin-orbit torques," *Nature Nanotechnology*, vol. 12, no. 10, pp. 980–986, Oct. 2017, doi: 10.1038/nnano.2017.151.

[17] E. P. Amaladass, *Magnetism of Amorphous and Highly Anisotropic Multilayer Systems on Flat Substrates and Nanospheres*, PhD Thesis, University of Stuttgart, Stuttgart, 2008. https://elib.uni-stuttgart.de/handle/11682/6705.

[18] Massimiliano d'Aquino, *Nonlinear Magnetization Dynamics in Thin-Films and Nanoparticles*. Napoli, 2004.

[19] B. Behin-Aein, D. Datta, S. Salahuddin, and S. Datta, "Proposal for an all-spin logic device with built-in memory," *Nature Nanotechnology*, vol. 5, no. 4, pp. 266–270, 2010, doi: 10.1038/nnano.2010.31.

[20] F. Katmis *et al.*, "A high-temperature ferromagnetic topological insulating phase by proximity coupling," *Nature*, vol. 533, no. 7604, pp. 513–516, 2016, doi: 10.1038/nature17635.

[21] I. E. Dzyaloshinskii, "Theory of helicoidal structures in antiferromagnets. I. Nonmetals," *JETP*, vol. 19, no. 4, pp. 960–971, 1964.

[22] I. E. Dzyaloshinskii, "Theory of helicoidal structures in antiferromagnets. II. Metals," *JETP*, vol. 20, no. 1, pp. 223–231, 1965.

[23] T. Moriya, "Anisotropic superexchange interaction and weak ferromagnetism," *Physical Review*, vol. 120, no. 1, pp. 91–98, Oct. 1960, doi: 10.1103/PhysRev.120.91.
[24] S. Seki, X. Z. Yu, S. Ishiwata, and Y. Tokura, "Observation of skyrmions in a multiferroic material," *Science*, vol. 336, no. 6078, pp. 198–201, Apr. 2012, doi: 10.1126/science.1214143.
[25] K. Shibata et al., "Towards control of the size and helicity of skyrmions in helimagnetic alloys by spin – orbit coupling," *Nature Nanotechnology*, vol. 8, no. 10, pp. 723–728, 2013, doi: 10.1038/nnano.2013.174.
[26] T. Seki et al., "Giant spin Hall effect in perpendicularly spin-polarized FePt/Au devices," *Nature Materials*, vol. 7, no. 2, pp. 125–129, 2008, doi: 10.1038/nmat2098.
[27] C. Pan and A. Naeemi, "Nonvolatile spintronic memory array performance benchmarking based on three-terminal memory cell," *IEEE Journal on Exploratory Solid-State Computational Devices and Circuits*, vol. 3, pp. 10–17, 2017, doi: 10.1109/JXCDC.2017.2669213.
[28] K. L. Wang, J. G. Alzate, and P. Khalili Amiri, "Low-power non-volatile spintronic memory: STT-RAM and beyond," *Journal of Physics D: Applied Physics*, vol. 46, no. 7, p. 074003, Feb. 2013, doi: 10.1088/0022-3727/46/7/074003.
[29] J. Kim et al., "Spin-based computing: Device concepts, current status, and a case study on a high-performance microprocessor," *Proceedings of the IEEE*, vol. 103, no. 1, pp. 106–130, 2015, doi: 10.1109/JPROC.2014.2361767.
[30] E. Chen et al., "Advances and future prospects of spin-transfer torque random access memory," *IEEE Transactions on Magnetics*, vol. 46, no. 6, pp. 1873–1878, 2010, doi: 10.1109/TMAG.2010.2042041.
[31] J. Puebla, J. Kim, K. Kondou, and Y. Otani, "Spintronic devices for energy-efficient data storage and energy harvesting," *Communications Materials*, vol. 1, no. 1, p. 24, 2020, doi: 10.1038/s43246-020-0022-5.
[32] B. Koopmans et al., "Explaining the paradoxical diversity of ultrafast laser-induced demagnetization," *Nature Materials*, vol. 9, no. 3, pp. 259–265, 2010, doi: 10.1038/nmat2593.
[33] C. Lucas et al., "Relativistic kinematics of a magnetic soliton," *Science*, vol. 370, no. 6523, pp. 1438–1442, Dec. 2020, doi: 10.1126/science.aba5555.
[34] M. Gibertini, M. Koperski, A. F. Morpurgo, and K. S. Novoselov, "Magnetic 2D materials and heterostructures," *Nature Nanotechnology*, vol. 14, no. 5, pp. 408–419, 2019, doi: 10.1038/s41565-019-0438-6.
[35] B. Huang, M. A. McGuire, A. F. May, D. Xiao, P. Jarillo-Herrero, and X. Xu, "Emergent phenomena and proximity effects in two-dimensional magnets and heterostructures," *Nature Materials*, vol. 19, no. 12, pp. 1276–1289, 2020, doi: 10.1038/s41563-020-0791-8.
[36] M. Kar, R. Sarkar, S. Pal, and P. Sarkar, "Engineering the magnetic properties of $PtSe_2$ monolayer through transition metal doping," *Journal of Physics: Condensed Matter*, vol. 31, no. 14, p. 145502, Apr. 2019, doi: 10.1088/1361-648X/aaff40.
[37] H. Zheng et al., "Visualization of point defects in ultrathin layered 1T-$PtSe_2$," *2D Materials*, vol. 6, no. 4, p. 041005, Sep. 2019, doi: 10.1088/2053-1583/ab3beb.
[38] A. Avsar, A. Ciarrocchi, M. Pizzochero, D. Unuchek, O. v Yazyev, and A. Kis, "Defect induced, layer-modulated magnetism in ultrathin metallic $PtSe_2$," *Nature Nanotechnology*, vol. 14, no. 7, pp. 674–678, 2019, doi: 10.1038/s41565-019-0467-1.
[39] A. Avsar et al., "Probing magnetism in atomically thin semiconducting $PtSe_2$," *Nature Communications*, vol. 11, no. 1, p. 4806, 2020, doi: 10.1038/s41467-020-18521-6.
[40] Nguyen, T. D., Jiang, J., Song, B., Tran, M. D., Choi, W., Kim, J. H., Kim, Y.-M., Duong, D. L., and Lee, Y. H. "Gate-tunable magnetism via resonant Se-vacancy levels in WSe2," *Advanced Science*, vol. 8, p. 2102911, Oct. 2021, https://doi.org/10.1002/advs.202102911.
[41] R. Chua et al., "Can reconstructed Se-deficient line defects in monolayer VSe_2 induce magnetism?" *Advanced Materials*, vol. 32, no. 24, p. 2000693, Jun. 2020, https://doi.org/10.1002/adma.202000693.
[42] X. Lou et al., "Electrical detection of spin transport in lateral ferromagnet – semiconductor devices," *Nature Physics*, vol. 3, no. 3, pp. 197–202, 2007, doi: 10.1038/nphys543.
[43] S. Lee et al., "Synthetic Rashba spin – orbit system using a silicon metal-oxide semiconductor," *Nature Materials*, vol. 20, no. 9, pp. 1228–1232, 2021, doi: 10.1038/s41563-021-01026-y.
[44] Y. Fukuma, L. Wang, H. Idzuchi, S. Takahashi, S. Maekawa, and Y. Otani, "Giant enhancement of spin accumulation and long-distance spin precession in metallic lateral spin valves," *Nature Materials*, vol. 10, no. 7, pp. 527–531, 2011, doi: 10.1038/nmat3046.
[45] J. F. Sierra, J. Fabian, R. K. Kawakami, S. Roche, and S. O. Valenzuela, "Van der Waals heterostructures for spintronics and opto-spintronics," *Nature Nanotechnology*, vol. 16, no. 8, pp. 856–868, 2021, doi: 10.1038/s41565-021-00936-x.
[46] A. Avsar, H. Ochoa, F. Guinea, B. Özyilmaz, B. J. van Wees, and I. J. Vera-Marun, "*Colloquium*: Spintronics in graphene and other two-dimensional materials," *Reviews of Modern Physics*, vol. 92, no. 2, p. 021003, Jun. 2020, doi: 10.1103/RevModPhys.92.021003.

[47] A. Avsar et al., "Gate-tunable black phosphorus spin valve with nanosecond spin lifetimes," *Nature Physics*, vol. 13, no. 9, pp. 888–893, 2017, doi: 10.1038/nphys4141.

[48] I. Žutić, J. Fabian, and S. das Sarma, "Spintronics: Fundamentals and applications," *Reviews of Modern Physics*, vol. 76, no. 2, pp. 323–410, Apr. 2004, doi: 10.1103/RevModPhys.76.323.

[49] C. D. Stanciu et al., "All-optical magnetic recording with circularly polarized light," *Physical Review Letters*, vol. 99, no. 4, p. 047601, Jul. 2007, doi: 10.1103/PhysRevLett.99.047601.

[50] E. Beaurepaire, J.-C. Merle, A. Daunois, and J.-Y. Bigot, "Ultrafast spin dynamics in ferromagnetic nickel," *Physical Review Letters*, vol. 76, no. 22, pp. 4250–4253, May 1996, doi: 10.1103/PhysRevLett.76.4250.

[51] J.-Y. Bigot, M. Vomir, and E. Beaurepaire, "Coherent ultrafast magnetism induced by femtosecond laser pulses," *Nature Physics*, vol. 5, no. 7, pp. 515–520, 2009, doi: 10.1038/nphys1285.

[52] R. Moreno, T. A. Ostler, R. W. Chantrell, and O. Chubykalo-Fesenko, "Conditions for thermally induced all-optical switching in ferrimagnetic alloys: Modeling of TbCo," *Physical Review B*, vol. 96, no. 1, p. 014409, Jul. 2017, doi: 10.1103/PhysRevB.96.014409.

[53] J. A. Fülöp, S. Tzortzakis, and T. Kampfrath, "Laser-driven strong-field terahertz sources," *Advanced Optical Materials*, vol. 8, no. 3, p. 1900681, Feb. 2020, https://doi.org/10.1002/adom.201900681.

[54] X. Chen et al., "Generation and manipulation of chiral broadband terahertz waves from cascade spintronic terahertz emitters," *Applied Physics Letters*, vol. 115, no. 22, p. 221104, Nov. 2019, doi: 10.1063/1.5128979.

[55] A. Kirilyuk, A. v. Kimel, and T. Rasing, "Ultrafast optical manipulation of magnetic order," *Reviews of Modern Physics*, vol. 82, no. 3, pp. 2731–2784, Sep. 2010, doi: 10.1103/RevModPhys.82.2731.

[56] S. M. Zanjani, M. T. Naseem, Ö. E. Müstecaplıoğlu, and M. C. Onbaşlı, "All optical control of magnetization in quantum confined ultrathin magnetic metals," *Scientific Reports*, vol. 11, no. 1, p. 15976, 2021, doi: 10.1038/s41598-021-95319-6.

[57] J. Kimling, J. Kimling, R. B. Wilson, B. Hebler, M. Albrecht, and D. G. Cahill, "Ultrafast demagnetization of FePt:Cu thin films and the role of magnetic heat capacity," *Physical Review B*, vol. 90, no. 22, p. 224408, Dec. 2014, doi: 10.1103/PhysRevB.90.224408.

[58] Z. Lin, L. v. Zhigilei, and V. Celli, "Electron-phonon coupling and electron heat capacity of metals under conditions of strong electron-phonon nonequilibrium," *Physical Review B*, vol. 77, no. 7, p. 075133, Feb. 2008, doi: 10.1103/PhysRevB.77.075133.

[59] M. Battiato, K. Carva, and P. M. Oppeneer, "Superdiffusive spin transport as a mechanism of ultrafast demagnetization," *Physical Review Letters*, vol. 105, no. 2, p. 027203, Jul. 2010, doi: 10.1103/PhysRevLett.105.027203.

[60] B. Koopmans, J. J. M. Ruigrok, F. D. Longa, and W. J. M. de Jonge, "Unifying ultrafast magnetization dynamics," *Physical Review Letters*, vol. 95, no. 26, p. 267207, Dec. 2005, doi: 10.1103/PhysRevLett.95.267207.

[61] S. M. Zanjani and M. C. Onbaşlı, "Ultrafast All Optical Magnetization Control for Broadband Terahertz Spin Wave Generation," May 2020, Accessed: Nov. 29, 2021. [Online]. Available: http://arxiv.org/abs/2005.03493

[62] P. B. Allen, "Theory of thermal relaxation of electrons in metals," *Physical Review Letters*, vol. 59, no. 13, pp. 1460–1463, Sep. 1987, doi: 10.1103/PhysRevLett.59.1460.

[63] J. P. Rogers, P. H. Cutler, T. E. Feuchtwang, and A. A. Lucas, "Quantum size effects in the fermi energy and electronic density of states in a finite square well thin film model," *Surface Science*, vol. 181, no. 3, pp. 436–456, 1987, https://doi.org/10.1016/0039-6028(87)90199-3.

[64] Z. Du, C. Chen, F. Cheng, Y. Liu, and L. Pan, "Prediction of deterministic all-optical switching of ferromagnetic thin film by ultrafast optothermal and optomagnetic couplings," *Scientific Reports*, vol. 7, no. 1, p. 13513, 2017, doi: 10.1038/s41598-017-13568-w.

[65] J. K. Chen, D. Y. Tzou, and J. E. Beraun, "A semiclassical two-temperature model for ultrafast laser heating," *International Journal of Heat and Mass Transfer*, vol. 49, no. 1, pp. 307–316, 2006, https://doi.org/10.1016/j.ijheatmasstransfer.2005.06.022.

Research Trends and Statistical-Thermodynamic Modeling the α″-Fe$_{16}$N$_2$-Based Phase for Permanent Magnets

18

Taras M. Radchenko, Olexander S. Gatsenko,
Vyacheslav V. Lizunov, and Valentyn A. Tatarenko

*G. V. Kurdyumov Institute for Metal Physics of the N.A.S. of Ukraine,
36 Academician Vernadsky Boulevard, Kyiv, Ukraine*

Contents

18.1	Introduction	344
	18.1.1 Rare-Earth-Free Materials for Permanent Magnets	344
	18.1.2 Discovery of α″-Fe$_{16}$N$_2$ and Its Unique Magnetic Properties	344
	18.1.3 Fabrication Techniques	345
	18.1.4 Motivation of the Study	346
18.2	Solid Solution of Non-Metal Atoms Occupying Interstices and Sites of b.c.c.(t). Lattice with Vacancies	347
	18.2.1 Configurational Gibbs Free Energy	347
	18.2.2 Tetragonality of Fe–N Crystal Lattice	352
	18.2.3 Interatomic Interaction Energy Parameters for the Paramagnetic State	354
18.3	Redistribution of Impurities Between the Interstitial and Substitutional Positions	357
	18.3.1 Change in the Interstitial Impurity Concentration	357
	18.3.2 Change in the Vacancy Content at the Host-Lattice Sites	358
18.4	Explicit Magnetic Contribution into the Gibbs Free Energy	359
18.5	Conclusions	362
Acknowledgements		362
References		363

DOI: 10.1201/9781003197492-18

18.1 INTRODUCTION

As known [1,2], currently permanent magnets are one of the strategic metal-based products in the world industry due to a wide range of their applications: from micro-electromechanical and nano-electromechanical systems to high-power electricity generators using many tons of magnetic materials containing critical rare-earth metals. Such non-renewable elements are the main factor that constrains or increases the cost of fabrication of magnetic products. Therefore, the challenge of obtaining new permanent magnets, which have comparable characteristics to existing magnets but are free from critical elements (such as rare-earth metals), is extremely important for industrial security in the world.

18.1.1 Rare-Earth-Free Materials for Permanent Magnets

The new tailored metallic phases that do not contain critical elements (non-renewable on the earth) may act as an alternative to the permendur and rare-earth permanent magnets. Among materials with interesting magnetoelectric, thermo-, and magnetomechanical effects, $L1_0$-Fe–Ni-type elinvars with a rather tetragonal crystal lattice (but not a cubic one due to the layered atomic ordering) [3,4] and cubic $L1_2$-Fe–Ni-type invars [5] are worth noticing. Due to their properties, they successfully complement a series of the best rare-earth intermetallics and permendur (e.g., $Fe_{65}Co_{35}$ based on a body-centered-cubic (b.c.c.) lattice), which possesses high magnetic saturation levels (at 2.45 T) [6]. Such metallic phases (particularly known as tetrataenite) are present in the meteorites, but their synthesis is very difficult in earth conditions.

In recent years, great progress in the world has been made toward studying and improving the microstructure and physical properties of non-rare-earth materials for permanent magnets [1–2]. Researchers investigated several new materials candidates. Some of them have shown realistic potential for replacing rare-earth permanent magnets for some applications [1–2]. The properties of these materials are described in the large published research articles and reviews. In the recent review [2], authors addressed such systems as Mn-based Mn–Al and Mn–Bi alloys with high magnetocrystalline anisotropy, spinodally decomposing Fe-based Al-, Ni-, Co-containing (Alnico) alloys, high-coercivity tetrataenite $L1_0$-phase in Fe–Ni and Fe–Co, Co-rich $HfCo_7$ and Zr_2Co_{11} intermetallic compounds, Co_3C and Co_2C carbides, and iron nitride α''-$Fe_{16}N_2$. The latter system attracts considerable interest due to its exceptionally high saturation magnetization, the low cost of Fe, and because its elements are the most earth-abundant among all magnetic materials [2].

18.1.2 Discovery of α''-$Fe_{16}N_2$ and Its Unique Magnetic Properties

A chemically ordered nitride α''-$Fe_{16}N_2$ (Figure 18.1) was discovered and characterized structurally by K. H. Jack in his classical works [7,8]. Despite discovery in the middle of the last century, the magnetic behavior of the chemically ordered α''-$Fe_{16}N_2$ nitride remained a mystery for over 40 years due to the questionable and even controversial results in a series of works concerning the saturation magnetization values (see review [9] and references therein). Nevertheless, all the values reported in the literature – 2.58 T [10], 2.6–2.8 T [11], 2.83 T [12], 2.9 T [13], and 3.1 T [14] – are far beyond Fe–Co and all other magnetic materials.

The article of Kim and Takahashi in 1972 [10] that reported on high magnetization in α''-$Fe_{16}N_2$ has inspired many groups of scientists all over the world to explore the material. Motivated and encouraged by the promising magnetic properties that the α''-$Fe_{16}N_2$ phase demonstrates, researchers undertook quite a few attempts to theoretically and experimentally study this phase and prepare (through different synthesis techniques) the samples with as larger its volume fraction as possible.

To explain the giant saturation magnetization behavior in α''-$Fe_{16}N_2$, a "cluster + atom" model was proposed [15]. This model is associated with partially localized electron states. Two years later, the

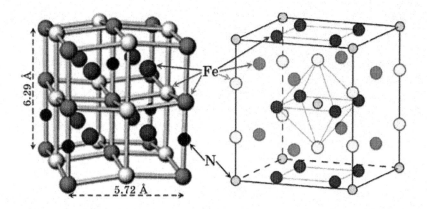

FIGURE 18.1 Structure of α″-Fe$_{16}$N$_2$, where the left and right unit cells are crystallographically equivalent: each of them contains 16 Fe and 2 N atoms. Figure left republished with permission of [8]. Copyright (2017) Royal Society. Permission conveyed through Copyright Clearance Center, Inc. Figure right republished with permission of [12]. Copyright (2017) AIDIC Servizi S.r.l.

partial localization behavior of 3d-electrons in α″-Fe$_{16}$N$_2$ was confirmed by means of the x-ray magnetic circular dichroism measurements [16]. Generally, the origin of the giant magnetic moment can be attributed to both the intrinsic effects such as partial localization of d-electrons in the Fe$_6$N octahedrons and the nature of the exchange interaction promoting the high-spin states through Hund's coupling [17]. The double-exchange interaction is the strongest and favors a high-spin state. The presence of the super-exchange interaction between the Fe sites within the Fe$_6$N octahedral region indicates the existence of localized states, which may also contribute to the occurrence of a giant magnetic moment in this material [17].

18.1.3 Fabrication Techniques

The earliest and most commonly used method so far to prepare a bulk (and, rarely, thin film) α″-Fe$_{16}$N$_2$ material is the process of nitriding, quenching, and annealing. The reaction-transformation sequence of this method includes four different phases (α, γ, α′, α″) and three phase transformations (including the γ→α′ martensitic one, which proceeds by shear without diffusion) [18,19]:

$$\alpha-\text{Fe} \xrightarrow[\approx 700-750°C]{\text{nitriding}} \gamma\text{Fe}-\text{N austenite} \xrightarrow[<0°C]{\text{quenching}} \alpha'\text{Fe}-\text{N martensite} \xrightarrow[\approx 120-150°C]{\text{tempering}} \alpha''-\text{Fe}_{16}\text{N}_2.$$

Here, at temperature $T \cong 750°C$, α-Fe is subjected to the nitriding procedure to form a nitrogen austenite phase, which then quenched below 0°C in order to obtain a martensitic α′-phase; then, it is annealed at $\cong 120°C$ (370–420 K) for ≈ 1–2 hours to obtain α″-Fe$_{16}$N$_2$ phase [18–19], where N atoms have an ordered distribution.

There are also several other methods to prepare α″-Fe$_{16}$N$_2$. They are: molecular beam epitaxy, ion implantation and beam deposition, facing target and reactive (magnetron) sputtering, the ball milling, and the so-called "strained-wire method" when uniaxial tensile stress is applied on the wire-shaped sample during the post-annealing stage (in details, see review [9] and references therein).

The α″-Fe$_{16}$N$_2$ structure is difficult for preparation and mass production. This structure is a metastable [20]: at $T > 463$ K [21] and with time (years) [22], it decomposes according to the decomposition reaction [21–22]: α″-Fe$_{16}$N$_2$ → 8(α-Fe) + 2(γ′-Fe$_4$N). Literature data on the decomposition temperature of α″-Fe$_{16}$N$_2$ strongly differ from ≈ 463 K to ≈ 673 K. To affect (improve) the thermal stability and soft magnetic problems, adding a small amount of the third element has been proposed. However, the doping (alloying) may degrade some other magnetic characteristics.

In addition to the metastability, the obtaining of α''-Fe$_{16}$N$_2$ is complicated by a wide variety of compounds detectable in the binary Fe–N system. Besides the α''-Fe$_{16}$N$_2$, they are [9]: α'-Fe$_8$N based on a body-centered-tetragonal (b.c.t.) lattice, γ'-Fe$_4$N based on a face-centered-cubic (f.c.c.) lattice, ε-Fe$_3$N based on a hexagonal-close-packed (h.c.p.) lattice, orthorhombic-based ζ-Fe$_2$N, (hypothetical) f.c.c.-based FeN, and tentative γ''-Fe$_8$N$_2$ of tetragonal symmetry. Nevertheless, in the literature, there are quite a lot of works where the α''-Fe$_{16}$N$_2$ phase is claimed to be synthesized mainly in (thin) films, rarer in bulk materials, powders, (nano)composites, (nano)particles, and (nano)ribbons.

Commonly, for generation of the α''-Fe$_{16}$N$_2$ phase in thin films and sometimes in low-dimensional (nanoscale) materials, a mixture of Fe and N atoms with a fixed atomic fraction has to be generated through the plasma or sputtering method in a high vacuum and then deposited on a single-crystalline substrate. A lattice mismatch between the substrate and the deposited Fe–N film causes the strain effect, which results in the martensitic α'-Fe$_8$N phase. Then, after a post-annealing process (below the fixed temperature in vacuum), the α'-Fe$_8$N phase transforms into the α''-Fe$_{16}$N$_2$ phase with an ordered distribution of N atoms.

Among dozens of articles, only a few works claimed pure α''-Fe$_{16}$N$_2$ phase preparation. Commonly, the prepared Fe–N samples consist of several possible phases with a partial volume ratio of an α''-Fe$_{16}$N$_2$ phase inside, but just a volume fraction of this phase defines magnetic properties: the higher volume fraction, the better properties [14]. In this respect, a key parameter to evaluate the quality of the prepared compound and its magnetic properties is information on the volume ratio of α''-Fe$_{16}$N$_2$ phase. To evaluate the amount of this phase in the prepared Fe–N system, the x-ray diffraction technique is most commonly used [13–14].

18.1.4 Motivation of the Study

Crucial properties of permanent magnets (both already widely used and acting as potential candidates), particularly coercivity and permanent magnetization, strongly depend on (micro)structure. That is why a deep understanding of metallurgical processing, phase stability, and (micro)structural changes is major for designing and improving permanent magnets as well as predicting their properties [1–2].

The key factors, which practically determine the phase magnetization, are external impacts (particularly strain and temperature), interatomic interactions, and order in the spatial distribution of nitrogen atoms, which can occupy both octahedral interstices and metal-lattice sites with vacancies. The N-atomic redistribution results in the partial (dis)ordering. The degree of this (dis)ordering correlates with a magnetization value: the latter increases (decreases) as the atomic long-range order parameter increases (decreases) [23].

So far, many research groups, including experimentalists and theoreticians, cannot reproduce or justify such unique magnetic characteristics of α''-Fe$_{16}$N$_2$. After a symposium on the topic of Fe$_{16}$N$_2$ at the Annual Conference on Magnetism and Magnetic Materials in 1996, where there was no decisive conclusion on the giant saturation magnetization origin, this research topic has been dropped by most magnetic researchers. To this day, therefore, there is no complete understanding and explanation of the magnetism of the α''-Fe$_{16}$N$_2$ phase. Thus, this problem remains topical and motivated and even somewhat of a mystery.

In spite of a lot of works on the synthesis of the α''-Fe$_{16}$N$_2$ phase, currently there is still a lack of full understanding of how one can change the external thermodynamic parameters (temperature, pressure, or deformation) and regulate the structure and thereby the properties of magnetic α''-Fe$_{16}$N$_2$-type martensite. To overcome such a theoretical gap, we are motivated to develop a microscopic model of a "hybrid" solid solution, in which the interstitial non-metal atoms can reside not only in octahedral interstices of the b.c.c. or b.c.t. metal (*Me*) lattice, but partially leave interstitial positions and move to the vacant sites of the lattice.

18.2 SOLID SOLUTION OF NON-METAL ATOMS OCCUPYING INTERSTICES AND SITES OF B.C.C.(T). LATTICE WITH VACANCIES

Commonly, the binary solid solutions are considered to belong to one of two geometrically model types, which are described as either interstitial or substitutional. However, currently it is quite clear that this simple differentiation is an approximation. In reality, solute atoms occupy both the lattice solvent-atom sites and one or more subsets of interstitial sites. Though the vast majority of solute (impurity) atoms in the overwhelming majority of solid solutions are located only in one subset of geometrically equivalent sites, this is not always the case. There are solutions where significant fractions of impurity atoms reside in both interstitial and substitutional positions [24,25].

In the literature, there is a viewpoint that the anomaly of the temperature dependence of the ratio of lattice parameters (a tetragonality degree) of (irradiated) b.c.c.(t.) α'-martensite is caused by a peculiar phase transformation, e.g., impurity C atoms from octahedral interstices are "captured" by vacant sites of b.c.c.(t.) Fe lattice. We can assume that this feature ("capture") is proper for not only irradiated martensites but also may occur in any (realistic) imperfect interstitial solid solution (alloy) that contains the vacant sites. McLellan attempted to construct a simple model of a binary "hybrid" interstitial–substitutional solid solution [26]; however, he has ignored the lattice uniform dilation effects and restricted the case to sufficiently low solute contents. In another work [27], the changes in the long-range $L1_2$-order parameter have been attributed to the introduction of deformation-induced interstitial atoms into vacant sites.

18.2.1 Configurational Gibbs Free Energy

Let us $c^\alpha(\mathbf{R})$ ($c_p^\beta(\mathbf{R})$) is a random quantity, which equals 1 if a site with radius-vector $\mathbf{R}$ (interstice with radius-vector $\mathbf{R} + \mathbf{h}_p$ in a primitive unit cell $\mathbf{R}$) is occupied by an "atom" of kind α = Fe, N_s, v (is occupied by an "atom" of kind β = N_i, $\varnothing$) and 0 otherwise (if not). Here, N_s and N_i denote the nitrogen atoms substituting sites s and octahedral interstices i in the b.c.c.(t.) Fe, respectively. To model the excess monovacancies on the sites or remaining unoccupied octahedral interstices, we consider them as "atoms" of additional "substitutional" (v) or "interstitial" ($\varnothing$) constituents, respectively. A set of octahedral interstices in the b.c.c. Fe consists of three interpenetrating sublattices $\{\mathbf{R} + \mathbf{h}_p\}$, where $p = 1, 2, 3$ (see Figure 18.2a–d). Each sublattice numbered with $p = 1, 2, 3$ is isostructural to the "mean" b.c.c. lattice of N_s sites $\{\mathbf{R}\}$ and displaced (as a whole) with respect to the origin site of the cubic conventional unit cell of b.c.c. crystal by the vector

$$\mathbf{h}_1 = \frac{a_0}{2}(1;0;0), \quad \mathbf{h}_1 = \frac{a_0}{2}(0;1;0), \quad \mathbf{h}_1 = \frac{a_0}{2}(0;0;1)$$

respectively, in a crystalline-physical system of coordinates, $Oxyz$, where a_0 is the basic lattice parameter of the impurity-free unit cell (Figure 18.2a–d). By this definition, $P^\alpha(\mathbf{R}) \equiv \langle c^\alpha(\mathbf{R}) \rangle$ and $P_p^\beta(\mathbf{R}) \equiv \langle c_p^\beta(\mathbf{R}) \rangle$ are single-site probabilities to find atoms α and β kind in the site $\mathbf{R}$ and the interstice $\mathbf{R} + \mathbf{h}_p$, respectively, and the symbol $\langle \ldots \rangle$ denotes the statistical averaging over all permitted atomic configurations in case of their canonical distribution. We assume a self-consistent mean-field approximation (neglecting correlation) without limiting an interatomic-interaction radius [28,29], i.e., take into account interatomic interactions in all coordination spheres. Then, the configurational contribution to the Gibbs free energy of such a solid solution (here and further, we consider the case of absence of external stresses) reads as [30,31]

$$G \cong G_0 + \sum_{\alpha=N_s,v} \sum_{\mathbf{R}} R^\alpha(\mathbf{R}) P^\alpha(\mathbf{R}) + \sum_{\beta=N_i,\varnothing} \sum_{p=1}^{3} \sum_{\mathbf{R}} R_p^\beta(\mathbf{R}) P_p^\beta(\mathbf{R}) +$$

$$+ \frac{1}{2} \sum_{\alpha=Fe,N_s,v} \sum_{\beta=Fe,N_s,v} \sum_{\mathbf{R},\mathbf{R}'} W^{\alpha\beta}(\mathbf{R}-\mathbf{R}') P^\alpha(\mathbf{R}) P^\beta(\mathbf{R}') +$$

$$+ \frac{1}{2} \sum_{\alpha=N_i,\varnothing} \sum_{\beta=N_i,\varnothing} \sum_{p,p'=1}^{3} \sum_{\mathbf{R},\mathbf{R}'} W_{pp'}^{\alpha\beta}(\mathbf{R}-\mathbf{R}') P_p^\alpha(\mathbf{R}) P_{p'}^\beta(\mathbf{R}') +$$

$$+ \sum_{\alpha=Fe,N_s,v} \sum_{\beta=N_i,\varnothing} \sum_{p=1}^{3} \sum_{\mathbf{R},\mathbf{R}'} W_p^{\alpha\beta}(\mathbf{R}-\mathbf{R}') P^\alpha(\mathbf{R}) P_p^\beta(\mathbf{R}') +$$

$$+ k_B T \sum_{\alpha=Fe,N_s,v} \sum_{\mathbf{R}} P^\alpha(\mathbf{R}) \ln P^\alpha(\mathbf{R}) + k_B T \sum_{\beta=N_i,\varnothing} \sum_{p=1}^{3} \sum_{\mathbf{R}} P_p^\beta(\mathbf{R}) \ln P_p^\beta(\mathbf{R}) \quad (1)$$

Here, k_B is the Boltzmann constant; T is the absolute temperature of the solution; $G_0 \ \forall \ T$ is the Gibbs free energy part insensitive to configurations of all atoms. $R^\alpha(\mathbf{R}) \cong \text{const} = R^\alpha$ and $R_p^\alpha(\mathbf{R}) \cong \text{const} = R_p^\alpha = R_1^\alpha = R_2^\alpha = R_3^\alpha$ are the specific (per atom α) energies related to the "expansion" or "compression" of each N_α atom uniformly introduced into the sites and/or crystallographically equivalent octahedral interstices of the b.c.c.(t.) Fe crystal [29–30]. $W^{\alpha\beta}(\mathbf{R}-\mathbf{R}')$, $W_{pp'}^{\alpha\beta}(\mathbf{R}-\mathbf{R}')$, or $W_p^{\alpha\beta}(\mathbf{R}-\mathbf{R}')$ are the effective pair interaction energies for the α and β atoms occupying sites with radius-vectors $\mathbf{R}$ and $\mathbf{R}'$, interstices with radius-vectors $\mathbf{R} + \mathbf{h}_p$ and $\mathbf{R}' + \mathbf{h}_{p'}$, or site $\mathbf{R}$ and interstice $\mathbf{R}' + \mathbf{h}_{p'}$, respectively [28–29].

$P_p^{N_i}(\mathbf{R})$ and $P^{N_s}(\mathbf{R})$ relate through conservation conditions for the total number N of nitrogen (N) atoms in interstices and sites $\left(N_N = N_{N_i} + N_{N_s} \right)$ during their redistribution:

$$N_N = N_s(3\kappa_{N_i} + \kappa_{N_s}) = \sum_{\mathbf{R}} \left(\sum_{p=1}^{3} P_p^{N_i}(\mathbf{R}) + P^{N_s}(\mathbf{R}) \right) \quad (2)$$

where $\kappa_{N_i} = N_{N_i}/(3\tilde{N}_s)$ and $\kappa_{N_s} = N_{N_s}/\tilde{N}_s$. Here and further, N_N denotes the total number of nitrogen atoms (and $\tilde{N}_s$ is the total number of sites), while N denotes the chemical element "nitrogen" (and N_s or N_i are nitrogen atoms in site and interstice, respectively). In addition, it is evident that any site (interstice) in any of $\tilde{N}_s$ primitive unit cells of the solvent has to be obligatory occupied by atoms of any allowable kind:

$$\forall \mathbf{R} \sum_{\alpha=Fe,N_s,v} P^\alpha(\mathbf{R}) = 1, \quad (3)$$

and

$$\forall \mathbf{R} \sum_{\beta=N_i,\varnothing} P_p^\beta(\mathbf{R}) = 1, \text{ where } p = 1, 2, 3. \quad (4)$$

Probabilities for interstices to be occupied by interstitial atoms N_i can be presented as the linear superpositions of the static concentration waves [29]:

$$P_p^{N_i}(\mathbf{R}) = \overline{P}_p^{N_i}(\mathbf{R}) + \delta P_p^{N_i}(\mathbf{R}) = \overline{P}_p^{N_i} + \sum_\tau \sum_{j_\tau} \delta \tilde{P}_p^{N_i}(\mathbf{k}_{j_\tau}) e^{i\mathbf{k}_{j_\tau} \cdot \mathbf{R}}$$

Here,

$$\overline{P}_p^{N_i} = \frac{1}{N_s} \sum_{\mathbf{R}} P_p^{N_i}(\mathbf{R})$$

is the relative concentration of N_i atoms in the p-th interstitial sublattice of two-component interstitial subsystem N_i–$\varnothing$. The amplitude of the plane wave $\exp(i\mathbf{k}_{j_\tau} \cdot \mathbf{R})$ can be represented as

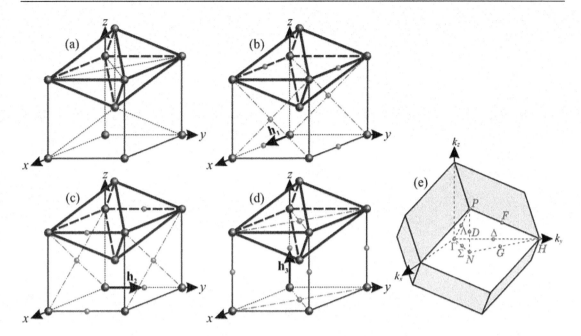

FIGURE 18.2 Unit cell (a) of b.c.c. crystal with sites (larger balls) and octahedral interstices (smaller balls) of the 1-st (b), 2-nd (c), and 3-rd (d) sublattices. The first Brillouin zone (*BZ*) of the b.c.c. crystal (e). The *BZ* boundary contains three high-symmetry points (*H*, *N*, *P*), while the symmetry axes (within the *BZ*) contain six ones (Δ, Σ, Λ, F, G, D). Γ point (**k** = 0) reside in the *BZ* center.

$$\delta \tilde{P}_p^{N_i}(\mathbf{k}_{j_\tau}) = \sum_{\omega=1}^{3} \eta_{\omega\tau}^{N_i} \gamma_{\omega\tau}^{N_i}(\mathbf{k}_{j_\tau}) \psi_{\omega p}(\mathbf{k}_{j_\tau}).$$

The symmetry coefficients $\{\gamma_{\omega\tau}^{N_i}(\mathbf{k}_{j_\tau})\}$ are constant within the T–κ_{N_i} range of thermodynamically stable existence of the phase. A set of these coefficients should be determined such that all linearly independent (non-negative by the definition) long-range order parameters $\{\eta_{\omega\tau}^{N_i}\}$ equal 1 for a totally ordered solution of N_i within the b.c.c.(t.) Fe. Such a superstructural state means a single-phase of a solid solution of a stoichiometric composition ($\kappa_{N_i} = \kappa_{N_i}^{st}$) described by the probabilities $\{P_p^{N_i}(\mathbf{R})\}$ possessing only two values, 0 or 1, and can be realized only at $T = 0$. $\psi_{\omega p}(\mathbf{k}_{j_\tau})$ is the p-th component of the ω-th orthonormalized column vector of "polarization" of the static concentration wave $\psi_{\omega p}(\mathbf{k}_{j_\tau})\exp(i\mathbf{k}_{j_\tau} \cdot \mathbf{R})$ ($\omega = 1, 2, 3$). In such an expansion of $\delta P_p^{N_i}(\mathbf{R})$, there exist only static concentration waves for those "stars" τ of non-zero (τ ≠ γ; see Figure 18.2e) wave vectors {**k**τ} (namely, rays $\{\mathbf{k}_{j_\tau}\}$ of the "star" τ) with non-zero factors $\{\gamma_{\omega\tau}^{N_i}(\mathbf{k}_{j_\tau})\}$ associated with atomic ordering.

Further, we shall consider non-stoichiometric b.c.t.-Fe–N solid solution. Suppose an ordered distribution of N_i atoms over the octahedral interstices [7–8]. Then, for the α″-Fe$_{16}$N$_2$-type superstructure, the second terms in the occupation probabilities (amplitudes of the plane waves) can be represented as follows [30]:

$$\delta P_1^{N_i}(\mathbf{R}) = \frac{1}{24}\left(\eta_{2H}^{N_i} - \eta_{1H}^{N_i}\right)e^{i\mathbf{k}_{1H}\cdot\mathbf{R}} + \frac{1}{16}\left(\eta_{1N}^{N_i} - \eta_{2N}^{N_i}\right)\left(e^{i\mathbf{k}_{3N}\cdot\mathbf{R}} + e^{i\mathbf{k}_{4N}\cdot\mathbf{R}}\right), \tag{5a}$$

$$\delta P_2^{N_i}(\mathbf{R}) = \frac{1}{24}\left(\eta_{2H}^{N_i} - \eta_{1H}^{N_i}\right)e^{i\mathbf{k}_{1H}\cdot\mathbf{R}} + \frac{1}{16}\left(\eta_{1N}^{N_i} - \eta_{2N}^{N_i}\right)\left(e^{i\mathbf{k}_{5N}\cdot\mathbf{R}} + e^{i\mathbf{k}_{6N}\cdot\mathbf{R}}\right), \tag{5b}$$

$$\delta P_3^{N_i}(\mathbf{R}) = \frac{1}{24}\left(2\eta_{1H}^{N_i} + \eta_{2H}^{N_i}\right)e^{i\mathbf{k}_{1H}\cdot\mathbf{R}} + \frac{1}{8}\eta_{3N}^{N_i}\left(e^{i\mathbf{k}_{1N}\cdot\mathbf{R}} + e^{i\mathbf{k}_{2N}\cdot\mathbf{R}}\right) + $$
$$+ \frac{1}{16}\left(\eta_{1N}^{N_i} + \eta_{2N}^{N_i}\right)\left(e^{i\mathbf{k}_{3N}\cdot\mathbf{R}} + e^{i\mathbf{k}_{4N}\cdot\mathbf{R}} + e^{i\mathbf{k}_{5N}\cdot\mathbf{R}} + e^{i\mathbf{k}_{6N}\cdot\mathbf{R}}\right), \quad (5c)$$

where (in the rectangular (Cartesian) coordinate system $Ok_xk_yk_z$) $\mathbf{k}_{1H}$ is a vector of the single-ray "star" $\{\mathbf{k}^H\}$, and $\mathbf{k}_{1N}$, $\mathbf{k}_{2N}$, $\mathbf{k}_{3N}$, $\mathbf{k}_{4N}$, $\mathbf{k}_{5N}$, $\mathbf{k}_{6N}$ are the vectors of the six-rays "star" $\{\mathbf{k}^N\}$ (see Figure 18.2e):

$$\mathbf{k}_{1H} = \frac{2\pi}{a_0}(0;1;0),$$
$$\mathbf{k}_{1N} = \frac{2\pi}{a_0}\left(\frac{1}{2};\frac{1}{2};0\right), \mathbf{k}_{2N} = \frac{2\pi}{a_0}\left(-\frac{1}{2};\frac{1}{2};0\right), \mathbf{k}_{3N} = \frac{2\pi}{a_0}\left(\frac{1}{2};0;\frac{1}{2}\right),$$
$$\mathbf{k}_{4N} = \frac{2\pi}{a_0}\left(-\frac{1}{2};0;\frac{1}{2}\right), \mathbf{k}_{5N} = \frac{2\pi}{a_0}\left(0;\frac{1}{2};\frac{1}{2}\right), \mathbf{k}_{6N} = \frac{2\pi}{a_0}\left(0;-\frac{1}{2};\frac{1}{2}\right).$$

A maximally ordered interstitial subsystem over the entire T-κ_{N_i} range of the existence of non-stoichiometric b.c.t. Fe–N solid solution satisfies the following relations:

$$\forall \mathbf{R}\ P_1^{N_i}(\mathbf{R}) \cong 0,\ P_2^{N_i}(\mathbf{R}) \cong 0,\ P_3^{N_i}(\mathbf{R}) \cong 3\kappa_{N_i}\left(1 + e^{i\mathbf{k}_{1H}\cdot\mathbf{R}} + \sum_{j_N=1}^{6} e^{i\mathbf{k}_{j_N}\cdot\mathbf{R}}\right) \quad (6)$$

Assume that the translational symmetry of N_s-atomic distribution over the available (vacant) b.c.c.(t.) lattice sites coincides with the symmetry of N_i-atomic distribution in the preferable (Figure 18.2d) interstitial sublattice, however, with the mutual arrangement of the firsts with respect to the latter so that the nearest neighborhoods N_s–N_i are absent. The case of the nearest-neighboring N_s–N_i atoms is excluded in order to be in accordance with strong ("blocking") but short-range "direct" "(electro)chemical" interatomic N–N interaction [32]. The occupation-probability function described such an ordered distribution of N_s atoms over the (single) sublattice of sites,

$$P^{N_s}(\mathbf{R}) = \kappa_{N_s} - \frac{1}{8}\eta_H^{N_s} e^{i\mathbf{k}_{1H}\cdot\mathbf{R}} + \frac{1}{8}\eta_N^{N_s}\sum_{j_N=1}^{6}(-1)^{j_N} e^{i\mathbf{k}_{j_N}\cdot\mathbf{R}}, \quad (7)$$

for the state with maximal long-range atomic order at the sites, which is possible at a fixed occupancy degree (relative concentration) κ_{N_s}, will have the following form:

$$P^{N_s}(\mathbf{R}) \cong \kappa_{N_s}\left(1 - e^{i\mathbf{k}_{1H}\cdot\mathbf{R}} + \sum_{j_N=1}^{6}(-1)^{j_N} e^{i\mathbf{k}_{j_N}\cdot\mathbf{R}}\right) \quad (8)$$

We assume that the sites v, which remained unoccupied by substituting atoms of Fe and N, are disorderedly distributed over the b.c.c.(t.) lattice: $P^v(\mathbf{R}) \cong \kappa_v = N_v/N_s$, where N_v is the number of such (residual) monovacancies. In addition, we suppose [33] that practically there are no mixed v–N_i complexes, v–N_s neighborhoods, or complexions of N atoms at the sites and that the number of these defects is small as compared to the number of lattice sites.

The long-range order of Fe and N_s atoms on the sites "inherit" the ordered N_i and $\varnothing$ "atoms" in the interstices of the third sublattice of octahedral interstices (Figure 18.2d). Such a "heritability" can be attributed to repulsive N_i–Fe and N_i–N_i interactions between the interstitial and substitutional subsystems rather than attractive (Fe–Fe) and repulsive (Fe–N_s, N_s–N_s) interactions in the substitutional Fe–N_s subsystem.

Taking into account Equations (1), (3), (4), (6), (8), we can easily obtain the expression of specific (i.e., with respect to the total number N_{Fe} of Fe ions) Gibbs free energy g for such an ordered phase of the b.c.t. Fe–N solution:

$$g \cong g_\varnothing + R^{N_s} c_{N_s} + R^v c_v + R_3^{N_i} c_{N_i} +$$

$$+ (1 + c_{N_s} + c_v)^{-1} \left\{ \frac{1}{2} \tilde{W}^{FeFe}(0) + \tilde{W}^{FeN_s}(0) c_{N_s} + \tilde{W}^{Fev}(0) c_v + \right.$$

$$+ \tilde{W}^{FeN_i}_3(0) c_{N_i} + \tilde{W}^{N_s v}(0) c_{N_s} c_v + \tilde{W}^{vN_i}_3(0) c_{N_i} c_v +$$

$$+ \left[\tilde{W}^{N_s N_i}_3(0) - \tilde{W}^{N_s N_i}_3(\mathbf{k}_{1_H}) + \tilde{W}^{FeN_i}_3(\mathbf{k}_{1_H}) + \sum_{j_N = 1}^{6} (-1)^{j_N} \left(\tilde{W}^{N_s N_i}_3(\mathbf{k}_{j_N}) - \tilde{W}^{FeN_i}_3(\mathbf{k}_{j_N}) \right) \right] c_{N_i} c_{N_s} +$$

$$+ \frac{1}{2} \left(\tilde{W}^{N_s N_s}(0) + \tilde{w}^{FeN_s}(\mathbf{k}_{1_H}) + 6 \tilde{w}^{FeN_s}(\mathbf{k}_{1_N}) \right) c_{N_s}^2 + \frac{1}{2} \tilde{W}^{vv}(0) c_v^2 +$$

$$+ \frac{1}{2} \left(\tilde{W}^{N_i N_i}_{33}(0) + \tilde{W}^{N_i N_i}_{33}(\mathbf{k}_{1_H}) + 4 \tilde{W}^{N_i N_i}_{11}(\mathbf{k}_{1_N}) + 2 \tilde{W}^{N_i N_i}_{33}(\mathbf{k}_{1_N}) \right) c_{N_i}^2 \right\} +$$

$$+ \frac{k_B T}{8} \left\{ 7(1 + c_{N_s}) \ln(1 + c_{N_s}) + (1 - 7 c_{N_s}) \ln(1 - 7 c_{N_s}) + \right.$$

$$+ 8 c_{N_s} \ln(8 c_{N_s}) + 8 c_v \ln(8 c_v) + 8 c_{N_i} \ln(8 c_{N_i}) + (1 + c_{N_s} + c_v - 8 c_{N_i}) \ln(1 + c_{N_s} + c_v - 8 c_{N_i}) -$$

$$- 9(1 + c_{N_s} + c_v) \ln(1 + c_{N_s} + c_v) \right\}, \tag{9}$$

where $g_\varnothing$ is a non-configurational part of g, $\tilde{W}^{\alpha\beta}(\mathbf{k})$, $\tilde{W}^{N_i N_i}_{pp}(\mathbf{k})$, and $\tilde{W}^{\alpha N_i}_p(\mathbf{k})$ are k-th ($\mathbf{k} = 0$, $\mathbf{k} = \mathbf{k}_{1_H}$, $\mathbf{k} = \mathbf{k}_{j_N}$) Fourier components of the effective pair interaction energies of atoms of corresponding constituents in the b.c.t. Fe–N solution, $\tilde{w}^{FeN_s}(\mathbf{k}_{1_H})$ and $\tilde{w}^{FeN_s}(\mathbf{k}_{1_N})$ are the Fourier components (in the $\mathbf{k}_{1_H}$ and $\mathbf{k}_{1_N}$ points of the 1-st Brillouin zone of b.c.c. Fe) of the mixing (also known as the interchange) energies

$$w^{FeN_s}(\mathbf{R} - \mathbf{R}') = W^{FeFe}(\mathbf{R} - \mathbf{R}') + W^{N_s N_s}(\mathbf{R} - \mathbf{R}') - 2 W^{FeN_s}(\mathbf{R} - \mathbf{R}').$$

The values

$$c_v = \frac{N_v}{N_{Fe}}, \quad c_{N_i} = \frac{N_{N_i}}{N_{Fe}}, \quad c_{N_s} = \frac{N_{N_s}}{N_{Fe}}$$

denote the relative concentrations of the residual vacancies v, and N atoms intruded into octahedral interstices and occupying the (smaller) part of sites in b.c.t. Fe, respectively. According to Equation (2),

$$c_{N_i} + c_{N_s} = c_N = \frac{N_N}{N_{Fe}}.$$

In Equation (9), we omitted the contributions of "soft" (Coulomb-type) α–$\varnothing$ interactions as negligibly small. The magnetic contribution (mainly due to the exchange interaction of Fe atoms) is implicitly contained in the mixing energies if temperature T is below the Curie point $T_C = T_C(c_{Ni}, c_{Ns} + c_v)$.

18.2.2 Tetragonality of Fe-N Crystal Lattice

Thermodynamically equilibrium values of two independent variables c_v and c_{N_i} can be obtained from the thermodynamic equilibrium conditions:

$$\frac{\partial g}{\partial c_v} = 0 \Rightarrow c_v\big|_T = c_v(c_N) \text{ and } \frac{\partial g}{\partial c_{N_i}} = 0 \Rightarrow c_{N_i}\big|_{c_N} = c_{N_i}(T).$$

Taking into account $c_{N_s} + c_{N_i} = c_N$, they result in the set of equations:

$$-R^v - \frac{1}{2}\tilde{W}^{vv}(\mathbf{0}) + (1 + c_{N_s} + c_v)^{-2}\left\{\frac{1}{2}\tilde{w}^{Fev}(\mathbf{0}) + \right.$$
$$+ \left(\tilde{W}^{FeN_s}(\mathbf{0}) + \tilde{W}^{vv}(\mathbf{0}) - \tilde{W}^{N_s v}(\mathbf{0}) - \tilde{W}^{Fev}(\mathbf{0})\right)c_{N_s} + \left(\tilde{W}_3^{FeN_i}(\mathbf{0}) - \tilde{W}_3^{vN_i}(\mathbf{0})\right)c_{N_i} +$$
$$+ \left[\tilde{W}_3^{N_s N_i}(\mathbf{0}) - \tilde{W}_3^{vN_i}(\mathbf{0})c_{N_s} - \tilde{W}_3^{N_s N_i}(\mathbf{k}_{1_H}) + \tilde{W}_3^{FeN_i}(\mathbf{k}_{1_H}) + \right.$$
$$+ \sum_{j_N=1}^{6}(-1)^{j_N}\left(\tilde{W}_3^{N_s N_i}(\mathbf{k}_{j_N}) - \tilde{W}_3^{FeN_i}(\mathbf{k}_{j_N})\right)\bigg]c_{N_s}c_{N_i} + \frac{1}{2}\left(\tilde{w}^{N_s v}(\mathbf{0}) + \tilde{w}^{FeN_s}(\mathbf{k}_{1_H}) + 6\tilde{w}^{FeN_s}(\mathbf{k}_{1_N})\right)c_{N_s}^2 +$$
$$\left. + \frac{1}{2}\left(\tilde{W}_{33}^{N_i N_i}(\mathbf{0}) + \tilde{W}_{33}^{N_i N_i}(\mathbf{k}_{1_H}) + 4\tilde{W}_{11}^{N_i N_i}(\mathbf{k}_{1_N}) + 2\tilde{W}_{33}^{N_i N_i}(\mathbf{k}_{1_N})\right)c_{N_i}^2\right\} \approx$$
$$\approx k_B T \ln\left\{c_v \sqrt[8]{\frac{1 + c_{N_s} + c_v - 8c_{N_i}}{(1 + c_{N_s} + c_v)^9}}\right\} \tag{10}$$

and

$$R^{N_i} - R^{N_s} - \frac{1}{2}\tilde{W}^{N_s N_s}(\mathbf{0}) + (1 + c_{N_s} + c_v)^{-1}\left\{\tilde{W}_3^{FeN_i}(\mathbf{0}) + \right.$$
$$+ \left[\tilde{W}_3^{N_s N_i}(\mathbf{0}) - \tilde{W}_3^{N_s N_i}(\mathbf{k}_{1_H}) + \tilde{W}_3^{FeN_i}(\mathbf{k}_{1_H}) + \right.$$
$$+ \sum_{j_N=1}^{6}(-1)^{j_N}\left(\tilde{W}_3^{N_s N_i}(\mathbf{k}_{j_N}) - \tilde{W}_3^{FeN_i}(\mathbf{k}_{j_N})\right)\bigg]c_{N_s} + \tilde{W}_3^{vN_i}(\mathbf{0})c_v +$$
$$+ \left(\tilde{W}_{33}^{N_i N_i}(\mathbf{0}) + \tilde{W}_{33}^{N_i N_i}(\mathbf{k}_{1_H}) + 4\tilde{W}_{11}^{N_i N_i}(\mathbf{k}_{1_N}) + 2\tilde{W}_{33}^{N_i N_i}(\mathbf{k}_{1_N})\right)c_{N_i}\right\} +$$
$$+ (1 + c_{N_s} + c_v)^{-2}\left\{\frac{1}{2}\tilde{w}^{FeN_s}(\mathbf{0}) + \left(\tilde{W}^{Fev}(\mathbf{0}) + \tilde{W}^{vv}(\mathbf{0}) - \tilde{W}^{FeN_s}(\mathbf{0}) - \tilde{W}^{N_s v}(\mathbf{0})\right)c_v + \right.$$
$$+ \tilde{W}_3^{FeN_i}(\mathbf{0})c_{N_i} - \frac{1}{2}\left(\tilde{w}^{FeN_s}(\mathbf{k}_{1_H}) + 6\tilde{w}^{FeN_s}(\mathbf{k}_{1_N})\right)\left(2c_{N_s}(1 + c_v) + c_{N_s}^2\right) -$$
$$- \left[\tilde{W}_3^{N_s N_i}(\mathbf{0}) - \tilde{W}_3^{N_s N_i}(\mathbf{k}_{1_H}) + \tilde{W}_3^{FeN_i}(\mathbf{k}_{1_H}) + \sum_{j_N=1}^{6}(-1)^{j_N}\left(\tilde{W}_3^{N_s N_i}(\mathbf{k}_{j_N}) - \tilde{W}_3^{FeN_i}(\mathbf{k}_{j_N})\right)\right]c_{N_i}(1 + c_v) +$$
$$\left. + \tilde{W}_3^{vN_i}(\mathbf{0})c_{N_i}c_v + \frac{1}{2}\tilde{w}^{N_s v}(\mathbf{0})c_v^2 + \frac{1}{2}\left(\tilde{W}_{33}^{N_i N_i}(\mathbf{0}) + \tilde{W}_{33}^{N_i N_i}(\mathbf{k}_{1_H}) + 4\tilde{W}_{11}^{N_i N_i}(\mathbf{k}_{1_N}) + 2\tilde{W}_{33}^{N_i N_i}(\mathbf{k}_{1_N})\right)c_{N_i}^2\right\} \approx$$
$$\approx k_B T \ln\left\{\frac{c_{N_s}}{c_{N_i}}\sqrt[8]{\frac{(1+c_{N_s})^7(1+c_{N_s}+c_v-8c_{N_i})^9}{(1-7c_{N_s})^7(1+c_{N_s}+c_v)^9}}\right\}, \tag{11}$$

where

$$\tilde{w}^{Fev}(\mathbf{0}) = \tilde{W}^{FeFe}(\mathbf{0}) + \tilde{W}^{vv}(\mathbf{0}) - 2\tilde{W}^{Fev}(\mathbf{0}) \equiv \tilde{w}(\mathbf{0}),$$
$$\tilde{w}^{N_s v}(\mathbf{0}) = \tilde{W}^{N_s N_s}(\mathbf{0}) + \tilde{W}^{vv}(\mathbf{0}) - 2\tilde{W}^{N_s v}(\mathbf{0}),$$
$$\tilde{w}^{FeN_s}(\mathbf{0}) = \tilde{W}^{FeFe}(\mathbf{0}) + \tilde{W}^{N_s N_s}(\mathbf{0}) - \tilde{W}^{FeN_s}(\mathbf{0}).$$

In the first simplest approximation, $0 < c_v = (1 + c_{N_s} - 8c_{N_i})$, and if in Equation (11) we neglect the terms containing the factor c_v or all the more c_v^2, the set of Equations (10) and (11) reduces to two equations. Equation (11) now does not contain c_v explicitly, and it serves to determine the equilibrium values of c_{N_i} and consequently $c_{N_s} = c_N - c_{N_i}$ (for the given T and c_N):

$$R^{N_i} - R^{N_s} - \frac{1}{2}\tilde{W}^{N_s N_s}(\mathbf{0}) + (1 + c_{N_s})^{-1}\left\{\tilde{W}_3^{FeN_i}(\mathbf{0}) + \right.$$
$$+ \left[\tilde{W}_3^{N_s N_i}(\mathbf{0}) - \tilde{W}_3^{N_s N_i}(\mathbf{k}_{1_H}) + \tilde{W}_3^{FeN_i}(\mathbf{k}_{1_H}) + \sum_{j_N=1}^{6}(-1)^{j_N}\left(\tilde{W}_3^{N_s N_i}(\mathbf{k}_{j_N}) - \tilde{W}_3^{FeN_i}(\mathbf{k}_{j_N})\right)\right]c_{N_s} +$$
$$+ \left(\tilde{W}_{33}^{N_i N_i}(\mathbf{0}) + \tilde{W}_{33}^{N_i N_i}(\mathbf{k}_{1_H}) + 4\tilde{W}_{11}^{N_i N_i}(\mathbf{k}_{1_N}) + 2\tilde{W}_{33}^{N_i N_i}(\mathbf{k}_{1_N})\right)c_{N_i}\right\} +$$
$$+ (1 + c_{N_s})^{-2}\left\{\frac{1}{2}\tilde{w}^{FeN_s}(\mathbf{0}) - \frac{1}{2}\left(\tilde{w}^{FeN_s}(\mathbf{k}_{1_H}) + 6\tilde{w}^{FeN_s}(\mathbf{k}_{1_N})\right)(2 + c_{N_s})c_{N_s} - \right.$$
$$- \left[\tilde{W}_3^{N_s N_i}(\mathbf{0}) - \tilde{W}_3^{FeN_i}(\mathbf{0}) - \tilde{W}_3^{N_s N_i}(\mathbf{k}_{1_H}) + \tilde{W}_3^{FeN_i}(\mathbf{k}_{1_H}) + \sum_{j_N=1}^{6}(-1)^{j_N}\left(\tilde{W}_3^{N_s N_i}(\mathbf{k}_{j_N}) - \tilde{W}_3^{FeN_i}(\mathbf{k}_{j_N})\right)\right]c_{N_i} +$$
$$+ \frac{1}{2}\left(\tilde{W}_{33}^{N_i N_i}(\mathbf{0}) + \tilde{W}_{33}^{N_i N_i}(\mathbf{k}_{1_H}) + 4\tilde{W}_{11}^{N_i N_i}(\mathbf{k}_{1_N}) + 2\tilde{W}_{33}^{N_i N_i}(\mathbf{k}_{1_N})\right)c_{N_i}^2\right\} \cong$$
$$\cong k_B T \ln\left\{\frac{c_{N_s}}{c_{N_i}}\sqrt[8]{\frac{(1 + c_{N_s} - 8c_{N_i})^9}{(1 - 7c_{N_s})^7(1 + c_{N_s})^2}}\right\}. \tag{12}$$

Equation (10) transforms into expression from which we can determine c_v at given $c_{N_i}(T, c_N)$ and $c_{N_s}(T, c_N)$:

$$c_v \cong c_v^{0'}\sqrt[8]{\frac{1 + c_{N_s} - 8c_{N_i}}{(1 + c_{N_s})^9}}\exp\left(-\frac{2R^v + \tilde{W}^{vv}(\mathbf{0})}{2k_B T}\right)\times$$
$$\times \exp\left(\frac{1}{2k_B T}(1 + c_{N_s})^{-2}\left[2\left(\tilde{W}^{FeN_s}(\mathbf{0}) + \tilde{W}^{vv}(\mathbf{0}) - \tilde{W}^{N_s v}(\mathbf{0}) - \tilde{W}^{Fev}(\mathbf{0})\right)c_{N_s} + \right.\right.$$
$$+ \left(\tilde{W}_3^{FeN_i}(\mathbf{0}) - \tilde{W}_3^{vN_i}(\mathbf{0})\right)c_{N_i} + \left\{\tilde{W}_3^{N_s N_i}(\mathbf{0}) - \tilde{W}_3^{vN_i}(\mathbf{0})c_{N_i} - \tilde{W}_3^{N_s N_i}(\mathbf{k}_{1_H}) + \tilde{W}_3^{FeN_i}(\mathbf{k}_{1_H}) + \right.$$
$$+ \sum_{j_N=1}^{6}(-1)^{j_N}\left(\tilde{W}_3^{N_s N_i}(\mathbf{k}_{j_N}) - \tilde{W}_3^{FeN_i}(\mathbf{k}_{j_N})\right)\right\}c_{N_s}c_{N_i} + \frac{1}{2}\left(\tilde{w}^{N_s v}(\mathbf{0}) + \tilde{w}^{FeN_s}(\mathbf{k}_{1_H}) + 6\tilde{w}^{FeN_s}(\mathbf{k}_{1_N})\right)c_{N_s}^2 +$$
$$+ \frac{1}{2}\left(\tilde{W}_{33}^{N_i N_i}(\mathbf{0}) + \tilde{W}_{33}^{N_i N_i}(\mathbf{k}_{1_H}) + 4\tilde{W}_{11}^{N_i N_i}(\mathbf{k}_{1_N}) + 2\tilde{W}_{33}^{N_i N_i}(\mathbf{k}_{1_N})\right)c_{N_i}^2\right]\right). \tag{13}$$

where the value

$$c_v^{0'} = \exp\left(-\frac{\tilde{w}^{Fev}(\mathbf{0})}{2k_B T(1+c_{N_s})^2}\right)$$

equals to the relative concentration c_v^0 ($=1$) of thermally activated vacancies in the b.c.c. Fe only in the hypothetical case of $c_{N_s} \equiv 0$.

The degree of tetragonality of a (weak) b.c.t. Fe–N solution (isostructural to iron–nitrogen martensite) is determined by the ratio of the main geometrical parameters $a(c_N)$ and $b(c_N)$ of the conventional lattice cell. In its maximally ordered state, it is approximately proportional to the concentration of N atoms accumulated on only one of three sublattices of octahedral interstices [29]:

$$\frac{b}{a} \cong 1 + (L_3^{N_i} - L_1^{N_i})c_{N_s}, \tag{14}$$

where

$$L_3^{N_i} \equiv L_{3ij}^{N_i} = L_{3zz}^{N_i} = L_{2yy}^{N_i} = L_{1xx}^{N_i} \cong \frac{1}{a_0}\frac{\partial b(c_N)}{\partial c_N}\bigg|_{c_N=0}$$

and

$$L_1^{N_i} \equiv L_{3xx}^{N_i} = L_{3yy}^{N_i} = L_{1yy}^{N_i} = L_{1zz}^{N_i} = L_{2zz}^{N_i} = L_{2xx}^{N_i} \cong \frac{1}{a_0}\frac{\partial a(c_N)}{\partial c_N}\bigg|_{c_N=0}$$

are both different values of non-zero (diagonal) elements of the tensor of concentration distortion coefficients $L_{3ij}^{N_i}$ ($i, j = x, y, z$) of b.c.c. Fe with interstitial N atoms in only one, the third (with $p = 3$) sublattice of the octahedral interstices (assuming the Vegard's rule, i.e., practically linear dependences of a and b on c_N at fixed T). Thus, Equation (12) will also determine the temperature dependence of the ratio b/a.

Thus, the spatial distribution of (nitrogen) atoms in the martensitic (b.c.t. iron–nitrogen) phase obviously affects its tetragonality degree. To analyze this effect adequately, we need to possess detailed information on the parameters of interionic interactions in the (b.c.c. Fe–N) solid solution.

18.2.3 Interatomic Interaction Energy Parameters for the Paramagnetic State

The effective pair interatomic interaction energy $W_{pp'}^{N_iN_i}(\mathbf{R}-\mathbf{R}')$ (and, therefore, its Fourier component $\tilde{W}_{pp'}^{N_iN_i}(\mathbf{k})$) includes the long-range strain-induced and short-range "(electro)chemical" contributions, $V_{pp'}^{N_iN_i}(\mathbf{R}-\mathbf{R}')$ and $\varphi_{pp'}^{N_iN_i}(\mathbf{R}-\mathbf{R}')$, respectively [32].

To calculate the Fourier components of the strain-induced interaction energies $\tilde{V}_{pp'}^{N_iN_i}(\mathbf{k})$ using the so-called lattice statics method [29], we used a quasi-harmonic model of the b.c.c.-crystal dynamics [29], taking into account dominating (in α-Fe [34]) interactions of metal ions within the first two coordination site spheres (central forces in both spheres and non-central ones in the first sphere only). The elasticity moduli values C_{11}, C_{12}, and C_{44} for b.c.c. Fe (with a corresponding lattice parameter $a_0 = a_0(T)$ [35]) depends on a temperature [36,37]; $L_3^{N_i} \cong 0.84$ and $L_1^{N_i} \cong -0.05$ at $T = 298$ K [38]. Assume that practically all interstitial N atoms still reside in one of the three sublattices of the b.c.t. Fe octahedral interstices (inheriting their mutual arrangement in homogeneous austenite). Other adapted parameters for b.c.c. Fe with vacancies (at $T = 298$ K): $C_{11} = 273$ GPa, $C_{12} = 150$ GPa, $C_{44} = 106.73$ GPa (i.e., with $\xi < 0$) [39], $a_0 = 2.8663$ Å [40], and $L^v = -0.016$ [41], where a_0 is a b.c.c.-lattice translation period, C_{11}, C_{12}, and C_{44} are the elastic moduli, $\xi \equiv (C_{11} = C_{12} - 2C_{44})/C_{44}$ is an elastic anisotropy factor, and L^v is a concentration dilatation coefficient of the b.c.c. Fe lattice due to the presence of vacancies.

To estimate roughly the "(electro)chemical" interactions N–N (i.e., N_i–N_i, N_s–N_s, and N_s–N_i) inside the b.c.c.(t.) Fe, we use a reasonable model [32]. The N ions at the octahedral interstices and/or sites of the b.c.t. Fe interact with each other in almost the same way as N atoms in different molecules containing these atoms (e.g., with the atom-atom N–N-potential for N atoms of the N_2 molecules). The potential of such a (direct) interaction of non-point N ions can be chosen as the Lennard-Jones potential with adopted parameters [42].

To calculate the Fourier components $\tilde{V}^{vv}(\mathbf{k}) = \tilde{W}^{vv}(\mathbf{k}) - \tilde{\varphi}^{vv}_{\text{el.chem}}(\mathbf{k})$ and $\tilde{V}^{vN_i}_{p'}(\mathbf{k})$ of the strain-induced interaction energies v–v and v–N_i ($V^{vv}(\mathbf{R}-\mathbf{R}')$ and $V^{vN_i}_{p'}(\mathbf{R}-\mathbf{R}')$, respectively), one can use theorem [43] for the potentials of "(electro)chemical" interactions of metal atoms (Me–Me) and vacancies (v–v): $\varphi^{MeMe}(r) = \varphi^{vv}_{\text{el.chem}}(r)$. The pair potential for the total pair interaction of metal atoms in b.c.c. Me crystal is [29]

$$W^{MeMe}(r) \cong \varphi^{MeMe}(r) \approx -\frac{A^{MeMe}}{r^4} + \frac{B^{MeMe}}{r^8}, \qquad (15)$$

where $r = |\mathbf{R} - \mathbf{R}'|$,

$$A^{MeMe} = -\frac{9\varepsilon^0}{4S_4}a_0^4(0\text{ K}) \cong -0.09939\varepsilon^0 a_0^4(0\text{ K})\ (S_4 \approx 22.63872),$$

$$B^{MeMe} = -\frac{81\varepsilon^0}{128S_8}a_0^8(0\text{ K}) \cong -0.0611\varepsilon^0 a_0^8(0\text{ K})\ (S_8 \approx 10.3552),$$

and ε^0 is the cohesive binding energy; for α-Fe, $\varepsilon^0 \cong -4.28$ eV/atom [44,45].

Thus, for example,

$$\tilde{w}^{Mev}(\mathbf{k}) = \tilde{W}^{MeMe}(\mathbf{k}) + \tilde{W}^{vv}(\mathbf{k}) - 2\tilde{W}^{Mev}(\mathbf{k}) \equiv \tilde{w}(\mathbf{k}) \cong$$

$$\cong \tilde{V}^{vv}(\mathbf{k}) + \tilde{\varphi}^{mix}_{\text{el.chem}}(\mathbf{k}) = \tilde{V}^{vv}(\mathbf{k}) + \tilde{\varphi}^{MeMe}(\mathbf{k}) + \tilde{\varphi}^{vv}_{\text{el.chem}}(\mathbf{k}) - 2\tilde{\varphi}^{Mev}(\mathbf{k}).$$

Further, we neglect the "(electro)chemical" (mainly Coulomb-type) contributions to the v–N_i and v–N_s interactions screened due to the presence of correlations between the distributed charges.

Using Equation (15), we can estimate the Fourier components $\tilde{W}^{FeFe}(\mathbf{k})$ and $\tilde{W}^{Fev}(\mathbf{k}) \cong -\tilde{W}^{FeFe}(\mathbf{k}) \cong -\tilde{\varphi}^{MeMe}(\mathbf{k}) = -\tilde{\varphi}^{vv}_{\text{el.chem}}(\mathbf{k})$ based on the model assumptions of Girifalco–Weizer [46] and Yamamoto–Doyama [47].

To evaluate the parameters of "(electro)chemical" Fe–N_s and Fe–N_i interactions, we assume that these interactions between N and Fe ions at a distance r are pairwise and centrally symmetric. Then, the interaction energy can be approximated as

$$\varphi^{FeN}(r) \cong \begin{cases} A^{FeN}e^{-d_{FeN}r}, & \text{if } r_I^0 = a_0/2 \le r \le r_{III}^0 = a_0\sqrt{5}/2; \\ 0, & \text{if } r > r_{III}^0. \end{cases} \qquad (16)$$

The conditions of a static mechanical equilibrium for the b.c.c.(t.)-lattice-based interstitial solution [29],

$$-\sum_{\mathbf{R}} \frac{\partial \varphi^{FeN_i}(r)}{\partial r} \frac{(R_x - h_{3x})^2}{r} \approx \frac{a_0^3}{4}\left[(C_{11} + C_{12})L_1^{N_i} + C_{12}L_3^{N_i}\right] \qquad (17)$$

$$-\sum_{\mathbf{R}} \frac{\partial \varphi^{FeN_i}(r)}{\partial r} \frac{(R_z - h_{3z})^2}{r} \approx \frac{a_0^3}{4}\left(C_{11}L_3^{N_i} + 2C_{12}L_1^{N_i}\right) \qquad (18)$$

and Equation (16) along with an assumption of the constraint N–Fe-interaction radius, we can find numerical values of the unified (for both N_i and N_s) fitting parameters A^{FeN} and d_{FeN} of function $\varphi^{FeN}(r)$. The Fe–N interaction potential parameters, obtained in this way, substantially depend on T due to the temperature-dependent experimentally determined C_{11}, C_{12}, and a_0.

The Fourier components $\tilde{W}^{FeN_s}(\mathbf{k})$ and $\tilde{W}_3^{FeN_i}(\mathbf{k})$ of the interaction energies $W^{FeN_s}(\mathbf{R}-\mathbf{R}') \cong \varphi^{FeN_s}(|\mathbf{R}-\mathbf{R}'|)$ and $\tilde{W}_3^{FeN_i}(\mathbf{R}-\mathbf{R}') \cong \varphi^{FeN_i}(|\mathbf{R}-\mathbf{R}'-\mathbf{h}_3|)$ can be calculated from Equation (16) within the assumption that the Fe–N_i and Fe–N_s interaction potentials are identical with $\varphi^{FeN}(r)$. In addition, based on the obtained values of the Fe–N interaction potential parameters, one can estimate the effective force with which the N atom substituting the b.c.c.(t.) Fe lattice site would act on the Fe ion occupying some other neighboring site. It means [28–29] that we possess a possibility to calculate the Fourier components $\tilde{V}^{N_sN_s}(\mathbf{k})$ and $\tilde{V}^{N_iN_i}(\mathbf{k})$ of the strain-induced N_s–N_s and N_i–N_i interaction energies.

The specific energy associated with the "expansion" ("compression") of the "impurity ion" $\alpha = v$, N_s (N_i) introduced into any site (octahedral interstice) of the b.c.c. Fe lattice is [28–29]

$$R^v \cong -\frac{1}{2}\tilde{V}^{vv}(\mathbf{0}) \text{ or } R^{N_s} \cong -\frac{1}{2}\tilde{V}^{N_sN_s}(\mathbf{0}) \ (R_3^{N_i} \cong -\frac{1}{2}\tilde{V}_{33}^{N_iN_i}(\mathbf{0}))$$

where $\tilde{V}^{vv}(\mathbf{0})$, $\tilde{V}^{N_sN_s}(\mathbf{0})$ (and $\tilde{V}_{33}^{N_iN_i}(\mathbf{0})$) are the Fourier components (at $\mathbf{k} = \mathbf{0}$) of the respective strain-induced interaction energies.

Thus, following the presented methodology, we can estimate all interaction energy parameters, which are contained in Equations (12) and (13). The curves in Figure 18.3 show the temperature dependences of the estimated values $\tilde{W}^{FeN_s}(\mathbf{0})$ and $\tilde{W}_3^{FeN_i}(\mathbf{0})$. These Fourier components reflect the total interaction energy of the selected nitrogen atom, which either substitutes a site, $\tilde{W}^{FeN_s}(\mathbf{0})$, or resides in the octahedral interstice, $\tilde{W}_3^{FeN_i}(\mathbf{0})$, of the b.c.t. Fe with all other Fe ions over the lattice sites. As the figure demonstrates, in note, these interaction energy parameters $\tilde{W}^{FeN_s}(\mathbf{0})$ and $\tilde{W}_3^{FeN_i}(\mathbf{0})$ increase as the temperature rises.

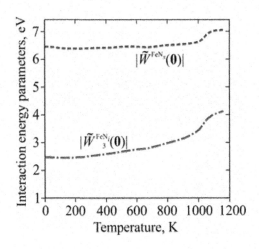

FIGURE 18.3 Absolute values of the Fourier components $\tilde{W}^{FeN_s}(\mathbf{0})$ and $\tilde{W}_3^{FeN_i}(\mathbf{0})$ of the interaction energies $W^{FeN_s}(\mathbf{R}-\mathbf{R}') \cong \varphi^{FeN_s}(|\mathbf{R}-\mathbf{R}'|)$ and $\tilde{W}_3^{FeN_i}(\mathbf{R}-\mathbf{R}') \cong \varphi^{FeN_i}(|\mathbf{R}-\mathbf{R}'-\mathbf{h}_3|)$, respectively, as the functions of temperature T.

18.3 REDISTRIBUTION OF IMPURITIES BETWEEN THE INTERSTITIAL AND SUBSTITUTIONAL POSITIONS

Here, firstly, we present the results of the calculation of the temperature-dependent concentration of that impurity (nitrogen) atoms residing in the interstitial (or substitutional) positions. Then, secondly, we show how such an atomic redistribution (from the interstitial to the substitutional sites) affects the vacancy content in the host lattice.

18.3.1 Change in the Interstitial Impurity Concentration

Since both the elasticity moduli $C_{IJ}(T)$ and lattice parameter $a_0(T)$ of the b.c.c. Fe are temperature-dependent, solving Equation (12) we take into account the temperature dependences of energy parameters, which enter into this equation. Figure 18.4 shows the $c_{N_i}\big|_{c_N} = c_{N_i}(T)$ curves at two different values of the total N-atomic content c_N. As seen from Figure 18.4, even for $T \to 0$ K ($\ll T_C$ – Courier temperature), part of the N atoms substitutes the b.c.t. Fe lattice sites in the equilibrium: $c_{N_i}\big|_{c_N=10\%}^{T\to 0\,K} \cong 9\%$ and $c_{N_i}\big|_{c_N=12\%}^{T\to 0\,K} \cong 9.9\%$. It means that, for the nitrogen concentration $c_N \in [10\%; 12\%]$ (i.e., close to the α''-Fe$_{16}$N$_2$ content), $\cong 82.5\%$–90% ($\cong 10\%$–17.5%) of N atoms may occupy the octahedral interstices (sites) at $T \to 0$ K. Further increase of the temperature leads to c_{N_i} decrease so that the "high-T" c_{N_i} value can be by more than 10% less than its "low-T" value. In part, $c_{N_i}\big|_{c_N=10\%}^{T\approx 1000\,K} \cong 8\%$ and $c_{N_i}\big|_{c_N=12\%}^{T\approx 1000\,K} \cong 9\%$, i.e., $\cong 75\%$–80% ($\cong 20\%$–25%) of N atoms can reside in the interstices (substitutional sites).

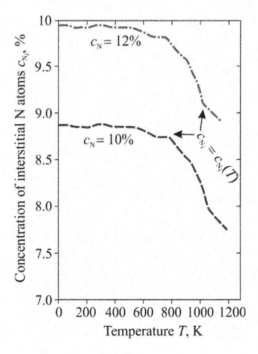

FIGURE 18.4 Temperature-dependent relative concentration c_{N_i} of nitrogen atoms, which remain to be interstitial occupying only one (the 3-rd in Figure 2d) sublattice of the octahedral interstices in the host b.c.t. Fe lattice, for two values of the total content of N atoms: c_N = 10% and 12%.

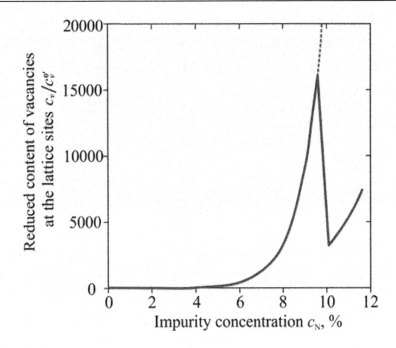

FIGURE 18.5 The ratio $c_v/c_v^{0'}$ (solid curve) as a function of the total concentration c_N of impurity nitrogen atoms (at $T = 1173$ K). Dashed line shows the hypothetical case $c_{N_s} \equiv 0$.

Thus (see Equation (14)), with increasing the temperature, the degree of tetragonality of the b.c.t. Fe–N solution can appreciably decrease. This is because of the redistribution of impurity nitrogen atoms between the 3-rd (Figure 18.2d) sublattice of octahedral interstices of b.c.t. Fe lattice and its vacant sites. Within the temperature range 4.2 K < T < 300 K, the function $c_{N_i} = c_{N_i}(T)$ can behave non-monotonically. This can be attributed to the non-monotonic temperature dependence of the energy parameters. However, these non-monotony effects are small and therefore may be insensible in experimental measurements of the tetragonality value for the Fe–N martensite within the $T \in (4.2$ K; 300 K).

18.3.2 Change in the Vacancy Content at the Host-Lattice Sites

As seen from Figure 18.5 (see also approximate expression (13)), the content of vacancies over the Fe–N lattice sites can substantially increase (even at a constant temperature) as the total concentration of interstitial N atoms increases. A huge kink in the solid curve ($c_v/c_v^{0'}$) in Figure 18.5 corresponds approximately to those values of the relative concentration c_N, when a spontaneous transition of part of the N atoms from octahedral interstices to b.c.t. Fe sites occurs (at 1173 K). Obviously, in this case, the N atoms occupy part of the vacant b.c.t. lattice sites and thus rapidly decrease the quantity of residual vacancies v in the solid solution. Assumption of $c_{N_s} \equiv 0$, which is when N atoms do not move from the interstices to the sites, results to the dashed line in Figure 18.5. This line exhibits a hypothetical dependence $c_v/c_v^{0'}$ on c_N. We can clearly see from comparison of both (solid and dash) lines in Figure 18.5 that most of the nitrogen atoms trend toward a movement to the b.c.t. iron lattice sites starting from a certain concentration c_N of interstitial atoms (and a little bit higher).

18.4 EXPLICIT MAGNETIC CONTRIBUTION INTO THE GIBBS FREE ENERGY

The presence of the ferromagnetism in the Fe–N system makes a theoretical analysis of this system more complicated. Fe–N alloy consists of only one magnetic component, Fe, with uncompensated spin s_{Fe} per one Fe atom. Within the framework of the Heisenberg model in the self-consistent "molecular"-field approximation [29] (similarly to [5,48,49]), the magnetic part of model configurational Hamiltonian of the Fe subsystem averaged over spin variables below the Curie point $T_C = T_C(c_{Ni}, c_{Ns} + c_v)$ is as follows:

$$H_{mag} \cong \frac{1}{2} \sum_{R,R'} J_{FeFe}(R - R') s_{Fe}^2 \sigma_{Fe}^2 c_{Fe}(R) c_{Fe}(R'), \qquad (19)$$

where $J_{FeFe}(R - R')$ is an exchange interaction parameter (exchange integral) for two Fe atoms at the sites R and R' and σ_{Fe} is a relative magnetization (per one atom) of the Fe subsystem.

The total model configurational Hamiltonian of the site Fe subsystem only includes both the paramagnetic interchange and magnetic contributions:

$$H_{tot} = \frac{1}{2} \sum_{R,R'} W_{tot}^{FeFe}(R - R') c_{Fe}(R) c_{Fe}(R'), \qquad (20)$$

where

$$W_{tot}^{FeFe}(R - R') = W_{prm}^{FeFe}(R - R') + J_{FeFe}(R - R') s_{Fe}^2 \sigma_{Fe}^2 \qquad (21)$$

is an effective total Fe–Fe interaction energy including exchange interactions of spins, $W_{prm}^{FeFe}(R - R')$ is Fe–Fe interaction energy in the paramagnetic state.

As follows from the last equations, the account of the exchange interactions of Fe spins in expression for the internal energy of alloy results, first of all, in the substitutions

$$\tilde{W}_{prm}^{FeFe}(k) \to \tilde{W}_{tot}^{FeFe}(k) \text{ and } \tilde{w}_{prm}^{FeN_s}(k) \to \tilde{w}_{tot}^{FeN_s}(k),$$

where, within the "molecular"-field approximation, for instance,

$$\tilde{w}_{tot}^{FeN_s}(k) = \tilde{w}_{prm}^{FeN_s}(k) + \tilde{J}_{FeFe}(k) s_{Fe}^2 \sigma_{Fe}^2; \qquad (22)$$

here, $\tilde{w}_{prm}^{FeN_s}(k)$ and $\tilde{w}_{tot}^{FeN_s}(k)$ are the Fourier transforms of the mixing energies in the paramagnetic and magnetic states, respectively, with

$$\tilde{J}_{FeFe}(k) = \sum_{R-R'} J_{FeFe}(R - R') e^{-ik \cdot (R-R')}$$

being the Fourier component of the exchange interaction parameters for Fe atoms.

In other words, the consideration of exchange interactions in the internal energy of Fe–N alloy leads only to a renormalization of the Fe–Fe interaction-dependent energy parameters entering into the internal energy expression and consequently into the Gibbs free energy: Equations (1) and (9).

The circumstances are more complex with the magnetic entropy contribution S_{mag}, which (following [5, 48–49]) can be calculated (in the "molecular"-field approximation [29]) as the entropy of the system of N_{Fe} noninteracting spins:

$$S_{mag} \cong k_B N_{Fe} \left[\ln \operatorname{sh}\left(\left(1+\frac{1}{2s_{Fe}}\right) y_{Fe}(\sigma_{Fe})\right) - \ln \operatorname{sh}\left(\frac{y_{Fe}(\sigma_{Fe})}{2s_{Fe}}\right) - \sigma_{Fe} y_{Fe}(\sigma_{Fe}) \right], \quad (23)$$

where $y_{Fe} = (s_{Fe} H_{eff}^{Fe})/(k_B T)$, $H_{eff}^{Fe} = -\mu_B g_{Fe} \Gamma \sigma_{Fe}$, $g_{Fe} (\cong 2)$ is the Landé factor, μ_B is the Bohr magneton, and γ is a coefficient of the Weiss "molecular"-field contribution.

With an account of the magnetic contribution, the specific Gibbs free energy (with respect to the total number N_{Fe} of Fe ions) in eq (9) is redefined as $g = g(\tilde{W}, \tilde{w}, \tilde{V}, c, T) - T S_{mag}/N_{Fe}$, where $\tilde{W}_{tot}^{FeFe}(0)$ along with $\tilde{W}_{tot}^{FeFe}(\mathbf{k}_{1_H})$ and $\tilde{W}_{tot}^{FeFe}(\mathbf{k}_{1_N})$ entering into $\tilde{w}_{tot}^{FeN_s}(\mathbf{k}_{1_H})$ and $\tilde{w}_{tot}^{FeN_s}(\mathbf{k}_{1_N})$, respectively, include the second σ_{Fe}-dependent term according to Equation (22):

$$\begin{aligned}
g \cong{} & g_\varnothing + R^{N_s} c_{N_s} + R^v c_v + R_3^{N_i} c_{N_i} + \\
& + (1 + c_{N_s} + c_v)^{-1} \Big\{ \frac{1}{2} [\tilde{W}_{prm}^{FeFe}(\mathbf{0}) + \tilde{J}_{FeFe}(\mathbf{0}) s_{Fe}^2 \sigma_{Fe}^2] + \tilde{W}^{FeN_s}(\mathbf{0}) c_{N_s} + \tilde{W}^{Fev}(\mathbf{0}) c_v + \\
& + \tilde{W}_3^{FeN_i}(\mathbf{0}) c_{N_i} + \tilde{W}^{N_s v}(\mathbf{0}) c_{N_s} c_v + \tilde{W}_3^{v N_i}(\mathbf{0}) c_{N_i} c_v + \\
& + \left[\tilde{W}_3^{N_s N_i}(\mathbf{0}) - \tilde{W}_3^{N_s N_i}(\mathbf{k}_{1_H}) + \tilde{W}_3^{FeN_i}(\mathbf{k}_{1_H}) + \sum_{j_N=1}^{6} (-1)^{j_N} \left(\tilde{W}_3^{N_s N_i}(\mathbf{k}_{j_N}) - \tilde{W}_3^{FeN_i}(\mathbf{k}_{j_N}) \right) \right] c_{N_i} c_{N_s} + \\
& + \frac{1}{2} \left[\tilde{W}^{N_s N_s}(\mathbf{0}) + (\tilde{W}_{prm}^{FeFe}(\mathbf{k}_{1_H}) + \tilde{J}_{FeFe}(\mathbf{k}_{1_H}) s_{Fe}^2 \sigma_{Fe}^2) + \tilde{W}^{N_s N_s}(\mathbf{k}_{1_H}) - 2\tilde{W}^{FeN_s}(\mathbf{k}_{1_H}) \right] c_{N_s}^2 + \\
& + 3 \left[\tilde{W}_{prm}^{FeFe}(\mathbf{k}_{1_N}) + \tilde{J}_{FeFe}(\mathbf{k}_{1_N}) s_{Fe}^2 \sigma_{Fe}^2 + \tilde{W}^{N_s N_s}(\mathbf{k}_{1_N}) - 2\tilde{W}^{FeN_s}(\mathbf{k}_{1_N}) \right] c_{N_s}^2 + \\
& + \frac{1}{2} \tilde{W}^{vv}(\mathbf{0}) c_v^2 + \frac{1}{2} \left[\tilde{W}_{33}^{N_i N_i}(\mathbf{0}) + \tilde{W}_{33}^{N_i N_i}(\mathbf{k}_{1_H}) + 4\tilde{W}_{11}^{N_i N_i}(\mathbf{k}_{1_N}) + 2\tilde{W}_{33}^{N_i N_i}(\mathbf{k}_{1_N}) \right] c_{N_i}^2 \Big\} + \\
& + \frac{k_B T}{8} \Big\{ 7(1 + c_{N_s}) \ln(1 + c_{N_s}) + (1 - 7 c_{N_s}) \ln(1 - 7 c_{N_s}) + \\
& + 8 c_{N_s} \ln(8 c_{N_s}) + 8 c_v \ln(8 c_v) + 8 c_{N_i} \ln(8 c_{N_i}) + (1 + c_{N_s} + c_v - 8 c_{N_i}) \ln(1 + c_{N_s} + c_v - 8 c_{N_i}) - \\
& - 9(1 + c_{N_s} + c_v) \ln(1 + c_{N_s} + c_v) \Big\} - \\
& - k_B T \left\{ \ln \operatorname{sh}\left[\left(1 + \frac{1}{2s_{Fe}}\right) y_{Fe}(\sigma_{Fe}) \right] - \ln \operatorname{sh}\left(\frac{y_{Fe}(\sigma_{Fe})}{2s_{Fe}} \right) - \sigma_{Fe} y_{Fe}(\sigma_{Fe}) \right\}.
\end{aligned} \quad (24)$$

Thermodynamically equilibrium value of the σ_{Fe} follows from the condition $\partial g/\partial \sigma_{Fe} = 0$, which results in the transcendental equation

$$\sigma_{Fe} \cong B_{s_{Fe}}(y_{Fe}) = B_{s_{Fe}}\left(-\frac{\left[\tilde{J}_{FeFe}(\mathbf{0}) + \left(\tilde{J}_{FeFe}(\mathbf{k}_{1_H}) + 6\tilde{J}_{FeFe}(\mathbf{k}_{1_N})\right) c_{N_s}^2 \right] s_{Fe}^2 \sigma_{Fe}}{k_B T(1 + c_{N_s} + c_v)} \right); \quad (25)$$

here, $B_{s_{Fe}}$ is the Brillouin function:

$$B_{s_{Fe}}(y_{Fe}) \equiv \left(1 + \frac{1}{2s_{Fe}}\right) \operatorname{cth}\left(\left(1 + \frac{1}{2s_{Fe}}\right) y_{Fe}(\sigma_{Fe}) \right) - \frac{1}{2s_{Fe}} \operatorname{cth}\left(\frac{1}{2s_{Fe}} y_{Fe}(\sigma_{Fe}) \right). \quad (26)$$

Let us finally dwell upon the simplest but realistic case. Suppose that after preparing an ordered Fe–N alloy and annealing it at some comparatively high-temperature $T_{annealing} > T_C = T_C(c_{Ni}, c_{Ns} + c_v)$, for which we still could assume $c_v = (1 + c_{N_s} - 8c_{N_i})$, the alloy is rapidly quenched down to comparatively low temperature $T_{quenching} \leq T_C = T_C(c_{Ni}, c_{Ns})$ ($c_{N_s} = c_N - c_{N_i}$) and then quickly transform to the ferromagnetic state with actually fixed distribution of N atoms over the octahedral interstices and sites with already fixed concentrations c_{N_i} = const and c_{N_s} = const corresponding to the temperature $T_{annealing}$. Accordingly, Equation (25) could be substantially simplified.

As an example, to solve Equation (25) numerically, we can assume the short-range pair ferromagnetic interaction of the magnetic moments in the b.c.c. lattice, namely, for the nearest-neighboring sites, $J_{FeFe}(r_1) < 0$, but for coordination spheres of more distant sites, $|J_{FeFe}(r_n)| = |J_{FeFe}(r_1)|$ ($n = 2, \ldots$). The Fourier transforms of the exchange integrals $\{J_{FeFe}(\mathbf{R} - \mathbf{R}')\}$ relate as [5, 29, 49]:

$$\tilde{J}_{FeFe}(\mathbf{0}) \cong 8J_{FeFe}(r_1) + 6J_{FeFe}(r_2) + 12J_{FeFe}(r_3) + 24J_{FeFe}(r_4) + 8J_{FeFe}(r_5) + \ldots, \quad (27a)$$

$$\tilde{J}_{FeFe}(\mathbf{k}_{1_H}) \cong -8J_{FeFe}(r_1) + 6J_{FeFe}(r_2) + 12J_{FeFe}(r_3) - 24J_{FeFe}(r_4) + 8J_{FeFe}(r_5) + \ldots, \quad (27b)$$

$$\tilde{J}_{FeFe}(\mathbf{k}_{1_N}) \cong -2J_{FeFe}(r_2) - 4J_{FeFe}(r_3) + 8J_{FeFe}(r_5) + \ldots, \quad (27c)$$

where $r_1 = a\sqrt{3}/2$ is the distance between two Fe atoms at the nearest neighboring sites $\mathbf{R}$ and $\mathbf{R}'$ in b.c.c. lattice with the lattice parameter a. (See also [49,50,51] and references therein.) Then, for instance, $\tilde{J}_{FeFe}(\mathbf{k}_{1_H}) \cong -\tilde{J}_{FeFe}(\mathbf{0})$ and $|\tilde{J}_{FeFe}(\mathbf{k}_{1_N})| \ll |\tilde{J}_{FeFe}(\mathbf{k}_{1_H})| \cong |\tilde{J}_{FeFe}(\mathbf{0})|$, and values for exchange-interaction energy parameters, $\tilde{J}_{FeFe}(\mathbf{k}_{1_H})$ or $\tilde{J}_{FeFe}(\mathbf{0})$, and spin number, s_{Fe}, contained in Equation (25) can be adopted (fitted) from the available literature (mainly experimental) data on the Courier temperature, T_C, for paramagnetic–ferromagnetic phase transition.

By the linearization of Equation (25), we can obtain expression of the Courier temperature, T_C, for such magnetic phase transition in hybrid b.c.c. Fe–N alloy at issue with one magnetic component Fe. If there are also two non-magnetic components (N_s and v) on sites, the T_C is defined by the formula:

$$T_C \cong -\frac{s_{Fe}(1 + s_{Fe})}{3k_B(1 + c_{N_s} + c_v)}\left\{\tilde{J}_{FeFe}(\mathbf{0}) + \left[\tilde{J}_{FeFe}(\mathbf{k}_{1_H}) + 6\tilde{J}_{FeFe}(\mathbf{k}_{1_N})\right]c_{N_s}^2\right\}. \quad (28)$$

Within the scope of the aforementioned rough approximation of the nearest-neighboring-spins' ferromagnetic coupling ($|\tilde{J}_{FeFe}(\mathbf{k}_{1_N})| \ll |\tilde{J}_{FeFe}(\mathbf{k}_{1_H})| \cong |\tilde{J}_{FeFe}(\mathbf{0})|$), the expression (28) for T_C becomes trivial, if $c_v = 1$:

$$T_C \cong -\frac{s_{Fe}(1 + s_{Fe})(1 - c_{N_s})}{3k_B}\tilde{J}_{FeFe}(\mathbf{0}) \cong -\frac{8s_{Fe}(1 + s_{Fe})(1 - c_{N_s})}{3k_B}J_{FeFe}(r_1) > 0. \quad (29)$$

From another side, taking into account inter-spin interactions (with exchange integrals "quasi-oscillating" by sign) within the even five site coordination spheres, applying Equation (27), we can easy see that

$$\tilde{J}_{FeFe}(\mathbf{k}_{1_H}) + 6\tilde{J}_{FeFe}(\mathbf{k}_{1_N}) \cong -\tilde{J}_{FeFe}(\mathbf{0}) + 64J_{FeFe}(r_5).$$

Therefore

$$T_C \cong -\frac{s_{Fe}(1 + s_{Fe})}{3k_B(1 + c_{N_s} + c_v)}\left\{\tilde{J}_{FeFe}(\mathbf{0}) - \left[\tilde{J}_{FeFe}(\mathbf{0}) - 64J_{FeFe}(r_5)\right]c_{N_s}^2\right\}, \quad (30)$$

and nonlinear behavior of the dependence of T_C on $c_{N_s} = c_N - c_{N_i}$ (along with c_v) can reflect in a noticeable non-monotonic dependence of T_C on the total nitrogen content c_N, if the contribution of exchange integrals on the fifth sphere into the static thermodynamic of the Fe atomic spins ordering is significant.

18.5 CONCLUSIONS

Among available in the literature attempts to interpret the giant saturation magnetization of the α''-$Fe_{16}N_2$ phase as a prospective material for fabrication of rare-earth-free permanent magnets, which do not contain the non-renewable metals, we stress on the structural contribution – the specific redistribution of nitrogen atoms within the host Fe crystal lattice.

Developing the statistical-thermodynamic model of "hybrid" b.c.t. lattice-based interstitial–substitutional solid solution containing non-metal impurity atoms at both the interstices and the sites of the metal lattice, we derived a set of equations. They allow calculation of the thermodynamically equilibrium concentrations of impurity N atoms in the interstices as compared to those moved to the sites, as well as of vacancies at the sites. To solve these equations, we have to possess information (mainly extracted from the available literature data) on the interaction energy parameters, including exchange interaction, which can depend on both the temperature and concentration.

Calculated results on the temperature-dependent content of impurity nitrogen atoms in the octahedral interstices of the b.c.t. Fe lattice confirm the expected predictions: even at a zero temperature, not all N atoms remain in the interstices; part of them move to the vacant sites of the host lattice. Further increasing the temperature just makes this trend stronger: for N concentration within the range 10%–12% (i.e., close to the $Fe_{16}N_2$ composition), the part of impurity N atoms found in the sites of the matrix Fe lattice rises from $\cong$ 10%–17.5% (at $T \approx 0$ K) to $\cong$ 20%–25% (at $T \approx 1000$ K), and even higher at higher T. Since the spatial distribution of N atoms in the b.c.t. Fe–N martensite correlates with its tetragonality, we can conclude that, by tuning (controlling) the temperature, one can manipulate the degree of tetragonality.

The non-monotonic dependence of the concentration of site vacancies on the total content of impurity nitrogen atoms in the Fe–N is observed. Initially, with increasing the N content, the vacancy concentration substantially increases (even at T = const). However, then (when part of N atoms spontaneously jump from the octahedral interstices into the host-lattice sites), the number of (residual) vacancies sharply decreases as a result of the occupation of vacant sites by the N atoms. Such an "abrupt" transition (of a significant part of N atoms to the b.c.t. Fe lattice sites) at the fixed temperature occurs at a certain concentration of interstitial impurity N atoms and continues for somewhat higher concentrations, however lower the stoichiometric $Fe_{16}N_2$ composition.

The methodology scheme for including the magnetic contribution into the interaction energy parameters is presented. Adopting available data from the (mainly experimental) literature on the spin-exchange interaction parameters, the proposed approach allows calculating the temperature-dependent curves of the magnetization of the Fe subsystem in the α''-$Fe_{16}N_2$-phase, which can contribute to much more clear understanding and explaining its magnetic properties.

The generation of excess vacancies in the iron sublattice and, actually, their simultaneous overwhelming filling of them with a noticeable number of ordering non-metallic (nitrogen) atoms promotes (contributes to) the transfer of electron charge to them (v and N) and the formation of energetically favorable "intermediate" compensated spin states near them, as well as facilitating the localization of high-spin configurations (with uncompensated spin) close to the iron atoms, including with the establishment of a double exchange ferromagnetic coupling between them.

ACKNOWLEDGEMENTS

All authors acknowledge the National Research Foundation of Ukraine (NRFU) for grant support of the project "Controlling the Atomic Distribution for the Functionalization of the Hybrid $Fe_{16}N_2$-Martensite

Phase Based Materials as an Alternative to Permanent Magnets Made of Rare Earth Intermetallics or Permendur" (State Reg. No. 0120U104061) within the NRFU Competition "Leading and Young Scientists Research Support" (application ID 2020.02/0191, agreements Nos. 53/02/2020 as from 27.10.2020 and 76/02/0191 as of 30.04.2021). The first and fourth authors are obliged to the National Academy of Sciences of Ukraine for partial support within the budget program КПКВК 6541230–1A "Support for the Development of Priority Areas of Scientific Research" for 2020–2021 (State Reg. No. 0120U000160). We thank the Armed Forces of Ukraine for providing security to perform this work. Royal Society (through Copyright Clearance Center, Inc.) and AIDIC/CET are acknowledged for permissions to adapt Figure 18.1.

REFERENCES

1. Wang J-P (2020) Environment-friendly bulk $Fe_{16}N_2$ permanent magnet: Review and prospective. *J Magn Magn Mater* 497:165962-1–12. https://doi.org/10.1016/j.jmmm.2019.165962
2. Cui J, Kramer M, Zhou L, Liu F, Gabay A, Hadjipanayis G, Balasubramanian B, Sellmyer D (2018) Current progress and future challenges in rare-earth-free permanent magnets. *Acta Mat* 158:118–137. https://doi.org/10.1016/j.actamat.2018.07.049
3. Radchenko TM, Tatarenko VA (2008) Atomic-ordering kinetics and diffusivities in Ni–Fe permalloy. *Defect Diffus Forum* 273–276:525–530. https://doi.org/10.4028/www.scientific.net/DDF.273-276.525
4. Radchenko TM, Tatarenko VA, Bokoch SM (2006) Diffusivities and kinetics of short-range and long-range orderings in Ni–Fe permalloys. *Metallofiz Noveishie Tekhnol* 28:1699–1720
5. Tatarenko VA, Bokoch SM, Nadutov VM, Radchenko TM, Park YB (2008) Semi-empirical parameterization of interatomic interactions and kinetics of the atomic ordering in Ni–Fe–C permalloys and elinvars. *Defect Diffus Forum* 280:29–78. https://doi.org/10.4028/www.scientific.net/DDF.280-281.29
6. O'Handley RC (2000) *Modern Magnetic Materials: Principles and Applications*. Wiley, New York
7. Jack KH (1951) The iron–nitrogen system: The preparation and the crystal structures of nitrogen–austenite (γ) and nitrogen–martensite (α') *Proc R Soc London Ser A* 208:200–215. https://doi.org/10.1098/rspa.1951.0154
8. Jack KH (1951) The occurrence and the crystal structure of α''-iron nitride; A new type of interstitial alloy formed during the tempering of nitrogen–martensite. *Proc Roy Soc London Ser A* 208:216–224. https://doi.org/10.1098/rspa.1951.0155
9. Radchenko TM, Gatsenko OS, Lizunov VV, Tatarenko VA (2020) Martensitic α''-$Fe_{16}N_2$-type phase of non-stoichiometric composition: Current status of research and microscopic statistical-thermodynamic model. *Prog Phys Met* 21:580–618. https://doi.org/10.15407/ufm.21.04.580
10. Kim TK, Takahashi M (1972) New magnetic material having ultrahigh magnetic moment. *Appl. Phys. Lett.* 20:492–494. https://doi.org/10.1063/1.1654030
11. Hang X, Matsuda M, Held JT, Mkhoyan KA, Wang J-P (2020) Magnetic structure of $Fe_{16}N_2$ determined by polarized neutron diffraction on thin-film samples. *Phys Rev B* 102:104402-1–8. https://doi.org/10.1103/PhysRevB.102.104402
12. Feng L, Zhanga D, Wang F, Dong L, Chen S, Liu J, Hui X (2017) A new structure of the environment-friendly material $Fe_{16}N_2$. *Chem Eng Trans* 61:1501–1506. https://doi.org/10.3303/CET1761248
13. Sugita Y, Takahashi H, Komuro M, Mitsuoka K, Sakuma A (1994) Magnetic and Mössbauer studies of single-crystal $Fe_{16}N_2$ and Fe–N martensite films epitaxially grown by molecular beam epitaxy. *J Appl Phys* 76:6637–6641. https://doi.org/10.1063/1.358157
14. Ji N, Lauter V, Zhang X, Ambaye H, Wang J-P (2013) Strain induced giant magnetism in epitaxial $Fe_{16}N_2$ thin film. *Appl Phys Lett* 102:072411-1–4. https://doi.org/10.1063/1.4792706
15. Ji N, Liu X, Wang J-P (2010) Theory of giant saturation magnetization in α''-$Fe_{16}N_2$: Role of partial localization in ferromagnetism of 3d transition metals. *New J Phys* 12:063032-1–8. https://doi.org/10.1088/1367-2630/12/6/063032
16. Wang J-P, Ji N, Liu X, Xu Y, Sanchez-Hanke C, Wu Y, de Groot FMF, Allard LF, Lara-Curzio E (2012) Fabrication of $Fe_{16}N_2$ films by sputtering process and experimental investigation of origin of giant saturation magnetization in $Fe_{16}N_2$. *IEEE Trans Magn* 48:1710–1717. https://doi.org/10.1109/TMAG.2011.2170156
17. Bhattacharjee S, Lee S-C (2019) First-principles study of the complex magnetism in $Fe_{16}N_2$. *Sci Rep* 9:8381-1–9. https://doi.org/10.1038/s41598-019-44799-8
18. Jack KH (1994) The synthesis, structure, and characterization of α''-$Fe_{16}N_2$. *J Appl Phys* 76:6620–6625. https://doi.org/10.1063/1.358482

19. Jack KH (1995) The synthesis and characterization of bulk α″->Fe$_{16}$N$_2$. *J Alloys Compd* 222:160–166. https://doi.org/10.1016/0925-8388(94)04901-7
20. Sugita Y, Mitsuoka K, Komuro M, Hoshiya H, Kozono Y, Hanazono M (1991) Giant magnetic moment and other magnetic properties of epitaxially grown Fe$_{16}$N$_2$ single-crystal films. *J Appl Phys* 70:5977–5982. https://doi.org/10.1063/1.350067
21. Fall I, Genin J-MR (1996) Mössbauer spectroscopy study of the aging and tempering of high nitrogen quenched Fe–N alloys: kinetics of formation of Fe$_{16}$N$_2$ nitride by interstitial ordering in martensite. *Metall Mater Trans A* 27:2160–2177. https://doi.org/10.1007/BF02651871
22. Shinpei Y, Gallage R, Ogata Y, Kusano Y, Kobayashi N, Ogawa T, Hayashi N, Kohara K, Takahashi M, Takano M (2013) Quantitative understanding of thermal stability of α″-Fe$_{16}$N$_2$. *Chem Commun* 49:7708–7710. https://doi.org/10.1039/C3CC43590C
23. Ji N, Allard LF, Lara-Curzio E, Wang J-P (2011) N site ordering effect on partially ordered Fe$_{16}$N$_2$. *Appl. Phys. Lett.* 98:092506-1–3. https://doi.org/10.1063/1.3560051
24. Hayashi Y, Sugeno T (1970) Nature of boron in α-iron. *Acta Metall* 18:693–697. https://doi.org/10.1016/0001-6160(70)90099-4
25. McLellan RB (1973) Interstitial solid solution of iron. In: *Chemical Metallurgy of Iron and Steel*. Iron and Steel Institute, London, pp. 337–343
26. McLellan RB (1989) The thermodynamics of hybrid binary interstitial–substitutional solid solutions. *J Phys Chem Solids* 50:49–54. https://doi.org/10.1016/0022-3697(89)90472-1
27. Starenchenko VA, Pantyukhova OD, Starenchenko SV, Kolupaeva SN (2001) Mechanisms of deformation-induced destruction of long-range order related to the generation of antiphase boundaries and point defects in alloys with the L1$_2$ superstructure. *Phys Met Metallogr* 91:85–93
28. Bugaev VN, Tatarenko VA (1989) *Interaction and Arrangement of Atoms in Interstitial Solid Solutions Based on Close-Packed Metals*. Naukova Dumka, Kiev, Ukraine
29. Khachaturyan AG (2008) *Theory of Structural Transformations in Solids*. Dover Publications, Mineola, NY
30. Tatarenko VA, Tsynman CL (1996) Temperature- and concentration-dependent tetragonality of a 'hybrid' binary solution in which non-metal atoms can occupy both interstices and sites of the b.c.t. metal lattice with vacancies. *Metallofiz Noveishie Tekhnol* 18(10):32–44
31. Levchuk KH, Radchenko TM, Tatarenko VA (2021) High-temperature entropy effects in tetragonality of the ordering interstitial–substitutional solution based on body-centred tetragonal metal. *Metallofiz Noveishie Tekhnol* 43:1–26. https://doi.org/10.15407/mfint.43.01.0001
32. Nadutov VM, Tatarenko VA, Tsynman CL, Ullakko K (1994) Interatomic interaction and atomic ordering in Fe–N martensite. *Metallofiz Noveishie Tekhnol* 16(8):34–40
33. Girifalco LA (1973) *Statistical Physics of Materials*. John Wiley and Sons, New York
34. Brockhouse BN, Abou-Helal HE, Hellman ED (1967) Lattice vibrations in iron at 296 K. *Solid State Commun* 5:211–216. https://doi.org/10.1016/0038-1098(67)90258-X
35. Kohlhaas R, Dünner Ph, Schmitz-Pranghe N (1967) Uber die temperaturabhangigkeit der gitterparameter von eisen, kobalt und nickel im bereich hoher temperaturen. *Z Angew Phys* 23:245–249
36. Rayne JA, Chandrasekhar BS (1961) Elastic constants of iron from 4.2 to 300 K. *Phys Rev* 122:1714–1716. https://doi.org/10.1103/PhysRev.122.1714
37. Dever DJ (1972) Temperature dependence of the elastic constants in α-iron single crystals: Relationship to spin order and diffusion anomalies. *J Appl Phys* 43:3293–3301. https://doi.org/10.1063/1.1661710
38. Cheng L, Boöttger A, de Keijser ThH, Mittemeijer EJ (1990) Lattice parameters of iron–carbon and iron–nitrogen martensites and austenites. *Scr Met Mater* 24:509–514. https://doi.org/10.1016/0956-716X(90)90192-J
39. Stoica GM, Stoica AD, Miller MK, Ma D (2014) Temperature-dependent elastic anisotropy and meso-scale deformation in a nanostructured ferritic alloy. *Nature Commun* 5:5178-1–8. https://doi.org/10.1038/ncomms6178
40. Basinski ZS, Hume-Rothery W, Sutton AL (1955) The lattice expansion of iron. *Proc R Soc Lond A* 229:459–467. https://doi.org/10.1098/rspa.1955.0102
41. Wolfer WG (2012) Fundamental Properties of Defects in Metal. In: Konings RJM, Allen TR, Stoller RE, Yamanaka Sh (eds) *Comprehensive Nuclear Materials*. Elsevier Science, Amsterdam, Kidlington, Oxford and Waltham, pp. 1–45. https://doi.org/10.1016/B978-0-08-056033-5.00001-X
42. Kuan T-S, Warshel A, Schnepp O (1970) Intermolecular potentials for N$_2$ molecules and the lattice vibrations of solid α-N$_2$. *J Chem Phys* 52:3012–3020. https://doi.org/10.1063/1.1673432
43. Harrison WA (1979) *Solid State Theory*. Dover Publ Inc, New York
44. Brewer L (1965) Prediction of high temperature metallic phase diagrams. In: Zackay VF (ed) *High-Strength Materials*. John Wiley and Sons, New York, pp. 12–103
45. Brewer L (1977) *The Cohesive Energies of the Elements*. Lawrence Berkeley Laboratory, Berkeley, CA

46. Weizer VG, Girifalco LA (1960) Vacancy–vacancy interaction in copper. *Phys Rev* 120:837–839. https://doi.org/10.1103/PhysRev.120.837
47. Yamamoto R, Doyama MJ (1973) The interactions between vacancies and impurities in metals by the pseudopotential method. *J Phys F Met Phys* 3:1524–1530. https://doi.org/10.1088/0305-4608/3/8/007
48. Tatarenko VA, Radchenko TM (2003) The application of radiation diffuse scattering to the calculation of phase diagrams of f.c.c. substitutional alloys. *Intermetallics* 11:1319–1326. https://doi.org/10.1016/S0966-9795(03)00174-2
49. Melnyk IM, Radchenko TM, Tatarenko VA (2010) Semi-empirical parameterization of interatomic interactions, which is based on statistical-thermodynamic analysis of data on phase equilibriums in b.c.c.-Fe–Co alloy. I. Primary ordering. *Metallofiz Noveishie Tekhnol* 32:1191–1212
50. Tatarenko VA, Radchenko TM, Nadutov VM (2003) Parameters of interatomic interaction in a substitutional alloy f.c.c. Ni–Fe according to experimental data about the magnetic characteristics and equilibrium values of intensity of a diffuse scattering of radiations. *Metallofiz Noveishie Tekhnol* 25:1303–1319
51. Radchenko TM, Tatarenko VA (2008) Fe–Ni alloys at high pressures and temperatures: statistical thermodynamics and kinetics of the $L1_2$ or $D0_{19}$ atomic order. *Usp Fiz Met* 9:1–170. https://doi.org/10.15407/ufm.09.01.001

Index

0–9

2D, 2–3, 11, 14, 18, 60–63, 66, 69, 74–76, 78–80, 81–82, 84–89, 97–98, 111–112, 115, 121–122, 126–127, 144–145, 147, 154, 155, 167, 186, 210, 217, 227, 327, 331–332, 339, 357
3D, 3, 9, 42, 48, 74–75, 78, 80, 87, 89, 94–95, 106–107, 126–127, 147–150, 154–155, 210, 215, 227–228, 252, 320, 330

A

AFM, 7, 27–29, 31, 34, 36, 37, 43, 76, 78, 80–84, 86, 88, 129, 133, 150, 176, 177, 179, 214, 215, 245, 246, 252, 253, 265, 290, 295, 296, 332
anisotropy, 2–4, 6–8, 11, 15, 18, 30, 45–47, 49–51, 53, 60, 78–79, 81, 84, 85, 95, 97, 113–117, 119–122, 126, 127, 138, 145, 150, 164, 175, 180, 185, 189, 191, 195, 196, 199, 209–213, 215, 219, 220, 226, 237, 241, 278, 280, 281, 285, 328–331, 344, 355
antiferromagnetism, 88, 183–185, 192, 213, 214, 216, 227
atomic force microscopy, 12, 27, 150, 215, 246, 252, 290

B

bacteria, 11, 16, 310–312, 316, 319–321
bandgap, 24, 25, 34, 43, 53, 80, 155, 160, 258–260
bimetallic, 55, 226–228, 235, 240
biogenesis, 312–314, 318
biological, 7, 8, 11, 13, 15, 16, 24, 153, 184, 318
biomedical, 2, 13, 15, 16, 19, 43, 54, 60, 126, 152, 155, 184, 185, 195
bioremediation, 321
biosensing, 16, 184
biosensors, 16, 54, 55, 320
BN, 175, 176
boron nitride, 60, 74, 85
bottom-up, 8, 10, 11, 61, 63, 66, 69, 75, 140, 141, 145, 147, 155
broken symmetry, 213, 273, 274, 276, 285
broken translation, 102, 106, 107, 209, 210

C

cancer, 15, 19, 24, 54, 152, 246, 320, 321
catalysis, 13, 54, 55, 126, 140, 154, 155, 199, 210
CFT, 26, 35
chalcogenides, 217
chemical, 3, 7–10, 12, 13, 18, 19, 25–27, 30, 42, 45, 54, 61–63, 66, 67, 69, 74, 75, 98, 115–119, 130, 131, 134, 141, 142, 145, 147, 148, 154, 155, 164, 169, 179, 192, 202, 209, 211, 217, 219–221, 226–229, 235, 237, 246, 251, 256, 259, 261, 262, 269, 311, 316, 319, 321, 337, 338, 348, 350, 354, 355
coercivity, 2, 3, 12, 13, 48, 60, 65, 179, 192, 211, 212, 218, 220, 346
coherent, 18, 48–51, 290, 291, 295, 298

combustion, 25, 128, 131, 155
computational, 19, 94, 95, 102, 148, 227–230, 232, 276, 277, 282, 327–329
confinement, 336
critical size, 193, 210, 218
crystal, 3, 8, 12, 25, 26, 28–30, 34, 36, 43, 45, 46, 53, 62–64, 67, 69, 74, 80, 81, 87, 114, 141, 167, 179, 184–187, 189, 199, 202, 209, 210, 213–221, 227, 246, 247, 249, 251, 260–262, 278, 279, 282, 297, 311–316, 319, 331, 344, 347–349, 352, 354, 355, 362
crystal field theory, 26
Curie temperature, 60, 61, 84, 89, 111–115, 121, 122, 128, 130, 145, 177–179, 192, 194, 195, 209, 219, 220, 226, 302, 331, 334, 335, 351
curling-vortex, 50

D

data, 13, 14, 17–19, 24, 36, 43, 52, 53, 55, 65, 93, 103, 105, 122, 126, 128, 129, 147, 150, 151, 162–164, 179, 185, 189, 202, 214, 221, 248, 252, 255, 260, 263, 278, 282, 290, 296–299, 301–305, 318, 326, 327, 345, 361, 362
defect, 86, 115, 162, 171, 217
devices, 3, 8, 14, 17–19, 25, 30, 31, 35, 43, 53–55, 69, 74, 87–89, 93–95, 106, 111, 112, 115, 116, 118, 126, 140, 141, 145, 147, 150–152, 155, 162–164, 184, 185, 191, 192, 208, 221, 247, 257, 278, 284, 290, 299, 306, 326, 331, 332, 339
diagnostic, 66, 158, 320
dielectric, 88, 95, 191, 246, 262–265
diluted magnetic semiconductor, 25, 31
dipolar, 47, 48, 50, 104, 153, 192, 198, 212, 213, 264, 282, 329
DMS, 25–27, 31, 36, 164, 166, 167, 170
domain, 3–5, 17, 18, 34, 48–51, 53, 55, 86, 95, 117, 126, 127, 148, 184, 190, 192, 193, 195–198, 209, 210, 215, 216, 221, 246, 254, 303, 326, 328, 331, 339
drug, 15, 24, 74, 152, 153, 184, 195, 246, 319–321

E

electrodeposition, 42, 43, 55, 299, 300
electroless, 42, 45
electronics, 14, 25, 151, 162, 194, 326, 331
electron microscopy, 248, 265, 311
environmental, 13, 16, 17, 66, 126, 154, 155, 185, 220, 316, 318, 321
epitaxy, 25, 60–63, 65, 66, 69, 75, 94, 112, 142, 345, 363
exchange, 3, 5–8, 18, 25, 26, 29–31, 34, 36, 43, 45–47, 49, 51, 62, 63, 66, 69, 74, 76–79, 84, 85, 89, 95, 104, 113, 120, 126, 127, 152, 169, 170, 178, 191, 192, 208–210, 212–215, 217–220, 233, 270–273, 276–278, 281, 282, 284, 328, 329, 334, 335, 345, 351, 359, 361, 362
exchange interactions, 26, 31, 34, 192, 208, 214, 217, 359
exfoliation, 60–63, 69, 75, 80, 115

F

ferrimagnetism, 114, 184, 191, 194, 213
ferroelectric, 66, 89, 200
ferromagnetic, 4–6, 12, 16, 18, 43, 47, 48, 54, 60, 63, 74–77, 84–86, 94, 95, 97, 98, 100, 112, 113, 115–118, 120, 121, 128, 129, 133, 139, 145, 147, 148, 150, 152, 163, 164, 167–171, 174, 179, 189, 192–197, 210, 212, 214, 217–219, 221, 226, 227, 233, 271, 275, 292, 299, 300, 302, 304, 328, 361, 362
ferromagnetism, 2, 60, 74, 75, 79, 114, 167–170, 172, 174, 184, 185, 191, 192, 199, 200, 213, 214, 216–221, 302, 304, 359, 363
Fourier transform infrared, 12, 246, 260
FTIR, 185, 187, 201, 202, 246, 260, 261, 265

G

gene, 54, 153, 184, 312, 319
glassy, 24–27, 29, 32–34, 198
graphene, 17, 60, 62, 68, 74, 85, 87, 88, 115, 116, 159, 169, 181, 199, 251, 252, 266, 267, 332, 341
grinding, 10, 25, 128, 130, 131, 155, 156

H

heterostructures, 74, 85, 87, 88, 117, 126, 152, 164, 226, 290, 299, 300, 302, 303, 305, 331
hydrothermal, 9, 10, 25, 68, 128, 134, 138, 139, 143, 144, 157

I

imaging, 2, 15, 16, 19, 86, 152–154, 218, 246, 249–252, 254, 255, 320

L

lithography, 8, 11, 42, 45, 94, 147, 148, 150

M

magnetic force microscopy, 48, 86, 221, 246, 252
magnetic moment, 80–82, 119, 139, 152, 167, 170, 172, 177, 185, 192, 195, 209, 212, 214, 215, 226–228, 233, 234, 237–241, 254, 278, 279, 290, 292, 297, 298, 304, 327–330, 345, 363
magnetism, 3, 7, 12, 18, 45, 53, 60, 61, 74–76, 78, 83, 84, 85, 116, 118–120, 139, 162, 168, 170, 171, 174, 184, 190, 192, 194, 195, 198, 199, 202, 209, 212, 214–219, 221, 269, 272, 284, 290, 292, 297, 300, 302, 304–306, 327, 328, 331, 332, 335, 339, 346, 363
magnetization, 2–6, 8, 11, 12, 14–16, 18, 45–55, 60, 74–76, 79, 80, 82–88, 94, 95, 97, 98, 100, 103–105, 113–115, 117, 119, 120, 126–128, 130, 131, 139, 143, 152, 185, 189–192, 194–196, 198, 199, 202, 209–220, 233, 240, 278–280, 285, 292, 295, 297, 299–302, 304, 305, 327, 328, 330, 332–339, 344, 346, 359, 362
magnetosome, 310–321
magnetostatic, 3, 6–8, 46, 47, 78, 100, 102, 104, 106, 126, 213
MBE, 25, 60, 66, 75, 112, 114
mechanical, 10, 13, 14, 18, 25, 45, 60–63, 69, 74, 75, 102, 104, 115, 129, 131, 139, 151, 184, 191, 192, 200, 201, 208, 254, 285, 328, 355

memory, 2, 14, 15, 43, 53, 94, 113, 126, 145, 147, 150, 151, 155, 163, 179, 184, 191, 217, 221, 278, 326, 328, 330, 331
metal-organic frame, 61, 69
metal oxides, 10, 74, 76, 185, 216
MFM, 27–29, 31, 33–36, 85, 86, 246, 252, 254, 265
microemulsion, 9, 139, 140, 155, 219
micromagnetism, 6, 47, 212, 213, 328–331, 333
microwave, 8, 15, 17, 27, 53, 54, 94, 98–102, 106, 138, 177
MO, 284
modeling, 48, 95–97, 104, 105, 154, 190, 195, 270, 280, 327, 331, 333, 334, 339
MOF, 61
molecular beam epitaxy, 25, 60, 61, 66, 75, 112, 142, 345, 363
monoatomic, 226, 228, 233, 234, 240
morphology, 2, 3, 5, 12, 69, 75, 128, 131, 133, 134, 138, 139, 153, 155, 157, 184, 185, 201, 208, 212, 215, 217, 220, 221, 246, 249, 252, 255, 300, 310–312, 315, 319
multilayer, 14, 54, 120, 151, 205, 292, 297, 298, 302–304
MXene, 76, 117

N

nanomagnetic, 48, 185, 190, 191, 196, 198, 202, 214, 216, 219, 221
nanostructure, 6, 7, 11, 52, 54, 148, 185, 187–189, 202, 210
nanowires, 7, 8, 11, 13, 42, 43, 48, 55, 60, 66, 68, 126, 127, 140–145, 147, 155, 184, 189, 200
neuromorphic, 55, 94

O

optical, 13, 24, 25, 27, 30, 31, 33–36, 42, 60, 82, 84, 89, 98, 140, 147, 168, 184, 187, 191, 199, 221, 246, 254, 257, 290, 291, 327, 333, 339
organic, 7–10, 16, 25, 54, 61, 63, 66, 68, 69, 112, 118, 119, 121, 122, 131, 133, 134, 148, 155, 184, 202, 218, 219, 311, 312, 314–316, 321

P

paramagnetism, 184, 185, 191, 193–196, 199, 213
perovskites, 60, 63, 76, 156, 217–221
photoluminescence, 36, 82, 246, 259, 260
physical, 2, 7, 8, 10, 12, 13, 18, 24, 25, 27, 30, 48, 55, 60–62, 66, 74, 89, 113, 116, 121, 126, 133, 140, 147, 154, 155, 163, 167, 172, 176, 177, 187, 190–192, 195, 198, 199, 209–211, 215, 217, 220, 227, 228, 246, 252, 259, 277, 284, 289, 290, 297, 300, 303, 327, 328, 344
PL, 26, 31, 32, 82, 84, 89, 259, 260, 265, 277
PLD, 64, 65, 69, 145, 148–150
precipitation, 25, 27, 45, 131
pulsed laser deposition, 64, 65, 69, 145, 304

R

racetrack, 17, 18, 43, 53
Raman spectroscopy, 27, 35, 84, 187, 201, 246, 259, 261, 262
redox, 17, 139, 315–318
reflection, 290, 293
reflectivity, 293–297
refrigeration, 126, 152, 155
RKKY, 77, 169

S

scanning electron microscope, 127, 215, 249
scanning probe microscopy, 252
scanning tunneling microscopy, 246, 255
SEM, 44, 67, 68, 127–129, 134, 141, 142, 145, 148, 149, 185–187, 215, 246, 249, 250, 265
semiconductor, 14, 24–26, 30, 34, 43, 60, 66, 75, 80–81, 88, 93, 94, 112, 150, 151, 162–164, 166–168, 170, 174, 259, 332
size effect, 4, 209
sol-gel, 9, 128, 131, 133, 134, 142, 145, 155, 263
solid-state, 14, 31, 128, 155, 208, 216, 251
solvothermal, 10, 25, 61, 63, 66, 69, 128, 134, 143, 155
space groups, 215, 218
spinels, 218
spinterface, 118–120, 122
spin-orbit couplings, 194
spintronic, 14, 25, 30, 35, 43, 44, 53, 66, 69, 111, 112, 114–116, 118, 119, 121, 122, 125, 126, 141, 145, 151, 155, 158, 162–164, 170, 179, 191, 326, 331, 333, 339
SPM, 128, 129, 152, 153, 195, 196, 198, 252
STM, 65, 87, 246, 255
Stoner, 49, 169, 170
storage, 13, 14, 17–19, 24, 43, 52, 53, 55, 74, 122, 150, 151, 184, 191, 208, 211, 218, 221, 226, 262, 278, 284, 302, 326, 327, 331
superconductor, 60, 72, 290, 303, 304, 307
superexchange, 77, 168, 169, 177, 270, 275, 281
superferromagnetism, 184, 185, 191, 196, 198, 202
superparamagnetic, 4, 13, 15, 53, 114, 193, 195, 196, 216, 220
superparamagnetism, 3, 45, 128, 184, 185, 191, 193, 195, 196, 200, 211, 216, 218, 219
surface effects, 126, 141, 209, 210, 219, 221, 259
synthesis, 2, 8–11, 13, 18, 19, 25, 27, 35, 43, 55, 60, 61, 63, 65–69, 128, 131, 133, 138–141, 143, 144, 147, 148, 155, 216, 218, 219, 221, 249, 251, 284, 309, 315–317, 344, 346

T

TEM, 27–31, 33, 35–37, 61, 127, 131, 134, 142–145, 185–188, 201, 215, 246, 249–251, 265, 290
thermal, 3, 8, 10–15, 17, 18, 25, 30, 43, 47, 53, 62, 64, 67, 68, 78–80, 113, 120, 152, 191, 194, 195, 199, 209, 211, 215, 219, 279, 290, 291, 295, 335, 337, 339, 345, 364
three-dimensional, 42, 95, 102, 104–106, 247
TMD, 117
top-down, 8, 10, 11, 62, 69, 75, 140, 141, 145
transition metal dichalcogenides, 60, 74, 75, 81, 115, 217
transition metal halides, 75
transition metal phosphorous, 45, 75
transmission electron microscopy, 27, 94, 127, 131, 185, 249, 290
transport, 18, 55, 60, 75, 85, 94, 95, 102, 104–106, 112, 115, 118, 119, 126, 156, 157, 162, 178, 221, 312, 314, 315, 327, 331–333
transverse, 17, 51, 85, 97, 99–102, 117, 153, 154, 280, 337
two-dimensional, 2, 11, 18, 60, 67, 74, 96, 97, 100, 111, 126, 163, 303

U

UV-Vis Spectroscopy, 36, 257, 258, 265

V

vacancy, 77, 86, 154, 162, 173–176, 217, 260, 357, 358, 362
valleytronics, 88
vibrational, 36, 184, 187, 191, 201, 202, 229

X

XPS, 185–187, 246, 256, 257
x-ray diffraction, 12, 27, 128, 185, 186, 215, 246, 251, 295, 297–299, 346
X-ray photoelectron spectroscopy, 246, 256
XRD, 27–29, 31, 34–36, 128, 131, 133, 147, 185–187, 189, 190, 215, 247, 248, 252, 259, 265, 299, 304

Z

Zeeman, 26, 47, 49, 86, 89, 191, 212, 213, 329

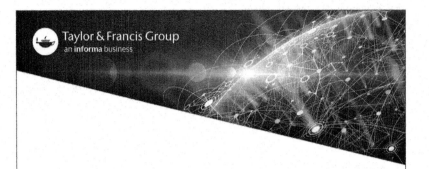